Lecture Notes in Computer Science 13094

Constantine Stephanidis ·
Marcelo M. Soares · Elizabeth Rosenzweig ·
Aaron Marcus · Sakae Yamamoto ·
Hirohiko Mori · Pei-Luen Patrick Rau ·
Gabriele Meiselwitz · Xiaowen Fang ·
Abbas Moallem (Eds.)

HCI International 2021 - Late Breaking Papers

Design and User Experience

23rd HCI International Conference, HCII 2021
Virtual Event, July 24–29, 2021
Proceedings

Springer

Editors
Constantine Stephanidis
University of Creteand Foundation for Research and Technology – Hellas (FORTH)
Heraklion, Crete, Greece

Elizabeth Rosenzweig
World Usability Day and Brandeis University
Newton Center, MA, USA

Sakae Yamamoto
Tokyo University of Science
Tokyo, Saitama, Japan

Pei-Luen Patrick Rau
Tsinghua University
Beijing, China

Xiaowen Fang
DePaul University
Chicago, IL, USA

Marcelo M. Soares
Department of Design
Federal University of Pernambuco
Recife, Brazil

School of Design
Hunan University
Changsha, China

Aaron Marcus
Aaron Marcus and Associates
Berkeley, CA, USA

Hirohiko Mori
Tokyo City University
Tokyo, Japan

Gabriele Meiselwitz
Computer and Information Sciences
Towson University
Towson, MD, USA

Abbas Moallem
San Jose State University
San Jose, CA, USA

ISSN 0302-9743 ISSN 1611-3349 (electronic)
Lecture Notes in Computer Science
ISBN 978-3-030-90237-7 ISBN 978-3-030-90238-4 (eBook)
https://doi.org/10.1007/978-3-030-90238-4

LNCS Sublibrary: SL3 – Information Systems and Applications, incl. Internet/Web, and HCI

This Springer imprint is published by the registered company Springer Nature Switzerland AG
The registered company address is: Gewerbestrasse 11, 6330 Cham, Switzerland

Foreword

Human-Computer Interaction (HCI) is acquiring an ever-increasing scientific and industrial importance, and having more impact on people's everyday life, as an ever-growing number of human activities are progressively moving from the physical to the digital world. This process, which has been ongoing for some time now, has been dramatically accelerated by the COVID-19 pandemic. The HCI International (HCII) conference series, held yearly, aims to respond to the compelling need to advance the exchange of knowledge and research and development efforts on the human aspects of design and use of computing systems.

The 23rd International Conference on Human-Computer Interaction, HCI International 2021 (HCII 2021), was planned to be held at the Washington Hilton Hotel, Washington DC, USA, during July 24–29, 2021. Due to the COVID-19 pandemic and with everyone's health and safety in mind, HCII 2021 was organized and run as a virtual conference. It incorporated the 21 thematic areas and affiliated conferences listed on the following page.

A total of 5222 individuals from academia, research institutes, industry, and governmental agencies from 81 countries submitted contributions, and 1276 papers and 241 posters were included in the volumes of the proceedings that were published before the start of the conference. Additionally, 174 papers and 146 posters are included in the volumes of the proceedings published after the conference, as "Late Breaking Work" (papers and posters). The contributions thoroughly cover the entire field of HCI, addressing major advances in knowledge and effective use of computers in a variety of application areas. These papers provide academics, researchers, engineers, scientists, practitioners, and students with state-of-the-art information on the most recent advances in HCI. The volumes constituting the full set of the HCII 2021 conference proceedings are listed in the following pages.

I would like to thank the Program Board Chairs and the members of the Program Boards of all thematic areas and affiliated conferences for their contribution towards the highest scientific quality and overall success of the HCI International 2021 conference.

This conference would not have been possible without the continuous and unwavering support and advice of Gavriel Salvendy, founder, General Chair Emeritus, and Scientific Advisor. For his outstanding efforts, I would like to express my appreciation to Abbas Moallem, Communications Chair and Editor of HCI International News.

July 2021 Constantine Stephanidis

HCI International 2021 Thematic Areas and Affiliated Conferences

Thematic Areas

- HCI: Human-Computer Interaction
- HIMI: Human Interface and the Management of Information

Affiliated Conferences

- EPCE: 18th International Conference on Engineering Psychology and Cognitive Ergonomics
- UAHCI: 15th International Conference on Universal Access in Human-Computer Interaction
- VAMR: 13th International Conference on Virtual, Augmented and Mixed Reality
- CCD: 13th International Conference on Cross-Cultural Design
- SCSM: 13th International Conference on Social Computing and Social Media
- AC: 15th International Conference on Augmented Cognition
- DHM: 12th International Conference on Digital Human Modeling and Applications in Health, Safety, Ergonomics and Risk Management
- DUXU: 10th International Conference on Design, User Experience, and Usability
- DAPI: 9th International Conference on Distributed, Ambient and Pervasive Interactions
- HCIBGO: 8th International Conference on HCI in Business, Government and Organizations
- LCT: 8th International Conference on Learning and Collaboration Technologies
- ITAP: 7th International Conference on Human Aspects of IT for the Aged Population
- HCI-CPT: 3rd International Conference on HCI for Cybersecurity, Privacy and Trust
- HCI-Games: 3rd International Conference on HCI in Games
- MobiTAS: 3rd International Conference on HCI in Mobility, Transport and Automotive Systems
- AIS: 3rd International Conference on Adaptive Instructional Systems
- C&C: 9th International Conference on Culture and Computing
- MOBILE: 2nd International Conference on Design, Operation and Evaluation of Mobile Communications
- AI-HCI: 2nd International Conference on Artificial Intelligence in HCI

Conference Proceedings – Full List of Volumes

1. LNCS 12762, Human-Computer Interaction: Theory, Methods and Tools (Part I), edited by Masaaki Kurosu
2. LNCS 12763, Human-Computer Interaction: Interaction Techniques and Novel Applications (Part II), edited by Masaaki Kurosu
3. LNCS 12764, Human-Computer Interaction: Design and User Experience Case Studies (Part III), edited by Masaaki Kurosu
4. LNCS 12765, Human Interface and the Management of Information: Information Presentation and Visualization (Part I), edited by Sakae Yamamoto and Hirohiko Mori
5. LNCS 12766, Human Interface and the Management of Information: Information-rich and Intelligent Environments (Part II), edited by Sakae Yamamoto and Hirohiko Mori
6. LNAI 12767, Engineering Psychology and Cognitive Ergonomics, edited by Don Harris and Wen-Chin Li
7. LNCS 12768, Universal Access in Human-Computer Interaction: Design Methods and User Experience (Part I), edited by Margherita Antona and Constantine Stephanidis
8. LNCS 12769, Universal Access in Human-Computer Interaction: Access to Media, Learning and Assistive Environments (Part II), edited by Margherita Antona and Constantine Stephanidis
9. LNCS 12770, Virtual, Augmented and Mixed Reality, edited by Jessie Y. C. Chen and Gino Fragomeni
10. LNCS 12771, Cross-Cultural Design: Experience and Product Design Across Cultures (Part I), edited by P. L. Patrick Rau
11. LNCS 12772, Cross-Cultural Design: Applications in Arts, Learning, Well-being, and Social Development (Part II), edited by P. L. Patrick Rau
12. LNCS 12773, Cross-Cultural Design: Applications in Cultural Heritage, Tourism, Autonomous Vehicles, and Intelligent Agents (Part III), edited by P. L. Patrick Rau
13. LNCS 12774, Social Computing and Social Media: Experience Design and Social Network Analysis (Part I), edited by Gabriele Meiselwitz
14. LNCS 12775, Social Computing and Social Media: Applications in Marketing, Learning, and Health (Part II), edited by Gabriele Meiselwitz
15. LNAI 12776, Augmented Cognition, edited by Dylan D. Schmorrow and Cali M. Fidopiastis
16. LNCS 12777, Digital Human Modeling and Applications in Health, Safety, Ergonomics and Risk Management: Human Body, Motion and Behavior (Part I), edited by Vincent G. Duffy
17. LNCS 12778, Digital Human Modeling and Applications in Health, Safety, Ergonomics and Risk Management: AI, Product and Service (Part II), edited by Vincent G. Duffy

18. LNCS 12779, Design, User Experience, and Usability: UX Research and Design (Part I), edited by Marcelo Soares, Elizabeth Rosenzweig, and Aaron Marcus
19. LNCS 12780, Design, User Experience, and Usability: Design for Diversity, Well-being, and Social Development (Part II), edited by Marcelo M. Soares, Elizabeth Rosenzweig, and Aaron Marcus
20. LNCS 12781, Design, User Experience, and Usability: Design for Contemporary Technological Environments (Part III), edited by Marcelo M. Soares, Elizabeth Rosenzweig, and Aaron Marcus
21. LNCS 12782, Distributed, Ambient and Pervasive Interactions, edited by Norbert Streitz and Shin'ichi Konomi
22. LNCS 12783, HCI in Business, Government and Organizations, edited by Fiona Fui-Hoon Nah and Keng Siau
23. LNCS 12784, Learning and Collaboration Technologies: New Challenges and Learning Experiences (Part I), edited by Panayiotis Zaphiris and Andri Ioannou
24. LNCS 12785, Learning and Collaboration Technologies: Games and Virtual Environments for Learning (Part II), edited by Panayiotis Zaphiris and Andri Ioannou
25. LNCS 12786, Human Aspects of IT for the Aged Population: Technology Design and Acceptance (Part I), edited by Qin Gao and Jia Zhou
26. LNCS 12787, Human Aspects of IT for the Aged Population: Supporting Everyday Life Activities (Part II), edited by Qin Gao and Jia Zhou
27. LNCS 12788, HCI for Cybersecurity, Privacy and Trust, edited by Abbas Moallem
28. LNCS 12789, HCI in Games: Experience Design and Game Mechanics (Part I), edited by Xiaowen Fang
29. LNCS 12790, HCI in Games: Serious and Immersive Games (Part II), edited by Xiaowen Fang
30. LNCS 12791, HCI in Mobility, Transport and Automotive Systems, edited by Heidi Krömker
31. LNCS 12792, Adaptive Instructional Systems: Design and Evaluation (Part I), edited by Robert A. Sottilare and Jessica Schwarz
32. LNCS 12793, Adaptive Instructional Systems: Adaptation Strategies and Methods (Part II), edited by Robert A. Sottilare and Jessica Schwarz
33. LNCS 12794, Culture and Computing: Interactive Cultural Heritage and Arts (Part I), edited by Matthias Rauterberg
34. LNCS 12795, Culture and Computing: Design Thinking and Cultural Computing (Part II), edited by Matthias Rauterberg
35. LNCS 12796, Design, Operation and Evaluation of Mobile Communications, edited by Gavriel Salvendy and June Wei
36. LNAI 12797, Artificial Intelligence in HCI, edited by Helmut Degen and Stavroula Ntoa
37. CCIS 1419, HCI International 2021 Posters - Part I, edited by Constantine Stephanidis, Margherita Antona, and Stavroula Ntoa

38. CCIS 1420, HCI International 2021 Posters - Part II, edited by Constantine Stephanidis, Margherita Antona, and Stavroula Ntoa
39. CCIS 1421, HCI International 2021 Posters - Part III, edited by Constantine Stephanidis, Margherita Antona, and Stavroula Ntoa
40. LNCS 13094, HCI International 2021 - Late Breaking Papers: Design and User Experience, edited by Constantine Stephanidis, Marcelo M. Soares, Elizabeth Rosenzweig, Aaron Marcus, Sakae Yamamoto, Hirohiko Mori, P. L. Patrick Rau, Gabriele Meiselwitz, Xiaowen Fang, and Abbas Moallem
41. LNCS 13095, HCI International 2021 - Late Breaking Papers: Multimodality, eXtended Reality, and Artificial Intelligence, edited by Constantine Stephanidis, Masaaki Kurosu, Jessie Y. C. Chen, Gino Fragomeni, Norbert Streitz, Shin'ichi Konomi, Helmut Degen, and Stavroula Ntoa
42. LNCS 13096, HCI International 2021 - Late Breaking Papers: Cognition, Inclusion, Learning, and Culture, edited by Constantine Stephanidis, Don Harris, Wen-Chin Li, Dylan D. Schmorrow, Cali M. Fidopiastis, Margherita Antona, Qin Gao, Jia Zhou, Panayiotis Zaphiris, Andri Ioannou, Robert A. Sottilare, Jessica Schwarz, and Matthias Rauterberg
43. LNCS 13097, HCI International 2021 - Late Breaking Papers: HCI Applications in Health, Transport, and Industry, edited by Constantine Stephanidis, Vincent G. Duffy, Heidi Krömker, Fiona Fui-Hoon Nah, Keng Siau, Gavriel Salvendy, and June Wei
44. CCIS 1498, HCI International 2021 - Late Breaking Posters (Part I), edited by Constantine Stephanidis, Margherita Antona, and Stavroula Ntoa
45. CCIS 1499, HCI International 2021 - Late Breaking Posters (Part II), edited by Constantine Stephanidis, Margherita Antona, and Stavroula Ntoa

http://2021.hci.international/proceedings

HCI International 2021 (HCII 2021)

The full list with the Program Board Chairs and the members of the Program Boards of all thematic areas and affiliated conferences is available online:

http://www.hci.international/board-members-2021.php

HCI International 2022

The 24th International Conference on Human-Computer Interaction, HCI International 2022, will be held jointly with the affiliated conferences at the Gothia Towers Hotel and Swedish Exhibition & Congress Centre, Gothenburg, Sweden, June 26 – July 1, 2022. It will cover a broad spectrum of themes related to Human-Computer Interaction, including theoretical issues, methods, tools, processes, and case studies in HCI design, as well as novel interaction techniques, interfaces, and applications. The proceedings will be published by Springer. More information will be available on the conference website: http://2022.hci.international/.

General Chair
Prof. Constantine Stephanidis
University of Crete and ICS-FORTH
Heraklion, Crete, Greece
Email: general_chair@hcii2022.org

http://2022.hci.international/

Contents

Design and Evaluation Methods, Techniques and Tools

Research on Information Design Matching with User's Need for Cognitive 3
Xuanyi Chen, Yanfei Zhu, and Chengqi Xue

Interaction Design for the Next Billion Users 16
Sumesh Dugar, Abhishek Mitra, Shweta Nandi, Biswajit Adhikary, Sonit Paul, and Madhav Manusuriya

The Crowd Thinks Aloud: Crowdsourcing Usability Testing with the Thinking Aloud Method 24
Edwin Gamboa, Rahul Galda, Cindy Mayas, and Matthias Hirth

Shaping AI as the Tool for Subconscious Design 40
Wentong Huang

An Experimental Study on "Consensus to Match" Game for Analyzing Emotional Interaction in Consensus Building Process 54
Kyoko Ito, Yoshiki Sakamoto, Rieko Yamamoto, Mizuki Yamawaki, Daisuke Miyazaki, Kimi Ueda, Hirotake Ishii, and Hiroshi Shimoda

Challenges and Workarounds of Conducting Augmented Reality Usability Tests Remotely a Case Study 63
Ted Kim, Santiago Arconada-Alvarez, and Young Mi Choi

Understanding Graphical User Interface (GUI) Trends Based on Kawaii (Cute) 72
Anirudh Kundu and Michiko Ohkura

Research on Interaction Design of Anti-addiction for Minor Games Based on Flow Theory 89
Xin Liang, Wei Yu, and Xueqing Zhao

Potential Design Strategies Based on Communication Design and Art Therapy for User Experience in COVID-19 101
Zhen Liu, Zulan Yang, and Ke Zhang

Preliminary Investigation of Methods for Graphic Simplification from Representation to Abstraction 116
Hui-Ping Lu

Suggestions for Online User Studies: Sharing Experiences from the Use of Four Platforms 127
Joni Salminen, Soon-gyo Jung, and Bernard J. Jansen

The Hidden Cost of Using Amazon Mechanical Turk for Research 147
Antonios Saravanos, Stavros Zervoudakis, Dongnanzi Zheng, Neil Stott, Bohdan Hawryluk, and Donatella Delfino

Research on Service Experience Design Framework Based on Semantics to Improve the Enterprise Service Capability 165
Kun Zhou, Xi Zhang, and Yuanlong Gui

Design, User Experience and Human Behavior Studies

Expectation, Perception, and Accuracy in News Recommender Systems: Understanding the Relationships of User Evaluation Criteria Using Direct Feedback 179
Poornima Belavadi, Laura Burbach, Stefan Ahlers, Martina Ziefle, and André Calero Valdez

Partial Consent: A Study on User Preference for Informed Consent 198
Sven Bock, Ashraf Ferdouse Chowdhury, and Nurul Momen

UI Development of Hardcore Battle Royale Game for Novice Users 217
Woo Jin Choi and Chang Joo Lim

The Reaches of Crowdsourcing: A Systematic Literature Review 229
Samantha Dishman and Vincent G. Duffy

A Bibliometric Analysis on Cybercrime in Nigeria 249
Monica Okwuchkwu Enebechi, Chidubem Nuela Enebechi, and Vincent G. Duffy

Differences in Product Selection Depend on Situations: Using Eyeglasses as an Example 270
Yuri Hamada, Atsuya Nagata, Naoki Takahashi, and Hiroko Shoji

Identifying Early Opinion Leaders on COVID-19 on Twitter 280
Zahra Hatami, Margeret Hall, and Neil Thorne

Impact of the Cyber Hygiene Intelligence and Performance (CHIP) Interface on Cyber Situation Awareness and Cyber Hygiene 298
Janine D. Mator and Jeremiah D. Still

The Effect of Social Media Based Electronic Word of Mouth on Propensity to Buy Wearable Devices 310
David Ntumba and Adheesh Budree

Applying Exploratory Testing and Ad-Hoc Usability Inspection to Improve the Ease of Use of a Mobile Power Consumption Registration App: An Experience Report 326
José Eduardo, Anderson Paiva, Victor Ferreira, Simara Rocha, Ítalo Santos, Luís Rivero, João Almeida, Geraldo Braz Junior, Anselmo Paiva, Aristofenes Silva, Hugo Nogueira, Eliana Monteiro, and Eduardo Fernandes

Euros from the Heart: Exploring Digital Money Gifts in Intimate Relationships 342
Freya Probst, Hyosun Kwon, and Cees de Bont

Impact of Social Media Marketing on University Students - *Peru* 357
Julissa Elizabeth Reyna González, Víctor Ricardo Flores-Rivas, and Irene Merino Flores

Dynamic Difficulty Adjustment Using Performance and Affective Data in a Platform Game 367
Marcos P. C. Rosa, Eduardo A. dos Santos, Iago L. R. de Moraes, Tiago B. P. e Silva, Mauricio M. Sarmet, Carla D. Castanho, and Ricardo P. Jacobi

Exploring the Effect of Resolution on the Usability of Locimetric Authentication 387
Antonios Saravanos, Dongnanzi Zheng, Stavros Zervoudakis, and Donatella Delfino

Usability Assessment of the GoPro Hero 7 Black for Chinese Users 397
Guo Sheng-nan, Chen Jia, Chang Le, Jiayu Zeng, and Marcelo M. Soares

A Study on Dual-Language Display Method Using the Law of Common Fate in Oscillatory Animation on Digital Signage 412
Takumi Uotani, Yuki Takashima, Kimi Ueda, Hirotake Ishii, Hiroshi Shimoda, Rika Mochizuki, and Masahiro Watanabe

Research and Analysis of the Office Socket Design Based on User Experience 424
Xiangrong Xu, Yuanlong Gui, Bo Fu, and Naizheng Liao

Research on Improving Empathy Based on the Campus Barrier-Free Virtual Experience Game 434
Junyu Yang, Yawen Zheng, Tianjiao Zhao, and Mu Zhang

Trust and Automation: A Systematic Review and Bibliometric Analysis 451
Zhengming Zhang, Vincent G. Duffy, and Renran Tian

Usability Assessment of Xiaomi Smart Band 4 465
Yiqing Zhou, Jiaqi Tang, Junchi Wu, Jiayu Zeng, and Marcelo M. Soares

Cross-Cultural Design

Factors Affecting e-Commerce Satisfaction in Qatar: A Cross-Cultural Comparison 481
Muth Mary Abraham and Pilsung Choe

Factors Influencing Trust in WhatsApp: A Cross-Cultural Study 495
Gabriela Beltrão and Sonia Sousa

The Research on the User Experience of Consultation Designed by China's Medical Mobile Media Platforms Under the Background of COVID-19 509
Lingxi Chen, Yuxuan Xiao, and Linda Huang

Research on the Attractive Factors and Design of Cultural Derivative Commodities Under Cultural Sustainability 522
Kuo-Liang Huang, Na Xu, Hsuan Lin, and Jinchen Jiang

Cross-Cultural Design in Consumer Vehicles to Improve Safety: A Systematic Literature Review 539
Priyanka Koratpallikar and Vincent G. Duffy

Cross-Cultural Differences of Designing Mobile Health Applications for Africans 554
Helina Oladapo and Joyram Chakraborty

"Tell Me Your Story, I'll Tell You What Makes It Meaningful": Characterization of Meaningful Social Interactions Between Intercultural Strangers and Design Considerations for Promoting Them 564
María Laura Ramírez Galleguillos, Aya Eloiriachi, Büşra Serdar, and Aykut Coşkun

Intercultural HMIs in Automotive: Do We Need Them? – An Analysis 584
Peter Rössger

Hybrid Kansei Research of Product's Interactive Design Experience Based on "Sensing" Technology. 597
Min Shi and Cheng-wei Fan

Author Index 609

Design and Evaluation Methods, Techniques and Tools

Research on Information Design Matching with User's Need for Cognitive

Xuanyi Chen, Yanfei Zhu, and Chengqi Xue(✉)

School of Mechanical Engineering, Southeast University - China Southeast University, Nanjing 211189, Jiangsu, People's Republic of China
ipd_xcq@seu.edu.cn

Abstract. Because of the influence of information presentation and individual cognitive traits, People's cognitive decision-making behaviors are different, and the current information interface design methods are not targeted, can not well meet the information needs of different users. In this environment, it is of great practical significance to study individual's cognitive decision-making behavior and explore its difference for optimizing interface information design and satisfying different users' information needs in the decision-making process. In recent years, the relationship between need for cognitive (NFC), a special cognitive trait, and behavior has attracted more and more attention from scholars in other fields. Need for Cognition (NFC) describe the tendency of individuals to engage in and enjoy certain cognitive activities that require effort and are often used to predict higher-order cognition, such as attitude formation and decision-making. Taking the decision-making scenario of waiting position for rail transit as an example, this study examines the effects of heuristic cues on decision-making behavior of users with different Need for Cognition (NFC), to verify the feasibility of using heuristic clues to design interface information to help improve the decision-making efficiency of users with low Need for Cognition (NFC). The findings are as follows: (1) the users with higher need for Cognition (NFC)s take longer time to make decisions under the four kinds of information interfaces, and the correct rate of decision-making results is higher. (2) the users with low need for Cognition (NFC) have longer decision-making time and lower correct rate in the information interface without heuristic clues, and have shorter decision-making time and higher correct rate in the information interface with heuristic clues. The level of Need for Cognition (NFC) has an effect on the efficiency of users' cognitive decision-making, and using heuristic clues to design interface information can improve the decision-making efficiency of users with low Need for Cognition (NFC).

Keywords: Need for Cognition (NFC) · Information visualization · Decision making

1 Introduction

With the rapid development of information technology and the sharp increase of the amount of information resources, we should pay attention to the problems brought by

C. Stephanidis et al. (Eds.): HCII 2021, LNCS 13094, pp. 3–15, 2021.
https://doi.org/10.1007/978-3-030-90238-4_1

the flood of information to users. Because the richer the sources of information and the more options available, the greater the conflict between the breadth and depth of information sharing and human cognitive ability, and the greater the uncertainty users feel in the decision making process, the "Psychological cost" of access to information will rise, leading to confusion, and ultimately difficult to choose. People's cognitive decision-making behavior is influenced by information presentation and individual cognitive traits, but the current information interface design methods are not targeted and can not meet the information needs of different users. In this environment, it is of great practical significance to study individual's cognitive decision-making behavior and explore its difference for optimizing interface information design and satisfying different users'information needs in the decision-making process.

In recent years, the relationship between need for cognitive (NFC), a special cognitive trait, and behavior has attracted more and more attention from scholars in other fields. Need for Cognition (NFC) describe the tendency of individuals to engage in and enjoy certain cognitive activities that require effort and are often used to predict higher order cognition, such as attitude formation and decision making. Studies show that high NFC users are more likely to use target specific perceptual cues to provide valid information for target selection, while low NFC users are more likely to use heuristic cues for judgment. The purpose of this study is to examine the influence of heuristic cues on decision-making behavior of users with different need for Cognition (NFC)s, to verify the feasibility of using heuristic clues to design interface information to help improve the decision-making efficiency of users with low Need for Cognition (NFC).

2 Background

2.1 Need for Cognition (NFC)

The concept of Need for Cognition (NFC) was first proposed in the 1950s by Cohen [2] and others, who described it as "The Need to organize the situation in a meaningful, integrated way". Stress the individual's sense of tension when this need is not met, thus prompting the individual to make a positive effort to organize the situation and improve understanding. Contemporary research on individual differences in Need for Cognition (NFC) began with Cacioppo and Petty (1982). Initial factor analysis studies by Cacioppo and Petty (1982) [4] show that many individual differences in people's propensity to participate in and enjoy cognitive effort can be expressed in a single factor, known as cognitive need. need for Cognition (NFC), which refers to the difference of cognitive motivation, is often used to predict higher-order cognition, such as attitude formation and decision-making.

Chaiken (1987) [5] used 20 projects from the NCS (Cacioppo and Petty 1982) and eight others to more specifically mine individual differences in heuristic processing trends. The items in each study were based on a single dominant factor: people with low need for Cognition (NFC)s were more likely to approve of items that described heuristics than to stay alert or struggle with information processing, people with high need for Cognition (NFC)s, on the other hand, are more likely to approve of items that describe goal related content: laborious rather than heuristic information processing. Individuals with cognitive deficits tend to avoid processing information arguments and rely on simple cues in the environment, but in some cases these individuals can be motivated to think about relevant information and avoid leading to biased cues. Cognitive ability and cognitive motivation are the key factors influencing individual decision-making behavior, and need for Cognition (NFC) affects decision-making through cognitive motivation. People with different Need for Cognition (NFC) have different cognitive involvement and cognitive liking, which leads to different behaviors when they face cognitive tasks that require effort. Compared with the individuals with lower level of Need for Cognition (NFC), the individuals with higher level of Need for Cognition (NFC) need to pay more efforts on specific cognitive tasks, contact more comprehensive information and information processing more efforts and more profound. Users with high cognitive ability are more likely to use target-specific perceptual cues to provide effective target recognition information, while users with low cognitive ability are more likely to use heuristic cues to make judgments.

Need for Cognition (NFC) reflects the difference of individual's cognitive motivation. Measuring need for Cognition (NFC) can help us understand the relationship between individual's internal characteristics and external behavior, and can help us optimize the way of information presentation, to make it more relevant to the psychology of different users.

2.2 Information Visualization

In recent years, information visualization has become increasingly important as a means of managing the vast amount of digital information available to users, and many applications need to be able to help users understand and manipulate data volumes of varying levels of complexity. The information visualization aims to provide graphical representations that help deal with this complexity.

Information visualization, sometimes referred to as abstract visualization, is a visual tool for visually representing abstract concepts that are not inherently spatial because there is no explicit physical representation of the data involved. The data represented can highlight visual patterns of concepts extracted from the data, such as clustering. Keim [8, 9] classifies these visualizations as exploratory analysis visualizations, the goal of which is to help users identify assumptions. The power of these tools lies in their ability to represent large amounts of information simultaneously, including attributes and internal relationships, such as encoding information with colors and shapes (Fig. 1). Because of people's innate perception, this information is readily apparent. These tools can be used by trained professionals as well as non-professional users.

However, recent research has shown that individual differences do have a significant impact on task efficiency and user satisfaction during information visualization. Some studies have found that personality traits affect user performance through different visual information designs. Vélez et al. [13] found that a user's spatial reasoning ability (for example, spatial orientation) is related to visual comprehension. Similarly, Conati & Maclaren [14] and Toker et al. [15] found that cognitive abilities such as perceived speed, visual/verbal working memory, and expertise can influence user performance or subjective preferences through a given visualization. These studies suggest that it is important to develop visual information interfaces that match the individual capabilities of the information visualization to suit individual differences.

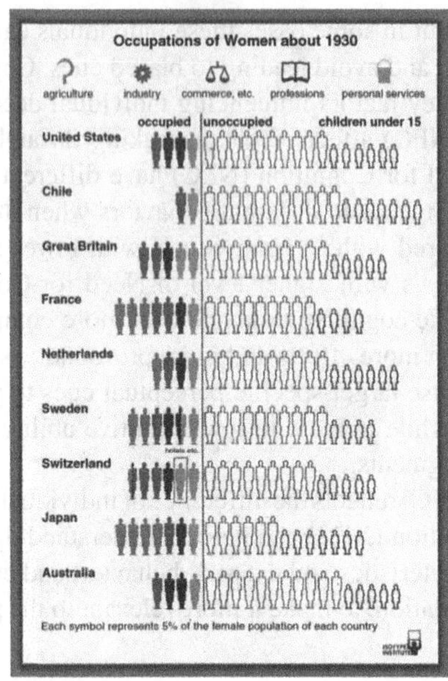

Fig. 1. Shows the proportion of women employed in different countries and their occupations in 1930 using colour and shape codes [10]

An information interface that matches a user's cognitive ability can help users understand complex information, guide their reasoning process, and reduce false reasoning in decision making to enhance their decision making. Therefore, in the design of information visualization, it should be taken into account that individuals often exhibit limited cognitive abilities when dealing with complex information, and that providing them with appropriate visualization tools can help minimize these limitations.

3 Experiment

3.1 Methodology

This paper mainly studies the influence of heuristic interface information design on the decision-making behavior of users with different Need for Cognition (NFC).

The first part of the experiment is the test of user's need for Cognition (NFC) scale, that is, the ability of users to understand and analyze the received information is measured by two levels (high need for Cognition (NFC) and low need for Cognition (NFC)). The second part of the experiment is the user decision behavior test, in which the participants browse the interface of information needed for decision making, these include an information interface that contains digital cues and three information interfaces that contain graphical heuristic cues (1 color as heuristic cue/2 shape as heuristic cue/3 color and shape as heuristic cue combination), after browsing the information interface, the subjects select the decision-making questions displayed on the screen, and record the subjects' decision-making results (decision-making time, decision-making accuracy).

The purpose of the task is to make the decision-making process as comprehensive and accurate as possible, and to satisfy the result.

3.2 Subjects

This experiment is an online experiment. We recruited 100 participants from the online experiment. The participants were all college students and master's degree students. Each experiment lasted about 30 min.

3.3 Materials

Need for Cognition (NFC) Scale. Thirty years after the concept of "Need for Cognition (NFC)"was proposed, Cacioppo and Petty published the first Need for Cognition (NFC) scale in 1982. Since then, Cacioppo, Petty and Kao revised the original Need for Cognition (NFC) scale in 1984 from 34 items to 18. All the subsequent studies were conducted on the basis of this scale. The most important one is the revision of the Need for Cognition (NFC) scale for college students by Kuang Yi and others. In this paper, the Need for Cognition (NFC) scale of Cacioppo abroad is modified by Chinese version, to make it more applicable to Chinese college students. In this experiment, NFC [16], using the Chinese version of Cacioppo and Petty Need for Cognition (NFC) scale revised by Kuang Yi of Department of Psychology of Peking University, had 17 items. The overall internal consistency coefficient of the scale was 0.89, the test-retest reliability coefficient is 0.86. The test was scored by a 5-point scale. 1 indicated that the statement of each question did not correspond to the individual situation of the subject at all, 5 indicated that it did. The total score of the scale is the need for Cognition (NFC) score of the subjects.

Passenger Information System Interface. In urban rail transit, the uneven distribution of passengers in train compartments and waiting platforms will affect the efficient operation of rail transit stations. When the passenger flow in the station is large, if a large number of passengers accumulate in the waiting area near the entrance of the station, it will hinder the passengers in the passage from entering the station quickly, and it is easy to form a bottleneck at the entrance of the station, resulting in safety accidents such as crowding and even trampling At the same time, the unbalanced passenger loading rate not only affects the passenger comfort, but also affects the passenger flow efficiency, and then affects the efficiency of rail transit. Therefore, it is necessary to present the relevant information through the passenger information system to guide the passengers in waiting position selection.

The waiting position selection behavior of platform passengers is affected by the passenger body, platform characteristics, train operation and other factors. The results of the field investigation and the L O G I T model analysis show that the train compartment full load rate and the waiting position queue length are the main explanatory variables of the non-passenger subjective factors when the passenger chooses the waiting position.

Take the Military Museum Station of Beijing Metro Line 1 (apple orchard direction) as an example. The platform is an island platform with 24 waiting positions on one side. The diagram of the platform is shown in Fig. 2.

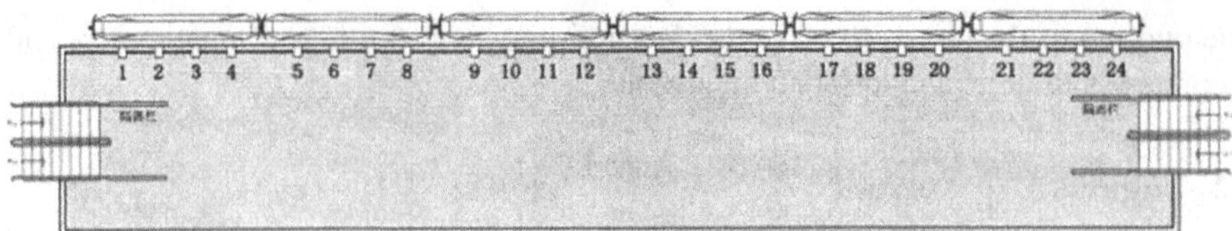

Fig. 2. Schematic view of the Military Museum Station 1 platform

The train is type A, each train has six cars, each car has four doors on one side, as shown in Fig. 2, the serial number is 24 waiting positions on the platform. The density of passengers in each compartment and the number of people waiting in the waiting area of four trains between 17:30 and 18:15 during the evening rush hour were selected as experimental materials. The passenger information system interface is designed as follows:

Number-coded clue interface design: in the digital cue interface, the number of cars and queues are displayed with digital information, as shown in Fig. 3.

Fig. 3. Number-coded clue interface design

Colour-coded clue interface design: according to the criteria for evaluating and recommending the density of standing passengers in vehicles [17], the crowding of cars can be divided into four grades according to the density of standing passengers, they are shown in green, yellow, red, and black, as shown in Table1.

The number of people waiting for a train is also divided into four grades according to the number of people. Below 10 people are shown in green, between 11 and 20 people are yellow, between 21 and 30 people are red, and above 31 people are black.

The color-coded design of the interface information is shown in Fig. 4.

Table 1. Class of compartment crowding

Class of compartment crowding	Crowded conditions	Average density of standing passengers (person/m^2)	Color
Comfort	Passengers can move freely, comfortable and satisfied	3 people/m^2 and below	Green
Slightly crowded	Passengers have contact with the body, the need for dislocation, more crowded	4–6 people/m^2	Yellow
Crowding	It is difficult for passengers to move around and it is very crowded. It is difficult for passengers waiting to board the train to get into the carriage	7–9 people/m^2	Red
It's terribly crowded	Extremely crowded, unbearable, not allowed to use	10 people/m^2 and above	Black

Fig. 4. Colour-coded clue interface design

Shape-coded clue interface design: the number of queues in the waiting area and the degree of congestion in the compartment are coded according to shape. The degree of congestion in the compartment is divided into four grades according to the evaluation standard and the recommended standard of the seat density in the vehicle, as shown in Fig. 5.

Fig. 5. Shape-coded clue interface design

Color and shape combinatorial coded clue interface design: color and shape coding of compartment congestion and queue size in the waiting area, as shown in Fig. 6.

Fig. 6. Color and shape combinatorial coded clue interface design

3.4 Process

The experiment consists of two parts. The first part of the experiment is the test of user's need for Cognition (NFC) scale, that is, the ability of users to understand and analyze the received information is measured by two levels (high need for Cognition (NFC) and low need for Cognition (NFC)). The second part of the experiment is the user decision behavior test, in which the participants browse the interface of information needed for decision making, these include an information interface that contains digital cues and three information interfaces that contain graphical heuristic cues (1 color as heuristic cue/2 shape as heuristic cue/3 color and shape as heuristic cue combination), after browsing the information interface, the subjects select the decision-making questions displayed on the screen, and record the subjects'decision-making results (decision-making time, decision-making accuracy). The purpose of the task is to make the decision-making process as comprehensive and accurate as possible, and to satisfy the result. Each participant completed a total of 34 experiments (2 practice experiments, 8 numbers + 8 shapes + 8 colors + 8 color shapes combination experiments). The whole experiment lasted about 30 min.

4 Results

According to the data obtained from the experiment, the correct rate and reaction time of the decision-making behavior experiment are the dependent variables, a two-factor analysis of variance (2) (need for Cognition (NFC) level: high vs low) × 4 (information presentation: Digital Cue/color-based heuristic cue/shape-based heuristic cue/3 color and shape-based heuristic cue) was performed. The results showed that the effect of information presentation mode on reaction time was significant ($f = 30.706$, $p = 0.000 < 0.001$), but not significant ($f = 3.099$, $p = 0.083 > 0.05$). The moderating effect of need for Cognition (NFC) on information presentation and decision reaction time was significant ($f = 6.130$, $p = 0.016 < 0.05$), but not significant ($f = 0.007$, $p = 0.933 > 0.05$).

4.1 Analysis of Need for Cognition (NFC) Test Results

Through the questionnaire survey and data processing (such as data entry, the reverse question into the positive, etc.) to obtain the total score of Need for Cognition (NFC). According to previous studies, there is only one dimension of need for Cognition (NFC), and finally, the scores of all items are added together and arranged from high to low. The items of the questionnaire are scored on a scale of 1 to 5. Choose 4 or 5 for high need for Cognition (NFC) and 1 or 2 for low need for Cognition (NFC), so a score above 68 is a high score, and a score below 34 is a low score. There were 56 people in the low NFC Group and 44 in the high NFC Group.

4.2 Correct Rate

For all users, the way of information presentation has some influence on the correct rate of decision-making. The user has the highest accuracy rate when making decisions in an information interface composed of heuristic cues of shape and color, as shown in Fig. 7. The accuracy of the information interface of the digital clue and the color clue was lower.

For all subjects, the level of need for Cognition (NFC) has a certain impact on the correct rate of decision-making. Users with high NFC scores make better decisions overall than users with low NFC scores, as shown in Fig. 8. The users with high NFC scores

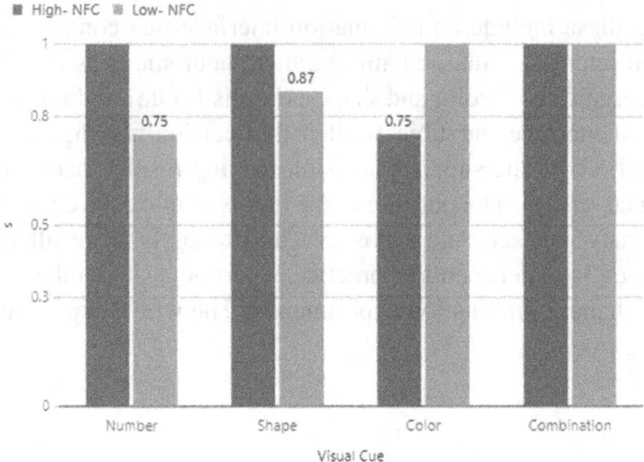

Fig. 7. Decision-making correct rate under visual clue grouping

were more consistent in the information interface of the four information presentation modes, and the users with high NFC scores were more accurate in the information interface of the digital clues and the shape clues than the users with low NFC scores. With a combination of shape and color cues, users with high NFC scores were more likely to be correct than those with low NFC scores. However, users with high NFC scores were less likely than those with low NFC scores to be correct in the color cue message interface.

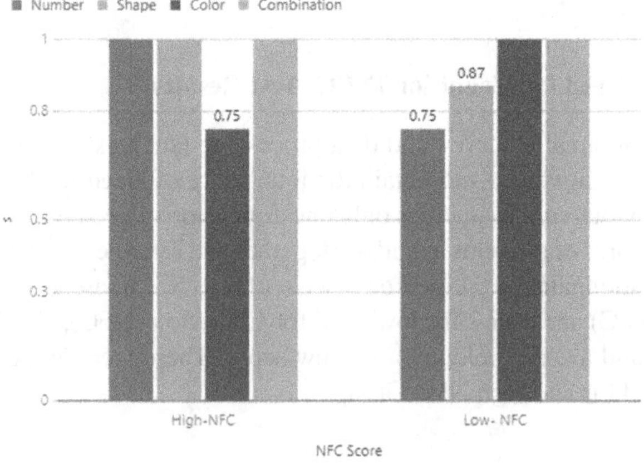

Fig. 8. NFC score grouping for decision correctness

4.3 Response Time

For all users, the way of information presentation has a significant impact on the decision-making reaction time. The decision-making time of users is the shortest and the highest under the information interface with color-based heuristic clues, and the decision-making time is the longest under the interface with digital clues, as shown in Fig. 9.

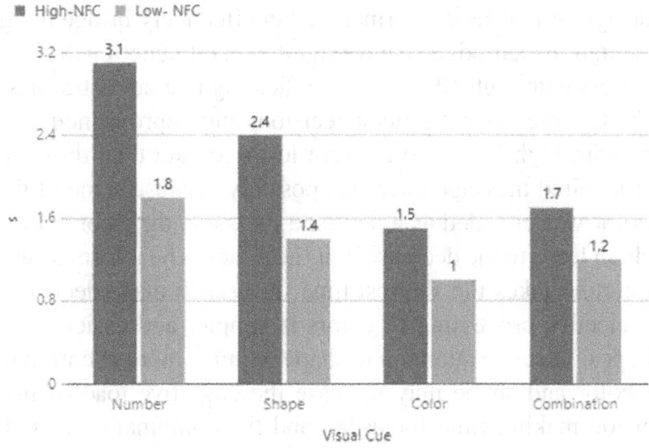

Fig. 9. Decision-making response time under visual clue grouping

For all subjects, the level of need for Cognition (NFC) has a certain impact on the correct rate of decision-making. Users with high NFC scores make decisions that are slightly higher overall than users with low NFC scores, as shown in Fig. 10.

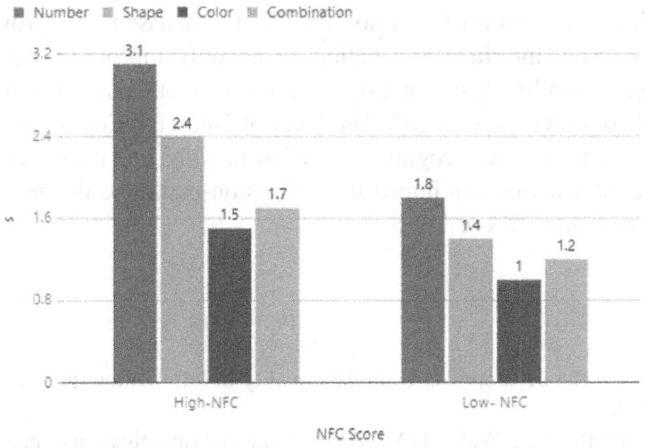

Fig. 10. Decision-making response time under NFC score grouping

5 Discussion

According to the analysis of the experimental results, the level of need for Cognition (NFC) has a certain impact on the efficiency of decision-making. Compared to users with high NFC scores, users with low NFC scores had lower decision accuracy in the digital cue interface, indicating that the ability of low NFC users to process digital information was different from that of users with high NFC scores. Under the heuristic information interface of shape and color, the accuracy of the two is consistent and high, which shows that the visual design of interface information can effectively bridge the gap. Different information presentation methods also have significant influence on the decision-making reaction time of users with high NFC scores, indicating that heuristic cues can also help users with high NFC scores to make quick decisions and improve their decision-making efficiency. Users with high NFC scores were less accurate than those with low NFC scores in a color-inspired message interface, possibly because some of the information was lost when color was encoded as a heuristic message, the information is expressed incompletely, which leads to the decrease of its accuracy. The information interface with color as heuristic clues takes the shortest time for user to make decisions, which may be because the selective processing of colors is simpler and easier than the selective processing of shapes, selective attention to shape requires more attention resources. The combination of color and shape may increase the cognitive load of users and result in a longer decision-making time for users, and the combination of color and shape may increase the cognitive load of users, but it is still the most efficient way to present information in this experiment.

6 Conclusion

In this study, taking the decision-making scene of waiting position for rail transit as an example, the effect of information presentation on the decision-making behavior of users with different need for Cognition (NFC)s is discussed, it is proved that using heuristic clues to design interface information can not only improve the decision-making efficiency of users with low need for Cognition (NFC)s, but also shorten the decision-making time of users with high NFC. The level of Need for Cognition (NFC) has an effect on the efficiency of users'cognitive decision-making, and using heuristic clues to design interface information can improve the decision-making efficiency of users with low Need for Cognition (NFC).

References

1. Song, L., Li, J.: The development from data quality to information quality. Inf. Sci. **28**(2), 182–186 (2010)
2. Cohen, A.R., Stotland, E., Wolfe, D.M.: An experimental investigation of need for cognition. J. Abnorm. Soc. Psychol. **1**(2), 291–294 (1995)
3. Fleischhauer, M., Miller, R., Enge, S., et al.: Need for cognition relates to low-level visual performance in a metacontrast masking paradigm. J. Res. Pers. **48**(Complete), 45–50 (2014)
4. Cacioppo, J.T., Petty, R.E.: The need for cognition. J. Pers. Soc. Psychol. **42**, 116–131 (1982)
5. Chaiken, S.: The heuristic model of persuasion. In: Zanna, M.P., Olson, J.M., Herman, C.P. (eds.) Socialin Fluence: The Ontario Symposium, vol. 5, pp. 3–39. Erlbaum, Hillsdale (1987)
6. Chen, Y., Wang, Y., Zhong, J.: The moderate effect of cognitive needs on the options time and decision aversion for consumers. Chin. J. Ergon. **18**(3), 41–44 (2012)
7. Shneiderman, B.: The eyes have it: a task by data type taxonomy for information visualizations. In: Proceedings of the IEEE Symposium on Visual Languages, 3–6 September 1996, pp. 336–343. IEEE Computer Society, Washington (1996)
8. Keim, D.A.: Visual exploration of large data sets. Commun. ACM **44**(8), 38–44 (2001)
9. Williams, G.: Women and Work. Nicholson & Watson, London (1945)

10. Velez, M.C., Silver, D., Tremaine, M.: Understanding visualization through spatial ability differences. In: Proceedings of Visualization, pp. 511–518 (2005)
11. Turner, M.L., Engle, R.W.: Is working memory capacity task dependent? J. Mem. Lang. **28**(2), 127–154 (1989)
12. Conati, C., Maclaren, H.: Exploring the role of individual differences in information visualization. In: Proceedings of the Working Conference on Advanced Visual Interfaces (AVI 2008), pp. 199–207. ACM, New York (2008)
13. Toker D., Conati C., Steichen B., Carenini G.: Individual user characteristics and information visualization: connecting the dots through eye tracking. In: Proceedings of the ACM SIGCHI Conference on Human Factors in Computing Systems, (CHI 2013) (2013, to appear)
14. Kwong, Y., Shi, J., Cai, Y.: Revision of the cognitive needs scale of college students. Chin. Ment. Health J. **19**(1), 57–60 (2005)
15. Liu, X., Qu, Y., Wang, W., Wu, J.: Analysis of passenger waiting position selection behavior and logit modeling on urban rail platforms. Shandong Sci. **33**(167.06), 81–88 (2020)
16. Shen, J.: Analysis on vehicle capacity and congestion. Urban Express Rail Transit **20**(5), 14–18 (2007)

Interaction Design for the Next Billion Users

Sumesh Dugar[1(✉)], Abhishek Mitra[2], Shweta Nandi[3], Biswajit Adhikary[4,5], Sonit Paul[6], and Madhav Manusuriya[4,5]

[1] CEO Webbies, Kolkata, India
sumesh@webbies.co
[2] Integrated Lifestyle Product Design, Kerala State Institute of Design, Chandanathope, Kollam, Kerala, India
[3] Aumni Techworks, 201, Suman Business Center, Kalyani Nagar, Pune 411006, India
[4] Fyle Technologies Pvt. Ltd., Princeton University, Princeton, NJ 08544, USA
[5] Paexskin Solutions Pvt. Ltd., Springer Heidelberg, Tiergartenstr. 17, 69121 Heidelberg, Germany
[6] Alinea Invest, Inc., New York, NY, USA
https://webbies.co

Abstract. As the first pandemic wave of Covid-19 hit India and the Indian Government announced a nationwide lockdown in March 2020, [1] it left millions of interstate migrant workers/daily-wage earners without a source of livelihood. Many of them had to return to their hometowns, often hundreds of kilometers away, with many walking a major part or the entirety of the distance [2].

This situation called for a design that could empower migrant workers to thrive in the new normal. An app was designed to enable migrant workers to search for jobs, learn about micro-entrepreneurship, get mental health assistance and develop skills.

In terms of usage and interaction, this app proposes several unique features - a result of researching about and understanding our target demographic. At the core of the interaction is a chatbot. It asks questions to the user through text and audio, but the user replies through clickable buttons only. This makes it easy, even for those who may not be literate. The keyboard has been removed from this app. Another unique feature is the usage of interactive videos to explain various things, such as micro-entrepreneurship, skill training, etc. These videos make use of buttons for the users to make choices.

In this paper, we discuss these two new interactions that will introduce a visual input method and an interactive video interface which will be inclusive of those less-privileged in terms of literacy in our ever-expanding tech world.

The contents of the paper are targeted toward helping the next billion users. However, some of the things proposed may go on to improve interactions for all.

Keywords: Interaction design · Migrant workers · App design

1 Introduction

1.1 Fundamentals

India has a large workforce of rural population that migrates from rural areas to urban areas in search for jobs. They are referred to as migrant laborer. These laborer work

C. Stephanidis et al. (Eds.): HCII 2021, LNCS 13094, pp. 16–23, 2021.
https://doi.org/10.1007/978-3-030-90238-4_2

on a daily wage, many keeping temporary or casual jobs. They make an average of $100–$200 if they are engaged on most working days.

Based on 2011 census data, it is estimated that currently there are about 65 million interstate migrants, across India and 33% of these migrants are workers. By conservative estimates, 30% of them are casual workers and another 30% work on regular basis but in the informal sector [3].

During the COVID-19 lockdown in India in 2020, millions of migrant workers lost their livelihood and approximately 10 million migrant workers returned to their native towns and villages, some even on foot.

With over 90% of the population working in the informal economy, ILO has predicted that as a result of the crisis and subsequent lockdown, about 400 million workers will fall deeper into poverty (this includes migrant as well as non-migrant workers) while forcing many of them to return to their places of origin in the rural areas [4].

This situation created a massive number of unemployed migrant workers and daily wage earners.

400 million workers in the informal economy, constituting 90% of India's workforce, were at risk of falling deeper into poverty. Outraged by the condition of the migrant workers, public policy academics have described the COVID-19 lockdown as 'the choice between virus and starvation' (Chen, 2020) [5].

In the context of this paper, low wage migrant workers are being focused upon. Such workers are usually engaged in factories, agriculture and construction industries. Many of them have very little education and usually work in urban centers that are far away from their rural homes.

Such migrant workers are economically weak, and often send a portion of their earnings back to their native place to their family members [6].

1.2 Personas

Based on our understanding of this target demographic, 2 user personas were developed.

Persona 1

(See Fig. 1).

Fig. 1. Persona 1 - A male construction worker

Raju Manjhi (Patna, Bihar)
29, Male
Construction Worker.

About Raju. Raju is from the state of Bihar, India. He was working as a construction worker in Mumbai, Maharashtra until COVID-19 lockdown was imposed. He lost his job.

Family Setting. Has a wife and 2 small children, a girl and a boy, lives far away from them in a city. His family lives back in his village.

Income (before losing his job). INR 12000 per month.

Education. Class 10 pass.

Values & Goals. Hard worker, wants to earn for family, hopes to give children better life, sends most of the money back home.

Worries. Not being able to provide for his family, not having food or shelter.

Influences. Popular culture (TV radio, songs), Local authorities, colleagues, employers.

Persona 2
(See Fig. 2).

Fig. 2. Persona 2 - A female house maid Meera (Howrah, West Bengal), 42, Female, House Maid

About Mira. Mira is from West Bengal. She was working as a house maid in Bangalore until COVID-19 came. Due to infection concerns, her employers discontinued her services.

Family Setting. Has a husband and a son. Lives with them in a city far away from her town.

Income (Before Losing His Job). INR 8000 per month.

Education. Class 12 pass.

Values & Goals. Hard worker, uses money to augment husband's income for running the household and for educating her son.

Worries. Not being able to provide for child's education.

Influences. Popular culture (TV radio, songs), members of the household where she works, neighbours.

In this context, designing an app for this demographic comes with its own set of challenges.

An app was designed to enable the migrant workers to do the following:

1. Find job opportunities
2. Learn about micro-entrepreneurship
3. Get mental health assistance
4. Learn new skills

It was important to create an interaction method that would be convenient to use for semi-literate and illiterate population. The interaction was heavily focused on clickable buttons for the user.

2 The Solution Design

Now that we have clearly defined the personas and problems, a feature set of a good solution needs to be demarked. A good solution would be one where:

1. The interaction is language/literacy independent.
2. The interface is simple.
3. The content and interface for learning content or skill upgrading is engaging.
4. The communication is personalized and contextual.
5. The keyboard needs to be ruled out and a new interaction mechanism needs to be imagined, that does not come with the assumption that the user recognizes character/alphabets.
6. Audio visual feedback will be a powerful tool (based on our findings that are targeted demography enjoys YouTube).
7. Interface announcements can serve as a virtual guide. This may go on to eliminate the current dependency on a more learned/literate operator.

2.1 The Interface

We propose that the keyboard is eliminated and replaced by visual inputs. Elaborating through an example, in cases where we are providing livelihood support, if the chatbot asks for 'What kind of jobs you are looking for?'. We make visual images of different options available like mechanics, Carpenters etc. Along with the visual identification, the system also makes on-screen announcements while highlighting each element in the options on the screen. The announcements will be made in the language of the user's choice, thus not only eliminating the need to understand an unknown language to put inputs but also becomes a virtual guide on how to run the system. This system also reduces the need for guidance and lowers the set-up barrier for operating tech products.

The Chatbot Experience in Brief

1. Open app
2. Chatbot sends a message and announces it
3. Lower half of the screen {keyboard area} is replaced by visual inputs
4. Announcements are highlighted one by one in a language of the user's preference
5. User selects the choice
6. Chatbot responds and makes announcements
7. Options for these questions are announced.
8. Option to replay announcements on touch for confirmation (Fig. 3).

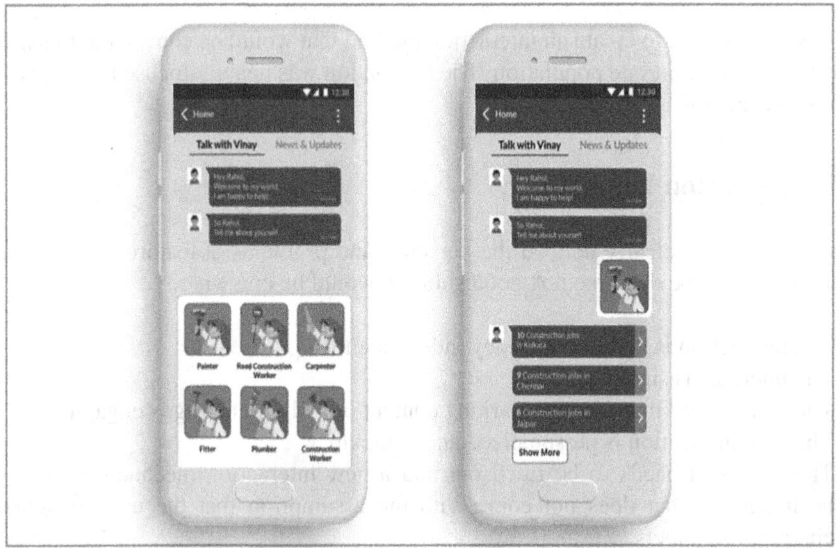

Fig. 3. Find the interactive video on titled Jobs.mp4 and Mental health.mp4: https://rb.gy/ydt9vg

The other interaction we recommend is an interactive video. Having understood that people we are catering to like spending time on YouTube, we came up with a tech interaction that is audio-visual yet at the same time allows easy navigation. Our proposed system allows for personalized educative content. To elaborate, for example if the concept of micro entrepreneurship in various industries needs to be explained, first the introduction of a video is played and then a question is asked,

What field of micro entrepreneurship would you like to learn more about?

1. Electronics
2. Farming
3. Religious products
4. Ayurveda

Once the user selects a particular choice the process will move forward. We retain the announcements on screen, similar to the visual chatbot instance. This makes the content engaging while personalizing it on the fly.

The Interactive Video Experience in Brief

1. Open app
2. Selects '*Micro entrepreneurship*'
3. Invited by a video message (in a language of the user's choice)
4. Video pauses and announces to make a choice on screen
5. Option to replay announcements
6. User selects an option and the video moves forward
7. Step 2 to step 5 is repeated (Fig. 4).

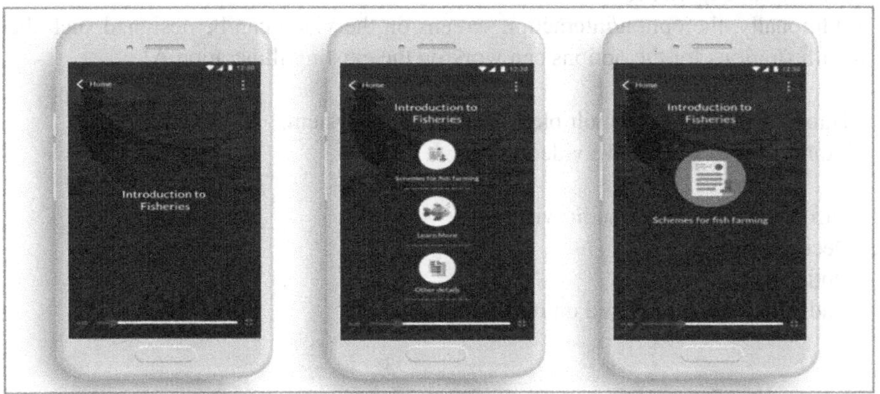

Fig. 4. Find the interactive video Titled Micro-entrepreneurship.mp4: https://rb.gy/ydt9vg

2.2 The Technology

The technology for a novel system like Vinay needs a correct approach and innovation mix. The technology will enable automated content creation as well as informed personalization to enable efficient decision making and increase spend time on the app.

Three places where we see usage of AI improving the scalability of the app.

AI for Illustration

- Creation of the app artwork that replaces keyboards:
- The simplest solution will be to manually draw up illustrations for inputs, however, this will lack scalability and act as friction to quick content creation.
- Thus, the advanced suggestion to enable scaling and quick content creation is the training of an AI system that can pick up keywords from admin's input, run's a google image search API, creates illustrations out of those image recommendations automatically. This will enable infinite scalability to questions/answers designed for the chat bot.

AI for Video Making

- Additionally, the interactive video making process can be automated. Content text can be created on the backend by the admin on the basis of slides he wants to show. The AI then goes ahead does a keyword analysis and creates the image illustration to create the slide and add music basis those keywords.
- The audio inputs will be a computer synthesized voice that is close to a human speaking.
- Finally, both the Slide creation AI's and audio AI's output can be merged to create an audio-visual experience.
- Additionally, the option/interaction screens on the video can be designed such that the illustrations for the options can work on the existing illustration AI.

Thus, the AI can serve multiple use cases in the system.
Example of an interactive video flow.

1. Introduction 3 slides (Audio-visual)
2. Decision screen
3. Option selection
4. Audio visual option based on response received.

Personalization AI

- This will be a key feature of the app to improve repeat usage.
- Once the system has understood basic liking and disliking, personalized notifications with audio announcements (Eliminating the need to read notifications) can be heard in a language of their choice.
- Once they tap on the notification, a popup will open with the option to replay announcements or to proceed with the desired action.
- Further personalization Ai will determine when the last time was that a successful action was received. Based on this, the AI will trigger content/notifications for increased engagement and interest development. The same can be based on opportunities of upskilling reskilling possibilities.

3 In Conclusion

In conclusion, our suggestions to expand the reach of technology through our proposed interfaces will not only reduce barriers to understanding interfaces, but it will also eliminate the need to have letters/alphabet recognition to input on devices. The paper sees this as the starting step to build a more equitable and inclusive system which recognizes more the needs of those still left out of availing the benefits of technological progress holistically.

Further the paper suggested AI models that have the potential to help create content at scale that maybe beneficial and used across various media generating and artwork generating platforms. The automation process will help creators focus on their content quality while the AI and systems can take care of representations and engagement. This again will democratize the content making process and enable automatic localization of content as it is image heavy.

References

1. Gettleman, J., Schultz, K.: Modi Orders 3-Week Total lockdown for all 1.3 billion Indians. The New York Times, 24 March 2020. ISSN 0362-4331
2. Slater, J., Masih, N.: In India, the world's biggest lockdown has forced migrants to walk hundreds of miles home. The Washington Post, 28 March 2020. Accessed 13 May 2020
3. https://indianexpress.com/article/explained/coronavirus-india-lockdown-migran-workers-mass-exodus-6348834/
4. https://www.ilo.org/wcmsp5/groups/public/---dgreports/---dcomm/documents/briefingnote/wcms_740877.pdf
5. https://journals.sagepub.com/doi/pdf/10.1177/2516602620933715
6. https://eprints.lancs.ac.uk/id/eprint/125072/1/MIGRATION_AND_POVERTY.pdf

The Crowd Thinks Aloud: Crowdsourcing Usability Testing with the Thinking Aloud Method

Edwin Gamboa[1(✉)], Rahul Galda[1], Cindy Mayas[2], and Matthias Hirth[1]

[1] User-Centric Analysis of Multimedia Data Group, TU Ilmenau, Ilmenau, Germany
{edwin.gamboa,rahul.galda,matthias.hirth}@tu-ilmenau.de
[2] u.works GmbH, Ilmenau, Germany
mayas@uworks.de

Abstract. Thinking aloud (TA) is a widely employed usability testing method and allows identifying usability problems based on testers' verbalization. Conducting TA studies is laborious and expensive due to the recruiting of participants, the required workforce and facilities, and the resulting monetary costs. Also, it is usually carried out in very controlled settings like laboratories, thereby limiting participants' diversity, test realism, and the quality of the results. Crowdsourcing, in contrast, provides access to a wide and diverse workforce at moderate costs. Hence, it might help to overcome the limitations of traditional TA. Although usability testing has been successfully crowdsourced with other methods before, there is still no evidence on the feasibility of conducting TA involving unsupervised workers as testers. Thus, we conducted a between-subjects user study using the example of an online web page, both via a traditional remote TA setup and a crowdsourcing setup with a self-developed platform. The results do not show a significant difference between both setups regarding participants' performance, and the quality and quantity of the identified usability problems. Therefore, we conclude that crowdsourcing is a feasible and cost-effective solution to conduct usability testing with the TA method, at least for selected use cases.

Keywords: Thinking aloud · Crowdsourcing · Usability testing

1 Introduction

Good usability is key to achieve user satisfaction and can be assured via usability testing. Thinking aloud (TA) is a method to capture what testers think while performing a task [24]. TA allows to uncover users' mental models and detect usability problems [9] and is probably one of the most used usability testing methods [8,20].

Nevertheless, traditional usability testing with TA is limited to local participants, settings, and scenarios [19]. Conducting TA sessions is usually expensive

C. Stephanidis et al. (Eds.): HCII 2021, LNCS 13094, pp. 24–39, 2021.
https://doi.org/10.1007/978-3-030-90238-4_3

and time-consuming [16,19], mainly due to the required resources to recruit participants and conduct the tests [12,28]. Similar drawbacks are also common in remote TA sessions [22]. These obstacles hinder usability practitioners' work, who are constrained by limited access to participants, time, and budget [12,25]. Thus, carrying out frequent TA sessions, as recommended in the literature [25], might be seen as impractical by practitioners [28].

Crowdsourcing might be a way to overcome some limitations of the traditional TA approach since it allows reaching a diverse, on-demand, and scalable workforce via the Internet in a time- and cost-effective manner. Indeed, other usability testing methods like the System Usability Scale (SUS) have been successfully crowdsourced [30]. However, crowdsourcing also creates new challenges. In a crowdsourcing setting, the participants are anonymous workers and there is in general not a possibility for direct communication with the workers, unlink to laboratory studies, traditional remote studies, and online panels. Further, the crowdsourcing workers are only recruited on a task-based via online market places, i.e., the workers complete a large variety of different tasks each day and are not trained for usability assessments. Previous works have detected a lower quality in the feedback provided by workers, e.g., via post-test questionnaires or scales [16,30]. Reasons for this might be the uncontrolled and unsupervised setting of crowdsourcing and the use of retrospective feedback. In this context, TA might improve participants' feedback quality since it does not rely on memory and avoid complex post-test tasks [9].

Therefore, we investigate the feasibility of crowdsourcing TA in usability testing by comparing a traditional synchronous remote concurrent TA (*Remote*) to a proposed asynchronous crowdsourced concurrent TA (*Crowd*). To aim this, we address four hypotheses. The first two compare both approaches regarding the identified usability problems and the last two the participant's behavior:

H1.1: The quantity of the identified usability problems per participant in the *Remote* setup is different from those in the *Crowd* setup.

H1.2: The quality of the identified usability problems per participant in the *Remote* setup is different from those in the *Crowd* setup.

H2.1: The performance of the participants in the *Remote* setup is different from those in the *Crowd* setup.

H2.2: The time the participants spent on the test in the *Remote* setup is different from those in the *Crowd* setup.

This paper is organized as follows: Sect. 2 presents the theoretical background of the TA method and gives a brief introduction to crowdsourcing, including its definition, relevance, benefits, and limitations. Moreover, we provide related work on crowdsourced usability testing as evidence for the potential to crowdsourced TA. Then, we detail the methodology and design of our study in Sect. 3. After that, in Sect. 4, we describe the demographics of the study participants. Further, we present and discuss the results of comparing both TA setups in Sect. 5. Section 6 summarizes our work and outlines future work.

2 Background and Related Works

In this section, we provide background information on the TA method and crowdsourcing. We also briefly outline works in which usability testing has already been successfully crowdsourced.

2.1 TA in Usability Testing

The TA method was initially proposed in [6] and allows identifying usability problems and their causes based on the tester's mental model [9,15,21,24]. During the test session, the testers speak their thoughts aloud while performing a given task. This allows understanding how they perform tasks, revealing when problems occur [15] and what must be fixed in the system under test [8]. TA is suitable for iterative development environments since it allows finding 75% of the problems with only 4 - 5 representative users [21].

Traditional TA can be *concurrent*, i.e., testers complete the given task and verbalize their thoughts simultaneously, or *retrospective*, in which participants work silently and verbalize their thoughts after finishing [9]. Furthermore, TA can be carried out *remotely*, i.e., evaluators are spatially separated from the testers. If evaluator and testers are additionally separated in time, it is known as *asynchronous* TA [13]. Research showed that concurrent TA is more effective than retrospective TA [9], and synchronous remote TA is as effective as concurrent TA [28].

Nevertheless, the demand to talk, the presence of a facilitator, and possible recording equipment create an artificial environment that can negatively influence the test realism and testers' behavior in traditional TA. TA studies are also laborious and expensive to conduct, due to the required logistics, equipment, and workforce. Even well-trained moderators can handle only four to six sessions in a day [24] and recruitment is usually limited to local testers, thereby limiting the diversity of the results [15]. These are crucial challenges for usability practitioners, who are usually under time and budget pressure [4,25].

2.2 Crowdsourcing

Crowdsourcing allows an individual or an organization, i.e., crowdsourcer, to propose the undertaking of tasks to a heterogeneous group of individuals, i.e., workers, via a flexible open call [7]. Performing a task results in a mutual benefit since the workers receive a kind of reward, e.g. economic or social recognition, and the crowdsourcer takes advantage of worker's work [7]. Typical crowdsourcing tasks include tagging content [3], conducting user surveys [2], collecting subjective user ratings [5], and completing and correcting data [17]. Crowdsourcing allows gathering feedback from a large and diverse group of workers [30] directly in their real-contexts [19] in a time- and cost-effective way [11,25,30].

Nevertheless, the uncontrolled crowdsourcing setting implies challenges mainly related to the anonymity of the workers [30] and the absence of a test facilitator [11,25]. This might result in low-quality results; thus, task design is a crucial activity in this context, which includes designing unambiguous and simple instructions [16], transforming complex tasks into micro-tasks [11], and including quality and reliability checks to identify and filter out unreliable workers before, during and after tasks execution [11]. Also, different experimental settings, such as browser versions, video and audio recording equipment, and language, must be controlled [11,12,19].

2.3 Crowdsourced Usability Testing

Usability testing has been already crowdsourced without using TA and there is evidence of its benefits and drawbacks. For instance, a university website was tested in laboratory TA sessions and via crowdsourcing in [16]. The crowdsourced sessions included after-test open-ended questions about the system and required participants to complete the SUS instrument. Liu et al. showed that crowdsourcing allows reaching more participants with various backgrounds while lowering test duration and cost. Nevertheless, the authors found that the quality of the feedback collected via their crowdsourcing approach was lower than the one from the laboratory sessions. Moreover, the workers were less focused and spammers had to be handled. In [30] an overestimated perceived usability from workers was identified. Thus, a modified version of the SUS instrument was proposed to identify spammers. Nebeling et al. [19], introduced CrowdStudy, a tool to crowdsource usability testing. The tool guides testers by highlighting screen elements that are relevant to conduct a task, such as texts, and images. Using the tool, the authors found that the lower costs and access to real context offered by crowdsourcing outweigh the lower quality of feedback from workers. Finally, in [18] crowdsourcing was shown to produce comparable results to traditional usability testing of the information architecture of a website using navigation stress tests. In general, these research studies have shown that crowdsourcing's benefits might outweigh its drawbacks.

In summary, TA is a relevant method for usability testing that has limitations, which might be overcome via crowdsourcing as shown by previous works with other usability methods like the SUS scale. Furthermore, concurrent TA could enhance the quality of workers' feedback since it avoids retrospective feedback, which in previous works have been proven ineffective. However, the feasibility of crowdsourcing TA is still open for investigation.

3 Study Description

In this section, we first present an overview of the study. Then, we describe the system under test and the TA sessions procedure. Finally, we detail how the usability problems were identified, and how the participants' recordings were evaluated.

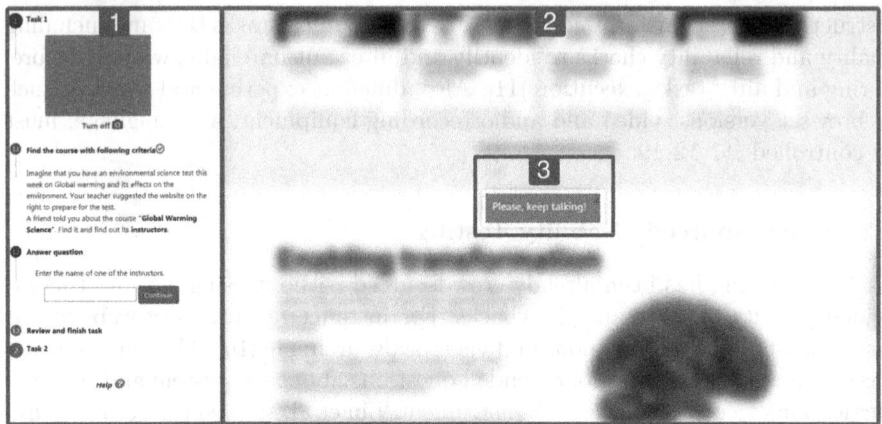

Fig. 1. The platform for the *Crowd* setup (1) presents tasks' description and question, (2) allows users interacting with the tested website, and (3) reminds them to verbalize when required.

3.1 Study Design

To investigate the feasibility of crowdsourcing for TA usability testing, we conduct a between-subjects study to compare a traditional *Remote* TA setup to the proposed *Crowd* TA setup. We use concurrent TA sessions for both setups as recommended in [13]. The *Remote* setup was realized synchronously via the Cisco Webex Meetings system[1], including a facilitator and an observer. The *Crowd* setup is asynchronous and a self-developed, web-based platform plays the role of the facilitator guiding workers through the sessions. No observer participated in the *Crowd* TA session. Figure 1 shows a screenshot of the developed platform. On the left (1), the platform presents the current task scenario, a question related to the task, the (un)completed steps, and a help option that gives access to the general instructions, the training information, and support contact details. On the right (2), it allows users to interact with the tested system, which is integrated via an *iframe*. Moreover, it shows the reminder (3) to speak their thoughts when needed. All texts and scripts in both setups employed simple non-technical English and followed the guidelines from [14]. We recorded the participants' audio for analyzing the verbalized thought and recorded the participants' screen while conducting the tasks. Optionally, the participants could also provide an additional webcam video of themselves. We also collected the participants' demographic information, experience with usability testing, and monitored the participants' task completion and the time spent in the tasks.

[1] https://www.webex.com/ Accessed June 2021.

3.2 Test Object and Tasks

The final test platform and task scenarios were selected after 3 pilot testing sessions with university students. We considered the following criteria to select the test system: (1) It should be targeted at a rather general audience, (2) allow conducting tasks without registering, and (3) allow integration within our platform via an *iframe* without hindering task completion. Therefore, we selected the *EDX learning platform*[2] as the system under test. Further, the tasks should be (1) representative of what actual users would do in the test platform and (2) diagnostic in relevant usability problems. The tasks used in our study and their respective completion criteria are presented in Table 1. In the *Crowd* setup, workers should answer an open-ended question to ensure task conduction and identify possible spammers.

Table 1. Description and completion condition of the conducted tasks.

Task	Description
1	Imagine that you have an environmental science test this week on global warming and its effects on the environment. Your teacher suggested the website on the right to prepare for the test. A friend told you about the course "Global Warming Science". Find it and find out its instructors *Completion condition:* Expected course instructors are found. Workers are asked to enter the name of one of the instructors
2	While studying global warming, you got interested in the topic "climate change". Now you want to find advanced courses about it, but only in the field of energy and earth sciences *Completion condition:* Courses on the topic are found using at least one of the filers. Workers are asked to enter the name of one of the found courses

3.3 TA Procedure

Both setups followed the same procedure divided into three parts depicted in Fig. 2. In the first part, the participants were welcomed and the test goal was explained highlighting that the system was evaluated and not the participant. Additionally, the *Crowd* platform presented payment information. Then, the participants provided consent for recordings, verbally for the *Remote* setup and by allowing the system to record in the *Crowd* setup. After that, the demographic data was collected. The second part included the TA training. In both setups, a demo video of a TA session was presented. The *Remote* setup included additionally a trial test, i.e., the participants should verbalize their first impression of the web page. In the third part, the participants read and completed the tasks in order. The only possible interaction with the facilitator was the reminder to think aloud to avoid affecting participants' verbalization as recommended in [6].

[2] https://www.edx.org/ Accessed June 2021.

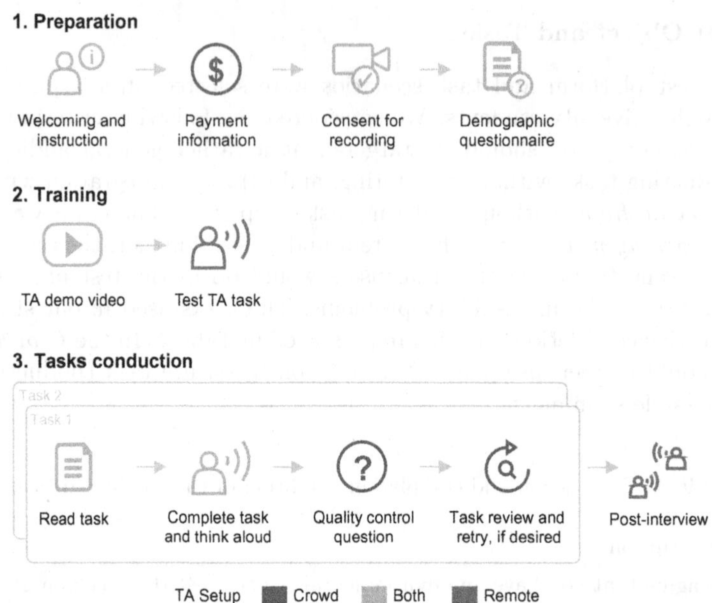

Fig. 2. Overview of the conducted procedure.

In both setups, the reminder was always "Please, keep talking." However, in the *Crowd*, it was displayed after the system detected 10 s of silence, while in the *Remote* setup, the facilitator reminded the participants after approximately 10 s to 20 s of silence. The *Remote* included also an observer for note-taking and monitoring. In both setups, the participants indicated when they finish, but in the *Crowd* setup, they were asked an additional open-ended question to ensure task conduction. Since the study depends on the quality of workers' verbalization, we allowed them to review their audio and screen recordings, and retry a task when desired. Finally, the *Remote* setup included a post-interview. The code of the developed platform, the employed training videos, and the instructions are open source and available online.[3]

3.4 Usability Problems Identification

A usability problem is defined as any aspect of the system that makes it unpleasant, inefficient, difficult, confusing, or impossible for participants to achieve given goals [28]. To identify potential usability problems observed in our test sessions, we coded the recordings to identify breakdown indicators, i.e., an event that may lead to some undesired outcome [29]. Breakdown indicators include, e.g., unexpected actions, interruption of the task, user indicating not to know how to complete a task. A complete list of breakdown indicators is presented in [29]. For

[3] https://github.com/nam-tuilmenau/crowd-ta.

our test sessions, two evaluators performed the coding independently, in random order to reduce the ordering effect and avoid biasing the results. After that, the evaluators discussed their results to produce a joint list of usability problems. Each problem is classified according to the scheme from [1] as:

- *Navigation*: indicates a problem to navigate between pages or identify a proper link to achieve a task.
- *Layout*: indicates a problem caused by layout, display, visibility, inconsistency, and structure issues.
- *Content*: indicates a problem related to the lack of information or understanding, or the presence of unnecessary information.
- *Functionality*: indicates a problem caused by the absence of necessary functions or the presence of problematic ones.

Additionally, the severity level of each problem is classified according to [26] as *critical* if it hinders the completion of a task, as *serious* if it delays the completion for several seconds, or as *cosmetic* if it does so for a few seconds.

4 Evaluation of the TA Setups

This section presents the results for the research hypotheses. Unless stated otherwise, we used the Wilcoxon rank-sum test to compare the data since it comes from two independent groups and is not normally distributed.

4.1 Participants

We recruited workers for the *Crowd* setup via the Microworkers[4] crowdsourcing platform. A pilot testing was initially performed from December 10 to 13, 2020. In this phase, we restricted participation to the US workers to test the correctness and clarity of the instructions with English native speakers. The workers were paid 1.70 USD to meet the minimum wage of 2020 according to the US Department of Labor. This pilot testing resulted in no changes, as no issues were identified in the submissions.

We published a second campaign internationally between December 14 and 17, 2020 to perform the actual usability study. The workers were paid 1.00 USD, as we expected most workers being from developing countries [10]. In total, 201 workers accessed our platform, but only 40 (19.90%) of them completed the session. To ensure the quality of the results, we excluded 15 (37.50%) workers' submissions for the evaluation, due to different reasons: 4 workers remained silent the whole session, 4 did not talk in English, 5 did not upload audio and screen recordings, 1 worker shared the wrong screen, and 1 did not watch the training video and did not verbalize during the tasks. Ultimately, 25 (62.50%) sessions remained for the study evaluation.

[4] https://www.microworkers.com/ Accessed June 2021.

The *Remote* usability tests were conducted between January 18 and 22, 2021. The participants did not receive any compensation. In total 11 volunteers participated, with 10 of them being university students and one a research assistant. Table 2 shows an overview of the participants' demographics of the *Crowd* and *Remote* usability test sessions.

Table 2. Overview of participants' demographic data.

Characteristic	Value	No. of participants	
		Crowd	*Remote*
Gender	Female	6 (24.00%)	6 (54.55%)
	Male	19 (76.00%)	5 (45.45%)
Age	19–25	12 (48.00%)	1(9.09%)
	26–30	6 (24.00%)	8 (72.73%)
	31–35	4 (16.00%)	2 (18.18%)
	36–40	3 (12.00%)	–
Location area	Africa	3 (12.00%)	–
	Asia	9 (36.00%)	–
	Europe	4 (16.00%)	11 (100%)
	North America	4 (16.00%)	–
	South America	5 (20.00%)	–
Native English speaker	Yes	6 (24.00%)	–
	No	19 (76.00%)	11 (100%)
Experience with usability test	Yes	9 (36.00%)	7 (63.64%)
	No	16 (64.00%)	4 (36.36%)

4.2 Quantity and Quality of Identified Problems

Regarding the the quantity of usability problems yielded by each setup (**H1.1**), the two evaluators identified 19 usability problems using the *Crowd* setup recordings, and 16 from the *Remote* setup ones. The *Crowd* setup yielded more unique problems (6) than the *Remote* setup (3), resulting in 22 unique usability problems combining both approaches. Although the *Crowd* setup resulted in more problems identified per participant (see Fig. 3a), we do not observe a statistically significant difference between both setups, as shown in Table 3. Similarly, when analyzing the number of problems of each type identified per participant (see Fig. 3b), we do not observe a statistically significant difference either. None of the setups yielded any navigation problems, which might be due to the pure searching nature of the tasks. Therefore, we do not observe evidence to support **H1.1**, meaning that both setups might not differ regarding the quantity of yielded usability problems.

Finally, we analyze the quality of the identified problems via the problems' severity level (**H1.2**). Figure 3c shows the number of times a participant identified a problem of a severity level. The number of cosmetic problems identified by participants in the *Crowd* setup is significantly higher than that from the *Remote* setup. This difference might be due to diversity and the higher number of participants in the *Crowd* setup. As stated in [13,21], TA allows identifying high relevant usability problems with a small number of subjects, and bigger samples result in detecting also less relevant problems. Furthermore, usually cosmetic problems are neglected due to time limitations in the development teams [13] Apart from that, the comparison does not result in a significant difference between both approaches. In both, the number of serious problems was higher than critical and cosmetic ones.

Table 3. Results of comparing both TA setups with the Wilcoxon rank-sum test.

Compared variable		W	p	r
No. of problems per participant (**H1.1**)		161.00	0.42	−0.13
No. of problems per participant by type (**H1.1**)	Content	129.50	0.77	−0.05
	Functionality	140.50	0.93	−0.02
	Layout	114.50	0.42	−0.14
No. problems per participant by severity level (**H1.2**)	Critical	128.50	0.72	−0.06
	Serious	150.50	0.66	−0.07
	Cosmetic	77.50	0.02*	−0.38
Participants' task performance (**H2.1 and H2.2**)	Completion rate	156.50	0.48	−0.12
	Time in Task 1 (s)	167.50	0.31	−0.17
	Time in Task 2 (s)	209.00	0.01*	−0.41

* $p < 0.05$. The effect size (r) is calculated using the z-score of the p-value as suggested in [23]

4.3 Participants' Performance and Spent Time

The Participants' performance is studied by comparing the task completion rate of the participants and the time they spent on each task. Figure 4a shows the share of participants who completed each task (**H2.1**). The share completion for Task 1 is higher in the *Remote* setup (18% more), but the completion share for Task 2 is the same. Similarly, we found that the number of tasks completed per participant does not differ significantly in both setups ($Median = 1.00$) as depicted in Table 3.

Figure 4b depicts results about the time spent by each participant in the tasks (**H2.2**). As Table 3 presents, the time spent by the *Crowd* and *Remote* participants in Task 1 does not differ significantly. However, we found that the participants from the *Remote* setup spent significantly more time in Task 2 than those from the *Crowd* setup. This can be expected since the presence of a facilitator and an observer might influence participants' behavior and they might spend more time seeking to complete the tasks [15]. Additionally, less engagement from workers was previously observed in [16].

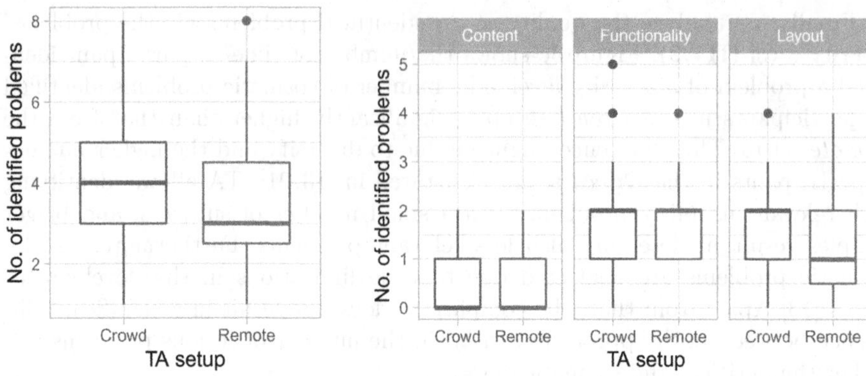

(a) Identified problems per participant.

(b) Identified problems per participant by type.

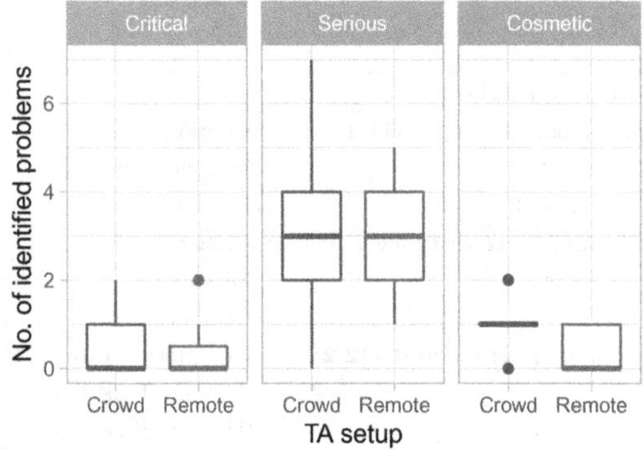

(c) Identified problems per participant by severity level.

Fig. 3. Number of problems identified per participant in each setup. The boxplots show the median, interquartile range, and maximum and minimum values.

Since the workers had the possibility to review their audio and screen recordings and retry the tasks when they were unsatisfied, we further analyze their behavior in both tasks, which is found to be similar. Approximately half of the workers reviewed their recordings before submission once or twice, 56% for Task 1 and 48% for Task 2. There were also 3 workers (12%), who reviewed the recordings of Task 1 at least 4 times. Similarly, 4 of them (16%) reviewed the recordings of Task 2 at least 3 times. Still, there was no significant statistical difference ($p = 0.63, r = -0.07$) between Task 1 ($Median = 1.00$) and Task 2 ($Median = 1.00$), when comparing the number of reviews per participant.

As expected, the results for the number of tries per task were similar. In Task 1, 48% of the workers tried the task once, and 40% did it twice. Also, 12% of them did it at least 4 times. Alike, 60% of the workers tried Task 2 once and 24% of them did it twice. Ultimately, 16% of them tried the tasks at least 3 times.

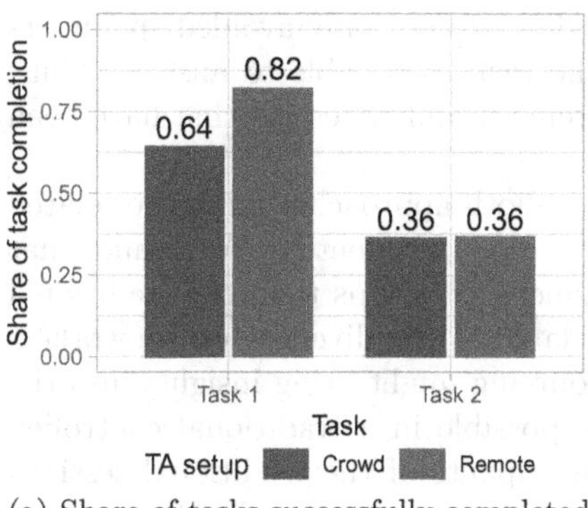

(a) Share of tasks successfully completed by all participants.

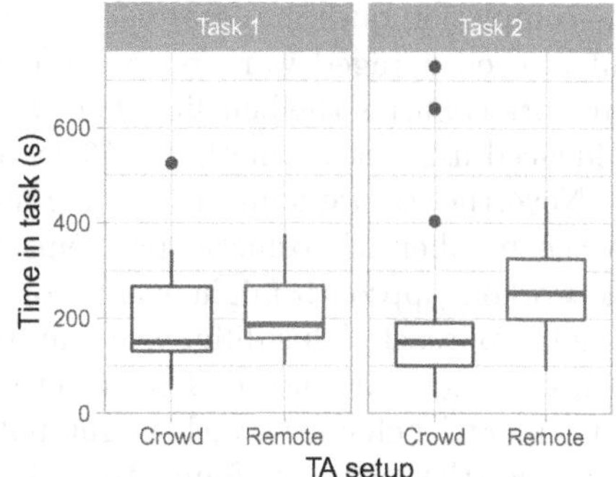

(b) Time spent in each task per participant.

Fig. 4. Overview of participants' performance on each task. The boxplot shows the median, interquartile range, and maximum and minimum values.

Again, we do not find any significant difference between the number of tries per participant in Task 1 ($Median = 2.00$) and Task 2 ($Median = 1.00$) after conducting a Wilcoxon signed-rank test ($p = 0.50, r = -0.14$). These results show evidence that the review opportunity might encourage some participants to deliver higher quality feedback. To avoid biasing the results due to multiple task tries, we only analyzed the first valid try of each worker, i.e., the first one in which they tried to conduct the task and verbalized their thoughts. Nevertheless, in a production scenario, all videos could be analyzed.

5 Discussion

In general, our results suggest that crowdsourcing usability testing with the TA method yields similar results to those of traditional remote sessions. We did not find significant differences when comparing the quantity (**H1.1**) and quality (**H1.2**) of the identified usability problems yielded by each approach. Similarly, the performance of the participants of the traditional sessions was not significantly different than the workers' performance (**H2.1**). Nevertheless, we found that the time invested in tasks might be higher in the traditional setup for more complex tasks (**H2.2**). These results might be due to the nature of the TA method since the problem identification process is based on the participants' verbalization and behavior while conducting a task. Thus, differences might come mainly from the quality of the collected verbalization. Contrary to [16,30], we did not detect lower quality feedback in the crowdsourced setup. Indeed, our work suggests that TA might enhance the quality of the feedback provided by workers. This could be due to the use of concurrent TA, which does not rely on participants' memory and does not require additional post-test tasks from the workers. Other reasons for this might be the possibility that we gave participants to review their audio and screen recordings before finishing each task. Moreover, the completion question after each task might have not only avoided spammers but also encouraged workers to conduct the task to be able to answer. Thus, our work confirms the benefits of crowdsourcing usability testing that have been evidenced using other methods [16,18,30].

Nevertheless, we found that differences in both approaches might be related to the number of cosmetic problems they allow to identify. We found that the workers approach might yield more cosmetic problems than the traditional remote approach. This difference may be related to the diversity of the workers and, thus, contexts of use. That is, crowdsourcing might bring insights into the participants' behavior which might not be possible in a traditional controlled setup. Furthermore, we found that the participants of the traditional sessions spent more time in Task 2 than the workers. This can be caused by the presence of a moderator and an observer [15], which can cause participants to spend more time trying to complete the task. This makes sense since Task 2 was more complex than Task 1. Contrary to this, workers approach the task without external pressure to come up with the correct path to task completion.

Although we did not quantify the effort invested in each setup, we estimate that the traditional setup was more time-consuming regarding recruitment and session conduction. The recruitment process included the invitation to participate, scheduling, and confirmation, and was carried out over a week. As stated in [27] this process requires on average 1.15 h per participant. Moreover, a facilitator and an observer were present and active in all traditional sessions. Some sessions took more time than expected since participants faced difficulties with screen sharing, internet connection, and audio. Therefore, at most 4 sessions were conducted in a day. In contrast, in the crowdsourced setup, we only needed to publish a job, which took some minutes for setting up. During conduction, a self-developed platform guided workers. We only monitored the platform occasionally to ensure correct functioning and approved or rejected submitted jobs in the crowdsourcing platform. Furthermore, we paid each worker at most 1.70 USD. On the contrary, a participant can cost 64 USD on average per session in traditional testing [27].

Therefore, our results are promising and demonstrate the feasibility of crowdsourcing TA at least for the selected use case. Our work can be used as a starting point for conducting cost-effective usability testing in real contexts overcoming traditional TA limitations and some challenges of collecting qualitative feedback from workers. Which might be relevant for development contexts in which resources are limited for Usability Testing.

This study is based on the use case of searching courses within a learning platform. Thus, the results cannot be generalized without studying further and more complex use cases. Additionally, we got a relatively low completion share in comparison with other crowdsourcing works [16]. A reason for this could be the requirement for recordings, which might raise privacy concerns. Also, TA in English might be considered difficult by non-native speakers, but this is still open for investigation. Furthermore, although remote TA has been found to be equivalent to classical laboratory TA [28], comparing the proposed approach with the classical one might be relevant. Additional future research includes the guaranty of privacy protection in the collected recordings, the handling of environmental noise that might affect speech quality, the usefulness of crowdsourcing to recruit specific audiences, and the effectiveness of different reminder to TA to avoid users describing what they see on screen.

6 Conclusion

TA is one of the most relevant usability testing methods in academia and the industry. However, conducting TA is highly cost- and time-consuming and limited to local testers or practitioners' recruitment capacity. Given this context, crowdsourcing might be useful to overcome such limitations. Thus, we investigated the feasibility of crowdsourcing TA for usability testing. To this aim, we compared a traditional remote concurrent TA setup to a proposed crowdsourced setup. Apart from differences in the number of identified cosmetic problems and the time spent by participants in one of the tasks, the comparison of both setups does not show further significant differences in the quantity and quality of the number of identified usability problems and the participants' performance. Additionally, we found that TA might enhance the quality of the feedback provided by workers. This confirms the potential of conducting cost-effective usability testing via crowdsourcing using the TA method.

References

1. Alhadreti, O., Mayhew, P.: To intervene or not to intervene: an investigation of three think-aloud protocols in usability testing. J. Usability Stud. **12**(3), 111–132 (2017)
2. Behrend, T.S., Sharek, D.J., Meade, A.W., Wiebe, E.N.: The viability of crowdsourcing for survey research. Behav. Res. Methods **43**(3), 800–813 (2011). https://doi.org/10.3758/s13428-011-0081-0
3. Bruggemann, J., Lander, G.C., Su, A.I.: Exploring applications of crowdsourcing to cryo-EM. J. Struct. Biol. **203**(1), 37–45 (2018). https://doi.org/10.1016/j.jsb.2018.02.006
4. Denning, S., Hoiem, D., Simpson, M., Sullivan, K.: The value of thinking-aloud protocols in industry: a case study at microsoft corporation. Proc. Hum. Factors Soc. Ann. Meet. **34**(17), 1285–1289 (1990). https://doi.org/10.1177/154193129003401723
5. Egger-Lampl, S., et al.: Crowdsourcing quality of experience experiments. In: Archambault, D., Purchase, H., Hoßfeld, T. (eds.) Evaluation in the Crowd. Crowdsourcing and Human-Centered Experiments. Evaluation in the Crowd. Crowdsourcing and Human-Centered Experiments, LNCS, vol. 10264, pp. 154–190. Springer, Cham (2017). https://doi.org/10.1007/978-3-319-66435-4_7
6. Ericsson, K.A., Simon, H.A.: Protocol Analysis: Verbal Reports as Data. The MIT Press, Cambridge (1984)
7. Estellés-Arolas, E.L., Guevara, F.G., Towards an integrated crowdsourcing definition: Towards an integrated crowdsourcing definition. J. Inf. Sci. **38**, 189–200 (2012). https://doi.org/10.1177/0165551512437638

8. Fan, M., Shi, S., Truong, K.N.: Practices and challenges of using think-aloud protocols in industry: an international survey. J. Usability Stud. **15**(2) (2020)
9. Haak, M.J., Jong, M.D., Schellens, P.J.: Evaluating municipal websites: a methodological comparison of three think-aloud variants. Gov. Inf. Qual. **26**(1) (2009). https://doi.org/10.1016/j.giq.2007.11.003
10. Hirth, M., Hoßfeld, T., Tran-Gia, P.: Anatomy of a crowdsourcing platform - using the example of Microworkers.com. In: International Conference on Innovative Mobile and Internet Services in Ubiquitous Computing (2011). https://doi.org/10.1109/IMIS.2011.89
11. Hossfeld, T., et al.: Best practices for QOE crowdtesting: QOE assessment with crowdsourcing. IEEE Trans. Multimedia **16**(2), 541–558 (2014). https://doi.org/10.1109/TMM.2013.2291663
12. Kittur, A., Chi, E.H., Suh, B.: Crowdsourcing user studies with mechanical Turk. In: Conference on Human Factors in Computing Systems (2008). https://doi.org/10.1145/1357054.1357127
13. Krug, S.: Rocket Surgery Made Easy: The Do-It-Yourself Guide to Finding and Fixing Usability Problems. New Riders (2009)
14. Krug, S.: Don't Make Me Think. Revisited - A Common Sense Approach to Web Usability, New Riders (2014)
15. Lewis, C.: Using the "Thinking Aloud" Method in Cognitive Interface Design. IBM TJ Watson Research Center Yorktown Heights, NY (1982)
16. Liu, D., Bias, R.G., Lease, M., Kuipers, R.: Crowdsourcing for usability testing. In: Proceedings of the American Society for Information Science and Technology **49**(1) (2012). https://doi.org/10.1002/meet.14504901100
17. Maier-Hein, L., et al.: Can masses of non-experts train highly accurate image classifiers? a crowdsourcing approach to instrument segmentation in laparoscopic images. Med. Image Comput. Comput.-Assist. Interv. - MICCAI **2014**(17), 438–445 (2014). https://doi.org/10.1007/978-3-319-10470-6_55
18. Meier, F.: Crowdsourcing als Rekrutierungsstrategie im Asynchronen Remote-Usability-Test. Information-Wissenschaft und Praxis, **63**(5) (2012). https://doi.org/10.1515/iwp-2012-0063
19. Nebeling, M., Speicher, M., Norrie, M.C.: CrowdStudy: general toolkit for Crowdsourced evaluation of web interfaces. In: SIGCHI Symposium on Engineering Interactive Computing Systems (2013). https://doi.org/10.1145/2494603.2480303
20. Nielsen, J.: Usability Engineering. Elsevier Science (1994)
21. Nielsen, J.: Estimating the number of subjects needed for a thinking aloud test. Int. J. Hum. - Comput. Stud. **41**(3) (1994)
22. Nielsen, L., Chavan, S.: Differences in task descriptions in the think aloud test. In: Aykin, N. (ed.) Usability and Internationalization. HCI and Culture, UI-HCII 2007. LNCS, vol. 4559, pp. 174–180. Springer, Heidelberg (2007). https://doi.org/10.1007/978-3-540-73287-7_22
23. Rosenthal, R.: Meta-Analytic Procedures for Social Research. SAGE Publications, Inc. (1991)
24. Rubin, J., Chisnell, D., Spool, J.: Handbook of Usability Testing: How to Plan, Design, and Conduct Effective Tests. Wiley, Hoboken (2011)
25. Schneider, H., Frison, K., Wagner, J., Butz, A.: CrowdUX: a case for using widespread and lightweight tools in the quest for UX. In: Conference on Designing Interactive Systems (2016). https://doi.org/10.1145/2901790.2901814
26. Skov, M.B., Stage, J.: Supporting problem identification in usability evaluations. In: Australia Conference on Computer-Human Interaction (2005)

27. Sova, D.H., Nielsen, J.: 234 Tips and Tricks for Recruiting Users as Participants in Usability Studies. Nielsen N Group (2003)
28. Thompson, K.E., Rozanski, E.P., Haake, A.R.: Here, there, anywhere: remote usability testing that works. In: Conference on Information Technology Education (2004). https://doi.org/10.1145/1029533.1029567
29. Vermeeren, A.P.O.S., den Bouwmeester, K., Aasman, J., Ridder, H., de Ridder, H.: DEVAN: a tool for detailed video analysis of user test aata. Behav. IT **21**(6) (2002)
30. Yuhui, W., Tian, L., Xinxiong, L.: Reliability of perceived usability assessment via crowdsourcing platform: retrospective analysis and novel feedback quality inspection method. Int. J. Hum.-Comput. Int. **36**(11) (2020). https://doi.org/10.1080/10447318.2019.1709339

Shaping AI as the Tool for Subconscious Design

Wentong Huang(✉)

Morpha Technology (Shenzhen) Co., Ltd., Shenzhen, China
wentong.huang@mail.polimi.it

Abstract. The whole process of human evolution is dynamic in the interaction with tools, as a designer, there is no doubt that we benefit from the digital tool - CAD (computer aid design) in the design process. But most of the tools are only the extension of our hand which reduce the repeating job, increase the efficiency, and help us create diverse outcome as quick as possible. This way of tools making bring us into the era of Mass-production and laid the foundation of modern design.

There is always a limitation for increasing the efficiency based on the First Principles [1], because the physical workflow is tangible. But the value of a designer is never about efficiency, it's about creativity and perception about the humanity. When the "design one and fit all" model hit its ceiling, we need to improve with another direction, which is the perception a deeper perception of the people and merging it into the early design process. Therefore, a new generation of tools will be required. Artificial Intelligent, especially the Neural Networks is a viable tool to generate the potential vision and form that driven by the data from the customers.

In this thesis, we take the machine learning model training project as an example. Proposing how to create Artificial intelligence tools/assistants, which helping bring the conscious/unconscious need of the customer into the beginning of the kitchen design and make the Mass-customization becoming a real-time experience. The outcome is a Machine Learning model which is suggestive assistant for the customer/designer to explore the form of the kitchen.

Keywords: AI · Machine learning · Generative adversarial neural networks · Mass customisation · Immersive design · Subconscious communication

1 Introduction

1.1 Preface

One of the most successful business model of the internet was the precise positioning of advertising by tracking the behaviour of the user online. The fundamental philosophy is, on the one hand, for the customers, it is difficult to tell what they want or show the preference with a very efficient way; on the other hand, when a company were facing a huge amount of user, the communication between people to people would become slow, ineffective and even misleading considering the feature of human. If we look backward to the history of how the information was distributed in the internet, from the portal website like Yahoo at the very beginning, to the search engines, content subscription

C. Stephanidis et al. (Eds.): HCII 2021, LNCS 13094, pp. 40–53, 2021.
https://doi.org/10.1007/978-3-030-90238-4_4

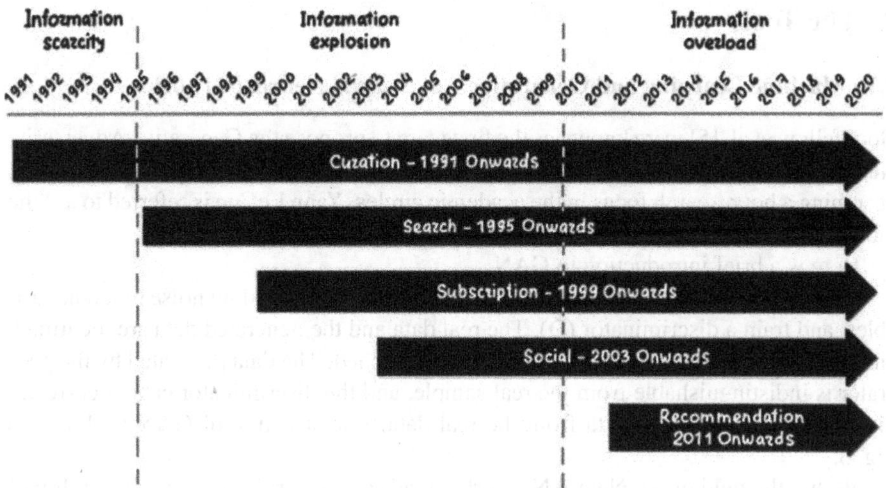

Fig. 1. Method of internet empowered informational distribution [2].

and recommendation, we can see the technology behind the phenomenon was heavily related to the amount of the information (Fig. 1).

2011 saw a breakthrough as Google started implementing a new machine learning system called Sibyl to make recommendations on YouTube [2], which made YouTube's suggested videos become the most accurate at guessing what would interest you at the time. For most of the people, we don't even notice it when we are using those platform and website to get the recommendation. Your behaviour was tracked silently and become the feed for the machine learning model.

"The beauty of relying on recommendations to improve engagement is that it creates a virtuous cycle of continual improvement over time, often referred to as a "data network effect" [3]. The more time spent using it, the better understanding user's preference and a more accurate content matches will be built for the better user experience. This naturally leads to more time being spent in the platform, which further enriches the user profile and so on. It is awkward that the machine understand the people better than the designer today. It is also a popular question to ask if AI will took the job of designer? "And the answer is absolutely yes, the Ai will take most of the Job of designer now, but not the Job of the designer in the future" [4] (Medium, 2019). But what is the designer in the future and what is the real value for designer to position in this society.

In the next chapter, the study will demonstrate how we can treat and discover the Machine learning model as a tool to build a virtuous cycle to empower the creativity from the early stage of the design.

2 The Tools

2.1 Machine Learning and Generative Adversarial Neural Networks

Goodfellow et al. [5] were known as the first team to propose the Generative Adversarial Network (GAN) in machine learning. This concept of generating against network is becoming a hot research focus in the academic circles. Yann LeCun is referred to as "the most exciting ideas in the field of machine learning over the past decade".

Here is a brief introduction to GAN.

Train a generator (G) to generate realistic samples from random noise or latent variables, and train a discriminator (D). The real data and the generated data are identified, and both are trained until a Nash equilibrium is reached. The data generated by the generator is indistinguishable from the real sample, and the discriminator cannot correctly distinguish the generated data from the real data. The structure of GAN is shown in Fig. 3.

Within the field of AI, Neural Networks stands as a key role. The creative ability of such models has been recently evidenced, through the further development of Generative Adversarial Neural Networks. As any machine-learning model, GANs learn statistically significant phenomena among data presented to them.

Consider the advantage of the neural networks model, the Generative Adversarial Neural Networks—or GANs- are here our weapon of choice. And the code of this project is strongly borrow from the Pix2Pix [6] in the GitHub, considered its requirement of small amount of data input and the ability of coping with the image.

2.2 Image Translation Trilogy: pix2pix, pix2pixHD

After the early work of Goodfellow et al. in 2014 [5], the field of Generative Adversarial Nets (GAN) has been keeping growing.

Image translation refers to the conversion from one image to another. It can be analogous to machine translation, one language to another.

The left figure shows some typical image translation tasks as the result from the Pix2Pix [6] paper: for example, the semantic segmentation map is converted to a real street view, the gray image is converted to a color map, and the daytime is converted to night, etc. (Fig. 2).

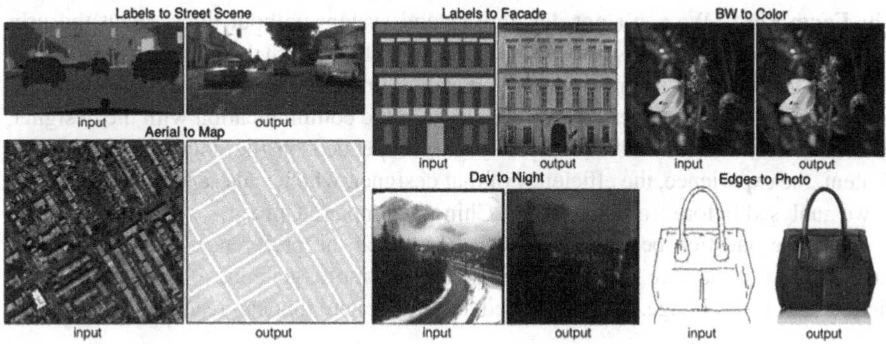

Fig. 2. The results of the method on several.

3 Shaping Neural Networks as the Tools of Mind

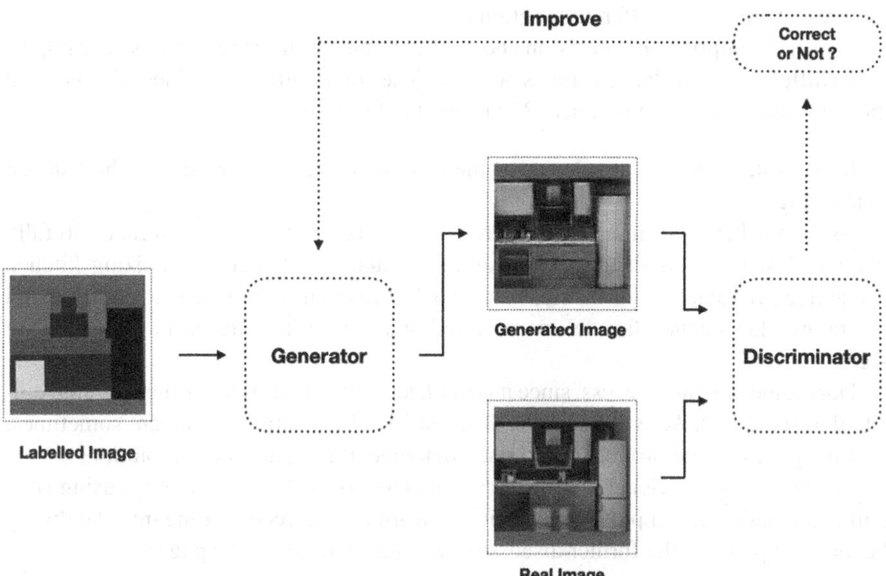

Fig. 3. Workflow of GAN model in my kitchen system

3.1 GANs (Generative Adversarial Neural Networks)

A. Neural Networks. The design of good Kitchen System is considered as the perfect combination of the Form & Function, in my project, after discovering the advantage of parametric design approach, it is proved to be a advanced tool for the function part which brings the bottom structure of the kitchen into the real function and production.

There is still a missing part between the system and the end user, which is the user experience. Because even though different kinds of of kitchen can be generated in a second, but the customer don't pick the kitchen by function. Function is invisible but the style is.

The Form First. We can't not develop a good system without think about the user experience. And the truth is, most of the people build their kitchen by asking a specific style and feeling. For example, normally they will find the style in the existing market or a brand and then pick some pictures they like for the communication with the designer. And in the reality, the final quality is highly depended on the understanding of the content, the experience, the efficiency of that designer, which cause a lot bad experience as we analysed before from the case of Chinese company Oppein.

Now the question become, how to better match the preference of need from the customer to the adaptive kitchen system. And even more, generate a style based on the picture they given image. So now, I need another system that can read your mind.

3.2 Training Process

Label Defined. First of all, a labeling rule should be created which uses different colors to represent areas with different functions.

In order to improve the precision and quality of the training process,it is very important to differentiate the label as far as possible. The color with RGB values of 0, 128 and 256 were selected, which products 27 different color label.

In our test, 11 color was selected as the label to define different area of the Kitchen system (Fig. 4).

As the result, 64 data image pairs were selected, sized to the same scale, and carefully marked. Based on this dataset, two trainings, kitchen-to-Label (recognizing kitchen facade and producing color labeled) and label-to-kitchen (inputting color labeled and generating plan kitchen facade), were tested and will be introduced in the following pages.

During the labeling process, since it is marked by human in a short time, it is unavoidable that some mistake will be made. But just like the creation of human, sometimes, error happened would produce a surprised outcome, it is a part of the evolution.

The beauty of the Generative Adversarial Network (GAN) is that, even using such a little amount of data input to train it, it still learn how to recorrect the mistake during the marking process, the further discussion will be show below the pages.

B. Data Pretreat. The GAN is a powerful and flexible tool in dealing with image data, its application in Kitchen facade design, especially in recognizing the function distribution and generating the mood boards, CMF and even the render is a great potential for further development and application (Fig. 5).

A process of training and evaluating between an image and its corresponding labeled map was carried out by the author in Python and Pytorch.

In addition, to simplify the study, all the image of the kitchen Facade was collected from the website of the Chinese Kitchen company - Oppein, after a series of Pretreatment, including eliminate 3D and perspective effects, in order to remove the influence of the impact of 3D effects on images. Here is one thing need to be noted is that, the purpose of pretreatment is just to simplify the test process, which offer a much simple and abstract understanding of how it works, it doesn't means that the GAN doesn't have the ability

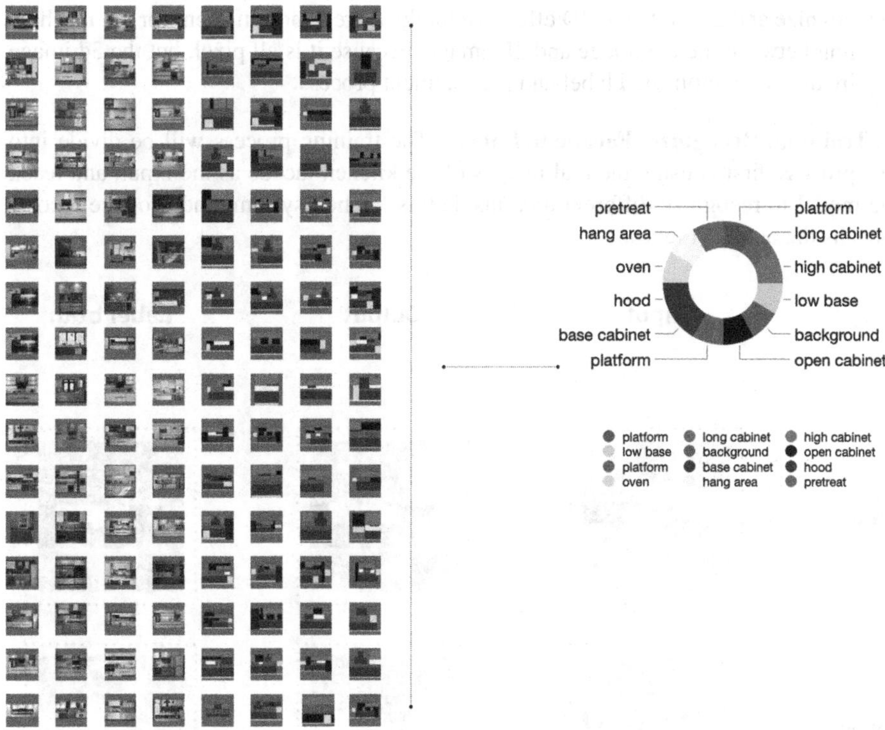

Fig. 4. Define label of data for the machine learning training process.

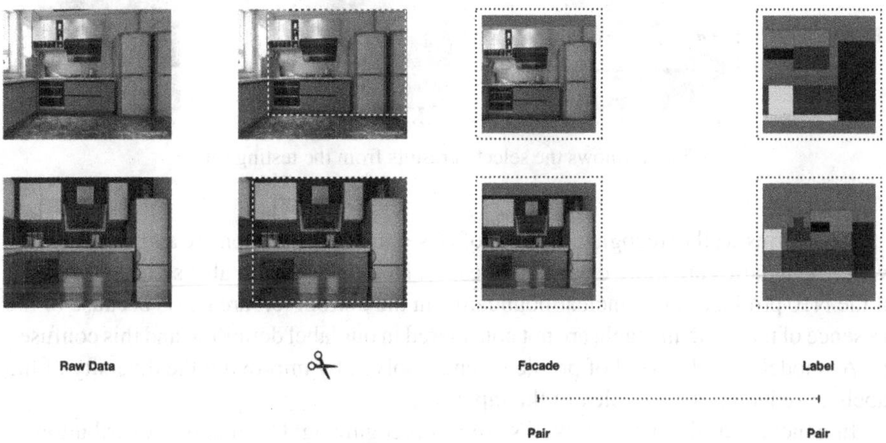

Fig. 5. Data Pretreatment for the machine learning training process.

to recognize and generate the 3D effect. Actually, there is not different for the machine learning between the 2D image and 3D image, because it is all pixel, but the 3d image require a more complicated label and pretreatment process.

C. Training: Recognize (Facade to Label). The training process will be divide into two process, first is using the real images of the kitchen facade as the input, and teach the model to recognize different area inside this kitchen system, and produce a color label as the output (Fig. 6).

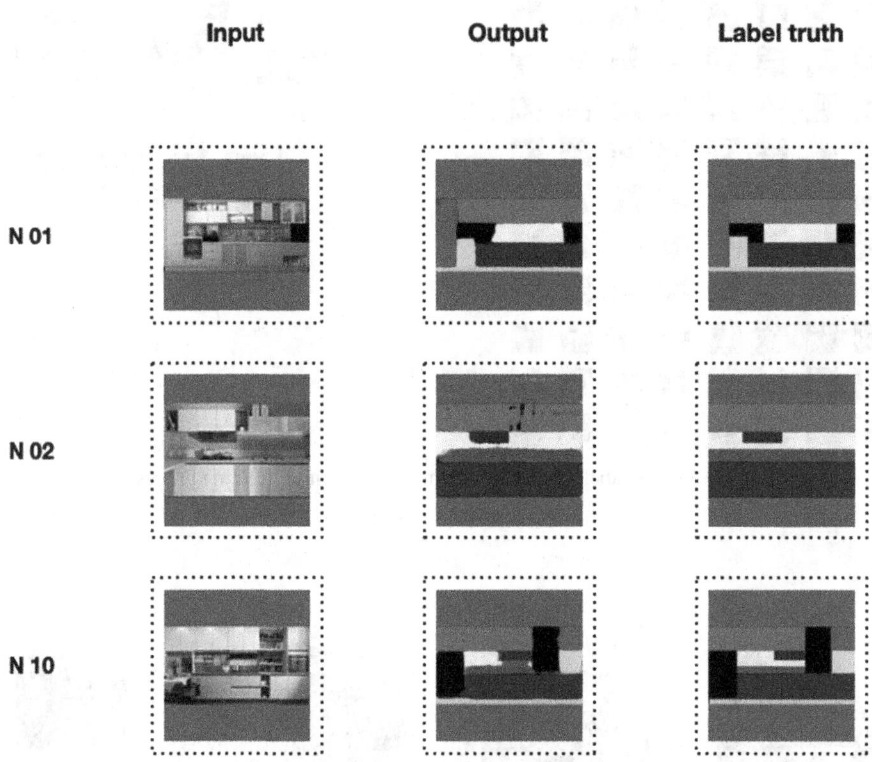

Fig. 6. Shows the selected results from the testing set.

It performs well in recognizing areas of Base cabinet, operation are and high cabinet, whose boundaries are more clear in the reality. However, we can also see that the result of the N10 produce some unclear boundaries in the bottom left area. It is because of the presence of the chair and table are not considered in our label definition and this confused the AI model. But this kind of problem can be solved by improving the diversity of the labels or reducing the complex of the input data.

In conclusion, the network works well in recognizing The function distribution of the kitchen facade. And the significant advantage is, instead of training set of thousands of images data commonly used in other machine learning model, a training set with 60 pairs images is enough for the network to learn and summarize the knowledge of kitchen system function distribution. More importantly, this advantage will play a key role to push Artificial Intelligent/Machine Learning, as a tools, from the big company and organization to the individual.

D. Training: Generate (Label to Facade). Next, instead of using the real images data as the input, I trained another network model using color labeled as input images and kitchen facade as output images. As what is expected, the program is going to generate a kitchen facade according to the input labeled map (Fig. 7).

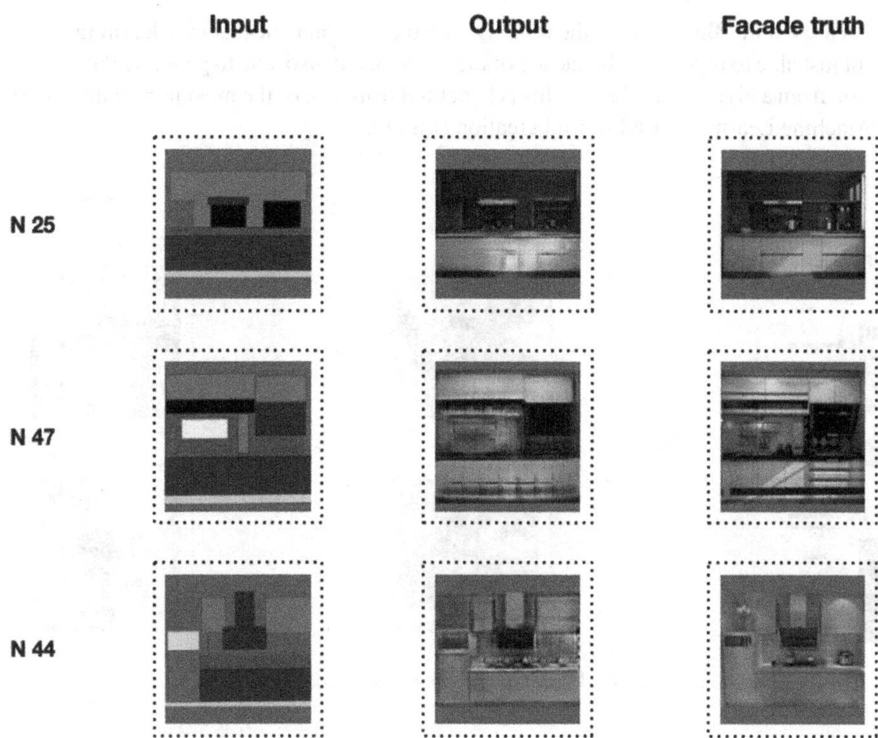

Fig. 7. Shows selected results from the testing set.

All three selected images show different result compare with the Facade truth. The N25, which represent 90% of result, show a clear and nice generation of the kitchen facade from tow aspects: the distribution of the function area and the color/materials.

During the analyze process, we can figure out there are some different effect like the N47 and N44, produce a different result. As we can see, the generations of them keep the distribution of the kitchen function, but the color and materials was totally changed, and if we compared them will the other 90% of the expected result, it seems that they try to keep the same feeling with them. In the machine learning model, the phenomenon is common because of the Weight Ratio of some of the fracture from the input training data. For example, all the source images data were collected from one Chinese company in one style classification - Europe Modern style, which common fracture is strong enough to shape the personality of this Network.

But instead of treating it as a mistake or failure, from the perspective of a designer, it will be very useful to shape the personality of the networks to get some specific result and this probably could change the way we work, instead of designing a specific kitchen, we can design the process of the designing.

3.3 The Outcome

A. Draw Color Block. After the training with the pair data, the machine learning model is not just able to recognize the facade of the kitchen, but also able to generate the kitchen facade from a giving color label. This is benefited from one of the most important feature of Machine Learning - End to End creation (Fig. 8).

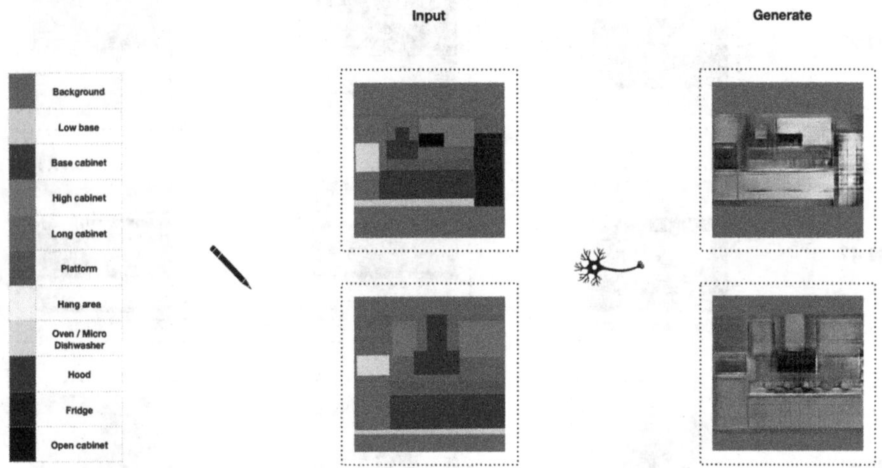

Fig. 8. Draw color block to generate kitchen

For the Creator. Compare with the parametric BIM [7] approach, which each step is clearly coded and defined by the creator, the ML model is like a black box and more humanity actually. Even though it produces lower quality of the image but the higher possibility of surprising outcome will be a nice tool for inspiring the creator.

For the Customer. Furthermore, just like the challenge we set in the parametric approach, drawing a line to generate a kitchen. This model will go further, the user can draw a color label to create the kitchen facade, which will be a starting point for the dynamic system to decide which type of the kitchen will be generated when a live is drew.

This will become a new way of communication between the customer and the kitchen brand, instead of communicating by language, which will cause misunderstanding and the lost of information. By giving the tool for the customer to sketch what they like with the most simple way, could significant improve the efficiency and customer participation.

B. Recognize & Statistical Analysis. Deep learning is to learn a variety of questions based on the way the human brain is thought. Through the relevant algorithms, the time spent on learning will be recorded, and the way of thinking pattern of human brain will be used to solve these problems.

After a certain amount of training, the computer with deep learning ability can automatically process some text or image information to realize the artificial intelligence (Fig. 9).

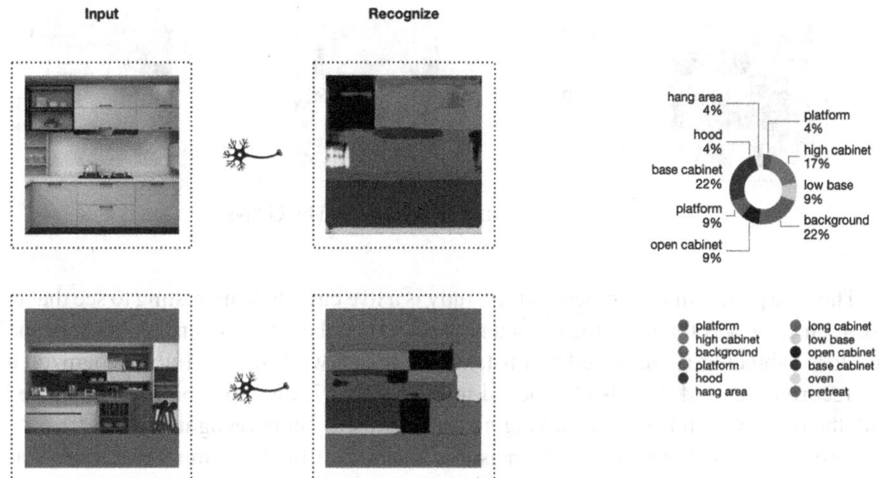

Fig. 9. Kitchen layout recognize & statistical analysis

In my project, after the training, the capability of recognizing the image and then labeling its distribution by following the principle defined by me, is working. Even though it look like a simple mission for the human brain, when it comes to deal with thousands and even million of the data, its advantage is obvious and that is the meaning of the ML.

In this case, Recognize the distribution of the kitchen means that the brand can rationally analyze the preference of the customer when some reference image are given. Furthermore, by analyzing the big data of the kitchen on the market will help the company to better understand the trend. If the definition of the color labels is changed from the function to the texture or CMF, then another door will be opened. The fundamental idea is not as complicated as we imagined but when it plug in the big data, a simple idea could be super powerful.

C. Error Correction. As a human, we make mistakes during the working process and life. And when it comes to the large number of datas, the mistakes inside the system is not easy to figure out by human's eye due to its biological limitation.

And of course sometime the error of human will produce some unexpectedly good outcome, especially for the algorithm for the generative graphic design, the mistake is always much more beauty than the plan. But we have to admit that most of the time, the small mistake will cause failure (Fig. 10).

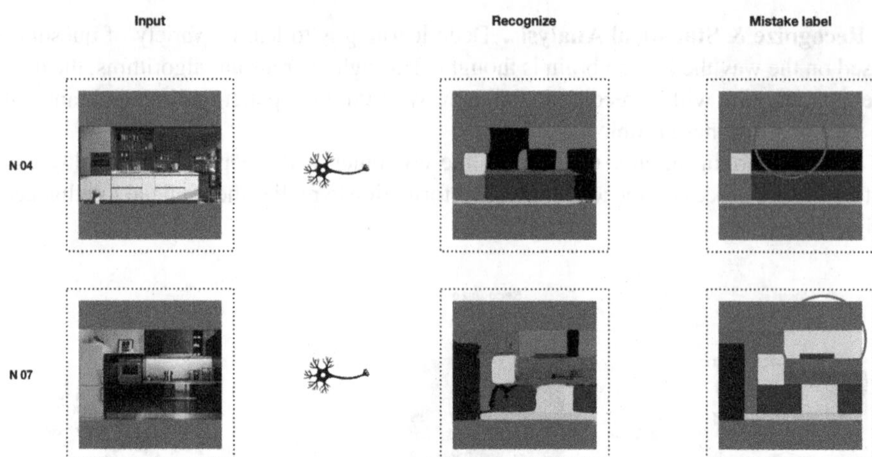

Fig. 10. How a mistake was corrected by GANs

The data pretreatment process in this study is a live case. It is interesting to see that in No. 4, an error was found by the trained network. The label of the open cabinet area and the close cabinet area are mixed by mistake, but the network successfully recognized it and recorrect it. And this also happened in the test N 07, the top cabinets was labeled with the oven, which is totally making no sense but still being recognized.

And when I look backward, the reason why most of the label mistake I made was in the beginning of the pretreatment process is part of the problem of the human brain. When a new workflow is defined, no matter how clear it is, our brain still need time to adapt it. This experience could be extended to the creativity management, by combining the Ai with workflow to improve the adaptability efficiency of human brain.

D. Style Shifting. The personality of every model is different even base on the same algorithm. If we take this advantage, we can create our own style "clone" as an assistant to generate the same style of image (Fig. 11).

Every machine learning model has its own Personality which means that every outcome from the same model with has some common feature or style. It depends on the kind of data we feed to the ML model and the way we train it.

For example, in this ML model, all the source data was collected from one brand and has a relatively unified style, material, color, and layout. In our case is modern style, as the result, when we use a traditional style data source as the input, we can shift the the traditional style into the modern style but keep the same function and distribution of the design.

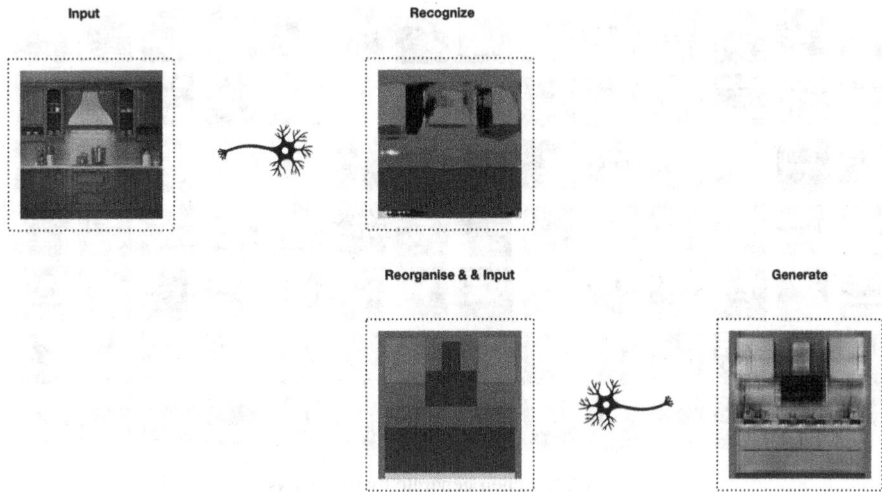

Fig. 11. From one style to another

The further step will be, by training different kinds of ML model with different way, we can generate different kind of outcome with the same data source, this will significantly change the way we design, create and improve the efficiency (Figs. 12 and 13).

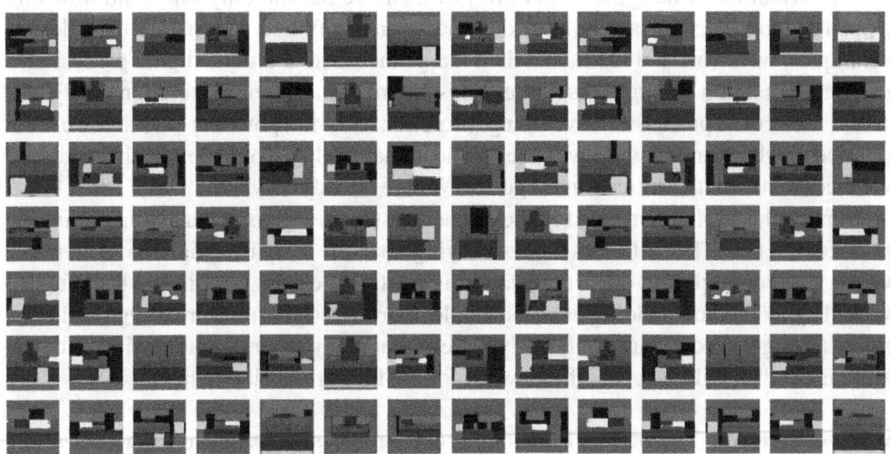

Fig. 12. Kitchen layout recognized by the GANs

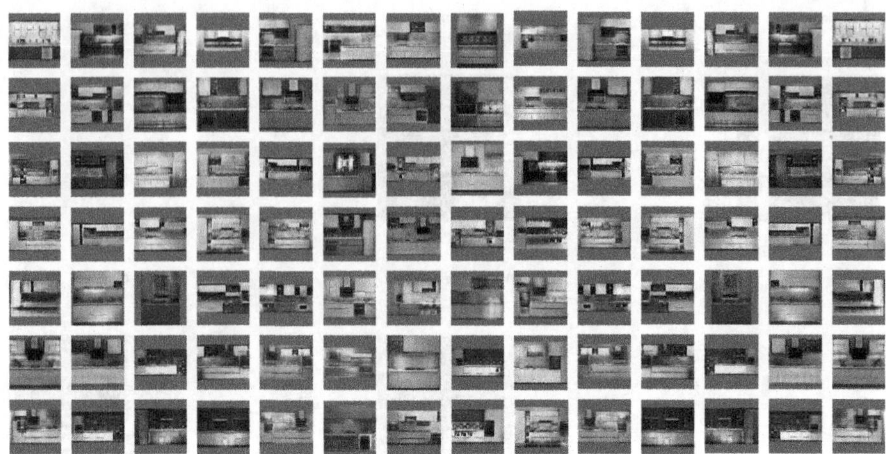

Fig. 13. Kitchen generate by the GANs

4 Conclusion

The practice demonstrated that the boundary of different subjects and technology is fading. And the barrier inside the human is much more stronger which stop us from becoming more powerful. We encourage the designer to embrace the power of the AI and bring the humanity to the world of technology.

The evolution of designer might not just a choice but the only one. And how this happen requires a process. For now, designers can verify and summarize their design techniques, workflow and ideas, and then merge them into a unique digital process and get further creation through this process. Instead of starting with a solution, designers should focus on a process of developing a solution. Still the power is ours and we have a choice of where and how we will use it. The human ability would be greatly expanded when combined with artificial intelligence. We are already a cyber.

Finally, the application of AI in the kitchen system design is just a starting point for the sake of focus. The real focus is about saying no at the early investigation. And when we dig deep enough, the other application is relatively easier. And in the bear future, we will apply this methodologies to the whole interior design, furniture design and even product design, And we are confidently believe that many other potential will be found. And this studies will become a framework for the further development.

Acknowledgement. The Machine Learning part of this paper is the continuing research of understanding and visualizing Generative Adversarial Networks in kitchen system design by the authors. And the code of this project is strongly borrow from the Pix2Pix in the GitHub.

References

1. First principle (2021). https://en.wikipedia.org/wiki/First_principle. Accessed 10 January 2021
2. Brennan, M.: Attention Factory: The Story of TikTok and China's ByteDance (2020)

3. Xiang, L.: putting into practice recommender systems. Beijing (2012)
4. Medium: Yes, AI Will Replace Designers (2019). https://medium.com/microsoft-design/yes-ai-will-replace-designers-9d90c6e34502. Accessed 24 Nov 2019
5. Goodfellow, I.J., et al.: Generative adversarial networks. arXiv:1406.2661 (2014)
6. Isola, P., Zhu, J.-Y., Zhou, T., Efros, A.A.: Pix2Pix, image-to-image translation with conditional adversarial networks. Berkeley AI Research (BAIR) Laboratory, UC Berkeley(2018)
7. Woodbury, R.: Elements of Parametric Design. Routledge, London (2010)

An Experimental Study on "Consensus to Match" Game for Analyzing Emotional Interaction in Consensus Building Process

Kyoko Ito[1](✉), Yoshiki Sakamoto[2], Rieko Yamamoto[2], Mizuki Yamawaki[2], Daisuke Miyazaki[2], Kimi Ueda[2], Hirotake Ishii[2], and Hiroshi Shimoda[2]

[1] Faculty of Engineering, Kyoto Tachibana University, 34 Yamada-cho, Oyake, Yamashina-ku, Kyoto 6078175, Japan
ito-ky@tachibana-u.ac.jp

[2] Graduate School of Energy Science, Kyoto University, Yoshida-honmachi, Sakyo-ku, Kyoto, Kyoto, Japan
{sakamoto,rieko,yamawaki,miyazaki,ueda,hirotake,shimoda}@ei.energy.kyoto-u.ac.jp

Abstract. For "consensus building" in which more than one people come to conclude, a variety of studies has been conducted on proposals of support methods using information technology, optimization simulations, analysis of consensus building for policy, etc. While their targets were the "outside" of the person, such as the opinions, positions, and conclusions issued by the person, these studies did not cover the "inside" of an individual person, that is, the process in the head such as the reasons leading to conclusions or remarks. Although human "emotion" is involved in the process, the involvement of emotion is not properly shown, and therefore it has not been an essential proposal to support consensus building or methodology. In this study, we conducted an experiment to consider the mechanism of consensus as a basic study to find the relationship between cognition and emotion in the process of consensus building. In order to conduct the consensus-building experiment, we developed a new "consensus to match" game and used "favorability" as the emotional aspect. We had 40 participants play the "consensus to match" game in pairs. The results showed that the sense of distance to consensus differed among the participants in the same dialogue, and that the "change rate in favorability to the other person" and the "change in favorability from the other person (estimated)" increased as a whole. In addition, "change rate in favorability to the other person" was larger than "change in favorability from the other person (estimated)" as a whole.

Keywords: Consensus building · Emotion · Cognition · Experiment · Communication · Process

1 Introduction

We build consensus to make group decisions in our daily lives. Consensus building has also been observed in the behavior of animals, such as honeybees, fish, and birds and is

C. Stephanidis et al. (Eds.): HCII 2021, LNCS 13094, pp. 54–62, 2021.
https://doi.org/10.1007/978-3-030-90238-4_5

the basis of social behavior. It has been studies in the fields of economics [1], biology [2], and social neuroscience [3]. Support methods using information technology have also been proposed [4–7]. However, it is unclear how consensus is formed among group members, especially how the emotional aspect is involved.

In this study, we conducted an online consensus-building experiment and analyzed the results in order to consider the influence of emotional aspects on human consensus building.

2 Method

In order to conduct an experiment to analyze the process of consensus building in humans, including emotions, the following experimental design was used. In order to examine the basic part of the consensus building process, we decided to conduct an online text chat from the perspective of reducing the human modality.

2.1 Consensus Building Theme (Task)

In this study, as a result of multiple preliminary experiments, we developed a new game named "consensus to match" game as a consensus-building theme (task). Figure 1 shows an overview of the game. There are four members (UserA, UserB, UserX, UserY) in a game. UserA and UserB are acquaintances, and UserX and UserY are acquaintances, while the pair of UserA and UserX and the pair of UserB and UserY are not acquaintances. With these four members, two pairs are created that are not acquaintances. The theme of consensus building is "Given three options, choose one option that matches the option between the pairs". If the options of each pair match, everyone will be presented with that option.

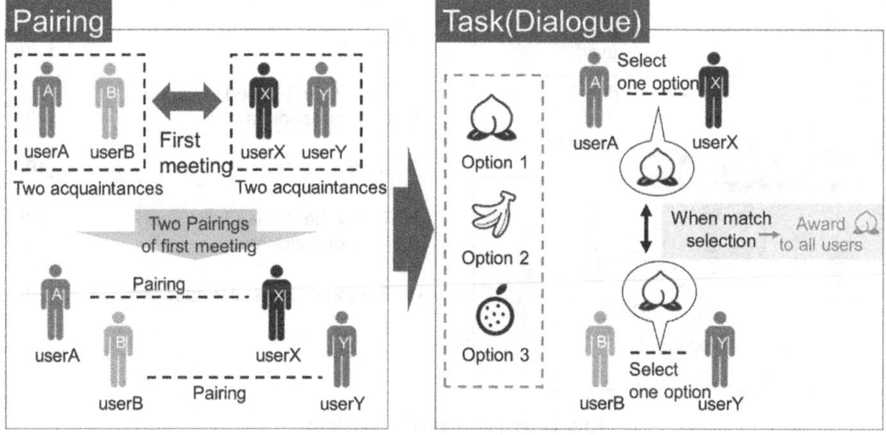

Fig. 1. Outline of "consensus to match" Game.

2.2 Measurement of Emotion

The participants were asked to select the "good feeling level" to the other person for each remark as the emotion during the consensus building. Specifically, they were asked to rate the "change rate in favorability to the other person" and the "change rate in favorability from the other person (estimated)" on a 9-point scale (1: became lower - 5: remained the same - 9: became higher), respectively.

2.3 Experimental Procedure

Figure 2 shows the flow of the experiment. The time required (minutes) is shown in []. The time of the experiment was about 2 h and 30 min and was conducted in two pairs of four participants. The experimenter provided the instructions online. For the task implementation, three options from six items were prepared based on a preliminary questionnaire. The six items are Japanese noodle, Japanese rice crackers, chocolate crashers, pre-packaged food, canned food, and apple juice. The task was conducted in pairs, and each participant was asked to perform the task twice. For Task 1 and Task 2, the pairs were changed as shown in Fig. 2. Figure 3 shows an example of the screen for the online chat. During the task, the participants were asked to select the change rate in favorability level for each remark. A participants was asked the change rate in favorability level for the other's message, while s/he was asked the change rate of the other's favorability level of her/himself (estimated). After the task was completed, the participants were asked to select the "distance to consensus" for each remark on a scale of 0 to 100 as a post questionnaire. In addition, personality questionnaire was conducted.

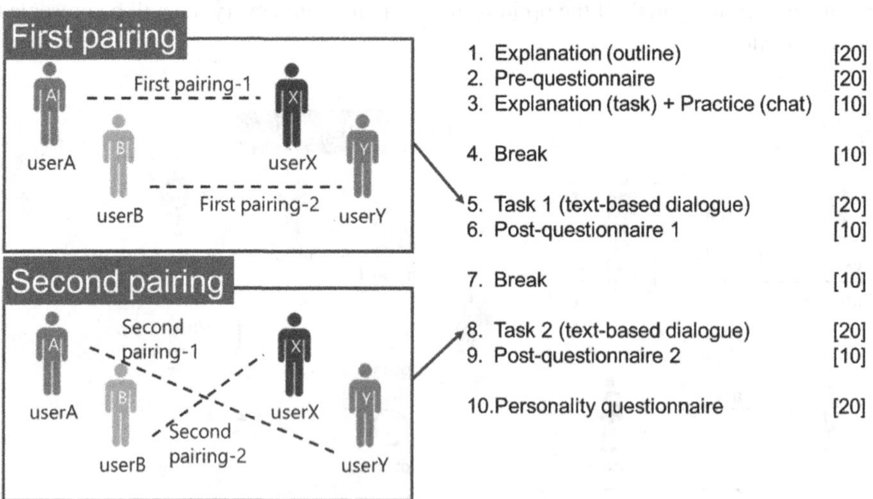

Fig. 2. Experimental procedure.

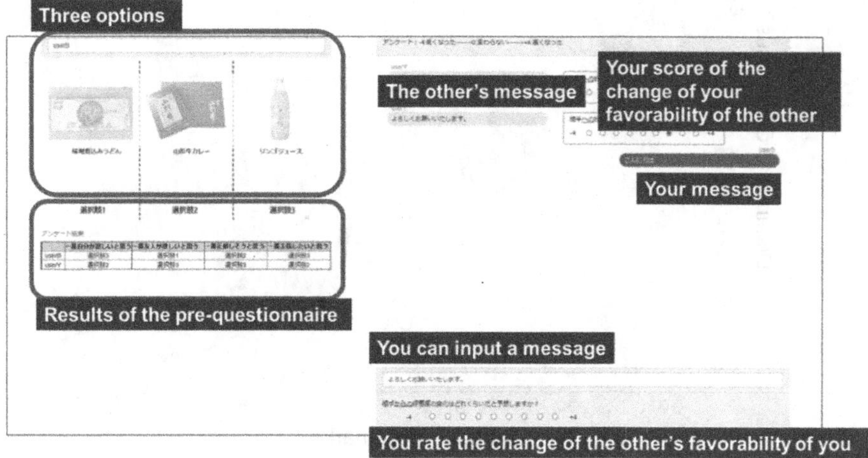

Fig. 3. An example of the screen for the online chat.

2.4 Outline of the Experiment

The number of participants in the experiment was forty. The implementation of the experiment was approved by the Ethics Committee of the Graduate School of Energy Science, Kyoto University.

3 Results

The number of the participants was forty (21 women and 19 men in their twenties). The total number of remarks in the 40 dialogues was 1028.

Figure 4, Fig. 5, and Fig. 6 show the results of the experiment. The order of remarks was normalized for each dialogue, and for each remark in each dialogue, the distance to consensus (0–100), the difference between two people in the distance to consensus, and the change rate in favorability are shown respectively. The lines show the average and the standard deviation are colored.

From Fig. 4, the overall distance to consensus decreased monotonically, with some increases. From Fig. 5, the difference in consensus between the two participants tended to increase until around 20% of the dialogue, and then decreased from around 60% of the dialogue. Figure 6 shows that the "change rate in favorability to the other person" and the "change rate in favorability from the other person (estimated)" appeared to increase as a whole. In addition, "change rate in favorability to the other person" was larger than "change in favorability from the other person (estimated)" as a whole.

In addition, we analyzed the remarks focusing on the process of consensus building. Specifically, we prepared the following six stages of consensus building.

1. Opening remarks
2. Confirmation of questionnaire results
3. Organization of reasons for selection

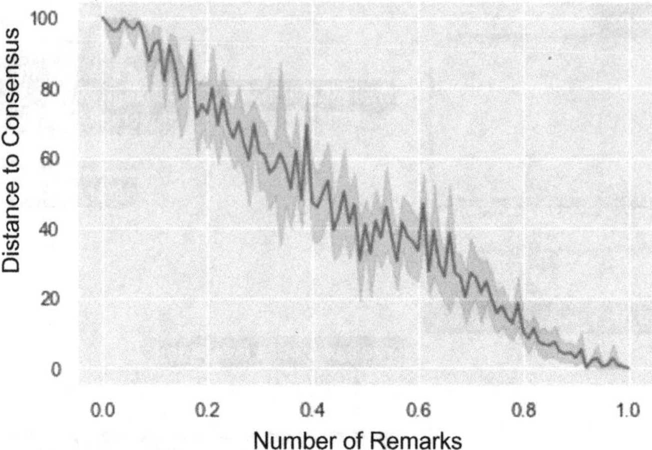

Fig. 4. Results (1): distance to consensus.

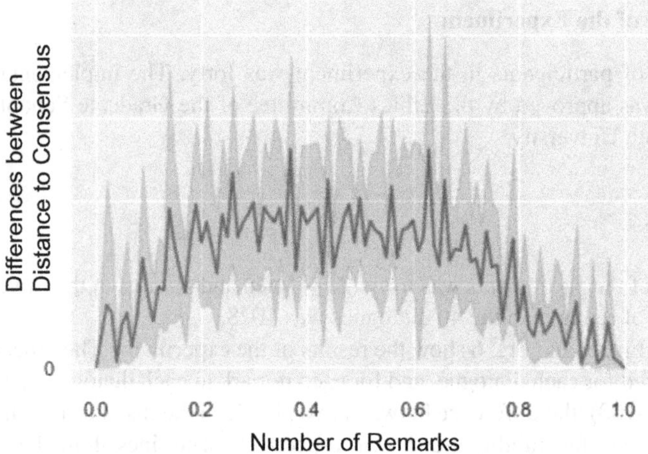

Fig. 5. Results (2): differences between distance to consensus.

4. Derivation of conclusions
5. Decision of conclusion
6. Ending remarks

Using the classification of the stages, we asked three people to classify all the remarks. As a result of the classification, 69.6% of the remarks agreed with all three, 28.8% agreed with two, and 1.6% disagreed with the classification of all three. Then, after classification, the participants were asked to discuss the results of the three classifications until they were in agreement.

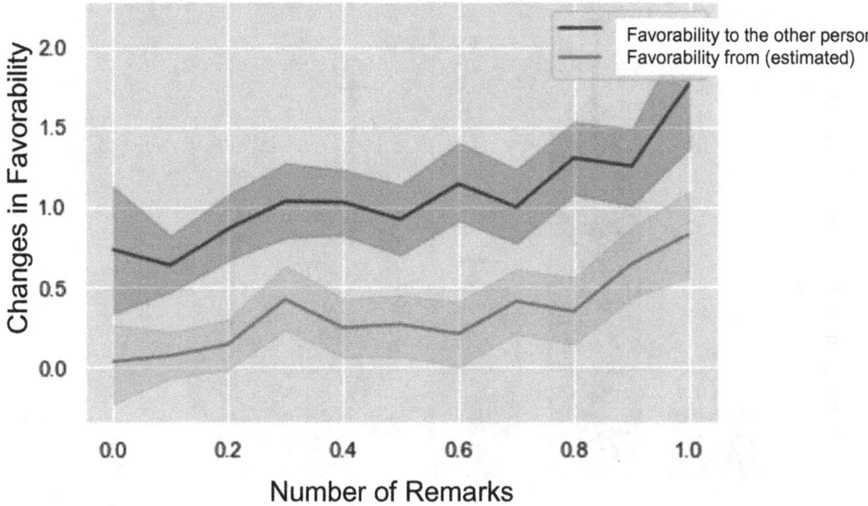

Fig. 6. Results (3): changes in the favorability.

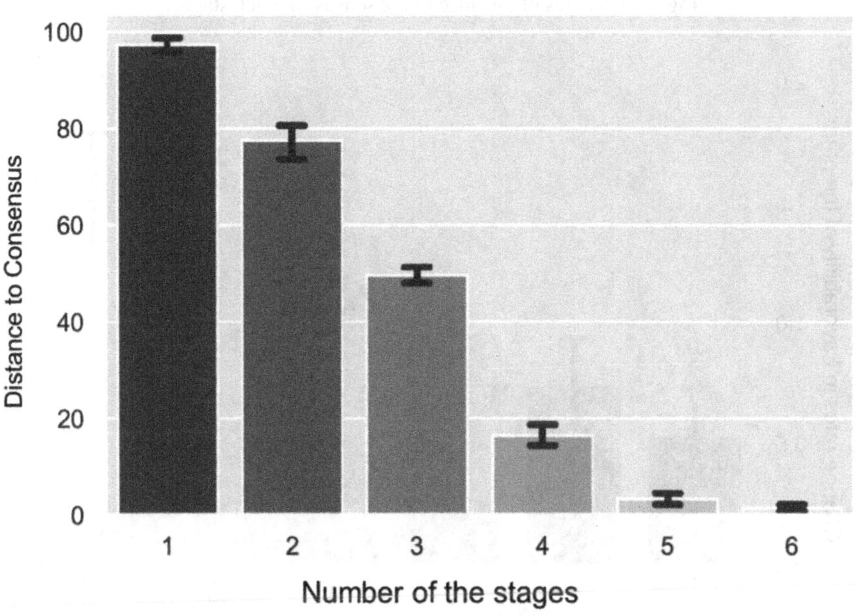

Fig. 7. Distance to consensus for each stage.

Based on the results of the classifications, the distance to consensus and the change in the distance to consensus are shown in Fig. 7 and Fig. 8, respectively. The error bars indicate the standard deviation.

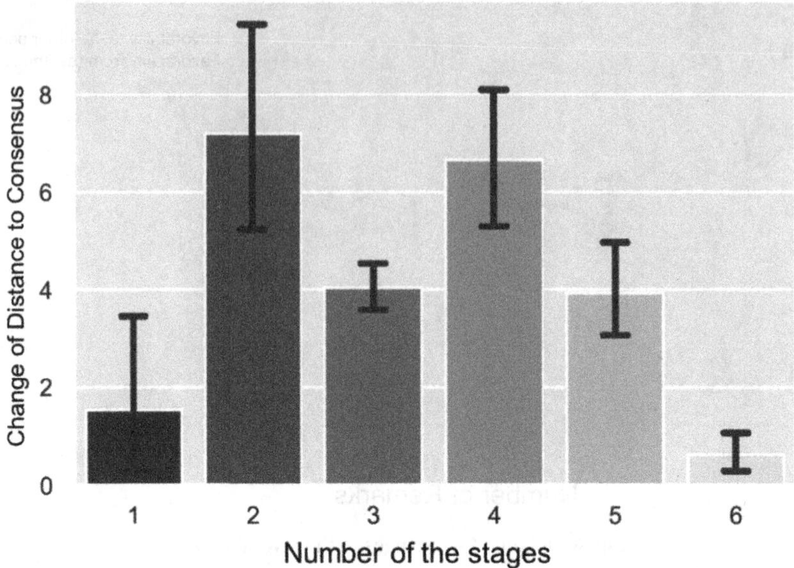

Fig. 8. Changes of distance to consensus for each stage.

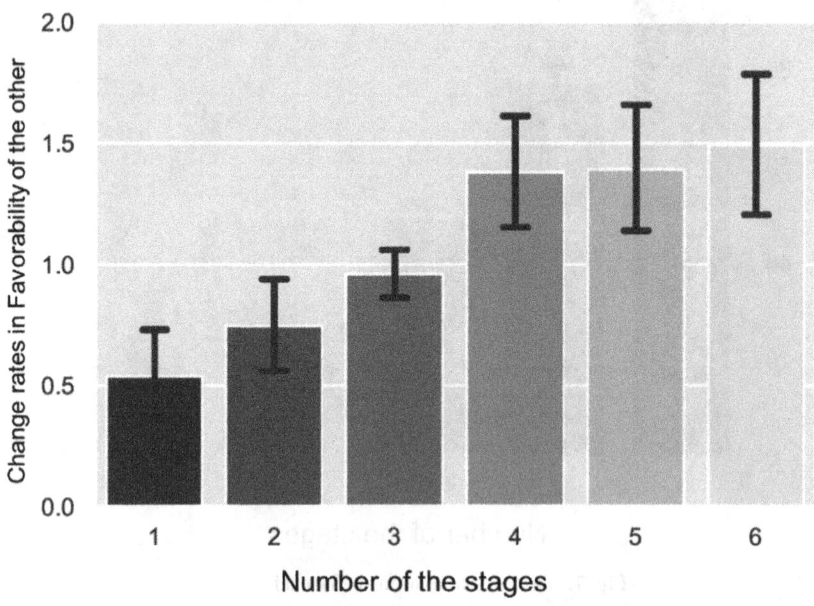

Fig. 9. Change rates in favorability of the other for each stage.

Figure 7 shows that the distance to consensus becomes smaller as the stage progresses. Figure 8 shows that the distance to consensus changes less in Stage 3 than in Stage 2 and Stage 4. It indicates that Stage 3 is not the stage where the distance to consensus changes because it is the phase where the reasons for the selection are being sorted out.

The change in favorability of the other person, and the change in favorability (estimated value) at each stage are shown in Fig. 9 and Fig. 10, respectively. The error bars indicate the standard deviation.

From Fig. 9, it is shown that as the stage progresses, the change in favorability of the other person becomes larger. Figure 10 shows that the estimated change in favorability from the other person in stage 2 is smaller than in Stage 1. It might be presented in Stage 2 that there are the differences in the selection.

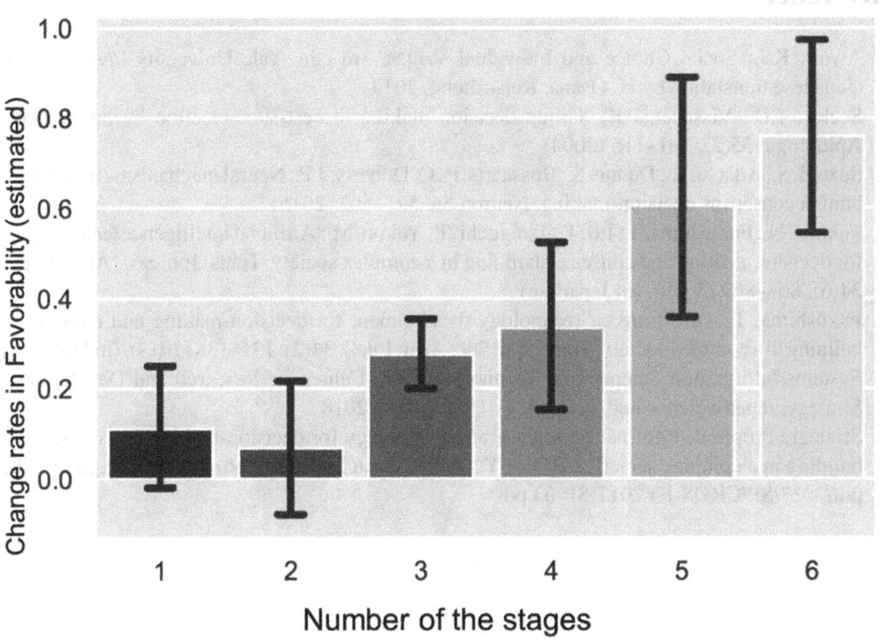

Fig. 10. Change rate in favorability (estimated) for each stage.

4 Conclusion

In this study, we conducted an online chatting experiment as a basic study to consider the human consensus-building mechanism, including its emotional aspects. In order to conduct the consensus-building experiment, we developed a new "consensus to match" game. We had 40 participants play the "consensus to match" game in pairs. The results showed that the sense of distance to consensus differed among the participants, and that the "change rate in favorability to the other person" and the "change in favorability from the other person (estimated)" increased as a whole. In addition, "change rate in favorability to the other person" was larger than "change in favorability from the other person (estimated)" as a whole.

In the near future, we will analyze the relationship between the emotional aspect and the consensus-building process, and would like to analyze the effects in conflict situations. Then, we will extract the causes of consensus building and classify the patterns of the process. Based on these findings, we aim to provide guidelines on how to provide information and how to create a place for consensus building.

Acknowledgements. This work was supported by JSPS KAKENHI Grant Number 20H01748.

References

1. Arrow, K.J.: Social Choice and Individual Values. 3rd edn. Yale University Press (2012). (Japanese translation by H. Osana, Keisoshobo, 2013)
2. Seeley, T.D., Visscher, P.K.: Group decision making in nest-site selection by honey bees. Apidologie **35**(2), 101–116 (2004)
3. Suzuki, S., Adachi, R., Dunne, S., Bossaerts, P., O'Doherty, J.P.: Neural mechanisms underlying human consensus decision-making. Neuron **86**, 591–602 (2015)
4. Fukuda, N., Fukushima, T., Ito, T., Taniguchi, T., Yokoo, M.: Artificial intelligence technologies for decision-making and consensusbuilding in a complex society. Trans. Jpn. Soc. Artif. Intell. **34**(6), 863–869 (2019). (in Japanese)
5. Fukushima, T.: Directions of technology development for decision-making and consensus-building in a complex society. Trans. Jpn. Soc. Artif. Intell. **34**(2), 131–138 (2019). (in Japanese)
6. Systems/Information Science and Technology Unit, Center for Research and Development Strategy, Japan Science and Technology Unit, March 2018
7. Strategic Proposal: Information science and technology for decision-making and consensus-building in a complex society. CRDS-FY207-SP-03 (in Japanese). https://www.jst.go.jp/crds/pdf/2017/SP/CRDS-FY2017-SP-03.pdf

Challenges and Workarounds of Conducting Augmented Reality Usability Tests Remotely a Case Study

Ted Kim(✉), Santiago Arconada-Alvarez, and Young Mi Choi

Georgia Institute of Technology, North Ave NW, Atlanta, GA 30332, USA
{tkim369,sarconada}@gatech.edu, christina.choi@design.gatech.edu

Abstract. It has been demonstrated that virtual prototypes can be used productively in usability testing. Virtual prototypes, Augmented Reality (AR) in particular, has received some focus as it combines both the real and virtual worlds and because platforms capable of supporting AR have become more ubiquitous.

This paper aims to explore the challenges of using AR to conduct usability testing in a remote delivery format. It presents a use case examination of the use of an AR tool to gather usability assessment data of a design prototype in a remote (distanced) format. An interview was conducted with those involved in conducting the test to understand issues related to the utilization of hardware provided by the usability test taker; the use of screen sharing to communicate instructions; challenges to efficiently gathering accurate observational data; and issues related to variability in hardware setup, consistency of the AR experience, and other environmental factors.

The results suggest that an approach for utilizing AR for remote usability testing was feasible, but many challenges and frustrations exist. Some of these related to communication with the participants, lack of motivation, difficulties extrapolating the prototype to a real setting, and technical issues. Usability test participants were often more interested in simply completing assigned tasks rather than exploring the tool or product in more detail that often happens in person. Other observations included some limitations in AR technology itself such as level of immersion (ability to accurately visualize the product). Challenges due to internet connectivity between the facilitator and tester at time affected the ability to effectively direct the testing session. Future work should investigate new tools and approaches for mitigating current challenges and experimentally compare in person versus remote testing outcomes.

Keywords: AR usability testing · Remote usability testing

1 Introduction

It is common during the development of new products for designers to feel that they do not have enough information about users' needs [1], particularly at the beginning of the process [2] when many different concepts are considered. The ability to gather and incorporate needs information is strongly linked to product success [3].

C. Stephanidis et al. (Eds.): HCII 2021, LNCS 13094, pp. 63–71, 2021.
https://doi.org/10.1007/978-3-030-90238-4_6

Usability testing is one of the most widely used and important methods for evaluating product design [4]. Usability is the effectiveness, efficiency, and satisfaction that users can achieve specific goals within a given environment [5]. In order to test, a product/service concept must be of an appropriate level of fidelity so that a user can use and assess how well a task or objective can be achieved. The more realistic the concept, the more accurately this evaluation can be done. Detailed physical prototypes are most often used and are especially helpful when evaluating subjective attributes like aesthetics and emotional appeal, ergonomics and usability, product integrity, or craftsmanship [6]. A big disadvantage is that these prototypes are expensive to produce (in both money and time). They are typically only available very late in the process after major design decisions are set. Substantial changes may be impractical if defects are found. Rapid prototyping techniques can speed up some aspects of constructing prototypes but does solve the inability to make late-stage changes.

Mixed reality technologies such as augmented reality (AR) offer a path to obtaining earlier user feedback. Augmented Reality refers to a view of the real or physical world in which certain elements of the environment are computer generated. This is generally referred to as mixed reality, since it combines the view of a real-world environment (such as a room) that has computer generated elements added in and seamlessly mixed into the view. This in contrast to virtual reality (VR), where the entire view and all elements within it are fully computer generated and do not contain any real-world elements.

There are a number of forms of AR technologies. vanKrevlan and Poleman [7] describe three main categories of AR: Handheld Displays (HHD), Head Mounted Displays (HMD), and Spatial Augmented Reality (SAR). The HHD is currently the most common methods to implement AR. The augmented view in these types of applications is often achieved through the use of a marker, such as a QR code. Software on the device is configured to detect the presence of the marker and replaces the marker with a computer-generated 3D object. These applications work much like a camera application, such as on a smartphone. As long as the environment is viewed through the device, the digital object appears just as it would appear in reality. HMDs typically integrate the device view into a headset. This frees the user's hands as the headset keeps the viewscreen in front of the eyes. An example of this kind of device would be the Microsoft HoloLens or Varjo XR-3. These devices are made up of a mostly see through display that sit in front of the eyes like the lenses of glasses. They are equipped with a camera for tracking and placing augmented objects into the user's view. SAR displays work by projecting a digital element into an environment. This is the kind of approach that is used to allow dead musicians to be integrated into live performances [8]. The projection is made onto a semi-transparent medium. It is even more specialized than HMD technology and requires custom equipment and care in setup to be effective.

The ubiquity of smartphones mean that many people own or have easy access to a device that can support AR. Some studies have already investigated how the real environment of AR and other factors affect the users through usability testing where an evaluator observes the interactions between the users and products to identify the user problems [9]. Choi explored Augmented Reality (AR) and Tangible Augmented Reality (TAR) as tools for the usability assessment of a space heater [10], and Ha, Chang and Woo investigated usability testing with multi-sensory feedback [11]. However, all these studies are conducted by lab-based usability testing, which typically does not build actual end-use environment for the test takers – it is impossible to reproduce specific environments. (e.g. construction site, bus, etc.) Due to this constraint, some usability problems are not identified in testing [9, 10]. On the other hand, remote usability testing where evaluators and test takers are separated in space, can resolve the limitations of lab testing. Many studies already verified that remote usability testing can be more effective than the lab-based usability testing [12], but when it comes to AR usability testing, none of the literature confirmed that the remote AR usability testing is more effective than the lab-based usability testing.

At a time when the covid19 pandemic has complicated in-person interactions, one of these being in-person usability testing, the motivation for this research is to explore the challenges of using AR to conduct usability testing in a remote delivery format. The new technological advancements in AR should be taken advantage of to carry out usability testing remotely; participants and evaluators are separated spatially and temporally. This paper will review the case of one such experiment.

Case Study: Efficiency of Authoring Interactions for Augmented Reality Experiences for Designers

The case study in this paper examines the process of evaluating the usability of a GUI based software designed to allow non-programmers to create augmented reality interactions without the need to write code. 22 participants tested the two different interface prototypes remotely. The software was implemented within a web browser to facilitate the remote testing. The usability test involved two groups. One group used the software to design an interaction with a car object. The other group designed an AR interaction utilizing a lamp object. The difference in the software interface is that one required developing the AR interaction by manipulating components of the virtual object. The other interface required developing an interaction by setting up events to drive changes in the view. The study ultimately found that the event-based prototype performed significantly better than the component-based prototype in both, the evaluation for efficiency in creating the AR interaction successfully and in the overall usability of the software [13].

The full study was originally designed to be conducted in person in a lab. The rest of the paper investigates the successes and challenges encountered while conducting the usability tests in a fully remote format (Figs. 2 and 3).

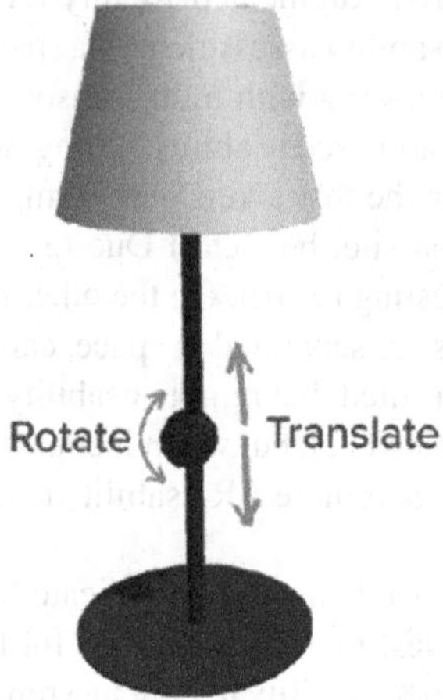

Fig. 1. AR prototype of table lamp (can be interacted with by swiping the finger up and down to translate and turn on by rotating the knob)

Fig. 2. AR prototype of toy car (can be translated by tapping on it)

2 Methodology

Retrospective interviews with those involved in the usability test were conducted. The interview format was semi-structured [8], where the respondents answer preset open-ended questions, and it was directed at dis-covering advantages and burdens that author might have experienced from the conducting the experiment. The interview questions covered the following themes:

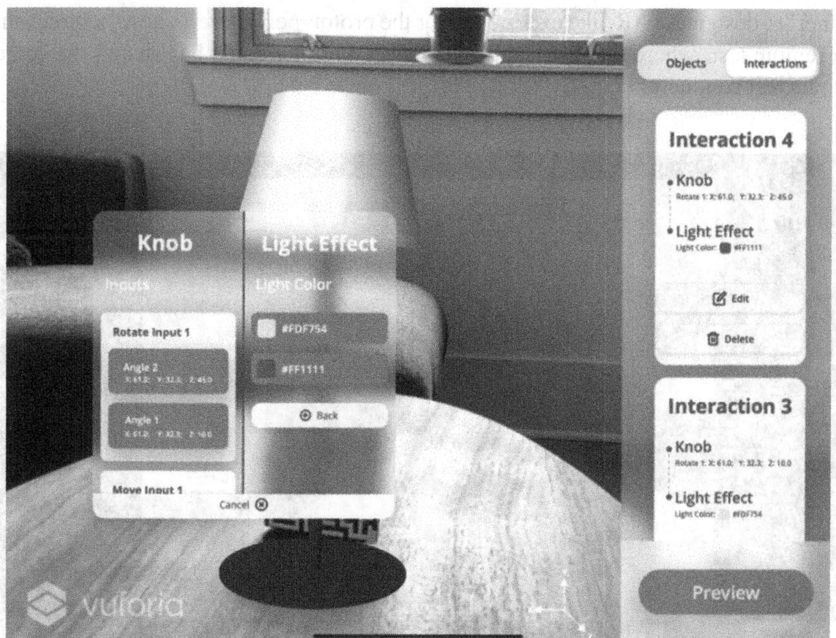

Fig. 3. Prototype used for the case study (shared with the permission of the author)

- Changes needed to collect data in a remote format
- Best and most frustrating experiences
- Technical issues
- Skills required to set up the experiment
- Newly emerged possibilities in general

The interviews were conducted by people who had not participated in the usability study. The interview data was transcribed and analyzed in two rounds focused around the main questions: 1) what are the changes that took place to conduct the experiment in remote 2) what were the advantages, and 3) what were the frustrations.

Changes Made to Support Remote Usability Testing

Though the prototypes (the car and lamp) were designed for mobile AR, participants tested the study prototypes on a desktop or laptop. Testing the prototypes on an iPad or mobile device would have required the participants to use their personal devices. This would have created inconsistencies in the setup due to various screen sizes and device capabilities. Moreover, using their personal devices makes it more challenging for those involved in conducting the usability test to observe and track the participant's actions. Through screen-sharing, they can examine what the test takers are viewing/doing on screen; though orientation, angle, and distance from the target is still hard to quantify.

Next, the software used for the prototype also had to be altered. The in-person usability tests would have utilized the game engine Unity coupled with Vuforia, which simplifies the process of creating AR experiences. The software needed had to be switched to 'framer' to design the AR-like experience for the prototype. Figure 1 shows a screenshot of a remote test, where the test taker is using the prototype and has shared the screen with the test conductor (Fig. 4).

Fig. 4. Remote AR usability study setting (shared with the permission of the author)

3 Results

The results suggest that an approach for utilizing AR for remote usability testing was feasible. In general, the evaluator felt that most of the participants were also very satisfied with remote testing. There were many benefits derived from a remote synchronous format for testing, but at the same time, a significant number of frustrations were also emerged.

Advantages

Easy to Recruit Participants

The first advantage was that recruiting participants in a remote format was a lot easier compared to recruiting participants for in-person testing. The best experience was described as: 'surprisingly, it was a lot easier to recruit participants. The fact that they did not have to come to the lab in-person was very fascinating to them, and some of the participants who are professionals conducted usability testing even during their company lunchtime'.

Frustrations

Communication

The first frustration mentioned was in-person communication is much easier for a number of tasks performed when running a study outside of simply collecting data. Collecting informed consent and other forms was much more difficult. 'Getting all the forms signed remotely takes more steps and longer time compared to in-person. I needed to email every participant one by one to communicate with them and some of the participants did not reply to me because they missed my emails. However, in-person, communication is much faster because you can just ask them to sign the documents once they walk into a lab'.

Lack of Motivation

There was frustration with participants not being actively engaged during the study. It was easy to demonstrate the technology remotely, but participants were more interested in moving along rather than exploring the tool in their own accord. 'AR is pretty novel and users are usually very fascinated by the technology, but at home, it wasn't the same because it wasn't true AR. In addition, in person, you came here to study, and you feel more obliged to play around and try, but remotely they just want to move quickly. The mindset of coming to a lab to conduct an experiment compared to doing from their bedroom is so much different'.

Understanding Output of a Prototype

Another area of frustration was users not understanding how the output of the prototypes would be used to represent the product/interaction. Since the prototypes were not providing a true AR experience, when the author explained how this tool would look like in a real AR setting, the participants could not imagine how this looked like. The evaluator also provided task-functioning prototype to the participants instead of fully functional prototype, and this made a lot harder for the test takers to imagine. 'I feel like this needs to be worked more… levels of imagination are different between participants, some of them fully understood the prototype without any problems, but some of them could not imagine'.

Technical Issues

The central frustration during the study was slow internet. Some participants had slow and unstable connectivity, which required a number of the use case scenarios to be restarted. This might be very critical because there is a clear difference between a user who is asked to conduct a task for the first time and a user who already knows what task they will be given. 'Internet could have been a big problem especially if your independent variable is the time taken to complete the task because when it's a bit lagging, the time would not be very accurate'.

4 Discussion

From the interview, some advantages and many frustrations emerged. The advantages the following: Easy to recruit participants (no restrictions on space and time) and learning about the possibilities and limitations when working with a new environment (remote AR setting). And the frustrations were the following: Communication, lack of motivation, inconceivability of the prototype, and technical issues (slow internet). In this section, evaluators discuss certain considerations regarding the experimental results. First, we explain the implications and recommendations of the prototype's level of immersion (ability to accurately visualize the product). Then we discuss how accurate the data is since the experimenter mimicked the AR experience. These discussions shall help other researchers who are considering conducting AR usability tests in a remote setting.

Experimental Mindset

One common theme in most remote studies is the variability between the users' frame of mind when joining the study. Some of the usability test participants were able to carry out the study during their lunch break, which helped with recruiting because some of his participants were working. One of the big differences here is that the participant has a shorter time frame to get "acclimatized" to the study, in the sense that there is no in between time their previous task and them starting the study. This in between time could help participants transition their mindset from their previous task. It would be interesting to study whether setting up a buffer time prior to the study to have the participant do a simple task or simple relax can influence in any way the results of the study.

Another key difference is that when participants come for an in-person study the number of distractions is diminished, often times lab environments are very minimalistic and void of attention-grabbing items so that participants focus on the task at hand; usually they are also asked to put their phone away. This is not the case in a remote environment. Participants tend to stay in the same environment where their computer is stationed. Email notifications, text messages, people walking by, etc., all these can grab their attention and cause them to drift away from the task at hand.

Internet Issues

Internet speeds during the task onboarding directly impacted the experiments as some users had to refresh the page to get the prototype to load. This could be mitigated by having participants download a native app on their computers, but it has the added difficulty of software development.

Level of Detail

Each user has a different ability to accurately visualize the product. Especially in this study, because the experimenter mimicked the AR experience, it was a lot harder for the users to imagine how the actual product would look like: 'Because I know how the actual AR is implemented, I did not expect that this mimicked prototype would be difficult for the users who do not have any AR experience to visualize the product. … I was looking at efficiency to complete tasks, so I had to ensure that the design of the interface did not hinder completion, no typos, everything is clear. … I did a validation study with usability experts as well. However, visualizing the actual product has nothing to do with completing the tasks'.

To alleviate this discovered limitation, providing an actual AR experience to the users in a remote setting also needs to be considered. In this case, the question remains as to how to provide the same AR medium to participants (same screen sizes), how the evaluator observes the participants, and how to manage and control the participants' surrounding environment.

References

1. Bruseberg, A., McDonagh-Philip, D.: Focus groups to support the industrial/product designer: a review based on current literature and designers' feedback. Appl. Ergon. **33**(1), 27–38 (2002)
2. Moultrie, J., Clarkson, P.J., Probert, D.: Development of a design audit tool for SMEs. J. Prod. Innov. Manag. **24**(4), 335–368 (2007)
3. Creusen, M.E.: Research opportunities related to consumer response to product design. J. Prod. Innov. Manag. **28**, 405–408 (2011)
4. Lewis, J.R.: Usability testing. In: Salvendy, G. (ed.) Handbook of Human Factors and Ergonomics, pp. 1275–1316 (2006)
5. ISO: ISO 9241-11. Ergonomic Requirements for Office Work with Visual Display Terminals (VDTs)-Part 11, Guidance on Usability (1998)
6. Srinivasan, V., Lovejoy, W.S., Beach, D.: Sharing user experiences in the product innovation process: Participatory design needs participatory communication. Creat. Innov. Manag. **16**, 35–45 (1997)
7. van Krevelen, D.W.F., Poelman, R.: A survey of augmented reality technologies, applications and limitations. Int. J. Virtual Reality **9**(2), 1–20 (2010)
8. Peddie, J.: Types of Augmented Reality. In: Peddie, J. (ed.) Augmented Reality, pp. 29–46. Springer, Cham (2017). https://doi.org/10.1007/978-3-319-54502-8_2
9. Folstad, A., Law, E.L.-C., Hornbaek, K.: Analysis in usability evaluations: an exploratory study. In: Proceedings of the 6th Nordic Conference on Human-Computer Interaction: Extending Boundaries, pp. 647–650 (2010). https://doi.org/10.1145/1868914.1868995
10. McFadden, E., Hager, D.R., Elie, C.J., Blackwell, J.M.: Remote usability evaluation: overview and case studies. Int. J. Hum.-Comput. Interact. **14**(3–4), 489–502 (2002). https://doi.org/10.1080/10447318.2002.9669131
11. Duh, H.B.-L., Tan, G.C.B., Chen, V.H.: Usability evaluation for mobile device: a comparison of laboratory and field tests. In: Proceedings of the 8th Conference on Human-Computer Interaction with Mobile Devices and Services, pp. 181–186 (2006). https://doi.org/10.1145/1152215.1152254
12. Martin, R., AlShamari, M., Seliaman, M.E., Mayhew, P.: Remote asynchronous testing: a cost-effective alternative for website usability evaluation. Int. J. Comput. Inf. Technol. **03**(January), 2279–764 (2014)
13. Jain, K.: Investigating the efficiency of authoring interactions for augmented reality experience for designers (unpublished master's thesis). Georgia Institute of Technology, Georgia, United States (2020)

Understanding Graphical User Interface (GUI) Trends Based on Kawaii (Cute)

Anirudh Kundu[1(✉)] and Michiko Ohkura[2]

[1] Department of Design, Delhi Technological University, Rohini, India
[2] SIT Research Laboratories, Shibaura Institute of Technology, Tokyo, Japan
ohkura@sic.shibaura-it.ac.jp

Abstract. In today's ever changing world, it is very important to keep a holistic track of a user's needs and wants. Maslow proposed that human needs can be organized into a hierarchy that ranges from base level needs such as food and water to top tier concepts such as self-fulfillment [1]. The notions of self fulfilment emerge from the emotional comfort that a user experiences whenever he/she interacts with any product or situation. Hence, a specialized understanding of various emotions and their effects is required. This brings us to a form of complicated human emotion that we have often elicited and experienced in our life. The emotion of 'kawaii' or 'cute' or the 'Aww emotion.' This study focuses on the kawaiiness of various design trends followed by different websites or graphical user interfaces (GUIs). It aims to evaluate the kawaii features that enhance or diminish the approachability and attractiveness of different web-based interfaces through self-devised experimentation and analysis. The results show clear notions of 'kawaiiness' in certain design trends that use pastel colors and softer animations. At the same time, evident characteristics for 'not kawaii' websites that used darker and bolder colors along with real images were also found.

Keywords: Kawaii (Cute) · User Experience (UX) · Graphical User Interface (GUI) · Usability · Emotion · Approachability

1 Introduction

1.1 Kawaii

'Kawaii' is a word of Japanese origin which means "cute," "lovable," or "charming." This has evolved to become a cultural concept or an emotional domain that relates to something or someone lovely, or someone or something that invokes the feeling of "wanting to protect" [2]. In the modern context, the notion of kawaii is embraced as a catalyst to evoke positive feelings [3], as can be seen in designs ranging from the branding industry in the form of 'Pokémon branded airplanes' and 'hello kitty branded shops' to the product industry in form of toys and road signs to name just a few examples. This emotion is primarily channelized by the approach-motivation towards things. The urge to interact, approach or to simply be with any product or situation can be classified as the influx of 'kawaii' [4]. Studies show that interacting with kawaii or cute stimuli increases the focus and attention span of a user drastically [3]. Such a response monumentally nourishes the user experience of any process.

C. Stephanidis et al. (Eds.): HCII 2021, LNCS 13094, pp. 72–88, 2021.
https://doi.org/10.1007/978-3-030-90238-4_7

1.2 Prior Work

Previous studies have examined various studies related to emotional understanding of interfaces and their psychophysiological responses. The larger classification of emotions in these studies are on the basis of a six emotional categorization [5]. Various physiological signals are measured in these studies along with simpler self-devised experimentation. Galvanic Skin Responses (GSR), Electroencephalograms (EEG), Electrocardiograms (ECG) and eye trackers are few of the equipment used for such psychophysiological data readings [6]. Softer and more complex emotions such as kawaii have not been addressed through a design perspective in depth. However, there have been several studies that are aimed at understanding the emotional and physical nature of kawaii. There exist studies that show the understanding of kawaii through colors [7]. Results of these studies show that softer colors in the spectrum of pink and purple are classified as kawaii. Furthermore, another interesting study shows the kawaii nature of basic shapes [8]. The outcomes of this study showed that rounded figures are generally classified as more kawaii. Applying the insights gained from these studies an exploration in the field of kawaii and its existence in screen design acts as a novel opportunity for the world of website and UI design.

1.3 GUI and Kawaii

In today's world, the usage of digital interfaces is on a rapid increase and it takes a user only 50 ms to build an impression about any Web based Graphical User Interface (GUI) [9]. These impressions are key driving factors that contribute to the successful usability and functionality of any GUI [10]. Hence, it is of utmost importance that designers who are working towards building such screen-based solutions learn the emotional and psychological experience their products provide. Only then can the entire UX be enhanced and humanized accurately. 'Kawaii' has the ability to make Web based GUI easy to approach, navigate and experience. In the last two decades, Kawaii has also been gaining audiences and customers worldwide as well as in Japan through kawaii products [11]. As such, kawaii design principles along with screen design principles are being incorporated into successful products that are being used globally including website and application design. This paper aims to evaluate and understand the kawaii-ness of current GUI design trends through self-devised experimentation.

2 Experimental Method

2.1 Stimuli Set

Customer wants can be broadly classified into three strata, Functionality, Usability and Desirability or Attractiveness, each of these strata are dependent on the level of their interactions. Functionality depends upon the physical level of interaction, Usability depends on the cognitive level of Interaction and lastly Attractiveness depends on the emotional level of interaction [12] (Fig. 1). While curating the stimuli set amongst a pool of various different websites, the visual language followed by the GUI was given precedence. Visual design is one of the most important aspects of screen design. It has the capacity to attract users, which is the first step towards building relationships. Furthermore, the visual/aesthetic appeal of any website is one of the key factors that induce an emotional response.

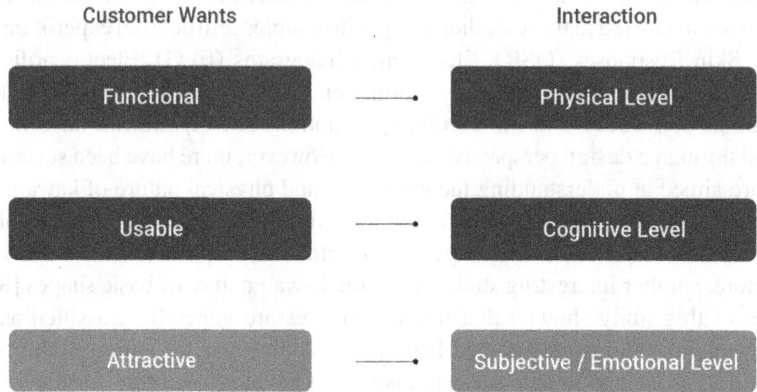

Fig. 1. Customer wants and their dependent interaction level. Highlighted: attractive customer wants and their dependency on the emotional level.

The colors used, the animations, the illustrations and graphics used amongst other factors contribute heavily towards a strong emotional response. Hence, in order to successfully develop an understanding of the emotional response of any GUI, it was of utmost importance to curate and classify different GUI accurately. A stimuli-set of 20 diverse gifs of different Web based GUI's were handpicked and curated based on 16 different GUI screen design visual trends.

The chosen GUI trends were:

a) Minimalism
b) Bold Typographic design
c) Retro design
d) White Space design
e) Dark Mode design
f) Usage of 3D objects
g) Usage of Micro Interactions
h) Pop design
i) Usage of Animated Illustrations
j) Neumorphism
k) Usage of Asymmetrical Layouts
l) Usage of Interactive Animations
m) Glass Morphism
n) Usage of Geometric Layouts
o) Material Design
p) Realism

These trends were selected on the basis of popularity, commonality in usage and accessibility to references. This list is not exhaustive and was made in order to cover the maximum number of different distinct trends used in web based screen design. Following this, a visual color palette was extracted from each website (Fig. 2). The trends followed by each website tallied with their URLs has been shown in Table 1.

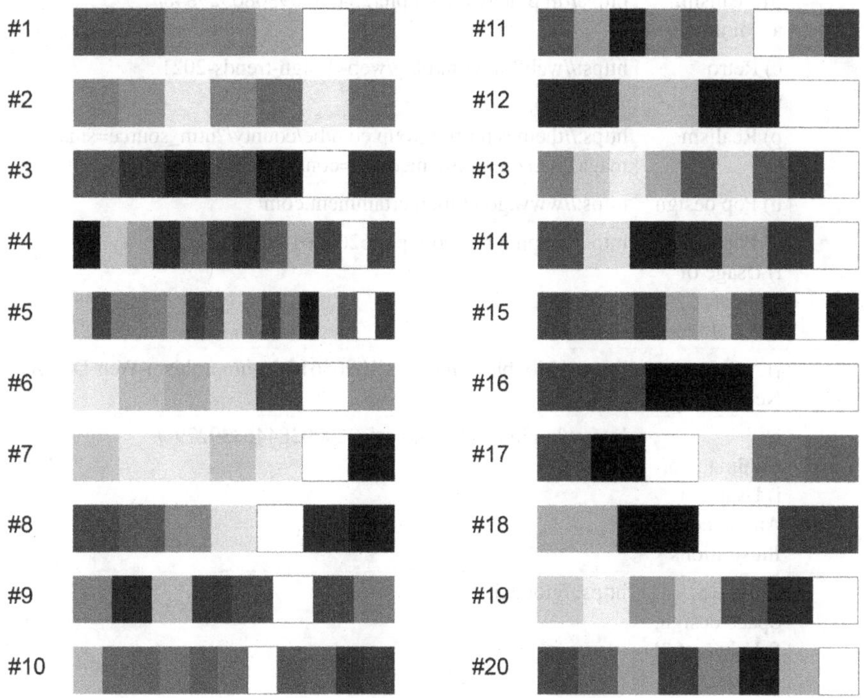

Fig. 2. Websites from stimuli set and their color palettes.

2.2 Materials and Method

The experiment was conducted with 30 undergraduate students aged 18 to 23 years old (15 males and 15 females), pursuing Design at Delhi Technological University, India. Due to the limitations posed by the Covid Pandemic in the form of a nationwide lockdown in India, the entire experiment was carried out through an online mode of communication. The chosen platform for discussion was 'Google meet'. A 'Google Forms' based questionnaire was generated containing the aforementioned stimuli set. Furthermore, a Visual Analogue Scale (VAS) [13] was generated that ranged from 'Not kawaii at all' to 'Extremely kawaii' (Fig. 3). The VAS scale was provided through an online collaborative workspace called 'Miro'.

Table 1. Websites used in the stimuli set and the respective design trends they follow. Along with their URL link.

Website Number	Trends Followed by the Website	URL to Website
1	p) Realism, a) Minimalism	https://in.pinterest.com/pin/234539093080529848/
2	c) Retro design	https://webflow.com/blog/web-design-trends-2021
3	p) Realism	https://themes.muffingroup.com/be/county/?utm_source=smashing magazine.com&utm_medium=content&utm_campaign=march19
4	h) Pop design	https://www.goliath-entertainment.com/
5	h) Pop design, i) Usage of Animated Illustrations	https://in.pinterest.com/pin/62698619801388815/
6	j) Neumorphism	https://dribbble.com/shots/10068614-Neumorphism-Web-Design
7	a) Minimalism, i) Usage of Animated Illustrations	https://in.pinterest.com/pin/792000284463392796/
8	d) White Space design, f) Usage of 3D objects, l) Usage of Interactive Animations	https://globekit.co/
9	l) Usage of Interactive Animations, p) Realism, g) Usage of Micro Interactions	https://in.pinterest.com/pin/219620919306514722/
10	o) Material design	https://github.com/creativetimofficial/material-kit

(*continued*)

Table 1. (*continued*)

Website Number	Trends Followed by the Website	URL to Website
11	l) Usage of Interactive Animations, g) Usage of Micro Interactions	https://soulless.medium.com/financial-website-design-15-best-examples-templates-tips-for-you-2d66328e9189
12	l) Usage of Interactive Animations, d) White Space design	https://www.awwwards.com/inspiration/julie-guzal-portfolio-reactive-cursor-animation-particles
13	m) Glass Morphism	https://dev.to/harshhhdev/ui-design-trend-of-2021-4fb7
14	n) Usage of Geometric Layouts	https://cyberchimps.com/blog/web-design-trends/
15	e) Dark Mode, g) Usage of Micro Interactions	https://dribbble.com/shots/11184058-Dashboard-Graph-Animation
16	b) Bold Typographic, g) Usage of Micro Interactions	https://in.pinterest.com/pin/366621225913470392/
17	a) Minimalism	https://www.sequoir.com/
18	a) Minimalism, d) White Space design	https://www.intercom.com/
19	Usage of 3D Objects, Usage of Micro Interactions, Minimalism	https://dribbble.com/shots/6036333-

(*continued*)

Table 1. (*continued*)

Website Number	Trends Followed by the Website	URL to Website
20	Usage of 3D Objects, Minimalism, Usage of Animated Illustrations	https://dribbble.com/shots/4753157-3D-Wall-Landing-Page

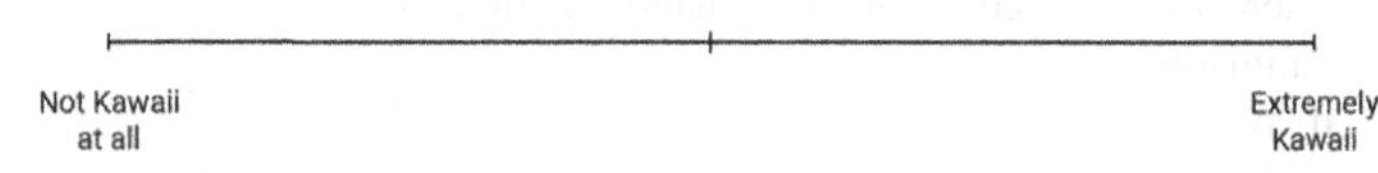

Fig. 3. Visual Analogue Scale (VAS) provided to the participants ranging from not kawaii at all to extremely kawaii.

The Entire experiment was carried out through 4 distinct steps as follows:

1) Discussion about 'Kawaii' and 'GUI' with participants

   Participants were briefed about kawaii as a concept, and their understanding of kawaii was discussed. Furthermore, a channelized line of thought related to website design principles and their emotional responses was generated. Lastly, any misconceptions about kawaii and or GUIs were cleared. This was done in order to bring all participants in a similar frame of mind in terms of understanding before the commencement of the tasks.

2) Rating in 'VAS' scale according to stimuli set from google form

   The participant was instructed to look at the stimulus containing the gif of a GUI and rate it's kawaii-ness on the VAS scale. The rating was done using the 'pen tool' available in the Miro workspace and by drawing a red | 'dash' on the given scale (Fig. 4).

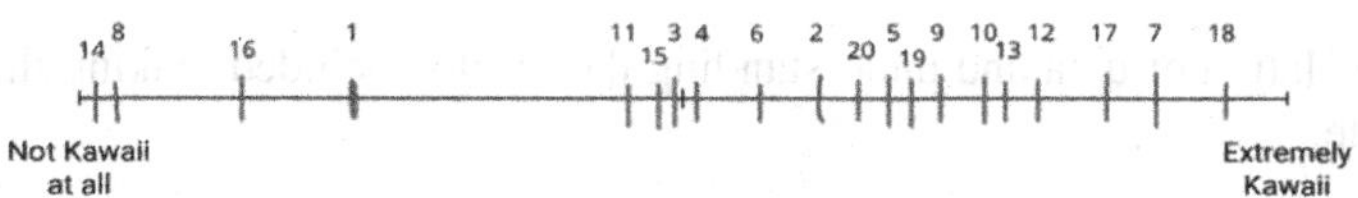

Name : Mithravinda KG
Age : 21

Fig. 4. Screenshot of VAS sheet marked with ratings by a participant.

3) Filling in response according to VAS scale rating in google form.
   The participant was asked to give a brief reason behind his rating in the VAS scale (Fig. 5).

6) According to you, how Kawaii is the GUI below? Rate it on the VAS sheet provided. Very briefly mention as to why you gave the above rating ? You can write specific keywords, any specific elements or a short paragraph.

30 responses

smooth, minimal, calming colours, friendly font

It gives me Kawaii feels By the Elements, Presentation and colors used.

Even though they haven't used contrasting colors, went with white and other muted shades but still because of the cognitive effort to process it, it's not kawaii.

Soft and Clean..has properties of kawaii

clean neat

not kawaii at all, doesn't have any elements that are cute in any way, no storytelling either. Very straightforward design

Neomorphism. V contemporary

the word hospital in itself doesn't feel like kawaii, tho looks ok

Fig. 5. Screenshot of 'Google Forms' based questionnaire provided to participant along with responses. Responses are the brief reasoning the participants have filled against a particular website stimulus.

4) Collecting and Extrapolating data.
   A 200 mm VAS scale was used. Post the experiment, the rated values were extrapolated on a scale of 100 mm for mathematical accuracy and ease of calculation.

3 Evaluation

Extrapolation of data and understanding the results included making the data readings to scale.

- The VAS Sheets were printed to scale on A4 size print paper.
- The initial measurement is taken on a 200 mm scale in cm. (a)
- The measurement is converted to mm units.
- The mm measurement is then converted for a 100 mm scale. (b)

$$b = (a \times 10)/2$$

The final scores (b) were used for drawing insights.

4 Results

4.1 Results in General

The 20 GUIs were ordered according to their averaged scores. A hierarchy of GUIs was created and graphical representation of various websites versus their average scores was formed (Fig. 6).

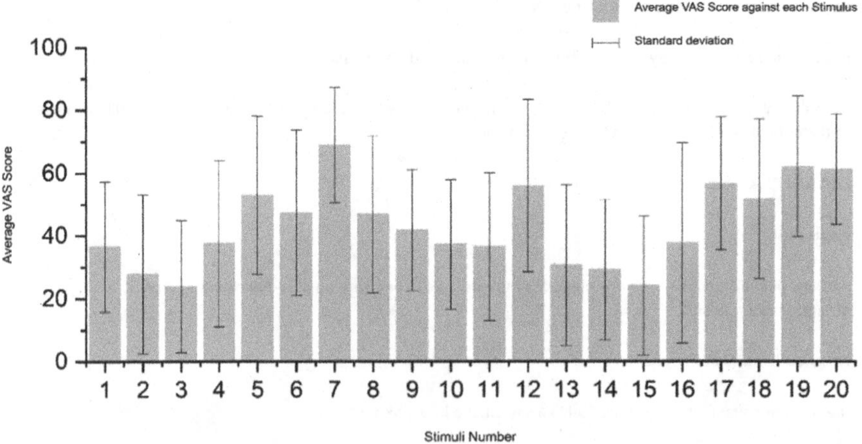

Fig. 6. Graphical representation of average VAS scores of twenty different stimuli.

With a variance in the intensity of 'kawaii-ness' experienced by the participant, there was a presence of large scatter in terms of absolute readings (Fig. 7). Some participants rated most websites highly kawaii. I.e. in the second half of the scale. Whereas there were some participants who found very few websites to have notions of kawaii in them, hence all their ratings were in the first half of the scale i.e. low scores. Lastly, there were

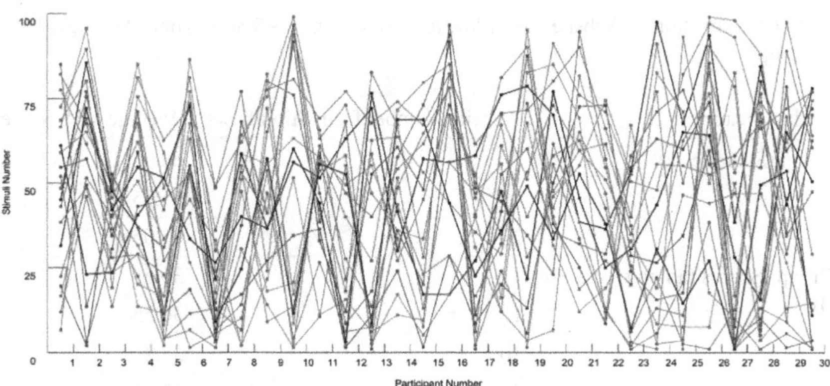

Fig. 7. Graphical representation of heavy scatter in absolute VAS scores against participant due to difference in opinion and level of understanding of kawaii.

some participants who moderated their scores, they were unable to decisively rate a website to be very kawaii or not kawaii at all, hence most of their scores were mediocre.

However, despite these differences in terms of opinion, there was a clear trend even in these ratings that pointed to some websites being rated more kawaii than others.

Hence, in order to verify the main effect of GUIs and gender of participants, two-factor ANOVA was carried out. The result showed a significant (1%) main effect of GUI's and no significant gender differences. The latter results showed that kawaii or approach motivation towards GUIs were not dependent on the gender of the user.

In addition, the post hoc tests of ANOVA showed significant differences (5%) between the following combinations:

- #1 < #7, #19, #20
- #2 < #5, #7, #12, #17, #18, #19, #20
- #3 < #5, #6, #7, #8, #12, #17, #18, #19, #20
- #4 < #7, #19, #20
- #8 < #7
- #9 < #7
- #10 < #7, #19, #20
- #11 < #7, #19, #20
- #13 < #5, #7, #12, #17, #19, #20
- #14 < #5, #7, #12, #17, #19, #20
- #15 < #5, #6, #7, #8, #12, #17, #18, #19, #20
- #16 < #7, #19, #20

4.2 Results by Classification

The entire website set was segregated and classified into three primary groups from the results of post hoc analysis, namely High score (H) group, Medium score (M) group, and Low score (L) group (Table 2).

Table 2. Classification of Websites according to their average VAS Scores into H group, M group and L group.

GUI Classification	Website Number from Stimuli Set	Average VAS Scores
H Group (Scores >60)	7	68.9
	19	61.95
	20	61.05
M Group (Scores ranging from 36 <60)	11	36.5
	1	36.55
	10	37.2
	4	37.55
	16	37.65
	9	41.75
	8	46.85
	6	47.25
	18	51.6
	5	52.95
	12	55.9
	17	56.5
L Group (Scores <31)	3	23.5
	15	24.15
	2	27.9
	14	29.2
	13	30.65

4.3 H Group

The H group contained websites that were rated the most 'Kawaii'. According to the recorded results, data plots were created that showed clear concentration of high ratings for certain websites (Fig. 8).

It can be clearly seen from Fig. 8 that there is a large concentration of high scores in website numbers 7, 19 and 20. This also verifies the high average scores of the three websites as discussed previously. Website number 7 has most of its scores higher than 50 with one exception, that has a low kawaii score. Similarly, websites 19 and 20 also have a large concentration of scores lying in the second half of the scale. I.e. high scores.

Common attributes among the H group were identified and a clear pattern in terms of kawaii-ness of different GUI was noted. Furthermore, some common attributes and features among highly kawaii interfaces were also identified.

1) The **color palette** used in websites was a primary factor which contributed to the kawaii-ness of GUIs. Some of the common colors identified are described further.

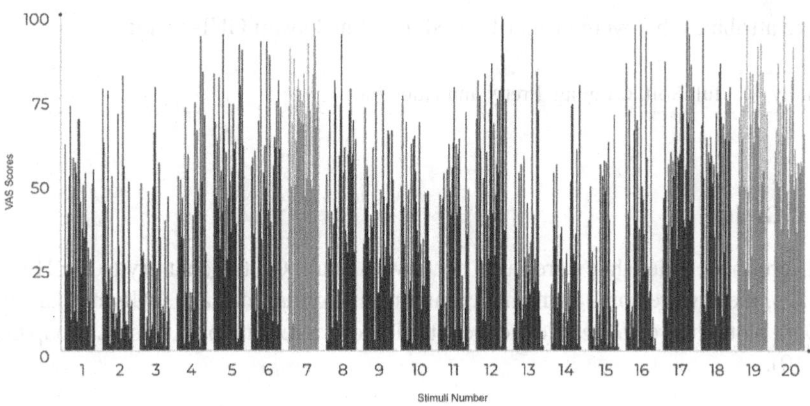

Fig. 8. Graphical representation showing higher concentration of high VAS scores in H group websites. i.e. stimuli number 7,19 and 20.

A color palette for websites number 7,19 and 20 have been identified (Fig. 9). The GUIs which used **pastel color** palettes were rated the most kawaii.

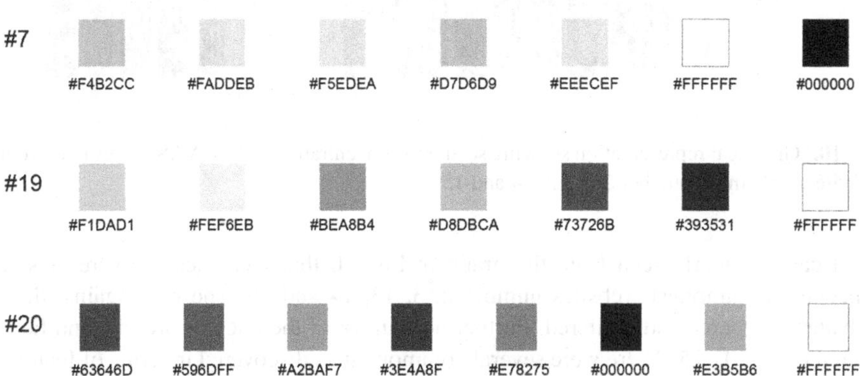

Fig. 9. Color palette used by websites from H group. i.e. Website number 7, 19 and 20

2) **Minimal illustrations** with large use of negative space were rated very kawaii.
   Illustrations having lighter line weights and usage of lesser and pastel colors as discussed in Fig. 9 along with a large amount of empty, white or background space were considered as 'cute', easy on the eyes and kawaii.
3) **Soft Animations** were also key factors that made GUIs more kawaii. The softness and smoothness of elements and transitions made GUIs appealing and inviting and eventually extremely kawaii.

The attributes that were primarily used to define kawaii GUIs were:

Friendly, Playful, Soft, Calming, Pretty and Adorable

4.4 L Group

The L group websites that were rated not kawaii at all. On the contrary to the H group websites, there were some elements common in L group to make GUIs not kawaii at all. Data plots showed a clear low concentration of scores amongst these L group GUIs (Fig. 10).

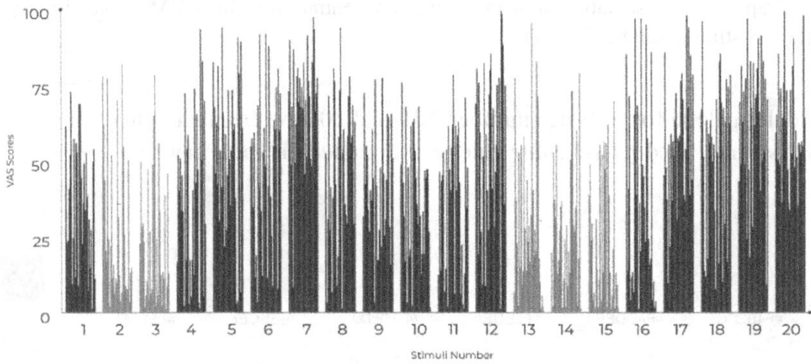

Fig. 10. Graphical representation showing scattered concentration of Low VAS scores in L group websites. i.e. stimuli number 2, 3, 13, 14 and 15

It can be clearly seen from the graph in Fig. 10 that there lies a severe lack of concentration amongst websites number 2, 3, 13, 14 and 15. The bars against these websites are scarce and scattered. Furthermore, most of their scores are low and lie in the interval of 0 - 25. There were several commonalities discovered in terms of features and attributes of these GUIs. They have been discussed ahead.

1) Colors played an important role here as well. The usage of **dark and primary colors** made some users find interfaces 'unwelcoming' and ultimately not kawaii at all (Fig. 11). The three primary colors Red, Yellow and Blue were clearly found to be synonymous to many of the colors that these websites had. Furthermore, the set of three secondary colors, I.e. Orange, Green and Violet were also amongst the shades in the palette from the 'not kawaii at all' websites. Website numbers 2, 14, and 15 showed the presence of these primary and secondary colors. On the other hand, websites 3 and 13 show the usage of a dark and dull color palettes that made these websites not kawaii at all.

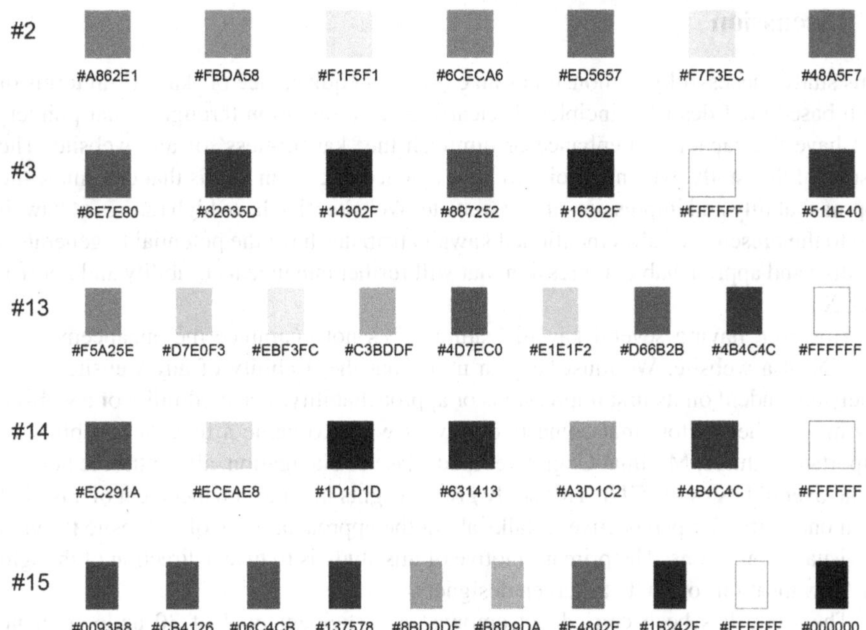

Fig. 11. Color palette used by websites from L group. i.e. Website number 2, 3, 13, 14 and 15

2) Another interesting insight that was found was the less kawaii-ness of GUIs that followed the trend of '**Realism**.' Real images and graphics made websites 'beautiful' and 'pretty' but not kawaii. Despite having positive notions of emotional feedback, realism was rated as not kawaii at all.

4.5 Insights and Outcomes

From the aforementioned results, a clear pattern can be observed in terms of the kawaiiness of websites. A hypothesis can be drawn in terms of the kawaiiness of GUIs using specific visual languages and animations. Websites and GUIs can be divided into five primary levels of segregation. These levels are the five dimensions of GUI design [14]. The first dimension being words and content oriented features. The second dimension discusses 2D graphics and Illustrations and other visuals. The third dimension talks about the platform of interaction with the media, i.e. web based or mobile application based. The fourth dimension is time based. Here, animations, transitions and other time dependent factors are discussed. The last dimension is the psychological and experiential dimension. With the help of this study, a clear direction of kawaii and not kawaii attributes of GUI design can be found out in terms of the second, fourth and fifth dimensions. Websites having pastel colors, softer animations and minimal illustrations will have higher notions of approachability and kawaiiness. On the contrary, websites with darker colors and real graphics following realism trends will be less kawaii in their nature.

5 Discussion

This study successfully demonstrates an emotional equivalence of 'kawaii' in terms of Web based GUI design principles. It clearly gives a direction through visual pointers that have the capacity to enhance or diminish the 'kawaii-ness' of any website. The results of this study evidently point to the existence of certain trends that determine the approachability and impression of any website. Websites that have high ratings of kawaii due to the presence of aforementioned kawaii attributes have the potential to generate a positive and approachable impression that will further enhance its usability and nourish its UX.

However, having several kawaii features does not guarantee the enhancement of the UX of a website. We must keep in mind that the usability of any website is not solely dependent on its first impressions or approachability. The credibility of a website has many other factors that come into play as well. To name a few, Accessibility of important features, Minimal Cognitive Load, Ease of Navigation, all contribute heavily to the overall UX of the GUI. This study gives insights into the enhancement of any GUI from one particular perspective. It talks about the approachability of a website through its visual impressions. The primary motive of this study is to give a direction of thought for 'kawaiisation' of GUIs to screen designers.

The study has been carried out amongst a participant pool of 30 undergraduate design students. Since a study in the lines of kawaii and its existence in digital design is a fairly novel field of research, the first step to successfully gain insights lie in the determination of the most proficient user set. Young people who are in their teens up until their late twenties are amongst the most avid users of digital media interfaces [15]. Hence, a pool of undergraduate students with a fair knowledge regarding screen design were chosen for the experiment. However, due to the limited age group of participants, the findings of this study are not exhaustive. The understanding of kawaii as a concept might change between different age groups of people. Furthermore, the proficiency of usage of websites might also differ from younger generations to older generations. However, despite these limitations the findings of this experiment are novel and the insights gathered are sufficient to point out certain trends of screen design and their interplay with kawaii as an attribute.

6 Conclusion

The experiment on the kawaiiness of GUI is a novel study in terms of the understanding the user experience and approachability of different websites. It helps provide a clear trend in terms of screen design principles for complex emotions and attributes of the likeness of kawaii. This study also takes a step in the direction of cross cultural user experience of digital products. It tests the notions of perception and understanding when boundaries of nationalities and cultures are pushed. Furthermore, with it being the first of its kind. It opens several opportunities for a series of future experimentation and research amongst a variety of participant pools and ever evolving website design trends. This paper helps gain novel insights in the field of screen design and its interplay with complex human emotions. The approach motivation towards GUIs was evaluated through the 'kawaii' emotion.

The results of the study show clear trends of kawaiiness in website designs that incorporate pastel color palettes and softer animations, paired with minimal illustrations. Furthermore, features of 'not kawaiiness' were also discovered in websites that used darker coloration and followed realistic visual graphics.

This study successfully explores the diverse fields of UX research, human psychology and Kawaii engineering to give insights in terms of accessibility and usability of digital product design.

Acknowledgement. We would like to thank the students of Department of Design, Delhi Technological University, who participated in the experimentation and helped support the research process with great enthusiasm and vigor. Furthermore, we would also like to thank Prof. Ravindra Singh and Prof. Partha Pratim Das from Delhi Technological University, India for their support and guidance.

References

1. Mcleod, S.: Maslow's Hierarchy of Needs. Simply Psychology 29 December 2020. https://www.simplypsychology.org/maslow.html#gsc.tab=0. Accessed 15 May 2021
2. Pellitteri, M.: Kawaii Aesthetics from Japan to Europe: Theory of the Japanese "Cute" and Transcultural Adoption of Its Styles in Italian and French Comics Production and Commodified Culture Goods. MDPI, 07 2018, Arts (2018)
3. Nittono, H., Fukushima, M., Yano, A., Moriya, H.: The power of kawaii: viewing cute images promotes a careful behavior and narrows attentional focus. Plos One Group (9) (2012)
4. Nittono, H., Ihara, N.: Psychophysiological responses to kawaii pictures with or without baby schema. SAGE Open (April 2017)
5. Damien, L., Nathalie, B.: Emotion and interface design. How to measure interface design emotional effect? In: KEER 2014, International Conference on Kansei Engineering and Emotion Research (2014)
6. Cyr, D., Head, M., Larios, H.: Colour appeal in website design within and across cultures: a multi-method evaluation. Int. J. Comput. Stud. **68**(1-2), 1–21 (2010)
7. Komatsu, T., Ohkura, M.: Study on evaluation of kawaii colors using visual analog scale. In: Smith, M.J., Salvendy, G. (eds.) Human Interface 2011. LNCS, vol. 6771, pp. 103–108. Springer, Heidelberg (2011). https://doi.org/10.1007/978-3-642-21793-7_12
8. Ohkura M., Aoto, T.: Systematic study of kawaii products: relation between kawaii feelings and attributes of industrial products. In: ASME, 2010. ASME 2010 International Design Engineering Technical Conferences and Computers and Information in Engineering Conference (2010)
9. Lindgaard, G., Fernandes, G., Dudek, C., Brown, J.: Attention web designers: you have 50 milliseconds to make a good first impression! Behav. Inf. Technol. **25**(2), 115–126 (2006)
10. Laja, P.: First Impressions Matter: Why Great Visual Design Is Essential. CXL. 25 September 2019. https://cxl.com/blog/first-impressions-matter-the-importance-of-great-visual-design/#:~:text=For%20first%20impressions%2C%20visual%20appeal%20even%20beats%20usability.&text=Users%20completed%20different%20tasks%20on,visual%20appeal%20of%20the%20site. Accessed 16 May 2021
11. Lieber-Milo, S., Nittono, H.: How the Japanese Term kawaii is perceived outside of Japan: a study in Israel. LEAN LIBRARY, SAGE Journals, August 2019

12. Collinge, R.: The Importance of Visual Appeal in Web Design. Usabillia by Survey Monkey. CX Insights, 28 June 2017. https://usabilla.com/blog/visual-appeal-web-design/#:~:text=Ensure%20that%2C%20as%20well%20as,for%20an%20exceptional%20user%20experience. Accessed 16 May 2021
13. Sung, Y., Wu, J.: The visual analogue scale for rating, ranking and paired-comparison (VAS-RRP): a new technique for psychological measurement. Behav. Res. Methods **50**(4), 1694–1715 (2018)
14. Devazya, A.: Interaction design and its 5 dimensions. UX Collective, Medium, 07 October 2019. https://uxdesign.cc/interaction-design-and-its-dimensions-39ca7e1d09f0. Accessed 16 May 2021
15. Keelery, S.: Distribution of internet users in India 2019, by age group. Statista 05 July 2021. https://www.statista.com/statistics/751005/india-share-of-internet-users-by-age-group/. Accessed 16 May 2021

Research on Interaction Design of Anti-addiction for Minor Games Based on Flow Theory

Xin Liang, Wei Yu(✉), and Xueqing Zhao

School of Art Design and Media, East China University of Science and Technology, Xuhui District, No. 130, Meilong Road, Shanghai, People's Republic of China

Abstract. With the development of Internet technology, minors' addiction to games has become more and more serious. How to improve this problem has become a focus of social attention. This paper introduces the theory of flow, and through its positive and negative applications, it takes the premise of causing as little rebelliousness as possible in minors, and subtly reduces the time they devote to online games, so as to achieve the purpose of preventing minors from getting addicted to games, and finally establishes the interaction model of preventing minors from getting addicted to games. Firstly, through literature research method, inductive method and user interview method, we analyze the main causes of minors' game addiction, combine them with nine features of flow theory, and summarize the inspiring principles of flow theory in the interaction design of minors' game anti-addiction. The interaction design model of anti-addiction interaction for minors' games is proposed by applying the flow theory in reverse and positive directions while minimizing the impact on the game experience, so that it can be a reference for the interaction design of anti-addiction interaction model for minors' games in several mobile game apps in the future.

Keywords: Minors game anti-addiction · Flow theory · interaction design · Reverse application

1 Research Background

At present, with the continuous development of Internet technology and game industry, minors' addiction to online games has become a social problem that cannot be ignored, and minors with weak self-control have become the biggest victims of online games. The survey shows that the proportion of online addiction among Chinese online minors is as high as 26%, and the proportion of online addiction tendency is as high as 12%. The strong involvement of online games in minors' life and study is considered an important factor affecting their psychological state and growth process. For minors, moderate online games can play a role of physical and mental pleasure to a certain extent, but on the other hand, Internet addiction syndrome formed by excessive addiction to games has negative parental effects such as damaging minors' physical and mental health, inducing violent psychology, deteriorating the relationship between minors and their

C. Stephanidis et al. (Eds.): HCII 2021, LNCS 13094, pp. 89–100, 2021.
https://doi.org/10.1007/978-3-030-90238-4_8

parents and wasting their studies. Therefore, it is extremely important to guide and assist children to establish good and moderate game usage habits, and a game anti-addiction system for minors has been created. At present, with the introduction of relevant national policies, major Internet companies have launched "minors' mode" and "growth guardian platform" and other cell phone anti-addiction systems for minors. However, since the measures to stop and control the game behavior of the anti-addiction system are rough and direct, it is easy to trigger a rebellious mentality among minors, which will lead to negative effects such as arguing with guardians and seeking ways to break the system. The purpose of preventing minors from getting addicted to games is achieved in a subtle way.

2 Analysis of Research Methods for the Design of Anti-addiction System for Minors Based on the Flow Theory

2.1 Flow Theory

Csikszentmihalyi, an American psychologist who first conducted systematic scientific research on the theory of flow, defined flow experience as an emotional experience when people are fully engaged in a certain activity or thing, when their attention is unprecedentedly focused, and when all external influences are filtered out. [1] Based on Csikszentmihalyi's research, researchers at the University of Milan added and adapted the concept of flow experience: Flow is a positive experience that occurs when people engage in challenging or skill- and difficulty-matched activities and tasks, in which they show fascination and dedication to the activity, concentrate and forget about In this experience, people show fascination and devotion to the activity, concentrate on it and forget about the passage of time, and achieve a state of integration of time and activity and forgetfulness. [2] During this period, people experience a high level of pleasure and fulfillment, and their perception of the external world is greatly diminished.

Although flow is not a continuous experience, it is an important psychological reason why people are willing to continue a certain activity.

Mindfulness Model

The Csikszentmihalyi model of flow is shown in Fig. 1, where the challenge is more difficult than the user's skills. Conversely, users will become bored and their excitement and exploration will decrease. Only when both challenges and skills reach a balanced state will the flow experience be generated. In game design, the balance of challenge and skill has always been an essential part of game design in order to improve the immersion and addiction of the game (See Fig. 1) [3].

Characteristics of Flow Experience

Csikszentmihalyi summarized nine characteristics of flow experience:

Clear and explicit goals, i.e. people carry out activities with clear goals;

Balance between challenge and skill, i.e. people's skill level matches the level of challenge of the activity;

Timely and valuable feedback, i.e. people can effectively understand the progress of the current activity and the direction of the later activity during the activity;

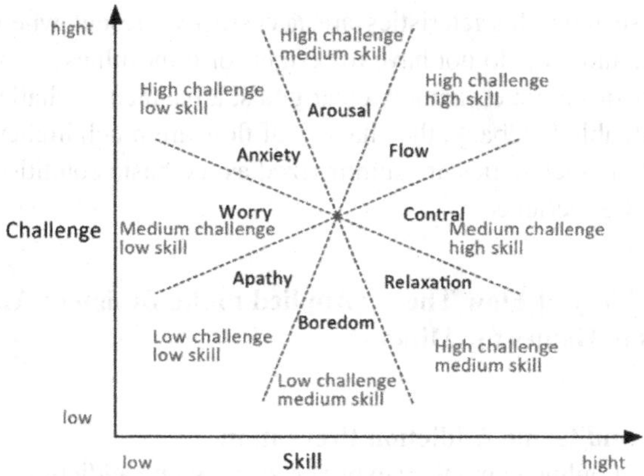

Fig. 1. The relationship between challenges and skills

Integration of awareness and action, i.e. awareness and action remain consistent;

Complete concentration, i.e. people are fully engaged in the activity;

A sense of control over the experiential activity, i.e. people are fully engaged in the activity.

Loss of individual self-awareness, i.e., people lose the perception of bodily functions and external influences, such as hunger and external sound or temperature changes;

An altered sense of time in life, such as the ability to engage in an activity for a long time without perceiving the passage of time.

A sense of involvement in experiencing an activity, i.e., the activity is both the process and the purpose of the experience.

Nocak and Michel grouped these nine characteristics into three phases based on the process of flow experience generation (See Table 1): antecedent conditions, characteristics, and sequences of experience.

Table 1. Nine characteristics of flow experience

Flow experience process	Flow experience characteristics
Antecedent conditions	Clear and explicit goals
	Balance between challenge and skill
	Timely and valuable feedback
Characteristics	Integration of awareness and action
	Complete concentration
	A sense of control over the experiential activity
Consequences of experience	Loss of individual self-awareness
	An altered sense of time in life
	A sense of involvement in experiencing an activity

Not all of these nine characteristics are necessarily present when a mindfulness experience occurs, and they do not have to be met for mindfulness to occur. If three of the nine characteristics are met, namely clear goals, a balance of challenges and skills, and timely and valuable feedback, the chances of flow are much higher, so to a certain extent, these three characteristics are summarized as the basic conditions that motivate users to have a flow experience.

2.2 The Applicability of Flow Theory Applied to the Design of Anti-addiction Interactions in Games for Minors

Flow Experience and Game Addiction Prevention

Research shows that online game flow experience and game addiction are strongly positively correlated, and the likelihood of game addiction increases with the intensity of flow experience. At the same time, online game flow experience indirectly influences game addiction through the fun perception dimension of value perception, and value perception plays a partly mediating role between flow experience and game addiction. [4] Chou et al. argued that flow experience brings online game players a sense of accomplishment and pleasure, which will prompt players to repeat the game behavior, and the interaction between flow experience and repeated use will lead to online game addiction [5].

Based on the previous discussion on the strong positive correlation between game addiction and flow experience, this paper, after literature research and user interviews, comes up with the main causes of game addiction among minors and corresponds them to the nine characteristics of flow experience.

There are many reasons for minors' game addiction, which is an inevitable and likely long-term problem in the Internet era, and its sources include society, government, schools, parents and minors' own irresponsibility. In this paper, we analyze the causes of minors' game addiction in terms of the objective reasons of game design that lead to addiction and their own subjective reasons.

Objective reasons (taking the King of glory handheld game as an example)

Online games have the feature of easy to start: most of the popular mobile games on the market have the feature of easy to start and simple operation, which allows users to easily master the initial skills and start the game. Users can easily control the game interface with their left and right thumbs and complete a series of cool game actions from simple to complex; this feature corresponds to the "sense of control in experiencing activities" and "balance of challenge and skill" in the mindstream experience. If the game skills and challenge requirements are too high and the user loses the sense of control, it will greatly reduce the user's motivation.

Diversification of modes: The diversified design of game modes improves user stickiness, brings users a constant sense of freshness, and avoids users' boredom caused by monotonous game modes, such as different challenge modes, live games, etc.; this feature can correspond to the "balance of challenge and skill" in the flow experience, when users play the same mode for a long time. When users play the same mode for a long time, their skills will continue to improve, but the challenge is not improved enough, and users will get bored to a certain extent. At this time, users will choose other modes of play to rebuild the balance of skills and challenges, and can get the satisfaction of freshness.

Reward system: When users reach a certain level of experience value or achievement in the game, they will get corresponding rewards or titles, such as rare skins, equipment and high ranking. For minors, these rewards or titles that make them gain high achievement are very tempting, and it is necessary to invest very much time, effort and money to get them. [6] This feature corresponds to the "timely and valuable feedback" and "clear and definite goals" in the mindstream experience, where each stage of the game is rewarded with virtual rewards and level advancement. These feedbacks can let users have a clear understanding of the current game progress and even the future direction of the game, which can make them establish clear goals.

Subjective reasons
Psychological needs: Maslow's hierarchy of needs divides human needs into five levels: physiological needs, security needs, social needs, respect needs, and self-actualization needs. After the low-level needs are satisfied, people will pursue higher-level pursuits, such as respect needs and self-actualization needs. People can obtain a certain degree of satisfaction of higher-level needs by defeating opponents, winning rewards or reaching high ranks in games. For students entering secondary school, studying occupies the majority of their life time. With the increasing academic pressure, many students are under a lot of pressure and distress, especially those with marginal academic performance, who have less "presence in the school class" and can easily shift their focus to games that can fulfill them. The virtual world of games provides a sense of existential satisfaction that is lacking in real life, and people tend to get addicted to it easily, not to mention the immature minors. [7] This feature corresponds to the "clear and explicit goals" and "loss of self-awareness" in the mindstream theory. The virtual environment of online games makes people easily addicted to them, and minors' minds are not yet fully mature, so the high sense of achievement of online games makes them easily bring themselves into the game characters and situations, and it is difficult for them to get out of this state because they are overly involved in them.

Group psychology: people who are in a group can get a sense of belonging from the group on the one hand, and get support from the group on the other hand to avoid being isolated. In 2019, among the various activities minors often engage in online, the proportion of minors who choose to play games reached 61%, which leads many minors who were not too interested in games to be easily influenced by their friends who play games. In order to integrate into the group and maintain classmate relationships they also start playing games. This characteristic corresponds to the "sense of participation in experiential activities" in the mindstream theory. Participation is both an experiential process and an experiential purpose, i.e., playing games both for the fun of the game and for the game itself as a form of social interaction.

Flow Experience and Interaction Design
The purpose of interaction design is to design efficient and useful interaction solutions through sufficient research and insight into users' product use experience, behavioral characteristics and psychological features. Eventually, users can get a good experience

of using the product and satisfy their emotional needs. [1] Flow experience is an optimal experience in user experience, which can greatly enhance users' pleasure and satisfaction when using the product, which is the same as the satisfaction of emotional needs and improvement of user experience pursued by interaction design.

3 The Construction of an Interaction Design Model for the Prevention of Addiction in Games for Minors Based on Flow Theory

3.1 Enlightening Principles of the Design of Anti-addiction of Games for Minors Based on the Flow Theory

The application of flow theory in this paper is divided into two aspects: reverse application and positive application.

Reverse Application
Through the reverse application of the three stages of flow experience: premise, characteristics and effect, the state of minors being addicted to computer games for too long and even producing flow experience is broken. For example, an important feature of the mindstream experience is the balance of challenge and skill, it is taken that after a certain amount of game time, the degree of challenge of the match matched to exceeds their own skills, thus leading to their inability to get a high sense of acquisition. The combination of persuasive behavior that triggers their self-reflection and interaction design that prompts them to passively end their gaming behavior achieves the goal of guiding minors to stop playing for too long while affecting their gaming experience within normal gaming time as little as possible.

The previous section combines the causes of minors' game addiction with the characteristics of flow experience. The six characteristics of the mindstreaming experience, are the balance of challenge and skill, clear goals, timely and valuable feedback, a sense of control over the activity, a loss of individual self-awareness, and a sense of involvement in the activity. In the following section, we will introduce the inspiring principles of the flow theory for the design of anti-addiction for minors (See Table 2), and construct a model of interaction design for anti-addiction for minors based on the flow theory from three aspects: user layer, behavior layer, and goal layer.

Positive Application
Minors have certain resistance to the anti-addiction system, but at the same time, they also have the initial self-reflection consciousness that long-time addiction to games has negative effects on them and desire to achieve a healthy and moderate time of game experience. The introduction of the flow theory can make users have a better interactive experience, which is very useful for reducing the resistance of minors and better realizing the anti-addiction function.

The first three features of the nine features of the flow theory, i.e., the features of the antecedent conditions, are the basic conditions for generating the best user experience - flow, while the characteristics stage and the consequences of experience stage mainly

Table 2. Inspiring principles of the flow theory for the design of anti-addiction for minors

	Reasons for minors' game addiction	Enlightening principles
Balance between challenge and skill	Online games are easy to start, diversify modes	The balance of challenges and skills is the most important condition for generating flow experience. Different users have different game skills, and they are matched with different skills. With the process of in-depth use of game products, users are matched to challenges of gradually increasing difficulty and can explore new modes and new functions, which constantly generate a sense of acquisition and freshness. In the game anti-addiction design, after the user has been addicted to the game for a long time, the balance of challenges and skills in the game is broken by applying the theory of flow in reverse, so that the boredom of matching low challenges with high skills or the frustration of matching high challenges with low skills is generated, and the purpose of stopping game addiction is achieved
Clear and explicit goals	Reward System	Clear and explicit goals are one of the necessary conditions for generating a flow experience, which can have an encouraging effect on users and increase their motivation. The clearer the goal, the more obvious the encouragement for users, and the stronger the motivation of users. In the design of anti-addiction games for minors, the design of goals can be blurred after the specified game time is exceeded, so as to reduce the various kinds of rewards and ranking system that tempt minors to be addicted to games, so that minors lack the goal and sense of gain to continue playing
Timely and valuable feedback	Reward System	Timely and valuable feedback is also one of the necessary conditions for the generation of flow experience, which can help users determine the progress of the activity and the route of future development, and enhance their sense of control and motivation towards the activity. The reward mechanism in the game can provide users with timely and valuable positive feedback, which can make them feel satisfied and enhance their sense of participation and challenge in the activity, which is an important reason to prolong the time of flow and addiction. Therefore, in the design of anti-addiction for minors' games, the positive feedback that encourages minors to continue to be addicted to games can be appropriately reduced

(continued)

Table 2. (*continued*)

	Reasons for minors' game addiction	Enlightening principles
A sense of control over the experiential activity	Online games are easy to start	The sense of control over the activity is the second stage of the flow experience, i.e., the "characteristics" stage, which indicates that the user can feel the operational autonomy and control over the activity. When the control is too low or the operational autonomy is small, it is easy to trigger anxiety and boredom. In the design of anti-addiction game for minors, when the user exceeds the set playing time, the controllability of the activity can be limited
Loss of individual self-awareness	Psychological needs	Most minor players are aware that online games have some negative effects on them. Minors have also begun to develop a sense of self-management. The research found that the factor that minors self-identify as limiting their gaming time accounts for the highest percentage is the fear that excessive gaming will affect their studies, which is different from our conventional idea that the main factor controlling minors' gaming time comes from the external environment (such as parental management). [8] In the process of gaming, minors, due to their lack of self-control, can easily be immersed in the game for too long without realizing it, but this does not mean that they are not self-reflective about it. The problem is that the immersion experience in the game process makes it difficult for minors to realize the passage of time, and it is feasible to improve this problem and achieve the awakening of minors' self-reflective consciousness to make them reduce their game time voluntarily
A sense of involvement in experiencing an activity	Social Needs	A sense of involvement in experiencing an activity is a characteristic of the "effect" stage of the mindstream experience, which emphasizes that the purpose of experiential activities is both the experience and the activity itself, and represents in the play behavior of minors: the purpose of minors' participation in games is both to experience the game process and to play the game itself, an activity with social properties. At present, online games, especially mobile games, have become a common form of social entertainment among minors collectively, and many students who were originally not interested in games are exposed to games or even addicted to games for social purposes. Therefore, in the design of anti-addiction of minors' games, after going beyond the prescribed game market, the social attributes can be restricted to reduce the influence of social attributes on game addiction

illustrate the features and effects of the flow experience process. The purpose of the positive application of the flow theory in this paper is to achieve a better experience, so the three conditional features of the prerequisite stage are proposed to inspire design principles for the interaction design of the anti-addiction system (See Table 3).

Table 3. Enlightening principles of anti-addiction interaction design

	Enlightening principles
Balance between challenge and skill	One of the reasons minors are addicted to games is their lack of self-control. If the anti-addiction efforts are too strong or progress too fast, the challenge will be much harder than the skill level, making minors anxious and frustrated. Therefore, the prevention of addiction needs to be gradual, and it is best to combine the strengths of parents, schools and society, rather than expecting a one-step solution
Clear and explicit goals	Setting clear goals at the beginning of an activity can increase the motivation of users and provide them with a great incentive. By setting clear anti-addiction goals on their own or with their parents, minors can be more motivated to prevent addiction and have a higher sense of satisfaction after completing their tasks
Timely and valuable feedback	Positive feedback is given to users' behaviors on the anti-addiction tasks they complete, leading them to feel the sense of gain from anti-addiction

3.2 Interaction Design Model of Anti-addiction System for Minors' Games Based on Flow Theory

Based on the previous analysis, combined with flow theory and interaction design methods, an interaction design model for anti-addiction for minors is constructed from three aspects: user layer, behavior layer, and target layer (See Fig. 2).

User Level

Set clear anti-addiction targets: Minors are explicitly informed of the maximum game time or set their own game time when they enter the game. During the game, the system informs the user of the process of game time at a certain point in time (choose the point in time that does not affect the ongoing game) to avoid minors from forgetting the time when they are immersed in the game. In addition, adding an incentive mechanism to goal setting, corresponding game anti-addiction with other things that have a strong sense of acquisition, such as designing incentive pop-ups, pages or videos, and linking the reduction of game time with positive activities such as increasing physical exercise, shopping and socializing, and study time to improve minors' sense of acquisition and motivation.

Realize Multi-level User Orientation: Users at different stages have different environments, purposes, needs and psychological states. Grading the game time of minors in different age groups can, on the one hand, avoid arbitrarily deciding the game time of all people without considering the age gap of minors and realize more accurate game anti-addiction. On the other hand, it can achieve the balance of minors' anti-addiction challenges and skills and improve the executability of anti-addiction tasks.

User Identification System: The setting of anti-addiction mode for minors depends on the users themselves, and if parents cannot play a supervisory role, minors can change their own passwords, etc., so that the anti-addiction system cannot be used effectively, and the additional user identification and authentication function can effectively detect the guardians of minors.

Goal Level

Reward mechanism: Users can get certain rewards after completing a series of anti-addiction tasks on time or in advance, which can be achieved through the guardians of minors (the guardians and minors agree on this reward, and the data is recorded on the platform and feedback is given at the end); when users do not complete the anti-addiction tasks and continue the game behavior after exceeding the maximum game time, the game is reduced rewards, such as reducing the increase in the energy value required for upgrades. For users who do not complete the anti-addiction task, certain penalties are imposed, such as reducing the time of the next game, increasing the difficulty of obtaining the energy value required for upgrading, etc.

Decompose the Process of Goal Achievement: Establish progressive goals, set short-term goals and final goals, and increase the controllability of minors' use of the anti-addiction system.

Balance Between Challenges and Skills: After the game time is exceeded, if minors do not end the game behavior, they will be matched with game challenges that are not suitable for their skill level, with high skills matching low challenges and low skills matching high challenges.

Social Functions: Add social functions such as friend reminding anti-addiction and ranking to meet both social needs and anti-addiction needs. After the game time is exceeded, the module of contacting friends to play against each other is canceled, and only the mode of randomly matching teammates can be chosen to play, so as to reduce the influence of social attributes on game addiction.

Behavioral Level

Provide feedback: After exceeding the game time, reduce the feedback that encourages users to continue to indulge in the game, such as displaying a page reminder for users to play the next stage of the game and a reward for game completion. After completing the anti-addiction task, actively provide positive feedback to make users understand the current anti-addiction result as well as the process and generate a high sense of acquisition.

Designing Multi-channel Interaction Experience: Providing users with anti-addiction service from three aspects: visual, auditory and tactile. For example, the visual aspect includes pop-up reminder, hover ball reminder, screen brightness change reminder, rest screen reminder, video reminder, etc.; the auditory aspect includes ringtone reminder, voice reminder, etc.; the haptic aspect includes voice vibration, etc. The combination of multi-sensory experience can break the state of minors being addicted to games for a long time. Conversely, in improving the acceptability of the anti-addiction system, i.e., the positive application of the flow theory, the combination of multi-sensory channels can be used to improve its interactive experience.

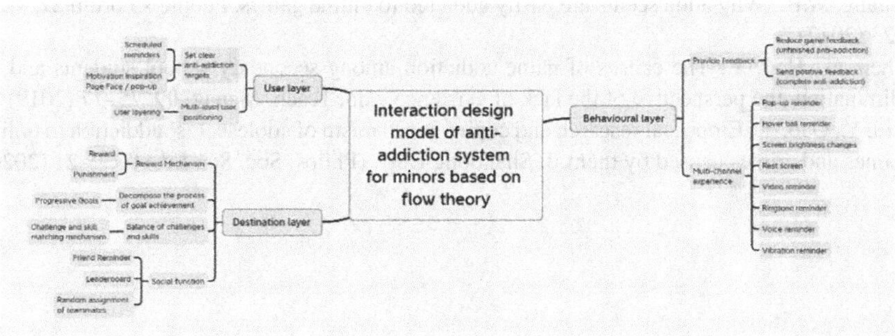

Fig. 2. Interaction design model of anti-addiction system for minors' games based on flow theory

4 Conclusion and Prospect

This study constructs an interaction design model of game anti-addiction for minors based on the flow theory. Through the reverse application of flow theory to crack the flow experience process that can cause game addiction and reduce minors' game time; through the positive use of flow theory, the acceptability of the anti-addiction system is improved and minors' rebelliousness to the system is reduced.

The biggest innovation of this paper is the reverse application of the flow theory, i.e., it discusses the properties of flow theory that have negative effects on addiction and applies them to the design of game anti-addiction, focusing precisely on the main causes of minors' game addiction and using the features of flow theory to crack them one by one.

The model of game addiction prevention for minors constructed in this thesis needs further research and development in the following aspects. (1) The causes of minors' game addiction come from the failure of responsibility of society, schools, families and their individuals in many aspects. This paper only discusses the factors of families and individuals, which can be improved in the future research. (2) The research on the design of anti-addiction is limited to qualitative research, and future research can include qualitative research to make the model more accurate.

References

1. Xiaoyun, X., Li, F., Yang, P.: A review of product interaction design research based on mind flow theory. Packag. Eng. **41**(24), 14–21 (2020)
2. Li, S.: Research on internet product experience design based on mind flow theory. Southeast University (2016)
3. Li, J., Tan, Q., Qiu, P., Zhang, W., Li, Z.: Research on interaction design of cultural and creative products based on mind flow theory. Packag. Eng. **41**(18), 287–293 (2020)
4. Huang, S., Zhu, D.: The relationship between adolescents' online game mind-flow experience and game addiction. Chin. Youth Soc. Sci. **40**(01), 79–89 (2021)
5. Chou, T.J., Ting, C.C.: The role of flow experience in cyber game addiction. Cyber Psychol. Behav. (6) (2003)
6. Kuang, W.B.: Why adolescents are easily addicted to online games. People's Forum **22**, 122–123 (2017)
7. Shen, G., Hou, Y.: The causes of game addiction among secondary school students and its elimination–the perspective of the lack of existence gain. Teach. Manag. **09**, 75–77 (2019)
8. Liu, Y., Gao, Y.: Empirical research and coping mechanism of adolescents' addiction to online games and crimes caused by them. J. Shandong Univ. (Philos. Soc. Sci. Ed.) **03**, 9–21 (2020)

Potential Design Strategies Based on Communication Design and Art Therapy for User Experience in COVID-19

Zhen Liu, Zulan Yang(✉), and Ke Zhang

School of Design, South China University of Technology, Guangzhou 510006, People's Republic of China

Abstract. At present, the 2019 novel coronavirus (COVID-19) is seriously affecting everyone's life, body and psychology. A review of relevant literature found that communication design (CD) and art therapy (AT) can bring immediate benefits to human health. However, the current CD focuses on the design of visual communication, and AT focuses on the methods and results of treatment, and rarely integrates the characteristics of the two to carry out healthy design interventions. Therefore, this paper aims to analyze the characteristics of CD and AT through the method of literature review, explore the potential positive effects of the integration of the two. And then start from the perspective of design and creation of cure, to design cures and think about potential interventions. Finally, a design strategy framework for experience under the epidemic will be constructed to enhance users' experience during treatment and reduce the impact of the COVID-19 pandemic on mental health. By integrating the analysis results of CD and AT, this paper constructs a broad framework of design strategy from three levels: users, design and elements. First, the design process is driven from top to bottom by stakeholders. Then, the CD process is constructed based on the design thinking of user experience by integrating visual information and artistic media. Finally, several design elements including senses, experience, humanity, rhetoric, new media and digital technology are integrated into the design process to realize the communication, exchange and interaction of information, and to construct the intervention measures to cure by design.

Keywords: COVID-19 · Communication design · Art therapy · User experience · Design strategy

1 Introduction

The outbreak of the 2019 novel coronavirus (COVID-19) has greatly affected everyone in life, body, and psychology. Many studies have found that the impact of COVID-19 on mental health is extensive and may have a negative impact on various groups of people, including medical service providers, children, adolescents, the elderly, LGBTQ groups, and individuals with pre-existing mental illness [1]. In terms of mental health, COVID-19 may cause emotional distress and anxiety in various groups [2], as well as easily

C. Stephanidis et al. (Eds.): HCII 2021, LNCS 13094, pp. 101–115, 2021.
https://doi.org/10.1007/978-3-030-90238-4_9

produce fear and stress [3]. Therefore, it is necessary to explore an effective intervention to deal with the negative impact of COVID-19.

In the field of design, communication design (CD) has a certain positive effect on personal health. A team at Yokohama City University proposed the concept of Advertising Medicine (AD-MED) for the first time in the world, which explores people's behavioral health communication from the perspective of advertising, and is considered to be a new academic system for achieving communication goals in medical health [4]. AD-MED aims to solve various medical problems from the perspective of consumers by studying communication, which integrates an easy-to-understand and influential advertising point of view, such as design and copywriting, to realize human behavior change [5]. Among them, CD can be applied in the field of pediatric medical care [6], the design of the medical communication system for the disabled [7], the guidance system of hospital space [8] and the transmission of health information [9]. In addition, among the many treatment methods, art therapy (AT) has been gradually used a long time ago. It can provide a medium for the emotional expression of the treated person, and can be used as a channel for emotional release and self-expression to achieve a good therapeutic effect [10]. Importantly, AT can effectively promote mental health [11, 12]. In addition, creative art therapy can also have a positive impact on emotional symptoms and behavioral difficulties [13]. Compared with drug therapy, AT has great advantages. In general, CD and AT have potential value in promoting health.

Although COVID-19 has brought challenges to health gap research, it has also brought unprecedented opportunities for innovation [14]. Therefore, in the face of the impact of the COVID-19 pandemic, thinking about the practical need for a truly design-smart response to a global public health emergency is not only an opportunity for innovation, but also a necessity. Therefore, this paper aims to analyze the characteristics of CD and AT, explore the potential positive effects of the integration of the two. And then start from the perspective of design and creation of cure, to design cures and think about potential interventions. Finally, a design strategy framework for experience under the epidemic will be constructed to enhance users' experience during treatment and reduce the impact of the COVID-19 pandemic on mental health.

2 Communication Design (CD)

Design is a way of understanding communication, because CD arises when people intervene in the interaction design of some ongoing activities through the invention of technologies, devices and programs [15]. CD influences people's knowledge, attitude and behavior by constructing visual information [16, 17], which involves a series of elements of conception, planning, projection, coordination, selection and organization to create text and visual communication [18]. CD may be related to technical communication, information design, and content development, and it not only represents a set of symbolic representation practices, but also reflects a deeper shift to sociological work [19].

In the field of CD, visual communication design (VCD) is an important branch. The concept of CD is sometimes used as a shorthand for VCD, which is a term closely related to graphic design [18, 20]. VCD is an important part of modern design and the main form of modern social information transmission. It should not only be satisfied with the visual experience of function and surface, but also convey a certain emotional experience to consumers and viewers through visual language [21]. With the development of the Internet and technology, VCD is no longer limited to the composition and appearance of two-dimensional elements for the purpose of formal beauty, but more and more emphasizes the user's reading experience and emotional design of the work [22]. In addition, in the information environment of contemporary society, VCD increasingly emphasizes the need for multi-sensory expression [23] and emotional experience [24]. Among them, graphic language [25, 26], shared communication [27], and design thinking [28, 29] have a positive impact on VCD. With the continuous development of VCD, it has been included in plastic arts, communication, marketing, visual physiology, visual psychology, semiotics, philosophy, aesthetics and many other disciplines, and it has become a special cultural phenomenon which is an important form of human information exchange, exchange and interaction [30].

In terms of experience, CD mainly focuses on user perception and cultural experience. Integrating perception and emotion into the studio education of contemporary CD can enable students to have a positive sensory experience in the learning process [31]. In the visualization of data and information, the results of CD can be dynamically displayed in a variety of sensory forms, so that users can obtain sensory experiences outside of the physical space when browsing the website [32]. In addition, considering that culture may be one of the most important aspects that communication designers need to consider when designing [33], adding a cultural perspective to CD can expand the potential range of user experience [34].

CD often incorporates art, rhetoric, and virtual reality technology to improve the effectiveness of communication. The CD of public space from the artistic level can improve the behavioral pressure and waiting experience of patients in the hospital environment [35]. It is worth noting that until the invention of written language, storytelling was considered to be one of the oldest art forms in human history and the main method of passing wisdom, knowledge, and information from generation to generation [36]. Therefore, using digital stories as an art form of communication design can effectively solve the stigma and discrimination associated with human immunodeficiency virus positivity [37]. Moreover, effective rhetorical communication has a profound impact on society, because it can better convey information and understand user needs in the process of CD [38]. Furthermore, CD can benefit from the rhetoric of health and medicine, on the contrary, health and medicine generally require CD and rhetoric of health and medicine [39]. At the technical level, although the correlation between virtual reality and CD needs to be further discussed, virtual reality as a tool of CD practice still has potential utility [40]. In the interactive CD system, creating a virtual environment that simulates the natural world to adapt to the user's experience can enable users to find satisfactory and valuable information [41].

CD can not only promote the transmission of risk information, but also has certain value for disease treatment and health promotion. Based on the narrative drive of the physical psychological model, CD can be used to build risk perception and risk CD for the elderly [42]. At the health level, by creating visual results of CD, people's awareness and understanding of Alzheimer's disease can be improved, and better interaction between patients with Alzheimer's disease, their families and carers, and provide patients with practicality communication tool [43]. Moreover, a multidisciplinary team composed of CD experts, pediatric urology experts and health service researchers can design a self-reporting toolkit for adolescent patients to promote participants' reflection on social and mental health [44]. In general, in health-related CD, user experience, localization, language and interdisciplinary collaboration can be combined to create a bilingual and localized vocabulary to transcend a single, static, and dominant language ideology [45]. In addition, based on the social ecological model of health, a CD evaluation framework considering stakeholders is designed, which will have some implications for the evaluation of child health programs, other public health, communication and international development interventions [46].

Value, humanity, and emotion are the favorable factors considered in the CD process. The value essence of CD is not only reflected in helping organization members perform tasks more effectively, but also in promoting products to better meet the needs of users, thereby contributing value to personal daily life [47]. In the process of CD, the language and activities of the three cultures of humanities, designers, and informatics can be combined to construct a collaborative research model, and then digital technology can be used to create an information visualization interface to better manipulate and interpret human experience [48]. In addition, in the learning of CD, feeling emotion is a form of feedback, and learners can use it to analyze and explain the impact of the surrounding learning environment [49]. Moreover, effective sentiment analysis can be used as the basis of CD to support the organization's strategic relationship goals [50].

In general, CD focuses on defining and solving problems in novel ways and responding to the urgency of highly different situations to highlight the importance of what is being done [51]. As a complex learning activity, CD requires a creative, multidisciplinary approach to the collection, analysis and interpretation of data [52]. The CD design practice framework provides a path to open the black box of stakeholder engagement and opens up a new way of thinking about stakeholder engagement, which has implications for cultivating professional practice and improving communication resources and infrastructure for organizational decision-making investment [53]. From the historical and theoretical overview of visual research methods, it can be concluded that visual methods are mainly developed in anthropology and sociology, and usually include the use of photography, video and painting in the qualitative study of life experience, although visualization methods are relevant fields such as human-computer interaction and computer science education have been applied, but these methods are not yet perfect in the research of CD [54]. It is worth noting that making full use of digital health communication channels for CD can improve health education, health promotion, and health behavior [55].

3 Art Therapy (AT)

AT consists of the physical creation of art works and the cognitive evaluation of painting, and it can bring many benefits to human cognition, such as the improvement of visual spatial ability, attention, working memory and executive function [56]. Creative AT and group AT are two types of art therapy. They have a certain potential in relieving symptoms of trauma, depression, fear, stress, anxiety, and behavioral disorders. Part of the literature reviewed the research scope and therapeutic value of creative AT applied to traumatized children through a systematic review [57], and also showed that creative AT can have a positive impact on the psychology of breast and gynecological cancer patients from scratch [58]. In addition, group AT can reduce the depression level of elderly patients with neurocognitive disorder (ND), improve their ability to express themselves [59], and also improve the fear of childbirth in late pregnancy [60]. Moreover, group AT combined with respiratory therapy can allow anxious patients to recognize their negative emotions, thereby reducing anxiety symptoms and improving subjective well-being [61]. It is worth noting that trauma is mainly a concept of non-verbal problems [62], and the value of non-verbal problems can be found through artistic creation [63]. Therefore, AT is an effective and non-verbal treatment method in solving the problems related to trauma [64].

The art media used in AT mainly include painting, image, collage, sculpture, music, dance/movement. Among the art media based on painting, bridge, self-portrait, mandala, and graffiti are different forms of painting. On the basis of painting, different painting themes can be constructed, as well as digital painting and other treatment methods that change with technology. By analyzing the different presentation forms of painting, it can be found that bridge drawing can provide an opportunity of AT for orphans who have experienced psychological trauma [65]. Four-drawing in the form of self-portrait can reduce the impact of traumatic events [66]. In addition, the drawing form of mandala has a positive impact on the lives of psychiatric inpatients, so further research on the impact of mandala art on super-subjective well-being and resilience, as well as other psychological constructs, will help to better understand the impact of mandala art on psychiatric inpatients [67]. At the technical level, although digital painting is not a creative process, it is an artistic process in itself, so it has a nostalgic, projection, and treatment or healing three purposes, it can be an art therapist and has a tendency to art therapists use, thereby bringing customers to self-acceptance and self-awareness of deeper [68].

The image is a useful therapeutic tool that reflects the therapeutic relationship, as well as a result of a guided self-image, or a visual translation of an oral statement or mental state [69]. A quantitative scoring tool for two-dimensional image expression is constructed by combining seven content areas of representation, color, shape, space, movement, composition and expression, which can quantitatively analyze the relationship between artistic works and artists' different constructs, and open up a new perspective for basic AT and psychological research [70]. Moreover, treatment with collage as the main art medium goes beyond artistic creation as a means of treatment, and can enter the world of metaphor and psychological imagery [71]. Through collage-based visual creative AT, patients can transition from dysfunction to learning, growth, and challenge themselves, and become more communicative and more comfortable in social interaction [72]. Sculpture-based AT allows visual and verbal expression as a form of treatment, which may help solve psychological and emotional problems [73], and can also improve the mental state, attention and other psychological attributes of mental patients [74].

Music therapy is often combined with drama, dance/movement, and cognition-behavior to study the therapeutic effects of AT on different diseases. On the one hand, creative AT using music as a medium can alleviate the emotional symptoms and behavioral problems of refugee youth [13]. On the other hand, it may also be valuable in the treatment of drug abuse disorders [75]. It is worth noting that the creative AT combined with music therapy and drama, dance/movement therapy (D/MT) can not only develop, implement, evaluate and improve AT interventions for juvenile offenders in safe care [76], but also be an effective intervention for patients with B group personality disorder in terms of perceptual effect and theme synthesis [77]. In addition, cognitive-behavior-based music therapy can improve anger management skills in forensic psychiatry [78], and may be an effective intervention in treating fatigue in blood and bone marrow transplant patients [79].

D/MT is a new field of creative AT [80]. Creative AT that combines dance/movement with art and music can expand the scope of D/MT treatment to the application of different diseases. D/MT is a creative psychotherapy method based on movement metaphor [81], and mirroring is the cornerstone of the D/MT process [82]. Through the kinesthetic experience of movement metaphor, patients with schizophrenia can express their emotions through words [83]. In addition, mirroring has the benefits of empathy in D/MT, enhancing the therapist's empathy for patients [84]. D/MT mainly focuses on mental illnesses, such as post-traumatic stress disorder (PTSD), mental disorders, emotions, and stress. D/MT can not only explore the psychological and behavioral changes of children suffering from PTSD after an earthquake, but also provide cohesion for a group of children with mental disorders [85]. At the same time, D/MT has certain positive effects in reducing negative emotions [86] and alleviating stress problems [87].

In general, AT is a non-pharmacological, non-medical intervention of non-verbal therapy tool that can help patients promote communication, manage emotions, and provide an opportunity to review life, which has a positive and direct impact on health [88]. In the face of the mental health challenges associated with COVID-19 due to stress and trauma, AT can provide multiple benefits. First of all, AT is a good self-care activity that can benefit individuals throughout their life by respecting their inherent need to have autonomy in expression, so as to reduce the sense of alienation of isolation [89]. In the context of the current COVID-19 pandemic, both emotion-based directional painting intervention and mandala painting intervention may help improve the mental health of primary school students [90]. Moreover, when viewing, producing, and sharing the performance art of music street art, painting, graphic art, film and digital video as a therapeutic tool to enhance power, unity, and collective action, it can be found that our digital world can be filled with momentum, resilience, and hope through these arts [91].

4 The Relationship of CD and AT

From the previous analysis and summary, it can be seen that CD mainly uses VCD to build a good sensory experience and promote emotional communication. And CD often combines art and technology to improve health education, health promotion, and healthy behavior. In addition, in the literature review of AT, it is found that AT using painting, image, collage, sculpture, music, dance/movement as the art medium can improve the happiness of life, especially mental health.

In order to more fully understand and confirm the relationship between CD and AT, this paper uses the core collection database of Web of Science to conduct an advanced search with 'AK = ("Communication Design" OR "Art Therapy")' as the search term. A total of 693 articles published online as of April 2021. Then the 693 articles were then imported into VOSviewer for a visual analysis of keyword co-appearance, in order to discover the connection between CD and AT. Keywords are an important part of the literature that reflects the core content. The analysis method of keyword co-occurrence can well reflect the current academic research hotspots, knowledge structure and development trends of certain disciplines [92, 93].

The network visualization generated by VOSviewer is shown in Fig. 1. It can be seen from Fig. 1. that the keywords between CD and AT have a certain overlap. In the field of CD, VCD is still its most closely linked branch. Moreover, new media is an important medium for VCD. In the process of VCD, creating information that attracts users is essential to enhance the influence and dissemination of information. Among them, social media provides unprecedented opportunities to strengthen health communication and health care. These findings have certain prospects for guiding the CD of health-related social media [94]. In addition, the keywords that are closely related to CD and AT include children, dementia, behavior, emotion, communication, students, design education, and new media, as shown in Fig. 2.

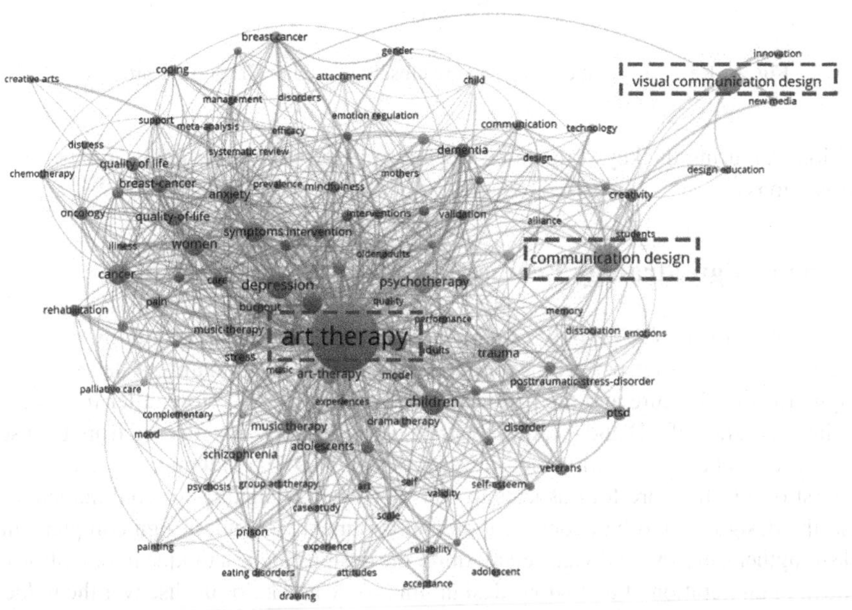

Fig. 1. Keyword analysis of communication design (CD) and art therapy (AT) related articles via VOSviewer.

In general, the visualization of keyword co-occurrence supports a certain connection between CD and AT. Moreover, the VCD from the perspective of visual senses is an important branch of research. Therefore, it is of potential value to combine the advantages of CD and AT and integrate the characteristics of both to promote health. In

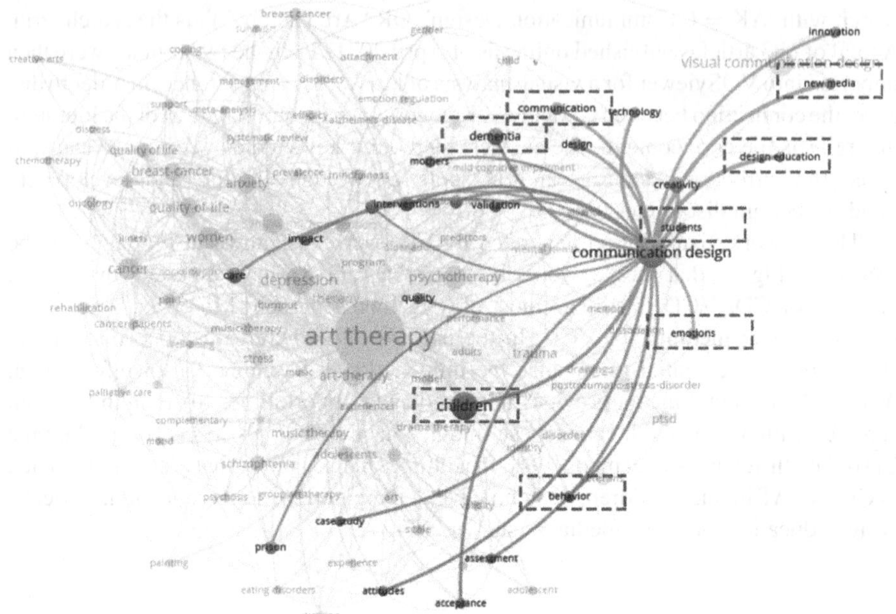

Fig. 2. Keywords that overlap and are closely related in CD and AT articles.

addition, capturing the keywords in between also helps build potential user experience interventions.

5 The Design Strategy Framework Combining CD and AT

In the context of the COVID-19 epidemic, when treatment is carried out by means of design, the relationship between experience and treatment is mainly reflected in the design effect of the cure and the patient's user experience. Therefore, after integrating the characteristics of CD and AT, design strategies are mainly constructed from the user experience level as intervention measures for the COVID-19 epidemic.

First of all, there are four aspects to consider when designing for communication. First, the design should be about creation and comment; second, design complements and strengthens theory (and vice versa); third, the design can be a unique intervention or context, or an iteration of a previous design; finally, design helps us discover the hidden properties of communication and increase opportunities for communication practice [95]. In addition, CD that mainly transmits information from a visual perspective should make full use of the advantages brought by new media. Secondly, in the process of AT, several art media including painting, pictures, collage, and sculpture are mainly used for therapeutic design, so as to improve the mental health problems caused by the COVID-19. Finally, in the face of the health impact of the epidemic, digital intervention is a general necessity. Increasing health communication with digital technology can provide a wide range of information acquisition channels for making health decisions

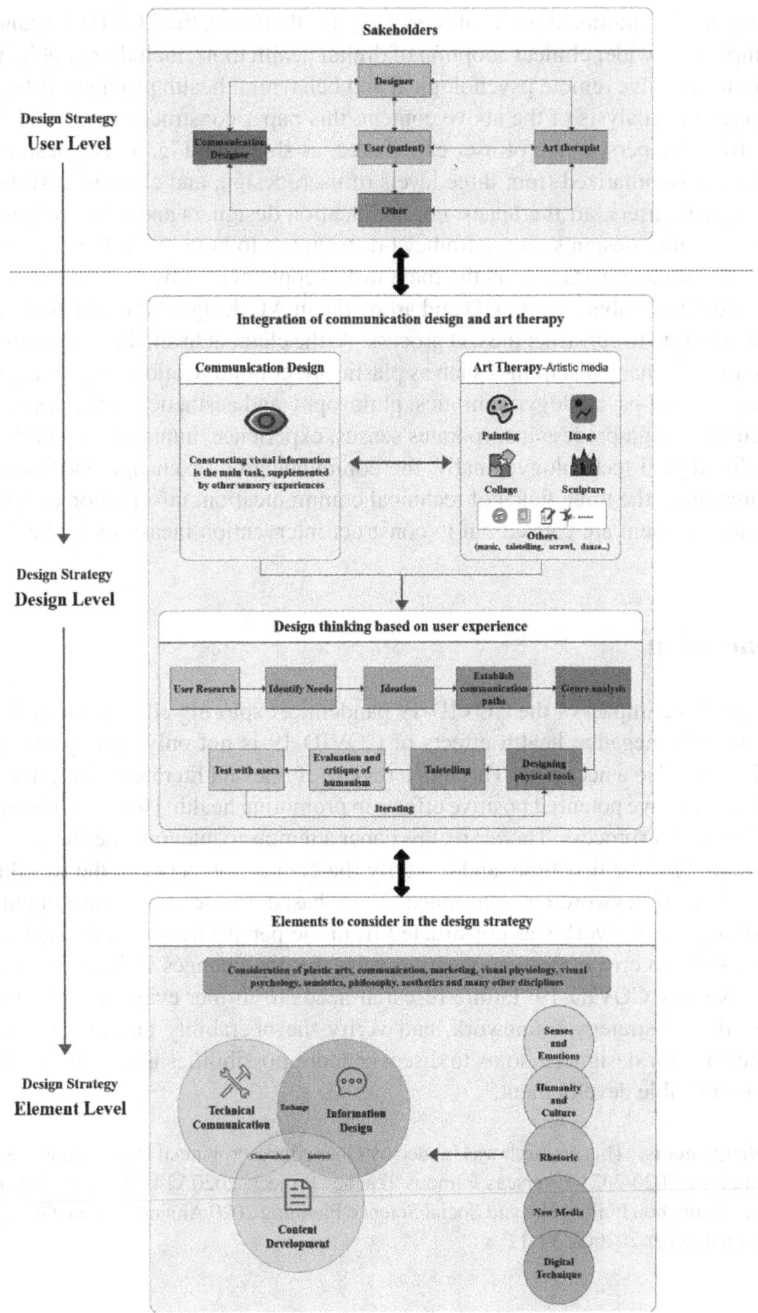

Fig. 3. The design strategy framework combining CD and AT.

and satisfying the emotional needs of users [55]. Furthermore, the COVID-19 pandemic has prompted the wider clinical adoption of digital health tools, including mobile health applications, to solve remote psychological and behavioral health problems [96].

Through the analysis of the above content, this paper constructs a potential design strategy from the perspective of user experience, as shown in Fig. 3. The design strategy is mainly summarized from three levels of user, design, and elements. At the user level, designers, users, art therapists, communication designers and other stakeholders promote the entire design strategy framework from top to bottom. At the design level, visual information construction is the main task, supplemented by other sensory experience. After the combination of CD and art media in AT, design thinking based on user experience is used to construct the CD process. At the element level, the main consideration is to include many disciplines such as plastic art, communication, marketing, visual physiology, visual psychology, semiotics, philosophy and aesthetics in the design analysis; then the design process incorporates senses, experience, humanities, and rhetoric, new media, digital technology; finally, the communication, exchange and interaction of information in the three fields of technical communication, information design, and content development are carried out to construct intervention measures for healing by design.

6 Conclusion

In the face of the impact of the COVID-19 pandemic, exploring effective interventions to deal with the negative health effects of COVID-19 is not only an opportunity for innovation, but also a necessity. Through a review of relevant literature, this paper finds that CD and AT have potential positive effects in promoting health. However, few articles apply AT in the CD process. Therefore, this paper attempts to integrate the characteristics of CD and AT for the first time, and to verify the connection between the two through visual analysis of keyword co-occurrence. Then, based on user experience, a potential design strategy framework was constructed from the perspectives of users, design and elements, so as to create more application value for the changes in health experience brought about by COVID-19. Future research needs to further evaluate and refine the proposed design strategy framework, and verify the operability and effectiveness of the framework for designers, so as to discover more possibilities and realistic paths to achieve sustainable development.

Acknowledgements. This research was funded by Guangdong Provincial Department of Science and Technology 2020-2021 Overseas Famous Teacher Project: 2020A1414010178. It was also funded by Guangzhou Philosophy and Social Science Planning 2020 Annual Project, Guangzhou, China, grant number 2020GZYB12.

References

1. Edwards, E., et al.: Preparing for the behavioral health impact of COVID-19 in Michigan. Curr. Psychiatry Rep. **22**(12), 1–19 (2020). https://doi.org/10.1007/s11920-020-01210-y

2. Pera, A.: Cognitive, behavioral, and emotional disorders in populations affected by the COVID-19 outbreak. Front. Psychol. **11**, 2263 (2020)
3. Tsur, N., Abu-Raiya, H.: COVID-19-related fear and stress among individuals who experienced child abuse: the mediating effect of complex posttraumatic stress disorder. Child Abuse Neglect **110**, 104694 (2020)
4. AD-MED Homepage. https://admed.jimdo.com/what-s-ad-med/. Accessed 29 April 2021
5. Iizuka, S., Takebe, T., Nishii, S., Kodaka, A.: Approach for communication design for motivation to health behavior. In: Yamamoto, S., Mori, H. (eds.) HCII 2019. LNCS, vol. 11569, pp. 425–436. Springer, Cham (2019). https://doi.org/10.1007/978-3-030-22660-2_31
6. Paulovich, B.: Clinicians as mediators in participatory design research: a communication design study in paediatric healthcare. J. Des. Res. **17**(1), 47 (2019)
7. Hossain, G.: Design analytics of complex communication systems involving two different sensory disabilities. Int. J. Healthcare Inf. Syst. Inform. (IJHISI) **12**(2), 65–80 (2017)
8. Giraldi, L., Maini, M., Meloni, D.: Way-finding and communication design as strategic systems to improve the well-being of children in paediatric hospitals. In: Bagnara, S., Tartaglia, R., Albolino, S., Alexander, T., Fujita, Y. (eds.) Proceedings of the 20th Congress of the International Ergonomics Association (IEA 2018). IEA 2018. Advances in Intelligent Systems and Computing, vol. 826, pp. 799–810. Springer, Cham (2019). https://doi.org/10.1007/978-3-319-96065-4_83
9. Harrison, S.: Health communication design: an innovative MA at Coventry university. J. Vis. Commun. Med. **30**(3), 119–124 (2007)
10. Levy, B.: Art therapy in a women's correctional facility. Art Psychother. **5**, 157–166 (1978)
11. Rankanen, M.: Clients' experiences of the impacts of an experiential art therapy group. Arts Psychother. **50**, 101–110 (2016)
12. Lee, J.H.: Effectiveness of group art therapy for mothers of children with disabilities. Arts Psychother. **73**, 101754 (2021)
13. Quinlan, R., Schweitzer, R.D., Khawaja, N., Griffin, J.: Evaluation of a school-based creative arts therapy programme foradolescents from refugee backgrounds. Arts Psychother. **47**, 72–78 (2016)
14. Harkness, A., et al.: Latinx health disparities research during COVID-19: challenges and innovations. Ann. Behav. Med. **54**(8), 544–547 (2020)
15. Aakhus, M.: Communication as design. Commun. Monogr. **74**(1), 112–117 (2007)
16. Frascara, J.: User-Centred Graphic Design: Mass Communications and Social Change, 1st edn. Taylor and Francis, London (1997)
17. Frascara, J.: Design and the Social Sciences: Making Connections, 1st edn. Taylor and Francis, London (2002)
18. Frascara, J.: Communication Design: Principles, Methods and Practice, 1st edn. Allworth Press, New York (2004)
19. Swarts, J.: Communication design. Q. Rev. **1**(1), 12–15 (2012)
20. Yates, D., Price, J.: Communication Design: Insights from the Creative Industries. Fairchild Books AVA, New York (2015)
21. Li, H., Wang, Q.: Research on the integration and performance of emotion concept in visual communication design. In: 2015 4th National Conference on Electrical, Electronics and Computer Engineering. Atlantis Press (2016)
22. Wang, Z.: Application of interaction design on visual communication design. In: 2018 4TH International Conference on Education, Management and Information Technology (ICEMIT 2018), pp. 247–250. Francis ACAD Press, London (2018)
23. Huang, H.: Research on the multisensory expression in the modern visual communication design. In: 2015 2nd International Conference on Education and Education Research (EER 2015), pp. 321–325. Singapore Management & Sports Science INST PTE Ltd., Singapore (2015)

24. Chang, D.: Application of emotional ideas on visual communication design. In: 2017 International Conference on Humanities, Arts and Language (HUMAL 2017), pp. 79–82. Francis ACAD Press, London (2017)
25. Zhang, W.R.: Application of the graphic creativity in the visual communication design. In: 2015 2nd International Conference on Education and Education Research (EER 2015), pp. 388–392. Singapore Management & Sports Science INST PTE Ltd, Singapore (2015)
26. Chen, H.-K., Guan, S.-S.: A perceptual evaluation of grating frequencies and velocities in the analysis of dynamic images. Displays **35**(1), 38–48 (2014)
27. Wu, H., Li, G.: Innovation and improvement of visual communication design of mobile app based on social network interaction interface design. Multimed. Tools Appl. **76**, 1–16 (2019). https://doi.org/10.1007/s11042-019-7523-6
28. Adiloglu, F.: Visual communication: design studio education through working the process. Procedia Soc. Behav. Sci. **28**, 982–991 (2011)
29. Peng, S., Liu, C., Wang, W.: The analysis of visual communication design of commonweal information through interactive design thinking - public commonweal information design and communication in urban traffic spatial environment as an example. In: Marcus, A., Wang, W. (eds.) DUXU 2018. LNCS, vol. 10920, pp. 351–362. Springer, Cham (2018). https://doi.org/10.1007/978-3-319-91806-8_27
30. Zhang, W.R.: Research on the aesthetic psychology of the audience in the visual communication design. In: 2015 2nd International Conference on Education and Education Research (EER 2015), pp. 393–397. Singapore Management & Sports Science INST PTE Ltd, Singapore (2015)
31. Marshalsey, L.: Investigating the experiential impact of sensory affect in contemporary communication design studio education. Int. J. Art Des. Educ. **34**(3), 336–348 (2015)
32. Hohl, M.: Beyond the screen: visualizing visits to a website as an experience in physical space. Vis. Commun. **8**(3), 273–284 (2009)
33. St. Amant, K.: Introduction to the special issue: cultural considerations for communication design: integrating ideas of culture, communication, and context into user experience design. Commun. Des. Q. Rev. **4**(1), 6–22 (2016)
34. McNely, B.J., Rivers, N.A.: All of the things: Engaging complex assemblages in communication design. In: Proceedings of the 32nd ACM International Conference on The Design of Communication CD-ROM, pp. 1–10. Association for Computing Machinery, New York (2014)
35. Klingemann, H., et al.: Public art and public space—waiting stress and waiting pleasure. Time Soc. **27**(1), 69–91 (2015)
36. Czarnecki, K.: Chapter 1: storytelling in context. Libr. Technol. Rep. **45**(7), 5–8 (2009)
37. Gray, B., Young, A., Blomfield, T.: Altered lives: assessing the effectiveness of digital storytelling as a form of communication design. Continuum **29**(4), 635–649 (2015)
38. Forlizzi, J., Lebbon, C.: From formalism to social significance in communication design. Des. Issues **18**(4), 3–13 (2002)
39. Meloncon, L., Frost, E.A.: Special issue introduction: charting an emerging field: the rhetorics of health and medicine and its importance in communication design. Commun. Des. Q. Rev. **3**(4), 7–14 (2015)
40. Laing, S., Apperley, M.: The relevance of virtual reality to communication design. Des. Stud. **71**, 100965 (2020)
41. Douma, M., Gritsay, P., Ligierko, G.: Intuitive browsing experiences with SpicyNodes. In: 4th International Conference on Cybernetics and Information Technologies, Systems and Applications/5th International Conference on Computing, Communications and Control Technologies, pp. 161–166. INT INST Informatics & Systemics, Orlando (2007)

42. Garg, V., Camp, L.J., Connelly, K., Lorenzen-Huber, L.: Risk communication design: video vs. text. In: Fischer-Hübner, S., Wright, M. (eds.) Privacy Enhancing Technologies. PETS 2012. Lecture Notes in Computer Science, vol. 7384, pp. 279–298. Springer, Berlin, Heidelberg (2012). https://doi.org/10.1007/978-3-642-31680-7_15
43. Branco, R.M.: How can communication design add value in the context of Alzheimer's disease? In: Proceedings of the 2nd European Conference on Design4Health, pp. 42–53. Art & Design Research Centre, Sheffield Hallam University, Sheffield (2014)
44. Chan, K.H., et al.: Community engagement of adolescents in the development of a patient-centered outcome tool for adolescents with a history of hypospadias repair. J. Pediatr. Urol. **15**(5), 448-e1 (2019)
45. Dura, L., Gonzales, L., Solis, G.: Creating a bilingual, localized glossary for end-of-life-decision-making in borderland communities. In: Proceedings of the 37th ACM International Conference on the Design of Communication, pp. 1–5. Association for Computing Machinery, New York (2019)
46. Schiavo, R., Basu Roy, U., Faroul, L., Solodunova, G.: Grounding evaluation design in the socio-ecological model of health: a logic framework for the assessment of a national routine immunization communication initiative in Kyrgyzstan. Glob. Health Promot. **27**(4), 59–68 (2020)
47. Amant, K.S.: Re-considering the nature of value in communication design. Commun. Des. Q. Rev. **4**(3), 4–8 (2017)
48. Caviglia, G., Ciuccarelli, P., Coleman, N.: Communication design and the digital humanities. In: Proceedings of the 4th International Forum of Design as a Process (2012)
49. Marshalsey, L., Sclater, M.: Supporting students' self-directed experiences of studio learning in communication design: the co-creation of a participatory methods process model. Australas. J. Educ. Technol. **34**(6) (2018)
50. McGuire, M., Kampf, C.: Using social media sentiment analysis for interaction design choices: An exploratory framework. In: Proceedings of the 33rd Annual International Conference on the Design of Communication, pp. 1–7. Association for Computing Machinery, New York (2015)
51. Johnson-Eilola, J., Selber, S.: Solving Problems in Technical Communication, 1st edn. University of Chicago Press, Chicago (2013)
52. Mehlenbacher, B.: Communication design and theories of learning. In: Proceedings of the 26th Annual ACM International Conference on Design of Communication, pp. 139–146. Association for Computing Machinery, New York (2008)
53. Aakhus, M., Bzdak, M.: Stakeholder engagement as communication design practice. J. Public Aff. **15**(2), 188–200 (2015)
54. McNely, B.J.: Visual research methods and communication design. In: Proceedings of the 31st ACM International Conference on Design of Communication, pp. 123–132. Association for Computing Machinery, New York (2013)
55. Kreps, G.L.: Online information and communication systems to enhance health outcomes through communication convergence. Hum. Commun. Res. **43**(4), 518–530 (2017)
56. Lee, R., et al.: Art therapy for the prevention of cognitive decline. Arts Psychother. **64**, 20–25 (2019)
57. Van Westrhenen, N., Fritz, E.: Creative arts therapy as treatment for child trauma: an overview. Arts Psychother. **41**, 527–534 (2014)
58. Hertrampf, R.-S., Wärja, M.: The effect of creative arts therapy and arts medicine on psychological outcomes in women with breast or gynecological cancer: a systematic review of arts-based interventions. Arts Psychother. **56**, 93–110 (2017)
59. Kim, H.-K., Kim, K.M., Nomura, S.: The effect of group art therapy on older Korean adults with neurocognitive disorders. Arts Psychother. **47**, 48–54 (2016)

60. Sezen, C., Ünsalver, B.Ö.: Group art therapy for the management of fear of childbirth. Arts Psychother. **64**, 9–19 (2019)
61. Kim, S., Kim, G., Ki, J.: Effects of group art therapy combined with breath meditation on the subjective well-being of depressed and anxious adolescents. Arts Psychother. **41**, 519–526 (2014)
62. Gantt, L., Tinnin, L.W.: Support for a neurobiological view of trauma with implications for art therapy. Arts Psychother. **36**, 148–153 (2009)
63. Walker, M.S., Kaimal, G., Koffman, R., DeGraba, T.J.: Art therapy for PTSD and TBI: a senior active duty military service member's therapeutic journey. Arts Psychother. **49**, 10–18 (2016)
64. Kometiani, M.K., Farmer, K.W.: Exploring resilience through case studies of art therapy with sex trafficking survivors and their advocates. Arts Psychother. **67**, 101582 (2020)
65. Darewych, O.: Building bridges with institutionalized orphans in Ukraine: an art therapy pilot study. Arts Psychother. **40**, 85–93 (2013)
66. Hass-Cohen, N., Bokoch, R., Findlay, J.C., Witting, A.B.: A four-drawing art therapy trauma and resiliency protocol study. Arts Psychother. **61**, 44–56 (2018)
67. Kim, H., Kim, S., Choe, K., Kim, J.-S.: Effects of mandala art therapy on subjective well-being, resilience, and hope in psychiatric inpatients. Arts Psychother. **32**, 167–173 (2018)
68. Rubin, L.C.: The use of paint-by-number art in therapy. Arts Psychother. **27**, 269–272 (2000)
69. Luzzatto, P.: Anorexia nervosa and art therapy: the double trap of the anorexic patient. Arts Psychother. **21**, 139–143 (1994)
70. Schoch, K., Gruber, H., Ostermann, T.: Measuring art: methodical development of a quantitative rating instrument measuring pictorial expression (RizbA). Arts Psychother. **55**, 73–79 (2017)
71. Greece, M.: Art therapy on a bone marrow transplant unit: the case study of a Vietnam veteran fighting myelofibrosis. Arts Psychother. **30**, 229–238 (2003)
72. Elkis-Abuhoff, D.L.: Art therapy applied to an adolescent with Asperger's syndrome. Arts Psychother. **35**, 262–270 (2008)
73. Brown, S.E., Shella, T., Pestana-Knight, E.: Development and use of the art therapy seizure assessment sculpture on an inpatient epilepsy monitoring unit. Arts Psychother. **9**, 6–9 (2018)
74. Seifert, K., Spottke, A., Fliessbach, K.: Effects of sculpture based art therapy in dementia patients–a pilot study. Heliyon **3**(11), e00460 (2017)
75. Megranahan, K., Lynskey, M.T.: Do creative arts therapies reduce substance misuse? A systematic review. Arts Psychother. **57**, 50–58 (2018)
76. Smeijsters, H., Kil, J., Kurstjens, H., Welten, J., Willemars, G.: Arts therapies for young offenders in secure care–a practice-based research. Arts Psychother. **38**, 41–51 (2011)
77. Havsteen-Franklin, D., Haeyen, S., Gante, C., Karkou, V.: A thematic synthesis of therapeutic actions in arts therapies and their perceived effects in the treatment of people with a diagnosis of cluster b personality disorder. Arts Psychother. **63**, 128–140 (2019)
78. Hakvoort, L., Bogaerts, S.: Theoretical foundations and workable assumptions for cognitive behavioral music therapy in forensic psychiatry. Arts Psychother. **40**, 192–200 (2013)
79. Fredenburg, H.A., Silverman, M.J.: Effects of cognitive-behavioral music therapy on fatigue in patients in a blood and marrow transplantation unit: a mixed-method pilot study. Arts Psychother. **41**, 433–444 (2014)
80. Milliken, R.: Dance/movement therapy as a creative arts therapy approach in prison to the treatment of violence. Arts Psychother. **29**, 203–206 (2002)
81. Lee, M.B.: Dance movement therapy: a creative psychotherapeutic approach. Arts Psychother. **30**, 53–57 (2003)
82. Berrol, C.F.: An introduction to medical dance/movement therapy: health care in motion. Arts Psychother. **33**(2), 153–155 (2005)

83. Ellis, R.: Movement metaphor as mediator: a model for the dance/movement therapy process. Arts Psychother. **28**, 181–190 (2001)
84. McGarry, L.M., Russo, F.A.: Mirroring in dance/movement therapy: potential mechanisms behind empathy enhancement. Arts Psychother. **38**, 178–184 (2011)
85. Erfer, T., Ziv, A.: Moving toward cohesion: group dance/movement therapy with children in psychiatry. Arts Psychother. **33**, 238–246 (2006)
86. Lee, H.-J., Jang, S.-H., Lee, S.-Y., Hwang, K.-S.: Effectiveness of dance/movement therapy on affect and psychotic symptoms in patients with schizophrenia. Arts Psychother. **45**, 64–68 (2015)
87. Bräuninger, I.: The efficacy of dance movement therapy group on improvement of quality of life: a randomized controlled trial. Arts Psychother. **39**, 296–303 (2012)
88. Johnson, C.M., Sullivan-Marx, E.M.: Art therapy: using the creative process for healing and hope among African American older adults. Arts Psychother. **27**, 309–316 (2006)
89. Braus, M., Morton, B.: Art therapy in the time of COVID-19. Psychol. Trauma Theory Res. Pract. Policy **12**(S1), S267–S268 (2020)
90. Malboeuf-Hurtubise, C., et al.: Online art therapy in elementary schools during COVID-19: results from a randomized cluster pilot and feasibility study and impact on mental health. Child Adolesc. Psychiatry Ment. Health **15**(1), 15 (2021)
91. Gupta, N.: Singing away the social distancing blues: art therapy in a time of coronavirus. J. Humanist. Psychol. **60**(5), 593–603 (2020)
92. Aimin, Z.: The cluster analysis of co-occurrence strength in the field of knowledge management in 2006. Mod. Inf. **28**(5), 30–33 (2008)
93. Liu, R., Lin, H., Zhao, H.: Research on field characteristics of shared earth system science data using keyword analysis and visualization. Procedia Environ. Sci. **10**, 561–567 (2011)
94. Rus, H.M., Cameron, L.D.: Health communication in social media: message features predicting user engagement on diabetes-related Facebook pages. Ann. Behav. Med. **50**(5), 678–689 (2016)
95. Harrison, T.R.: Enhancing communication interventions and evaluations through communication design. J. Appl. Commun. Res. **42**(2), 135–149 (2014)
96. Psihogios, A.M., Stiles-Shields, C., Neary, M.: The needle in the haystack: identifying credible mobile health apps for pediatric populations during a pandemic and beyond. J. Pediatr. Psychol. **45**(10), 1106–1113 (2020)

Preliminary Investigation of Methods for Graphic Simplification from Representation to Abstraction

Hui-Ping Lu(✉)

Department of Visual Communication Design, Asia University, Wufeng, Taichung 41354, Taiwan
jasmine@asia.edu.tw

Abstract. The ability to graphic simplification from representation to abstraction is considered a basic skill in the field of design. Therefore, how to train novice designers systematically to learn this skill has become an important subject in teaching. Creative process has always been regarded as a black box operation, because we often only see the results but can't understand the thinking modes, which makes it a challenge between teaching and learning. This study hopes to explore the thinking modes of novice designers and construct a systematic method of graphics simplification through the transformation perspective of works from representation to abstraction. The study focuses on the freshmen of the Department of Visual Communication Design of Asia University. They are given a design task - transforming personal image from representation into semi-representation and abstract images. And through work analysis and interviews to summarize the differences in thinking modes and creative methods of novice designers in different creative stages. We attempt to make the black box operation clear. If the instructors can understand the thinking modes and creative methods of the novice designers, not only can give appropriate guidance in the creative process, but also help establish a systematic teaching in the future.

Keywords: Design method · Modes of thinking · Graphic simplification · Representation to abstraction

1 Introduction

1.1 Research Background and Motives

"Less is more," an aphorism adopted by Ludwig Mies Van Der Rohe (1974), a master in Modernist architecture, epitomizes the spirit of modern design. A design is perfect not because nothing more remains to be added but because nothing is superfluous. The essence is extracted through the process of turning complexity into simplicity, which involves design that is more restrained. The more abstract a presence seems, the more boundless the imagination it contains. The ability to simplify forms is a basic yet basic skill in design. Thinking modes have a decisive impact on design work. To enhance

C. Stephanidis et al. (Eds.): HCII 2021, LNCS 13094, pp. 116–126, 2021.
https://doi.org/10.1007/978-3-030-90238-4_10

their skills, designers should be able to master the generalization and abstract extraction of forms. However, the creative process tends to be a type of black-box operation that escapes simple understanding. Making the creative process transparent facilitates the two-way communication between teaching and learning, further enabling designers to resolve problems more efficiently.

Design inspiration can be drawn from text, figures, or other emblems (Cheng 2010), but in general, graphics or images are more useful types of visual information. As supported by a substantial body of literature, designers tend to search and retrieve a considerable amount of visual data before generating ideas through mental synthesis (Schön and Wiggins 1992; Goldschmidt 1994; McGown et al. 1998; Verstijnem et al. 1998; Suwa et al. 2000; Dorst and Cross 2001). Referring to relevant data in the early stages of design is beneficial to the results (Herbert 1993; McGown et al. 1998). However, the process in which a designer transforms imagery into design ideas is linked to brain function with regard to individual intuitions and preferences. Senior designers, who have accumulated substantial experience, typically grasp the elements in an image more easily. By contrast, novice designers often experience challenge or failure in conception in the absence of effective auxiliary methods. The evidence indicates that human activities, including the creation of forms, can be performed through the plans and systemic procedures used by scientists—specifically, generalized principles that the ordinary individual can follow provided appropriate training is given.

Creative ideas are formed step by step, comparable to the design of scientific experiments (Tischler 2009). The difficult intellectual choices and analysis in the design development process are partially dependent on form and experience, but all can be transmitted through specific methods of instruction (Bayazit 2004). In a similar vein, Young (2003) asserted that the generation of creativity is as precise and specific as the process by which a Ford vehicle is manufactured; for example, both require a production line. To produce creativity, the mind must follow—and eventually master through practice—a series of controllable operative techniques.

1.2 Research Purpose

Taken together, these findings suggest that some design problems encountered by novice designers may be resolvable by building a series of systematic graphic simplification methods. Therefore, the present study investigated the transition from representation to abstraction by assigning a design task to and interviewing novice designers. In future, such a series of methods may serve as a reference for training novice designers on graphic simplification and ideas of form. The study objectives are presented as follows: to (1) complete a design task and produce work reflecting semi-representation and abstraction and (2) trigger the thinking modes and creative approaches novice designers commonly used through the analysis of their work as well as in-depth interviews.

2 Literature Review

The goal of simplification in art differs from that in design. Art values originality and experimentation, whereas design considers the viewer's perception and recognition of the design. Simplification approaches, which are highly practical, are often used to clarify graphic communication and facilitate memorization (Gombrich 1982). Relevant terms are defined and graphic simplification methods in various fields are examined as follows.

2.1 Definition of Simplification and Abstract

The term simplification designates the shift in form from complexity to simplicity. In art, this process involves placing the visual information of a material form in order (Hsu 2007b). After the removal of trivial details, a picture is modified accordingly; this is a pivotal skill. The visual factors and concrete details involved in representations limit the imagination and lead to preoccupation with details; in this context, the simplification of abstract forms is an attempt to seek the truth of form. Abstract and concise forms leave a deep impression and better embody the meaning and essence of the work. Simplification is an approach commonly adopted in the representation–abstraction transformation of a figure. In general, information and symbolic meaning can be better derived from simplified figures. Figures extracted through abstraction leave a strong visual impression. A mode of graphic simplification proposed by Hsu (2007b), extraction from the overall shape, is divided into three methods, namely depiction of the external contour, preserving structural relations, and planification. The second mode proposed, extraction of traits of a component, is divided into the methods of emphasizing visual features, emphasizing functional features, and keeping the texture and material features.

The term abstraction has two meanings. In one sense, abstraction refers to painting with complete reliance on intuition and imagination by disregarding the influence of the idea of representational figures from the outset. In the other sense, abstraction begins with representational figures and evolves into an action. Analogies concerning the aforementioned description and expressions of abstract painting are divided into two types. One is pure construction that is not based on natural things, and the other begins with natural phenomena, the expressive characteristics or factors of which are simplified or extracted into relatively generalized forms. These two types correspond to two paths: the internal psychological mechanism, which emphasizes the embodiment of the internal abstraction, and the external visual imitation, which highlights the abstraction of the external material form (Hsu and Wang 2007a). The two paths are often used in combination in the actual design process.

The external visual imitation was investigated in the present study, beginning with representation and gradually evolving into abstraction through simplification. In such processes, the structure of an object is first analyzed; next, the object is generalized into the final abstract form. Such a transformation reflects the creator's depth of feeling and appreciation for things, as well as various associations and a sense of beauty.

2.2 Graphic Simplification in Art

Picasso's Bull (1945) is a suite of 11 lithographs that faithfully records the evolution of a realistic bull into a cubist one. Picasso visually dissected the image of a bull, determined the essence of things through gradual analysis, and reclaimed the soul of the beast (Artyfactory 2019). Although the final bull appears almost devoid of meaning, its characteristics are present in the preserved lines. Picasso gradually extracted the essential information from a realistic image, obtaining an almost abstract shape through gradual simplification (Fig. 1). Roy Lichtenstein's Bull profile series (1973) starts with the original bull drawn from the real world to the final geometric form through gradual simplification and transformation (Fig. 2), the painting style vacillating between pure and applied art. In short, the series constitutes images of a bull that were abstracted in several stages (Teel and Claudia 2011).

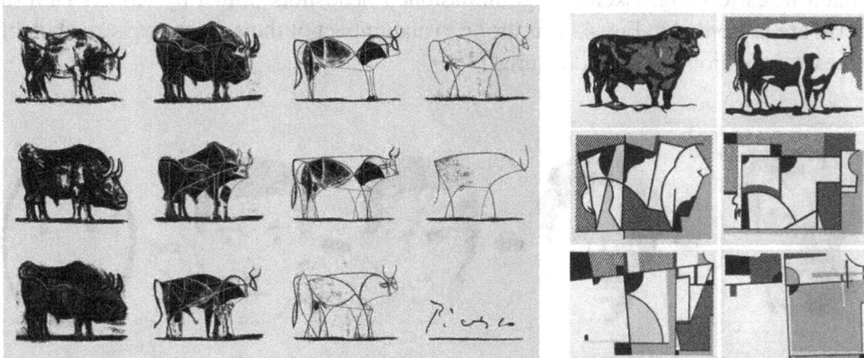

Fig. 1. Picasso's Bull (1945)

Fig. 2. Lichtenstein's Bull (1973)

2.3 Graphic Reduction in Design

In design, simplification refers to submitting visual information to a process of orderly organization, which involves the removal of trivial details, followed by various adjustments. Simplified graphics have wide applications, including illustration and the design of signage, posters, and logos. As noted by Hsu and Wang (2007a), numerous studies have indicated that designs featuring simplified figures draw greater attention from and improve the design recognition of the viewer. Figure 3 presents the five stages of graphic simplification developed by the cartoonist Scott McCloud (1993) presented five stages of graphic simplification (Figure 3). From left to right, the images show the process from realism to figuration, clearly demonstrating two typical orientations. Specifically, the images on the left are more detailed, realistic, and easy to recognize, whereas those on the right are more abstract and require advanced perception, leaving more room for interpretation. The use of different modes of expression immerses the reader in the world of comics and make the interpretation experience more enjoyable.

Moreover, as mentioned, simplification has a stronger effect on the viewer. By contrast, the structures of human faces are practically identical, with comparatively small variations in features. According to visual scientists, our brains seek to recognize such external features. To overcome this problem of homogeneity, cartoonists place the essential elements in the foreground of each panel, devoting relatively less attention to the rest (Austen 2011). Thus, just as our brains strive to determine the elements that make a person look unique, the visual differences in human faces are often amplified and exaggerated in cartooning. Exclusive commissioned headshots drawn by Robert DeJesus (2019) are presented in Fig. 4. The strong visual impact of the caricatures is attributable to the exaggeration of certain features.

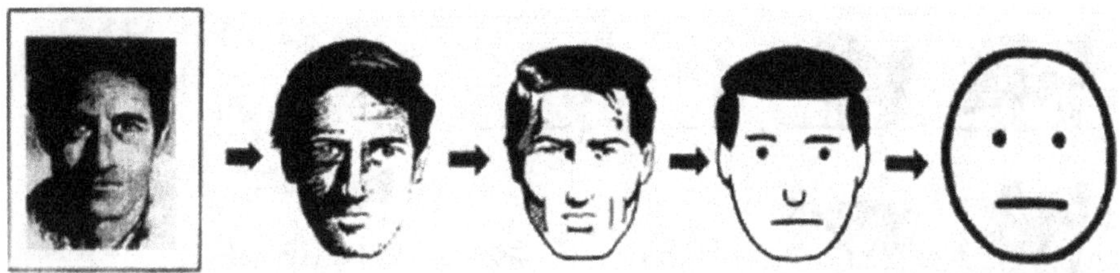

Fig. 3. McCloud's five stages of graphic simplification (1993)

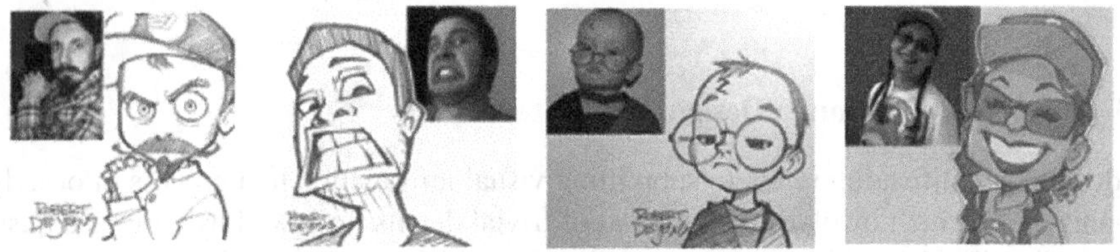

Fig. 4. Caricatures by Robert DeJesus (2019) Fig. 4 (From: FaceBook of Rebert DeJesus)

Italian designer Bruno Munari's *Variations on the Theme of the Human Face* (1971) exemplified another graphic simplification method: making representations of the human face from the front in as many forms as possible, taking into account the creativity and various characteristics manifested in each face. This exercise helps graphic designers determine the most suitable image for a given theme, with each figure or technique exhibiting certain qualities and delivering certain information (Fig. 5).

A concept rolled out by the Portuguese designer Eduardo Aires (2020; stylized as Porto.) was developed to reflect the identity of the city of Porto, the second largest in the country. After being inspired by the story of the blue-and-white tiles on the facades of several local ancient buildings, he designed the geometric icons of more than 70 representative cities and their citizens, simplifying natural landscapes and cultural characteristics into tile-like patterns. A continuous, tile-like network of interconnected patterns was built (Eduardo Aires 2020; Fig. 6).

As mentioned, the literature review revealed that the goal of simplification in art differs from that in design. Art values originality and experimentation, whereas design takes into account the viewer's perception and recognition of the work. The present

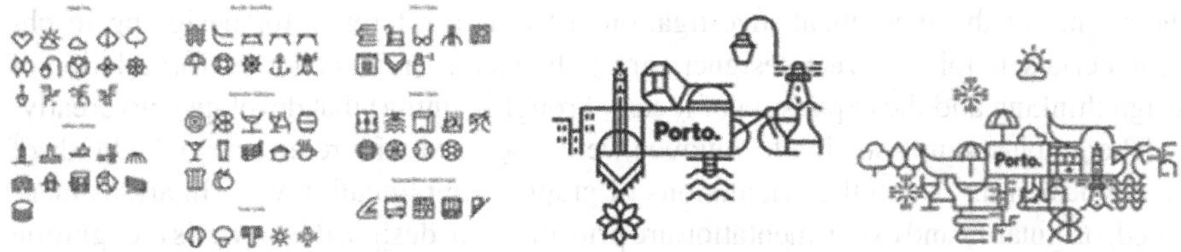

Fig. 5. Munari's variations on the theme of the human face (1971)

Fig. 6. Unit Forms of the city identity of Porto and the effect of their combinations

study considered novice designers' lack of knowledge and experience, including that concerning the viewer's recognition and perception. At this stage, originality and experimentation are the objectives and the participants were only required to transform objects featuring concrete figures into semi-representational and abstract figures. Design work of higher difficulty, including signage and posters, remains to be explored in the future.

3 Method

The novice designers were instructed to complete a design task. This was followed by an induction of the common thinking modes and design approaches they used, which served as a reference for constructing instructional methods. The present study was conducted through three stages: theoretical investigation, design work, and methodological induction. Regarding the theoretical investigation, design thinking modes were understood and methods of graphic simplification (from representation to abstraction) were examined. Design work involved two stages; representational objects were first transformed into semi-representational figures, which were then transformed into abstract ones. As for methodological induction, the designers were interviewed and their work was analyzed to determine their thinking modes and creative approaches.

3.1 Theoretical Investigation

The theoretical investigation mainly concerned the implications of design thinking and techniques of graphic simplification. In design, both ability and creativity are grounded on specific thinking abilities or thinking habits. Design thinking represents a designer's cognitive activity in a design process and thus has a decisive impact on design work.

The results of the theoretical investigation serve as a reference for designing teaching materials, helping novice designers grasp the factors influencing the formulation of design thinking and the expansion of ideas (through training) that develops into creativity. The graphic simplification techniques were discussed with reference to the fields of art and design, in which the orientations of graphic simplification vary. In art, as mentioned, originality and experimentation are prioritized; in design, the viewer's recognition and perception are at the forefront. Thus, the simplification technique used depends on the content of different designs. Shared through teaching materials, these techniques provided clear directions for novice designers, thus shortening the long search process.

3.2 Design Work

The design task was completed through three stages, as follows:

1. Collecting representational data (within a group): determine the themes and collect the photographs of the representational objects.
2. Transformation into semi-representation: transform the photographs of the representational objects into semi-representational forms through simplification.
3. Transformation into abstraction: simplify and transform the semi-representational forms into abstract ones.

The task was completed in groups of three designers. Group discussion was conducted to determine several themes that could be transformed and to select the object most suitable for transformation. Each individual produced two works; one involved the transformation of representational objects into semi-representational figures, and the other involved transforming semi-representational figures into abstract ones. The grouping had two objectives: first, using group discussion to provide novice designers with peer opinions as a reference in the early stages to design and thereby expand their design thinking; second, comparing (through the analysis of the works) the differences in the design thinking and design behavior underlying various designs on the same theme.

3.3 Methodological Induction

Once the design task was completed, the participants were interviewed. The focus was the transition process between ideas at different stages and the representations in the works. According to the author's teaching experience, in a creative process in which a designer creates forms and colors, they not only continue to use existing data but also tend to integrate their intuition and experience. If we were to analyze a work solely from the visual angle, by examining the forms and colors and with subjective awareness and imagination, the meaning of the work and the designer's thoughts at the moment of creation would not be sufficiently captured. Thus, the present study focused on the interviews; the designers were asked to describe their creative process in detail, including the construction of the forms and the use of colors.

4 Discussion

In this study, four groups of works were selected for discussion and analysis. Those that were more representative and distinct in expression are presented in Tables 1 and 2.

1. Representation: on the basis of the literature, three types of framing were derived from the original photographs, namely the close-up, the medium range, and the large range. Different types of framing had varying effects on the work produced. When a close-up composition was used, because the subject occupied a larger area, the designer tended to add numerous details to enhance variety. By contrast, when large-range composition was used, meaning that the subject occupied a larger area, subject-related details were disregarded in favor of simplifying the shape. Most participants used the compositions in the original photographs to develop semi-representational forms. However, greater change began to emerge at the stage of abstraction (Lu and Lin 2019).
2. Semi-representation: most participants opted to retain the original compositions of the photographs and only simplify the forms and colors; few (cases 4 and 10) altered the compositions to construct a new picture. Regarding color scheme, most participants drew from the original object. However, some participants (cases 1, 2, 4, 7, 8, 10, and 11) added color according to their own imagination and experience to enrich the picture if the original color palette was relatively monochromatic. Some participants added forms and colors through associations. For example, regarding the whale in case 9, chainlike forms were added to the original shape of the whale, and red was added to the original blue-black color scheme. The interview with the participant revealed that the whale theme was meant to evoke the human predation of these animals; the chainlike forms represented whaling tools, and the red represented the spillage of blood into the ocean. By contrast, the little bird in case 10 was drawn from the participant's own preference and experience. The interview revealed that the designer was practicing Zentangle drawing in the design process, incorporating relevant elements and colors.
3. Abstraction: the color schemes in the semi-representation and abstraction stages differed only slightly; by contrast, the compositions underwent greater changes. The color schemes were characterized both by removing mid-tones and constructing a picture with more distinct color planes. Regarding form, some participants deconstructed their work from the semi-representation stage and then recombined it into a new picture (cases 4, 7, 11, and 12). Others retained their original compositions but resimplified them (cases 1, 2, 3, 5, 6, 8, 9, 10, and 12). Most designers moved away from cubic representation, producing a graphic effect with distinct color planes. Still others used their imagination to add decorative patterns to their pictures (cases 1, 4, 8, and 10).

Table 1. From representation to abstraction (Cases 1–6)

Case	1	2	3	4	5	6
Representation					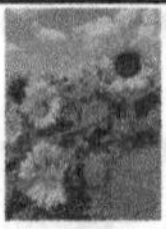	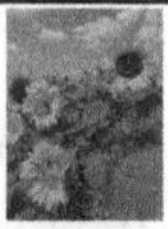
Semi-representation					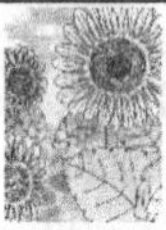	
abstraction	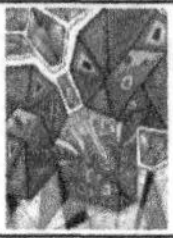			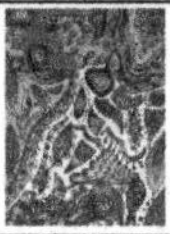		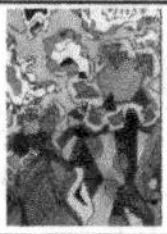

Table 2. From representation to abstraction (Cases 7–12)

Case	7	8	9	10	11	12
Representa-tion						
Semi-representation	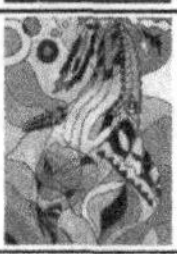					
abstraction	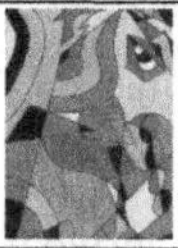					

Notably, the expressive approaches taken in cases 1, 2, and 3 were highly similar; splitting was used to construct the pictures. Interviews with the three participants revealed that, coincidentally, they all graduated from design-related departments of vocational high schools. Furthermore, they all regarded splitting as the simplest and most rapid method through which semi-representational and abstract compositions could be formed. This belief was associated with their design experience; splitting had allowed them to achieve certain design goals in technical examinations quickly and easily. These works were indicative of the participants' design education (specifically learning modes) and the decisive impact of their experience on their current work.

In the semi-representation stage, the designers tended to retain the original composition and alter the forms and colors. However, in general, the original forms and colors were retained when the colors in the original pictures were vivid. If the original palette was monochromatic, some designers enriched them, whereas others simplified them. In the abstraction stage, expression was more diverse. In one expression, the original composition was altered, and the design components were scattered and recombined to form a new composition. In another expression, the original composition was preserved, and the attributes of the details were partially magnified or minimized.

Overall, various thinking modes and creative approaches were adopted at different stages of design. Notably, in the two stages of transformation, some works were not only reflective of ideas concerning visual figures but also integrated the participants' individual feelings regarding the objects. In essence, the atmosphere was rebuilt through imagination and the incorporation of personal experience. These works serve as a valuable reference for constructing methods of instruction and design.

5 Research Results

As mentioned, design work is generally considered a black-box process; the design process did not become transparent until various design methods emerged. The present study examined the thinking transformation process in design and the creative approaches taken by novice designers. Some limitations are addressed as follows. Because the scope of this investigation was limited to the interviews and the analysis of their work, as well as the development of design methods (by combining relevant statements drawn from the literature), the present findings can only allow for generalization of and inferences concerning certain phenomena.

In general, the creative process of design is complex and involves a considerable amount of imagination and experience. The whole picture cannot be captured from only the visual angle. Thus, future research will involve interviews with senior designers. The principal aim of interviewing novice designers was to determine whether factors other than visual ones affected the work produced through the two stages. By contrast, interviews with senior designers facilitate the understanding of the stage-by-stage transformation process, as well as differences in thinking (i.e., from that of novice designers). The present findings serve as a reference for the construction of systematic design methods. Novice designers can benefit from the introduction of expert thinking in the early stages of the creative process.

Acknowledgements. We would like to thank the Ministry of Sciences and Technology for funding us under the "Individual Research Scheme" (MOST 109-2410-H-468-007), the Visual Communication Design Department at Asia University for their support, and all students enrolled in the course.

References

Artyfactory: Animals in art-Pablo Picasso (2019). http://www.artyfactory.com/art_appreciation/animals_in_art/pablo_picasso.htm

Austen, B.: What Caricatures Can Teach Us About Facial Recognition (2011). https://www.wired.com/2011/07/ff_caricature/

Bayazit, N.: Investigating design: a review of forty years of design research. Des. Issues **20**(1), 16–29 (2004)

Cheng, P.J.: A study on designers' searching-retrieving behavior in the ideation process. (PhD), National Yunlin University of Science & Technology (2010)

DeJesus, R.: The Art of Robert DeJesus (2019). https://www.facebook.com/pg/RobertDeJesusArt/photos/?ref=page_internal

Dorst, K., Cross, N.: Creativity in the design process co-evolution of problem-solution. Des. Stud. **22**(5), 425–437 (2001)

Eduardoaires: Studio Eduardo Aires (2020). https://www.eduardoaires.com/studio/

Kickinson, J.: MMART 165 Abstraction in Design (2018). https://medium.com/@joshuadickinson/mmart-165-abstraction-in-design-54bff62f95c6

Goldschmidt, G.: On visual design thinking: the vis kids of architecture. Des. Stud. **15**(2), 159–174 (1994)

Herbert, D.M.: Architectural Study Drawings. Nostrand Reinhold, New York (1993)

Hsu, C.C. Wang, W.Y.: Redefining abstraction in visual art and design. J. Des. **10**(3), 81–100 (2007a)

Hsu, C.C.: Cognitive study on the Design Operation of Graphic Simplification. Unpublished doctoral dissertation, National Taiwan University of Science and Technology, Taipei City (2007b)

Lu, H.P., Lin, H.H.: Exploration of novice designers' modes of thinking. In: International Conference on Knowledge Innovation and Invention 2019 (ICKII) (2019)

McCloud, S.: Understanding Comics: The Invisible Art. HarperCollins, New York (1993)

McGown, A., Green, G., Rodgers, P.: Visible ideas: Information patterns of conceptual sketch activity. Des. Stud. **19**(4), 431–453 (1998)

Munari, B.: Design as Art. (Creahg, P. Trans.). London: Penguin Group (1971). (Original work published 1966)

Schön, D.A., Wiggins, G.: Kinds of seeing and their function in designing. Des. Stud. **13**(2), 135–156 (1992)

Suwa, M., Gero, J., Purcell, T.: Unexpected discoveries and S-invention of design requirement: important vehicle for a design process. Des. Stud. **21**(6), 539–567 (2000)

Teel, S. Claudia, B.: Drawing: A Contemporary Approach. Wadsworth Pub Co., U.S.A. (2011)

Verstijnem, I.M., Hennessey, J., Leeuwen, C., Hamel, R., Goldschmidt, G.: Sketching and creative discovery. Des. Stud. **19**(4), 519–546 (1998)

Young, J.W.: A Technique for Producing Ideas. McGraw-Hill, New York (2003)

Suggestions for Online User Studies
Sharing Experiences from the Use of Four Platforms

Joni Salminen[1,2(✉)], Soon-gyo Jung[1], and Bernard J. Jansen[1]

[1] Qatar Computing Research Institute, Hamad Bin Khalifa University, Ar-Rayyan, Qatar
{jsalminen,sjung,bjansen}@hbku.edu.qa

[2] Turku School of Economics at the University of Turku, Turku, Finland

Abstract. During exceptional times when researchers do not have physical access to users of technology, the importance of remote user studies increases. We provide recommendations based on lessons learned from conducting online user studies utilizing four online research platforms (Appen, MTurk, Prolific, and Upwork). Our recommendations aim to help those inexperienced with online user studies. They are also beneficial for those interested in increasing their proficiency, employing this increasingly important research methodology for studying people's interactions with technology and information.

Keywords: Remote user studies · Online user studies · Platforms

1 Introduction

User studies in information science and related research are essential for enhanced understanding of people, technology, and information. The yearlong - and onwards! - global covid pandemic has caused tremendous hindrance to people's lives, including physical and mental health and well-being, professional coping and stress, and lack of access to others [1]. One of the other consequences has been that it is nearly impossible to carry out physical (in-person) user studies under these conditions, as social distancing restrictions are imposed for organizations and individuals. This hinders the scientific progress in fields such as human-computer interaction (HCI) and information science (IS) that have traditionally relied mainly on in-person user studies to conduct experiments regarding novel interaction techniques [2].

The lack of physical access to participants has required HCI researchers to seek alternative paths and methods for data collection in user studies, including crowdworker and freelancer platforms. Researchers have conducted online user studies [3] and have used these platforms in the past for data collection [4] with techniques such as survey and questionnaires; however, these platforms are now being employed for more complex user studies that previously were often conducted in a lab setting. Hence, the capabilities of these platforms are being pushed. The crowdworkers and freelancers taking part in user studies are confronted with more complex tasks. Researchers new to online user studies are using these online platforms out of necessity.

C. Stephanidis et al. (Eds.): HCII 2021, LNCS 13094, pp. 127–146, 2021.
https://doi.org/10.1007/978-3-030-90238-4_11

By *online user study*, we refer to user research, where study participants are users of a system, interact with information, or some similar task, regardless of their physical location, typically accessing the studied system via an internet browser. Guidelines for conducting these online user studies are scarce, imposing a challenge for researchers both less experienced with online tools for user studies and those experienced but leveraging these platforms for more complex research. To help address this need, we report our extension experiences with online (remote) user studies based on four well-known online research platforms (referred to as "ORPs" going forward). Our goal is to provide valuable recommendations for researchers concerning, designing, or considering a remote user study using crowdworkers and freelancers to accommodate data collection needs. This manuscript is particularly targeted to those new to online user studies, although those more experienced may also find it useful. The recommendations presented here are actionable for researchers interested in using ORPs now, and they can serve as the foundation for more rigorous guideline development in the future.

Our key tenet is that, with efficient use of ORPs, researchers can conduct high-quality research even during exceptional times, with innovative approaches for using the capabilities of the ORPs. As such, we first briefly touch on related work and methodology for comparisons. Then, we address our analysis of the four platforms included in this study and present recommendations for their employment. Additionally, we present some novel possible ongoing trends in remote user studies via online technologies. We end with a short discussion and conclusion.

2 Related Work

Tools and technologies for remote user studies are developing fast; perhaps too fast for the academic field to keep up. In the current day, researchers can use ORPs for surprisingly sophisticated studies involving complex participant judgments with excellent quality control at a reasonable cost in an astonishingly short amount of time for data collection. Yet, there is scarce scholarly work providing recommendations for conducting *remote* (i.e., virtual, online) user studies. The body of literature that applies remote user studies is predominantly descriptive, typically using an anonymous group ("crowd") of individuals for tasks that are often created to be simple and not requiring special expertise to be carried out. This is despite that fact that more advanced tasks requiring professional skills are relevant in many HCI and IS use cases. Examples of simple tasks that require little to no specialized expertise include layout experiments [5, 6], annotating datasets for machine learning [7–9], human computation [10], and the study of the crowd itself [11–13]. In contrast, an example of a user study requiring professional expertise would be a tele-medicine tablet for nurses.

Remote user studies often deploy surveys [14], although remote usability studies that these specific systems have also been taking speed [15]. Nonetheless, despite technological progress, researchers have not yet fully adopted remote user study tools and platforms, due to reasons such as lack of training, lack of institutional support, and due to multitude of options that can make choosing and learning new tools challenging. Therefore, providing experiences from the use of these platforms and tools can serve researchers across multiple fields. As previous work that deploys remote user techniques tends to be descriptive, i.e., communicating how the methods were used, rather than prescriptive (how they *should* be used) or instructive (*what* the best practices are). In the current work, we focus on sharing recommendations (i.e., best practices) based on our experience conducting online user studies over several years with four ORPs.

3 Research Objective

Our research objectives are: *Identify the strengths and weakness of four major ORPs for use in online user studies*, and *Provide recommendations based on lessons learned for employing these platforms.*

The four ORPs, described in detail below, are Appen, MTurk, Prolific, and Upwork. While other ORPs are available, we have found these platforms useful for online user studies in our field. We evaluate these four platforms along six dimensions defined as:

- **Diversity of reach**: the variety of and volume of participant populations based on nationalities, occupations, or other aspects.
- **User attributes available**: ease of accessing and range of available participant characteristics
- **Ease of use**: how straightforward the platform is to employ for an online user study as measured by complexity of the interface, coding required, etc.
- **Cost**: the amount that has to be paid or spent to conduct a user study, which is usually based on the time required for a participant to complete the task, number of participants, and amount of payment.
- **Modifiability**: the degree of ease at which one can make changes to the platform and the platform's flexibility adapting to such changes.
- **Blacklisting**: the ability of the platform to exclude certain participants from the study. This exclusion can be for reasons such as preventing repeated participation in the study (i.e., limiting to one session).

The metrics of evaluation for each dimension are on a scale of (*High, Medium, Low*) of relative degree that is compared to the complete set of platforms, and/or a binary measure (*Yes* or *No*) concerning whether the dimension applies to a given platform.

4 Approach

Our methodological approach is grounded in experience-based practice [16], in which the researchers immerse themselves in the contexts, with concepts, experiences, questions, and problem-solving are mixed together [17]. In this complex environment, research premises, ideas, data, and perspective are stress-tested in actual practice [18]. Given our research objectives, experience-based practice is an appropriate foundation for this exploratory work. Using this approach, for the recommendations presented here, we draw from our experience with more than a dozen online user studies, which have involved the use of multiple research designs and data collection platforms.

We provide examples from four major data collection platforms to illustrate specific points. The studies we base these guidelines on have been published in various journal and conference venues, including ACM Conference on Human Factors in Computing Systems, International Journal of Human-Computer Studies, International Journal of Human-Computer Interaction, and so on. Overall, the guidelines are based on the authors' experience conducting more than half a dozen user studies online, reported in various research papers [19–25], mostly related to testing an interactive persona system or designing data-driven personas [26]. The included platforms have been used in these studies, and we draw from experiences and best practices provided by each.

5 Remote User Study Platforms

The evaluation of the four data collection platforms based on our assessment in multiple user studies along the six dimensions is provided in Table 1, with a detailed discussion of each of the four platforms following. There are other remote user study platforms as well (e.g., Maze, UserTesting and UseBerry, Preely) – these are not included in the current work as we have no experience with these platforms.

Table 1. Evaluation of data collection platforms for remote user studies. N/A – means the dimension is not applicable to the platform.

	Diversity of reach	User attributes	Ease of use	Cost	Modifiability	Blacklisting
Appen	High	**No**	High	Low	Low	**No**
MTurk	Medium	Yes	**Low**	Low	**High**	Yes
Prolific	**Low**	Yes	High	Medium	Medium	Yes
Upwork	Medium	Yes	High	**High**	Medium	N/A

Appen (formerly known as Figure-Eight, CrowdFlower) is a data annotation platform. According to their website, Appen has more than a million workers (called contractors) in 130 countries, covering 180 unique languages and dialects[1]. The Appen platform is easy to use, providing several templates for different task types, such as Sentiment Analysis, Search Relevance, Data Categorization, Image Annotation, Transcription, and Content Moderation. In addition, one can create a study using a customizable empty template. The functionalities of Appen make it ideal for data annotation tasks, such as the creation of training sets for machine learning, and the cost is low for these tasks. The applicability to user studies is hindered by the fact that, apart from the workers' location, the platform does not provide other background information (e.g., age, gender) of the workers. As such, the sample characteristics cannot be reported, and the data cannot be analyzed by user attributes. Additionally, in Appen one cannot apply a custom blacklist to prevent the repetitive entry of the same participant in different experiment flows. Despite these shortcomings, Appen is easier to use than platforms that require

[1] https://www.appen.com (retrieved January 11, 2021).

advanced programming skills (e.g., MTurk), as it provides an easy-to-use drag and drop interface for some types of online user studies, as shown in Fig. 1.

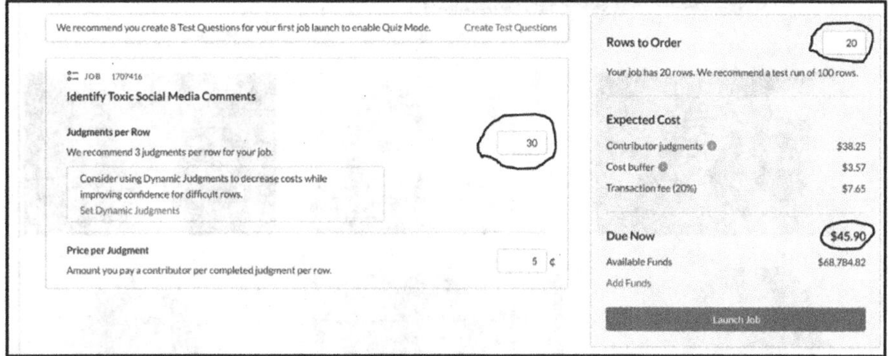

Fig. 1. Manipulation check study using crowdsourcing via Appen. The platform enables easy configuration and launch of the task.

The participants in this user study sees two types of social media comments, toxic and non-toxic comments. These comments were used in creating two types of persona profiles: toxic and non-toxic ones. To ensure that the assigned toxicity in the persona profiles matches with users' general perceptions of toxicity, a manipulation study was conducted (Fig. 1). Unlike typical crowdsourcing labeling tasks where thousands of rows are annotated for machine learning, here 20 rows are adequate as that corresponds to the number of comments used in the user study. For each comment (i.e., "row"), 30 ratings are obtained. The cost ($45.90 US) can be seen as reasonable for a manipulation check, and the results are typically available within minutes from launching the data collection. As such, the example illustrates the fast set-up and data collection process via Appen for simple tasks supporting user studies, such as manipulation check.

MTurk (full name: *Amazon Mechanical Turk*) is the first large-scale crowdsourcing platform, launched in 2005. Amazon has stated the platform has more than 500K registered workers from more than 190 countries[2], although the number of active workers is considerably lower, perhaps less than a fifth [27]. There has been observed a geographic bias, with approximately 60% of the workers located in the US and 30% in India. The cost per task low. The advantage of MTurk is its malleability [28]; through its programming interface, any sort of user study can be programmed (see Fig. 2). This modifiability comes with the heavy fee of technical skills; MTurk is the hardest to use out of the four platforms. In Fig. 2, the implementation required the creation of a dynamic HTML template that related to an underlying database that recorded for each participant the selected pictures and the participant's country. A certain sample size from each country was recruited. This study design is complex enough to be only carried out via customized programming and scripting, features that are most developed in the MTurk

[2] https://forums.aws.amazon.com/thread.jspa?threadID=58891.

platform. This study design could not have been possible with Appen's user interface, illustrating MTurk's superiority for more complex research designs.

Fig. 2. A study in which the participants were required to select (by clicking) pictures that could represent individuals from their own country.

Prolific is a research platform for recruiting subjects for user studies and sruveys [29]. As of June 2021, Prolific has more than 70,000 participants from various countries[3]. It has the fewest number of users out of the four platforms, but the platform itself is easy to use, and the quality of responses is good, as discussed below. The challenge is that the pool of participants is demographically biased: more than 60% are located in two countries (the UK and US). Perhaps, as a result, the cost is relatively high but reasonable for the quality of the responses. The participant pool is also about 75% aged between 20–40, and around 70% are ethnically White/Caucasian. The gender distribution, however, is close to 50%. The platform allows blacklisting of participants.

Upwork has more than 10 million users in 180 countries[4]. The advantage of Upwork is that it enables researchers to find highly skilled professionals [30], including designers, marketers, developers, public health experts, and so on (see Fig. 3). Unlike other platforms examined in this study, Upwork provides a way to recruit individual participants and view their work history. This feature is beneficial for recruiting specialized experts, such as public health professionals, who would be difficult to find on other platforms. Finding similarly skilled individuals in other platforms is much harder. The challenge is

[3] https://prolific.ac.

[4] https://www.prnewswire.com/news-releases/snagajob-appoints-former-upwork-ceo-to-board-of-directors-300417689.html.

that finding and recruiting professionals tends to require considerable communication costs. Typically, Upwork participants pose questions and expect more communication than participants in other platforms. However, the range of possible studies is broad, and participant selection is exact, so there is no need for blacklisting.

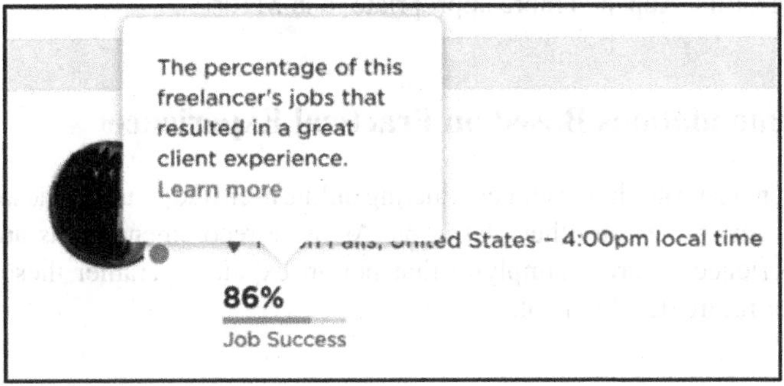

Fig. 3. Upwork profiles provide a metric of successful client encounters for each freelancer.

Comparison of Key Platform Features: When using Appen, data collection and participant recruitment occur on the same platform. When using MTurk, one can collect data either from outside or using MTurk itself. When using Prolific or Upwork, there is a separation of data collection and recruitment, which means these platforms are for participant recruitment only. The study itself will be run outside the platform, e.g., in Google Forms, Qualtrics, or a custom-made HTML page (we provide a short discussion of these tools shortly). *When the goal is to recruit professionals with specific skills,*

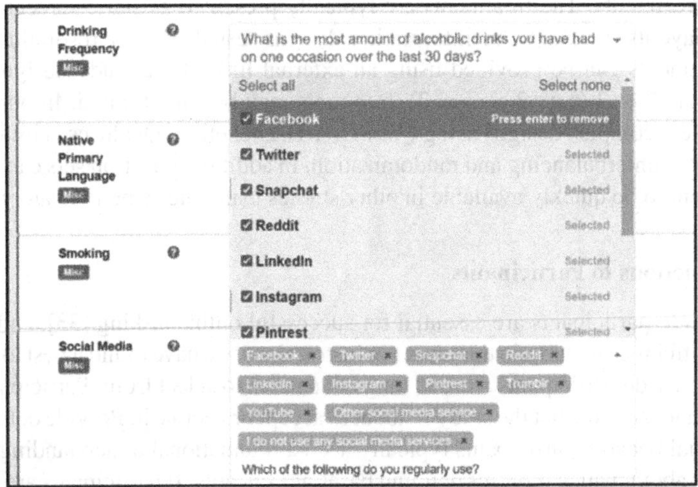

Fig. 4. Cloudresearch, an example of a user-friendly interface to MTurk. This service can be used to conveniently sample participants from MTurk, without the need of programming skills. Without an external service, using MTurk for data collection requires programming skills.

Prolific or Upwork are recommended, as these two platforms provide more information about users. MTurk's sampling features require more technical skills to access unless using an external service provider such as Cloudresearch (see Fig. 4). *When the goal is to annotate a large volume of data*, MTurk or Appen are commended, as these platforms have been designed with data annotation tasks in mind. When the research team lacks programming skills, Appen is more appropriate than MTurk.

6 Recommendations Based on Practical Experience

We here present some guidelines for conducting online user studies using these platforms, with insights for studies on other platforms. As these recommendations are based on practical experience, we are not implying that they are exhaustive; rather, these guidelines are a basis for future development.

6.1 Participant Attributes

Typical background information to ask in user studies includes demographic information (e.g., age, gender, country), job position, and the user's experience in their jobs or in the industry [31, 32]. Some platforms, such as Prolific, provide basic demographics (age, gender, country) for each participant in the data export without the need for asking them. In other platforms, such as Appen, it is extremely difficult to obtain these variables, as the platforms are based on the anonymity of crowdworkers.

However, ORPs can be supplemented with other tools to increase their robustness. For example, 2×2 experiments can be configured easily within *Google Forms*, as the tool enables one to duplicate and move items. In a 2×2 experiment, one can create four study sequences in Google Forms that are identical in every way except the order of presenting the study treatments. The treatments are typically presented as static screenshots that represent layouts, mockups, or information designs tested. Higher resolution images of the treatments can be provided using an external link. If the study design is more complex than 2×2 (e.g., $2 \times 2 \times 2$), free tools may become limited. In our studies, we implement complex designs using *Qualtrics*. The benefit of Qualtrics is its advanced features for counterbalancing and randomization. In addition, question sets can be saved in the system to be quickly available in other studies using the same indicators.

6.2 Instructions to Participants

Instructions to participants are essential for successful online tasking [33], and all four platforms offer nice features in this area. For instructions, we have found it best to be brief; platforms provide participants a broad range of studies to select from. Participants need to quickly understand what the study is about to accept or decline it. Provide definitions – this is crucial because participants typically need a foundational understanding of what the study is about, which requires defining baseline concepts. If participants are required to categorize data, provide examples. In all your instructions, use plain language; avoid professional or scientific lingo unless the study is particularly targeting people capable of grasping such lingo. Example instructions are shown in Fig. 5.

The default instruction template of Appen illustrates central information sections to include (see Fig. 5). Also, note that the study topic's level of subjectivity or interpretation influences how detailed the guidance should be. For simple human-intelligence tasks (HITs) such as "Is there a car in this picture?", no complex explanations are required, but for many social science topics, such as "Does this social media comment contain toxicity?", elaboration and explanations are required.

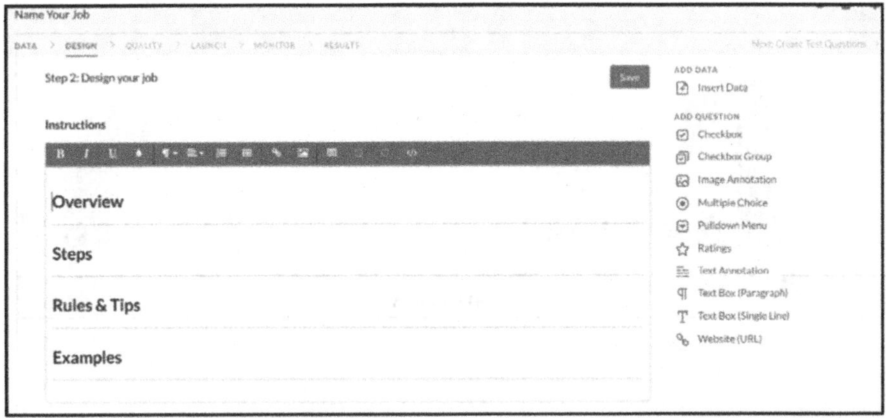

Fig. 5. Appen's user interface. The default template includes four sections for instructions. After uploading data, researchers can insert data items (e.g., text, pictures) and add questions, including multiple choice, checkbox, rating, and annotation.

Appen also enables the creation of *test questions* with known ground-truth values to train the participants. In case the participant answers wrong to a test question, the researcher can write an explanation that clarifies why the choice was wrong. Naturally, studies measuring attitudes such as trust or other user perceptions would not have ground truth values, as the responses are individually determined. If the task has a degree of subjectivity, monitoring the response patterns to test questions is important, as well as trying to select clear examples to the researchers' best ability.

6.3 Participant Recruitment

In our experience, user studies may employ an arbitrary number of participants, ranging from 30 (or lower) to a few hundred participants typically. However, the sample size can be computed in a data-driven way, using the pre-screened number of total candidate participants as a "population" to estimate the required sample size. For example, given 4,878 possible participants in a user study we conducted, with 95% confidence level and 5% confidence interval, 356 participants are required to make generalizable claims of this user population. (The sample size can easily be calculated using online tools[5].)

Virtually all ORPs struggle with providing a geographically and demographically balanced pool of participants. MTurk is heavily centered on participants from the US and

[5] https://www.surveysystem.com/sscalc.htm.

India [27]. Prolific is focused on young Caucasians from the US and UK (see Fig. 6). A general problem in all platforms seems to be that they are skewed to younger populations, making it harder to recruit participants of middle-aged and elderly age groups. Although the ORPs provide global coverage *in theory*, in practice this promise does not generally come into fruition, and researchers that need cross-cultural samples or representative samples from many countries may need to carry out data collection jobs in several platforms and then combine the results.

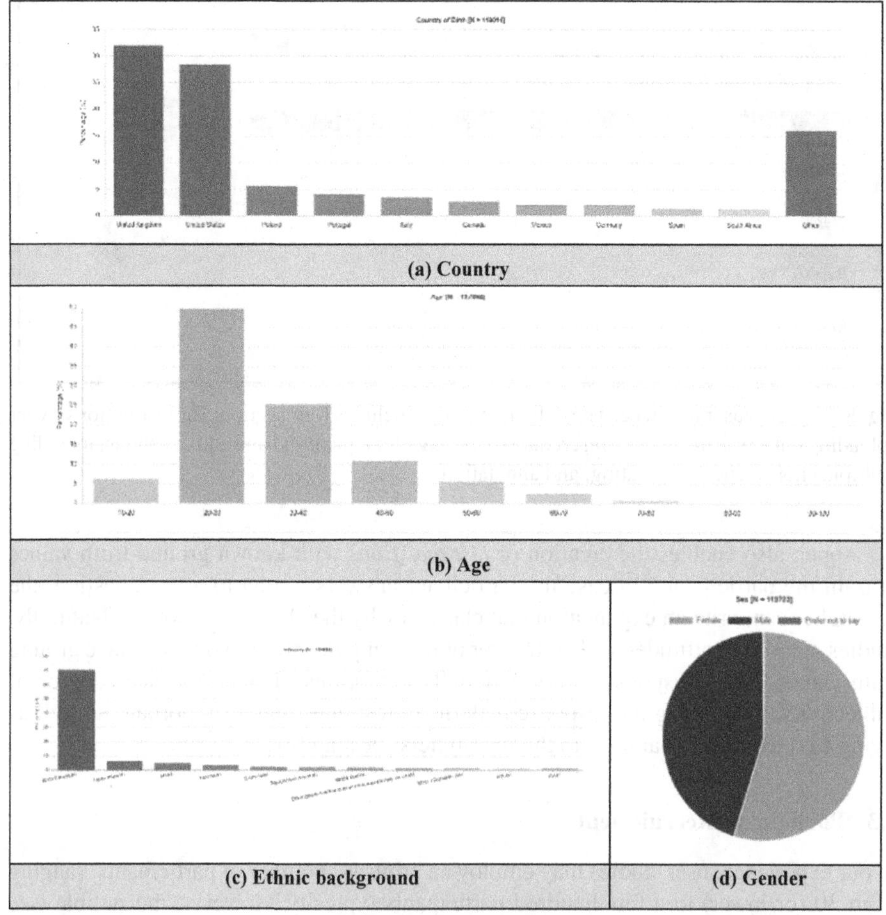

Fig. 6. Participant pool in Prolific: (a) UK and US are the dominating bars on the left. The last bar is 'Other'. (b) 20–30 and 30–40 are the two highest age groups. (c) White/Caucasian is the dominating bar. (d) The gender ratio is roughly balanced.

Most ORPs do not provide adequate tools for representative sampling or cross-cultural studies. Nonetheless, Prolific has a feature that provides representative population samples in two countries: the UK and the US. Using a representative sample, researchers can increase the correspondence of age, gender, and ethnicity in their sample to the characteristics in given national population. The estimates in Prolific are informed by census data from the UK Office of National Statistics and the US Census Bureau[6].

When not using geographic sampling criteria (as may often be the case in crowdsourcing), it is necessary to assess if cultural variability will affect the responses obtained. Diversity may not always be positive. Consider having 1 participant from 30 countries, in total N = 30 participants. The results do not necessarily inform us about the views of people from 30 countries. Still, they may instead involve hidden cultural biases that are impossible to analyze, as the sample for each country is only N = 1. Therefore, if there is any doubt of culture or other demographic variables affecting the participants' responses or behavior, the researchers should aim to apply a stratified sample. For example, in the above case, it would be more valuable to have 15 participants from 2 countries, as this would yield *some* data to analyze if the country affects the dependent variables. The more cultures or regions one wants to cover, the higher the sample size should be obtained.

6.4 Fair Compensation

Researchers should adhere to the fair compensation of online workers [34]. According to our experience, online workers are willing to work for less than minimum wages. This willingness makes it researchers' responsibility to maintain wage standards. Although the primary concern one may hear about is underpaying, prior research has shown possible downsides to overpaying [35]. Prolific provides as easy slider in the user interface to adjust the participants' compensation (see Fig. 7).

Prolific also shows the hourly rate based on (a) estimated completion time and (b) reward per completion that the researcher provides. Currently, researchers can "game" the system by providing low-ball estimates of the completion times, in which case the calculated hourly rate seems artificially good. However, this cannot actually mislead participants, if the platform shows them hourly rates based on actual completion times in a given job. One related issue is, however, that the variation among completion times tends to be high among participants; it is not uncommon that one participant can spend three times more on a survey than another. This is an issue because the effective hourly rate is set on the average; therefore, the quicker participants are able to earn better than slower ones. This leads into two issues: (a) the realized hourly rate of slower participants can drop below minimum wage, even when the researcher has set the reward to exceed the minimum wage (based on the researcher's estimate made in good faith); and, (b) if the speed of completion is positively correlated to quality of work, then the participants that take longer to provide quality outputs are effectively penalized for doing a good job, whereas those that rush through the job are rewarded at the expense of quality data. As such, ORPs still need to work considerably more to develop fairness mechanisms that prevent gaming from both the researcher and the participant.

6.5 Quality Control

The quality control of the participants can be done in two primary ways. First, platforms provide options for allowing only participants with a proven track record to participate.

[6] https://researcher-help.prolific.co/hc/en-gb/articles/360019236753-Representative-Samples-on-Prolific.

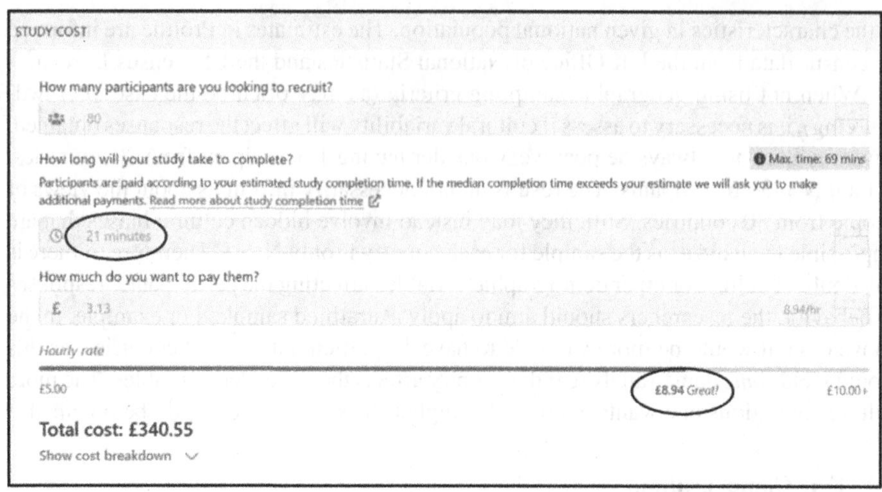

Fig. 7. Determining study compensation. The study completion time can be estimated by conducting a pilot study or (as in this case) relying on Qualtrics' automatic survey duration assessment. The hourly rate should be set fairly, at least to cover the sampled country's minimum wage. In this case, the minimum wage was verified from the UK government website[7]. If participants come from various countries, it may be advisable to replicate the study and change only the targeting and the compensation to satisfy local wage levels. Note that it is possible to releases bonuses for specific (or all) participants after checking the results.

In Appen, there are three quality levels, with "Level 3" indicating the smallest number of the participants with the highest accuracy rate based on their past jobs. Similarly, MTurk has three pre-defined quality levels: Masters (those with consistently high quality across tasks), "Number of HITs Approved" (the minimum number of accepted HITs needed to apply for a job), and "HIT Approval Rate" (the share of accepted HITs a worker has from all submitted ones). In Prolific, one can use approval rate (the percentage of studies where the participant has been approved) or the number of prior studies the person has taken part in. Upwork provides the applicant's historical success rate.

Second, in addition to using the platform's pre-defined categorization, one should employ attention checks – randomly placed items that ask for a specific response ("Please select 'Agree strongly'."). The researcher can then review the participants' answers and reject those that failed the attention check. Prolific has a high quality of participants in our experience, as attention checks are typically passed by more than 95% of the participants. The quality of Prolific is also supported in independent evaluation studies [29, 36]. Combining several techniques for quality control is suitable. As the crowd may learn and adapt to specific methods [37] – to be truly efficient at validating attentiveness, the checks need to have a degree of surprise.

[7] https://www.gov.uk/government/publications/the-national-minimum-wage-in-2020.

7 Current and Future Opportunities

As leveraging online platforms is currently a work-in-progress, we now discuss opportunities that could become trends in HCI, IS, and related domains due to their potential for online user research. Although there are many conceivable opportunities for remote user studies, we focus on two: behavioral tracking via live user sessions, and longitudinal online user studies, as these are somewhat underexploited in the current practice, but their implementation is technically possible and might result into insightful datasets.

7.1 Behavioral Tracking via Live User Sessions

There are at least two prominent methods for capturing user behavior data on system usage: mouse-tracking and eye-tracking. Mouse-tracking requires no calibration, and it can be implemented to record different user actions (hovering and clicking). Therefore, mouse-tracking is an unobtrusive form of recording user behavior. Another benefit is that mouse-tracking has no measurement error – the browser is able to precisely determine the exact screen coordinates the user's cursor is placed on at a given point in time. The recording can also be done based on HTML elements. The purpose of using HTML elements for tracking is to enable a direct interpretation of the user's interactions with a system – e.g., what menu items, views, and features were used (see Fig. 8). Such tracking systems can be developed by using standard HTML and JavaScript techniques. A consent

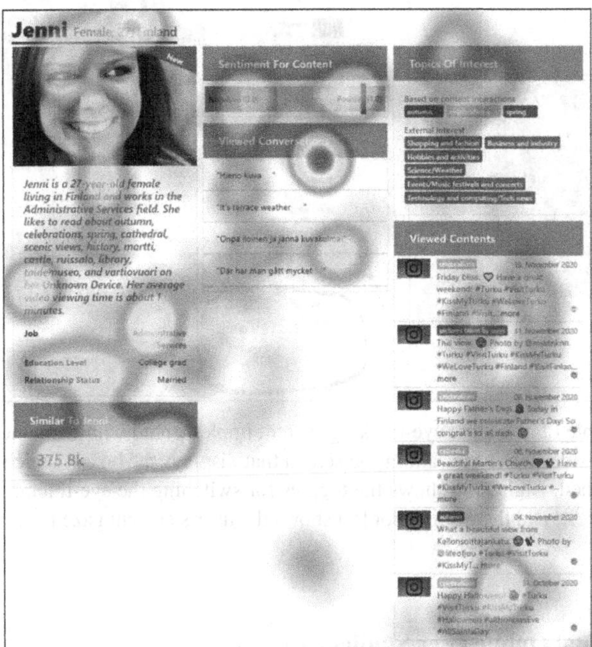

Fig. 8. An example of a heatmap generated with a mouse-tracking system [38] that tracks an interactive persona system. The color intensity indicates more movement.

from users is required for monitoring their behavior, and it might be appropriate to seek approval from an Institutional Review Board (IBR) and ensure compliance to privacy protection measures such as General Data Protection Regulation (EU GDPR) prior to launching remote-tracking studies.

An issue with mouse-tracking is that it does not necessarily correlate with a user's attention [39, 40] – that is, the user can be fixated on an item on the screen while placing their cursor on a different item. As such, information about how users focus their attention on a system can be lost. A potential solution is applying eye-tracking. Online eye-tracking has developed a lot over the years, but its use in research is still rare relative to physical eye-tracking systems that provide higher accuracy and more control. The challenges of online eye-tracking include, e.g., users' varying lighting conditions and webcams' ability to track gaze relative to infrared trackers. Perhaps the most advanced online eye-tracker is WebGazer.js [41–43] that can be integrated into web applications using JavaScript (see Fig. 9), and that is available as open source.

Because online eye-tracking currently imposes quality challenges, researchers are advised to take precautions, such as defining an accuracy threshold– participants failing to meet this threshold can be dropped from the final analysis. The accuracy values can be obtained from the online eye-tracking tool's calibration step. Because the previous technique likely results in losing a non-negligible number of participants from the final dataset, researchers should pre-emptively consider this by increasing sample sizes.

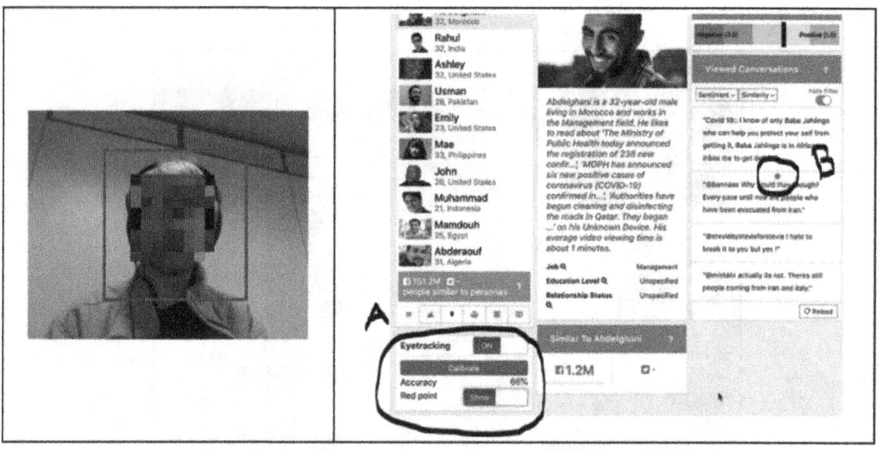

Fig. 9. An example of an online eye-tracking system implemented within an interactive persona system [44]. Left side shows the calibration screen that displays the box in which the user's head should be positioned. Right side shows the toggles for switching the eye-tracking and on-screen trailing on or off (A), and the trailing dot [B] shows the user's current gaze position.

7.2 Longitudinal Online User Studies

Longitudinal studies are extremely important for understanding how users' relationship and usage of a system evolves in time [45, 46]. Yet, longitudinal remains as a rarely exploited opportunity in HCI and IS user studies. In theory, such studies are simple to set up; given that the ORP stores the User IDs of each participant, researchers can use this ID to target the participants with future rounds of data collection.

An important consideration is attrition. It is unlikely that 100% of the participants remain active throughout a longitudinal study. Instead, a certain percentage will churn at each stage. Therefore, the initial pool of participants has to be adequate to suffer this loss. For example, assuming the attrition rate of 80%, out of 100 participants, only 51 remain after three extra rounds of data collection. In studies running more than a year, the attrition rate can decrease to less than 25% [47].

Ways to mitigate the dropout rate include decreasing the temporal distance of the data collection rounds, designing and communicating a reward structure that incentivizes attrition (i.e., increasing payment after each round), and screening for participants with a relatively long history with the platform (increased chance of commitment over the long term). Finally, if the payments can be delayed (e.g., in Prolific, this can be done up until 21 days), then one can make the full compensation contingent on participation in all the data collection rounds.

8 Discussion

Commercial ORPs provide three main advantages for researchers: (a) *tools and features* that are designed to accommodate specific research needs, such as rapid data annotation (e.g., Appen, MTurk), running large-scale surveys (e.g., Prolific) or locating expert freelancers (e.g., Upwork); (b) *commercial maturity and support* that lower the barrier of trying the tools and finding help for specific questions, and (c) critical mass of respondents/participants available to take part in a study at almost any given time.

However, despite the prominence of ready-made tools, it is worthwhile discuss the dilemma of "build or buy?" – meaning, should researchers use commercial or open-source tools, or build their own research tools. While commercial tools and platforms undoubtedly are the favorite choice of most researchers due to their low barrier of entry (they do not require much learning, and the development cost/time/effort tends to be low), in some cases it can make sense to pursue custom-made solutions. For example, in the persona analytics case illustrated in Figs. 8 and 9, it would have been possible to build the user tracking via Google Analytics (GA) instead of programming a customized solution. However, the tailored implementation of persona analytics has several advantages over using an industry standard tool such as GA:

- **Data Ownership:** the data resides in the research institute's own serves, not in 3rd party servers such as Google's. There are obvious reasons as to why storing delicate research data to commercial parties' servers might not be ideal.
- **Appropriate Event and Information Tagging:** When designing the data collection system from scratch, the researcher can determine the precise structure and functionality of *what* is collected and *how*; these requirements result in tracking only the information elements and events that matter for the research goals.
- **Data Availability:** Unlike in the standard GA installation, the researchers building a tailored solution can extract raw log data of every action a user took in the tracked

system. The standard GA installation only gives access to aggregated reports that are often not sufficient to analyze user behavior in depth.
- **Meaningful Reports:** the reports in GA are designed for website business analytics, not for academic research systems. Therefore, the metrics and reports available in GA may not be meaningful for addressing serious research questions. When building their own reports, researchers can consider specific research questions for which the data would be used. Hence, the reports can serve empirical user research better than the reports provided by industry tracking solutions like GA.

These advantages support building research-dedicated tracking solutions for systems. For example, persona analytics does a better job than GA at the specific purpose for which it was developed, which is empirically tracking persona user behavior.

9 Practical Suggestions for Researchers

Concerning recommendations for specific platform use, we offer the following:

- **Simple Tasks**: If you want users to do simple tasks where user attributes are not needed, Appen is a good platform of choice.
- **Complex Tasks**: If you want users to do complex user task flows and need user attributes, MTurk is a good choice, albeit with some upfront set-up efforts.
- **Specific Segments**: If you need to target specific population segments, such as professionals, Prolific may be your only reasonable choice.
- **High Quality, Complex Task, & Small Sample**: If you need high-quality responses to complex tasks from a small sample of skilled participants, then Upwork may be your only reasonable choice.

While remote user study techniques might not work for all conceivable user studies – for example, haptic technologies strictly require the human presence and "touch" [48] – user studies of many sorts can be carried out effectively online. Researchers' needs drive selecting and configuring platforms:

- *Are you testing a static vs. live system?* If static, survey-based data collection can be optimal. If a live system, behavioral (mouse- and eye-)tracking might be more sensible, ideally coupled with a survey.
- *Would the topic benefit more from a cross-sectional or longitudinal study design?* Both can be implemented fairly easily with online platforms by leveraging platforms' User IDs for repeated data collection.
- *Is your a HIT or an interpretative task?* For interpretative tasks, designing instructions and verifying quality is more difficult; a combination of attention checks and heuristic quality thresholds based on the platform's quality ratings can be applied.

An aspect increasing in research interest but still not effectively deployed in the field is fair remuneration [34]. Researchers typically set one price per response, which may be too high for a given region and too low for another. More dynamic reward-setting would be required, considering the country-specific variation in minimum and average wages. Relatedly, there are ethical issues [49] that must be considered when using these platforms, just as with other user studies approaches, such as the use of students [50]. Similarly, considerations of representativeness are typically overlooked in remote user studies. Even a small sample can include participants from more than a dozen countries: *does the underlying cultural variation matter for the findings?* HCI and IS researchers conducting user studies should consider applying stratified sampling more often and discuss the potential effects of cultural variability on their findings [51].

10 Conclusion

In this exploratory research, we present suggestions for employing online platforms for user studies. Although our experience is extensive in using these platforms, a limitation is that there are insights outside of our experiences. These other experiences, platforms, and uses are future research areas. Future research also needs to take these recommendations and apply a more rigorous methodology of platform evaluation to develop rigorous guidelines for researchers, including detailed comparisons of online studies to those in the lab. However, the evaluations of the platforms and resulting recommendations are practically actionable for researchers desiring to employ these services for online user studies and related research.

References

1. Ford, D., et al.: A Tale of Two Cities: Software Developers Working from Home During the COVID-19 Pandemic. arXiv:2008.11147 [cs]. (2020)
2. Yee, N., Bailenson, J.N., Rickertsen, K.: A meta-analysis of the impact of the inclusion and realism of human-like faces on user experiences in interfaces. In: Proceedings of the SIGCHI Conference on Human Factors in Computing Systems, pp. 1–10 (2007)
3. Toms, E.G., Freund, L., Li, C.: WiIRE: the Web interactive information retrieval experimentation system prototype. Inf. Process. Manag. **40**, 655–675 (2004). https://doi.org/10.1016/j.ipm.2003.08.006
4. Simons, R.N., Gurari, D., Fleischmann, K.R.: I hope this is helpful: understanding crowdworkers' challenges and motivations for an image description task. Proc. ACM Hum.-Comput. Interact. **4**, 105:1–105:26 (2020). https://doi.org/10.1145/3415176
5. Heer, J., Bostock, M.: Crowdsourcing graphical perception: using mechanical turk to assess visualization design. In: Proceedings of the SIGCHI Conference on Human Factors in Computing Systems, pp. 203–212 (2010)
6. Komarov, S., Reinecke, K., Gajos, K.Z.: Crowdsourcing performance evaluations of user interfaces. In: Proceedings of the SIGCHI Conference on Human Factors in Computing Systems, pp. 207–216. Association for Computing Machinery, New York, NY, USA (2013). https://doi.org/10.1145/2470654.2470684
7. Barbosa, N.M., Chen, M.: Rehumanized crowdsourcing: a labeling framework addressing bias and ethics in machine learning. In: Proceedings of the 2019 CHI Conference on Human Factors in Computing Systems, pp. 1–12 (2019)
8. Davidson, T., Warmsley, D., Macy, M., Weber, I.: Automated hate speech detection and the problem of offensive language. In: Proceedings of Eleventh International AAAI Conference on Web and Social Media, pp. 512–515. Montreal, Canada (2017)

9. Weber, I., Mejova, Y.: Crowdsourcing health labels: inferring body weight from profile pictures. In: Proceedings of the 6th International Conference on Digital Health Conference, pp. 105–109 (2016)
10. Quinn, A.J., Bederson, B.B.: Human computation: a survey and taxonomy of a growing field. In: Proceedings of the SIGCHI Conference on Human Factors in Computing Systems, pp. 1403–1412 (2011)
11. Gadiraju, U., Kawase, R., Dietze, S., Demartini, G.: Understanding malicious behavior in crowdsourcing platforms: The case of online surveys. In: Proceedings of the 33rd Annual ACM Conference on Human Factors in Computing Systems, pp. 1631–1640 (2015)
12. Ikeda, K., Bernstein, M.S.: Pay it backward: Per-task payments on crowdsourcing platforms reduce productivity. In: Proceedings of the 2016 CHI Conference on Human Factors in Computing Systems, pp. 4111–4121 (2016)
13. Ross, J., Irani, L., Silberman, M.S., Zaldivar, A., Tomlinson, B.: Who are the crowdworkers? Shifting demographics in Mechanical Turk. In: CHI 2010 Extended Abstracts on Human Factors in Computing Systems. pp. 2863–2872 (2010)
14. Kittur, A., Chi, E.H., Suh, B.: Crowdsourcing user studies with Mechanical Turk. In: Proceedings of the twenty-sixth annual SIGCHI Conference on Human Factors in Computing Systems, pp. 453–456. ACM, New York, NY, USA (2008). https://doi.org/10.1145/1357054.1357127
15. Hill, J.R., Harrington, A.B., Adeoye, P., Campbell, N.L., Holden, R.J.: Going remote—demonstration and evaluation of remote technology delivery and usability assessment with older adults: survey study. JMIR mHealth and uHealth. **9**, e26702 (2021)
16. Cook, M.: Evidence-based medicine and experience-based practice-clash or consensus. Med. L. **23**, 735 (2004)
17. Kozleski, E.B.: The uses of qualitative research: powerful methods to inform evidence-based practice in education. Res. Pract. Pers. Sev. Disabil. **42**, 19–32 (2017). https://doi.org/10.1177/1540796916683710
18. Albarqouni, L., Hoffmann, T., Glasziou, P.: Evidence-based practice educational intervention studies: a systematic review of what is taught and how it is measured. BMC Med. Educ. **18**, 177 (2018). https://doi.org/10.1186/s12909-018-1284-1
19. Salminen, J., Jung, S., Santos, J.M., Jansen, B.J.: Does a smile matter if the person is not real? The effect of a smile and stock photos on persona perceptions Int. J. Hum.-Comput. Interact. 1–23 (2019). https://doi.org/10.1080/10447318.2019.1664068
20. Salminen, J., Jung, S., Santos, J.M., Jansen, B.J.: The effect of smiling pictures on perceptions of personas. In: UMAP 2019 Adjunct: Adjunct Publication of the 27th Conference on User Modeling, Adaptation and Personalization. ACM, Larnaca, Cyprus (2019). https://doi.org/10.1145/3314183.3324973
21. Salminen, J., Santos, J.M., Kwak, H., An, J., Jung, S., Jansen, B.J.: Persona perception scale: development and exploratory validation of an instrument for evaluating individuals' perceptions of personas. Int. J. Hum. Comput. Stud. **141**, 102437 (2020). https://doi.org/10.1016/j.ijhcs.2020.102437
22. Salminen, J., Kwak, H., Santos, J.M., Jung, S., An, J., Jansen, B.J.: Persona perception scale: developing and validating an instrument for human-like representations of data. In: Extended Abstracts of the 2018 CHI Conference on Human Factors in Computing Systems - CHI 2018, pp. 1–6. ACM Press, Montreal QC, Canada (2018). https://doi.org/10.1145/3170427.3188461
23. Salminen, J., Jung, S., Santos, J.M., Kamel, A.M., Jansen, B.J.: Picturing it!: the effect of image styles on user perceptions of personas. In: In the Proceedings of ACM Human Factors in Computing Systems (CHI 2021). ACM, Virtual Conference (2021)
24. Salminen, J., Jung, S., Kamel, A.M.S., Santos, J.M., Jansen, B.J.: Using artificially generated pictures in customer-facing systems: an evaluation study with data-driven personas Behav. Inf. Technol. 1–17 (2020). https://doi.org/10.1080/0144929X.2020.1838610

25. Salminen, J., Santos, J.M., Jung, S., Eslami, M., Jansen, B.J.: Persona transparency: analyzing the impact of explanations on perceptions of data-driven personas Int. J. Hum. Comput. Interact. 1–13 (2019). https://doi.org/10.1080/10447318.2019.1688946
26. Jansen, B., Salminen, J., Jung, S., Guan, K.: Data-Driven Personas. Morgan & Claypool Publishers, San Rafael (2021)
27. Difallah, D., Filatova, E., Ipeirotis, P.: Demographics and dynamics of mechanical Turk workers. In: Proceedings of the Eleventh ACM International Conference on Web Search and Data Mining, pp. 135–143. Association for Computing Machinery, New York, NY, USA (2018). https://doi.org/10.1145/3159652.3159661
28. Sheehan, K.B., Pittman, M.: Amazon's Mechanical Turk for academics: The HIT handbook for social science research. Melvin & Leigh, Publishers (2016)
29. Palan, S., Schitter, C.: Prolific. ac–a subject pool for online experiments. J. Behav. Exp. Financ. **17**, 22–27 (2018)
30. Popiel, P.: Boundaryless in the creative economy: assessing freelancing on upwork. Crit. Stud. Med. Commun. **34**, 220–233 (2017)
31. Salminen, J., Nielsen, L., Jung, S., An, J., Kwak, H., Jansen, B.J.: Is more better? Impact of multiple photos on perception of persona profiles. In: Proceedings of ACM CHI Conference on Human Factors in Computing Systems (CHI2018). ACM, Montréal, Canada (2018). https://doi.org/10.1145/3173574.3173891
32. Salminen, J., Jung, S., Chowdhury, S.A., Sengün, S., Jansen, B.J.: Personas and analytics: a comparative user study of efficiency and effectiveness for a user identification task. In: Proceedings of the ACM Conference of Human Factors in Computing Systems (CHI 2020). ACM, Honolulu, Hawaii, USA (2020). https://doi.org/10.1145/3313831.3376770
33. Alonso, O., Marshall, C.C., Najork, M.A.: A human-centered framework for ensuring reliability on crowdsourced labeling tasks. In: First AAAI Conference on Human Computation and Crowdsourcing (2013)
34. Gleibs, I.H.: Are all research fields equal? Rethinking practice for the use of data from crowdsourcing market places. Behav. Res. Methods **49**(4), 1333–1342 (2016). https://doi.org/10.3758/s13428-016-0789-y
35. d'Eon, G., Goh, J., Larson, K., Law, E.: Paying Crowd Workers for Collaborative Work. Proc. ACM Hum.-Comput. Interact. **3**, 1–24 (2019)
36. Peer, E., Brandimarte, L., Samat, S., Acquisti, A.: Beyond the Turk: alternative platforms for crowdsourcing behavioral research. J. Exp. Soc. Psychol. **70**, 153–163 (2017)
37. Hauser, D.J., Schwarz, N.: Attentive turkers: mturk participants perform better on online attention checks than do subject pool participants. Behav. Res. Methods **48**(1), 400–407 (2015). https://doi.org/10.3758/s13428-015-0578-z
38. Jung, S., Salminen, J., Jansen, B.J.: Persona analytics: implementing mouse-tracking for an interactive persona system. In: Extended Abstracts of ACM Human Factors in Computing Systems - CHI EA 2021. ACM, Virtual conference (2021). https://doi.org/10.1145/3411763.3451773
39. Chen, M.C., Anderson, J.R., Sohn, M.H.: What can a mouse cursor tell us more? Correlation of eye/mouse movements on web browsing. In: CHI 2001 Extended Abstracts on Human Factors in Computing Systems, pp. 281–282. ACM (2001)
40. Navalpakkam, V., Churchill, E.: Mouse tracking: measuring and predicting users' experience of web-based content. In: Proceedings of the SIGCHI Conference on Human Factors in Computing Systems, pp. 2963–2972 (2012)
41. Papoutsaki, A., Gokaslan, A., Tompkin, J., He, Y., Huang, J.: The eye of the typer: a benchmark and analysis of gaze behavior during typing. In: Proceedings of the 2018 ACM Symposium on Eye Tracking Research & Applications, pp. 1–9. Association for Computing Machinery, New York, NY, USA (2018). https://doi.org/10.1145/3204493.3204552

42. Papoutsaki, A., Laskey, J., Huang, J.: SearchGazer: Webcam eye tracking for remote studies of web search. In: Proceedings of the 2017 Conference on Conference Human Information Interaction and Retrieval, pp. 17–26. Association for Computing Machinery, New York, NY, USA (2017). https://doi.org/10.1145/3020165.3020170
43. Papoutsaki, A., Sangkloy, P., Laskey, J., Daskalova, N., Huang, J., Hays, J.: Webgazer: scalable webcam eye tracking using user interactions. In: Proceedings of the Twenty-Fifth International Joint Conference on Artificial Intelligence, pp. 3839–3845. AAAI Press, New York, New York, USA (2016)
44. Jung, S.-G., Salminen, J., Jansen, B.J.: Implementing eye-tracking for persona analytics. In: ETRA 2021 Adjunct: ACM Symposium on Eye Tracking Research and Applications, pp. 1–4. ACM, Virtual Conference (2021). https://doi.org/10.1145/3450341.3458765
45. Lee, G., Xia, W.: A longitudinal experimental study on the interaction effects of persuasion quality, user training, and first-hand use on user perceptions of new information technology. Inf. Manag. **48**, 288–295 (2011)
46. Yuan, W.: End-user searching behavior in information retrieval: a longitudinal study. J. Am. Soc. Inf. Sci. **48**, 218–234 (1997)
47. Kothe, E., Ling, M.: Retention of participants recruited to a one-year longitudinal study via Prolific (2019). https://psyarxiv.com/5yv2u/. https://doi.org/10.31234/osf.io/5yv2u
48. Luzhnica, G., Veas, E., Pammer, V.: Skin reading: encoding text in a 6-channel haptic display. In: Proceedings of the 2016 ACM International Symposium on Wearable Computers, pp. 148–155. ACM, New York, NY, USA (2016). https://doi.org/10.1145/2971763.2971769
49. Shklovski, I., Vertesi, J.: Un-Googling: research technologies, communities at risk and the ethics of user studies in HCI. In: The 26th BCS Conference on Human Computer Interaction 26, pp. 1–4 (2012)
50. Ferguson, L.M., Yonge, O., Myrick, F.: Students' involvement in faculty research: ethical and methodological issues. Int. J. Qual. Methods **3**, 56–68 (2004)
51. Häkkilä, J., et al.: Design sensibilities-designing for cultural sensitivity. In: Proceedings of the 11th Nordic Conference on Human-Computer Interaction: Shaping Experiences, Shaping Society, pp. 1–3 (2020)

The Hidden Cost of Using Amazon Mechanical Turk for Research

Antonios Saravanos[1(✉)], Stavros Zervoudakis[1], Dongnanzi Zheng[1], Neil Stott[2], Bohdan Hawryluk[1], and Donatella Delfino[1]

[1] New York University, New York, NY 10003, USA
{saravanos,zervoudakis,dz40,bh54,dd61}@nyu.edu
[2] Cambridge Judge Business School, Cambridge, Cambridgeshire CB2 1AG, UK
n.stott@jbs.cam.ac.uk

Abstract. In this study, we investigate the attentiveness exhibited by participants sourced through Amazon Mechanical Turk (MTurk), thereby discovering a significant level of inattentiveness amongst the platform's top crowd workers (those classified as 'Master', with an 'Approval Rate' of 98% or more, and a 'Number of HITS approved' value of 1,000 or more). A total of 564 individuals from the United States participated in our experiment. They were asked to read a vignette outlining one of four hypothetical technology products and then complete a related survey. Three forms of attention check (logic, honesty, and time) were used to assess attentiveness. Through this experiment we determined that a total of 126 (22.3%) participants failed at least one of the three forms of attention check, with most (94) failing the honesty check – followed by the logic check (31), and the time check (27). Thus, we established that significant levels of inattentiveness exist even among the most elite MTurk workers. The study concludes by reaffirming the need for multiple forms of carefully crafted attention checks, irrespective of whether participant quality is presumed to be high according to MTurk criteria such as 'Master', 'Approval Rate', and 'Number of HITS approved'. Furthermore, we propose that researchers adjust their proposals to account for the effort and costs required to address participant inattentiveness.

Keywords: Amazon Mechanical Turk · MTurk · Attention checks · Inattentive respondents · Worker reputation · Worker quality · Data quality

1 Introduction

Over time, online services for participant recruitment by researchers have increased in popularity [30]. Amazon Mechanical Turk (MTurk; also known as Mechanical Turk [15]) is one of the oldest and most frequently selected tools from a spectrum of web-based resources, enabling researchers to recruit participants online and lowering the required time, effort, and cost [24, 39]. A Google Scholar search for the term 'Mechanical Turk' reveals continuing growth in its use, with 1,080, 2,750, and 5,520 items found when filtering the results for 2010, 2012, and 2014 respectively [24]. The technology facilitates "an online labor market" where "individuals and organizations (requestors)" can "hire

C. Stephanidis et al. (Eds.): HCII 2021, LNCS 13094, pp. 147–164, 2021.
https://doi.org/10.1007/978-3-030-90238-4_12

humans (workers) to complete various computer-based tasks", which they describe as "Human Intelligence Tasks or HITs" [24]. The MTurk platform's suitability for use in research has been extensively evaluated [12], with most published studies describing it as a suitable means for recruitment [8], although a few also state reservations [9]. This paper builds on that work by investigating the reliability of the top crowd workers that can potentially be sourced from the MTurk platform, while concurrently motivating them through offers of high compensation. Specifically, we focus on the attentiveness exhibited by US based workers classified as 'Master', who had completed at least 98% of the tasks they committed to completing (i.e., an 'Approval Rate' of 98% or more). Additionally, the number of these activities was required to exceed 999 (i.e., the 'Number of HITS approved' had a value of 1,000 or more). To the best of our knowledge, this segment of the platform has yet to be studied with a focus on participant attention.

2 Background

Amazon does not disclose real-time data on the total number of workers available for hire via their MTurk service, or those online at any particular moment. Several researchers offer insight into what those values might be [9, 13, 36]. Ross et al. [36] report that in 2010 the platform had more than 400,000 workers registered. Likewise, there were anywhere between 50,000 to 100,000 HITs at any given time. In 2015, Chandler, Mueller, and Paolacci [9] wrote: "From a requester's perspective, the pool of available workers can seem limitless, and Amazon declares that the MTurk workforce exceeds 500,000 users". Stewart et al. [40] report that the turnover rate is not dissimilar to what one would experience in a university environment, with approximately 26% of the potential participants on MTurk retiring and being replenished by new people. More recently (2018), Difallah, Filatova, and Ipeirotis [13] found that at least 100,000 workers were registered on the platform, with 2,000 active at any given time. The authors also state that a significant worker turnover exists, with the half-life for workers estimated to be between 12 and 18 months [13]. Such numbers as those reported by Stewart et al. [40] and Difallah, Filatova, and Ipeirotis [13] indicate that recruiting the same worker more than once for a given experiment is highly unlikely.

MTurk is probably the most thoroughly studied of the available platforms for online participant recruitment through crowdsourcing. The literature on the suitability of MTurk for research presents a somewhat 'rosy' picture, labeling it as adequate for use with experiments. This includes the work of Casler, Bickel, and Hackett [8], who compared the data obtained through the recruitment of participants on MTurk with data collected from participants recruited through social media, and those recruited on an academic campus [8]. The authors found that the data was similar across all three pools and highlight that the MTurk sample was the most diverse [8]. Moreover, the authors [8] reveal that the results were similar irrespective of whether the experiments had been completed in-lab or online. Both the replicability and reliability of data collected through the MTurk platform have been established. Rand [34] found that participant responses across experiments were consistent, allowing for replication of results and Paolacci, Chandler, and Ipeirotis [31] found increased reliability of data. Hauser and Schwarz [17] found that participants recruited from MTurk exhibited superior attention to the assigned task compared to participants recruited using traditional approaches.

Despite all these reassuring findings, a growing body of literature raises warnings that must be addressed [2, 5, 20, 25, 50], and "significant concerns" remain [15]. Such concerns are not new. For example, the use of attention checks to identify inattentive participants was standard practice for a large group of research communities, and prior work from past decades shows that many participants (from 5% to 60%) answer survey questions carelessly [6, 21, 28]. However, to some extent, this would not be expected with MTurk, as it would be assumed that individuals are essentially workers, and as such, they would be devoted to the task and paying attention. The main concern is the assertion that the low remuneration attracts workers with limited abilities who cannot find better employment [15]. Stone et al. [42] note that participants recruited through MTurk "tend to be less satisfied with their lives than more nationally representative samples", although they comment that "the reasons for this discrepancy and its implications are far from obvious". The reliability of crowd workers has been widely discussed and studied by investigating the impact of attentiveness on the reliability of the crowd worker responses (e.g., [37]).

Researchers are increasingly concerned that participants sourced through MTurk "do not pay sufficient attention to study materials" [15]. A prominent example is the work of Chandler et al. [9], who reveal that participants were not always entirely devoted to the assigned task and were instead multitasking. Litman et al. [25] identified that the practice of multitasking while participating in a research study is problematic, as it can lead to inattentiveness and reduce participants' focus on details [25]. Consequently, studies relying on participants devoting their full attention to the current work are at risk [25]. The authors also state that "these findings are especially troubling, considering that the participants in the Chandler et al. study were some of the most reliable of MTurk respondents, with cumulative approval ratings over 95%". The current research seek to understand this inattentiveness, trusting that our research will be useful to others who source participants for research using this tool.

3 Methodology

This section outlines the methodology used for the study.

3.1 Experimental Design

To investigate participant attention an experimental approach was adopted. Participants solicited through MTurk were forwarded to the Qualtrics web-based software where they were randomly presented with a vignette describing one of four hypothetical technology products. Subsequently, they were asked questions on their intention to adopt that technology. MTurk has been used on numerous occasions to understand user intention to adopt technology [29, 38, 41, 48, 49, 51]. Participants were then given one of two technology acceptance questionnaires to share their perceptions of the technology presented in their respective vignettes. The questionnaires were adaptations of the most popular models used to study user adoption of technology [35]. The first questionnaire (short) reflected the instrument for the second version of the unified theory of acceptance and use of technology (UTAUT2) [45] model and comprised 52 questions in total. The second questionnaire (long) reflected the instrument for the third version of the technology acceptance model (TAM3) [46] and comprised 74 questions. Both questionnaires also included 10 demographic and experience questions. Aside from the demographic questions, each question was rated through a 7-point Likert scale ranging from 'strongly disagree' to 'strongly agree'.

Assessing Attention

Three forms of attention check, derived from the work of Abbey and Meloy [1], were used to gauge participant attention. The first was a logical check based on logical statements. It required participants to demonstrate "comprehension of logical relationships" [1]. An example of such a question might be 'at some point in my life, I have had to consume water in some form'. This check comprised two such logical statements to answer, as shown in Table 1. The second was an honesty check to "ask a respondent directly to reveal their perceptions of their effort and data validity for the study" [1]. An example was, 'I expended effort and attention sufficient to warrant using my responses for this research study'. As part of this check, participants were asked two questions regarding their perception of the attention invested in the experiment. Table 1 shows the questions used. These questions were also rated using a 7-point Likert scale ranging from 'strongly disagree' to 'strongly agree'. Participants who did not respond to both questions by selecting the 'strongly agree' choice were deemed to have failed their respective attention checks.

Table 1. Attention check questions.

Code	Type	Original question	Adapted question
ATT1	Logical Statement	I would rather eat a piece of fruit than a piece of paper	I would rather eat a piece of fruit than a piece of paper
ATT2		At some point in my life, I have had to consume water in some form	At some point in my life, I have had to consume water in some form
ATT3	Honesty check	On a scale of 1–10, with one being the least attention and 10 being the most attention, please indicate how much attention you applied while completing this study	I applied sufficient attention while completing this study
ATT4		Did you expend effort and attention sufficient to warrant using your responses for this research study?	I expended effort and attention sufficient to warrant using my responses for this research study

The third form of attention check was a time check, which used "response time" to ascertain attention, employing the concept that response times might be "overly fast or slow based on distributional or expected timing outcomes" [1]. Participants who were unable to complete the experiment within a reasonable time were deemed to have failed that check. To estimate the response time, we totaled up the number of words that participants would read as part of the informed consent document, instructions, and vignette. We then used the conservative reading rate (200 words per minute) described by Holland [18] to estimate the time participants would require to read that material. To determine the time participants would need to complete each Likert question, we used the estimate provided by Versta Research [47], which is 7.5 s on average. Table 2 summarizes our calculations. Participants who spent less than 70% of the estimated time on the survey were deemed to have failed the time check.

Table 2. Task composition and participant estimated effort and compensation.

Group	Vignette	Consent (# of words)	Vignette (# of words)	# of Likert questions	Participant total compensation	Estimated completion time (in seconds)	Effective wage (in USD)
1	1	265	80	52	$3.41	494	$24.85/hr
2	2	265	97	52	$3.41	499	$24.60/hr
3	3	265	106	52	$3.41	501	$24.50/hr
4	4	265	124	52	$3.41	507	$24.21/hr
5	1	265	80	74	$4.19	659	$22.89/hr
6	2	265	97	74	$4.19	664	$22.72/hr
7	3	265	106	74	$4.19	666	$22.65/hr
8	4	265	124	74	$4.19	672	$22.45/hr

Compensation

A factor that was considered important and that needed to be controlled for was compensation. The concern was that the level of compensation might influence participant attention. However, numerous studies have investigated how compensation influences the quality of data produced by MTurk workers [4, 7, 25]. Most found that the quality of results is not linked to the rate of compensation, with Litman, Robinson, and Rosenzweig [25] stating that "payment rates have virtually no detectable influence on data quality". One example of such a study was conducted by Buhrmester et al. [7], who offered participants 2 cents (i.e., $0.25/hour), 10 cents, or 50 cents (i.e., $6 per hour) to complete a five-minute task. The authors found that while recruiting participants took longer when lower compensation was offered, the data quality was similar irrespective of the offered compensation. A similar study was conducted by Andersen and Lau [4], who provided participants with either $2, $4, $6, or $8 to complete a task. They found that the remuneration did not influence participants' performance, writing that there was "no consistent or clear evidence that pay rates influenced our subject behavior".

A smaller number of studies show that the quality of work produced by those on MTurk is influenced by the compensation size. An example is seen in Aker et al. [3]

who compensated participants for a task at rates of $4, $8, and $10 per hour. Their "results indicate that in general higher payment is better when the aim is to obtain high quality results" [3]. Overall, most tasks on MTurk offer a minimal level of compensation [7, 14, 33]. In 2010, the mean and median wages were $3.63/hour and $1.38/hour, respectively [19]. In 2019, the median compensation in the United States was $3.01/hour [16]. Paolacci, Chandler, and Ipeirotis [31] comment that "given that Mechanical Turk workers are paid so little, one may wonder if they take experiments seriously".

The rate offered to participants in this study surpassed $22/hour in order to ensure that participants were adequately motivated and thereby control for compensation (Table 2). The size of this compensation could be considered excessive when considering what is traditionally offered to participants on MTurk, the federal minimum wage in the United States of $7.25/hour, and what is presented in studies examining the effect of compensation on performance. For example, Aker et al. [3] describe $10/hour as high. Offering an extremely generous wage was expected to negate undesirable effects and induce participants to devote their full attention to our study.

3.2 Participants

Participants were selected to represent the top workers the MTurk platform offers. This was accomplished by using a filtering mechanism, allowing only workers satisfying certain criteria to participate in the study [10]. The filters used were: 1) located in the United States; 2) classified by Amazon as 'Master' level; 3) had completed at least 98% of the tasks they committed to completing (i.e., an 'Approval Rate' of 98% or more); and 4) had completed at least 1,000 tasks (i.e., their 'Number of HITS approved' rating was 1,000 or more). The data was collected through two batches over 16 days (between December 22nd, 2019, and December 30th, 2019, and again between February 1st, 2020, and February 7th, 2020). Participation in this study was voluntary and all our participants were first asked to confirm that they were willing to participate before being allowed to begin the experiment. The privacy of participants was protected using confidential coding.

The sample was comprised of 564 participants, 293 (51.95%) identified as male, and 271 (48.05%) identified as female. Most participants were in the 31–55 age range (73.94%), had some form of undergraduate (74.47%) or postgraduate (9.93%) training, and earned below $60,000 (60.29%) per year. Most of the participants (480, equating to 85.11% of the sample) identified as white. Finally, most participants were either never married (284, or 50.35%) or married (215, or 38.12%). Table 3 shows the participants' demographics for the sample in greater detail. Figure 1 shows the locations of participants within the United States. All states were represented except for Wyoming, with the five most prevalent in the sample being California, Florida, Pennsylvania, Texas, and Michigan.

Table 3. Participant demographics.

Characteristic	Category	N	Percentage
Age	18–25	18	3.19%
26–30	78	13.83%	
31–55	417	73.94%	
56 or older	51	9.04%	
Gender	Male	293	51.95%
	Female	271	48.05%
Income	Less than $10,000	17	3.01%
	$10,000–$19,999	46	8.16%
	$20,000–$29,999	70	12.41%
	$30,000–$39,999	83	14.72%
	$40,000–$49,999	63	11.17%
	$50,000–$59,999	61	10.82%
	$60,000–$69,999	59	10.46%
	$70,000–$79,999	44	7.80%
	$80,000–$89,999	24	4.26%
	$90,000–$99,999	32	5.67%
	$100,000–$149,999	47	8.33%
	$150,000 or more	18	3.19%
Marital status	Never married	284	50.36%
	Married	215	38.12%
	Separated	3	0.53%
Divorced	52	9.22%	
Widowed	7	1.24%	
No response	3	0.53%	
Race	Asian	32	5.67%
	Black or African American	27	4.79%
	Other	25	4.43%
	White	480	85.11%
Schooling	<High school degree	2	0.35%
	High school graduate	85	15.07%
	Some college - no degree	119	21.10%
	Associate's degree	72	12.77%

(continued)

Table 3. (*continued*)

Characteristic	Category	N	Percentage
	Bachelor's degree	229	40.60%
	Master's degree	44	7.80%
	Professional degree	8	1.42%
	Doctoral degree	4	0.71%
	No response	1	0.18%

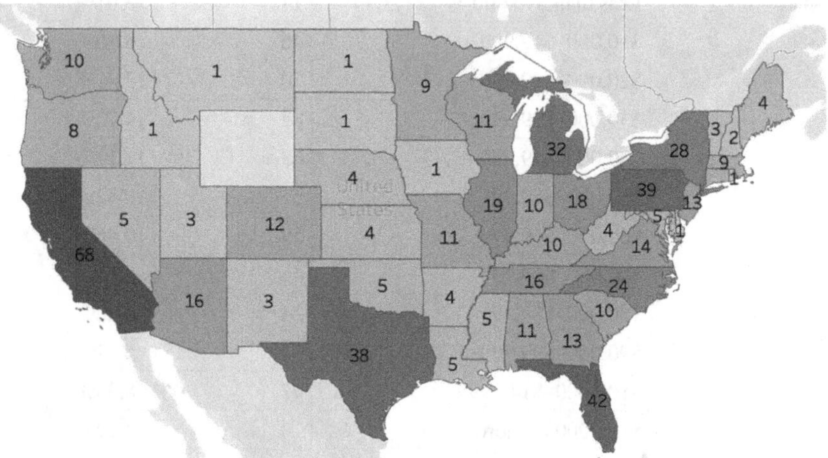

Fig. 1. Participants by State.

4 Analysis and Results

To analyze our data, we relied on three techniques. The first examined the frequency with which attention checks were passed or failed by participants; this revealed that 126 of the 564 participants (22.34%) failed at least one form of attention check. The attention check that most participants failed was the honesty check (94/564), followed by the logic check (31/564), and the time check (27/564). Some participants failed more than one check, with 14/564 (2.48%) failing both logic and honesty checks and 6/564 (1.06%) failing both time and honesty checks. Finally, 6/564 (1.06%) participants failed all three attention checks (logic, honesty, and time). Figure 2 illustrates the numbers of participants who failed and passed each form of attention check. As expected, participants who passed the time check were more likely to pass the other attention checks (logic and honesty).

The second technique used Spearman rank-order (rho) correlations to assess the correlation between the characteristics of age, gender, income, race, and prior experience with the technology and each of the three forms of attention checks (i.e., logic, honesty, and time checks). No significant correlation was found except in two instances. First, prior experience of using the technology being studied was positively correlated with the logic check ($r_s = 0.192$, $p = 0.000$). Second, prior experience of using the technology being studied was positively correlated with the honesty check ($r_s = 0.213$, $p = 0.000$). Therefore, the more familiar participants were with the technology, the more likely they were to pass the logic and honesty checks.

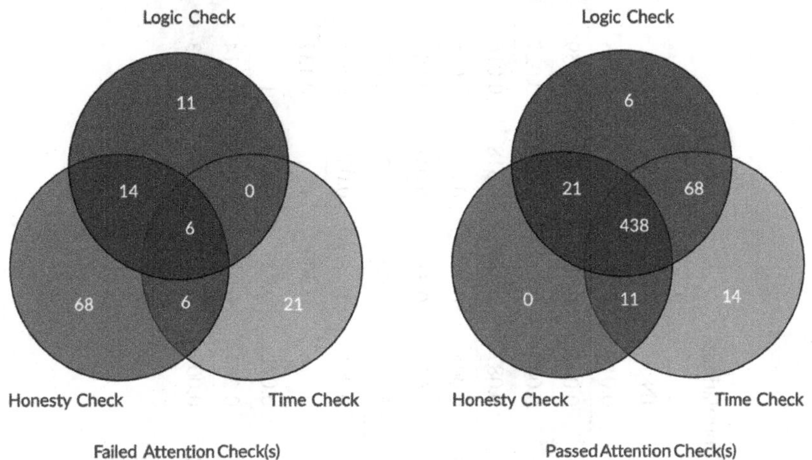

Fig. 2. Venn diagram depicting participant failure and passing of attention check by type.

Spearman rho correlations were also used to assess the relationship between the three different forms of attention checks and the passing of all three checks. A positive correlation was found between participants passing the logic check and passing the honesty check ($r_s = 0.310$, $p = 0.000$), failing the time check and failing the honesty check ($r_s = 0.132$, $p = 0.002$), and failing the time check and failing the logic check ($r_s = 0.139$, $p = 0.001$). That is, participants who pass one of the three attention checks are more likely to pass the other two attention checks. Table 4 shows the results of the correlations.

Finally, a logistic regression analysis was used to investigate whether age, income, gender, prior experience, and time on task influenced participant attention. All four assumptions required for logistic regression were satisfied [23]. Only prior experience with the technology in the logistic regression analysis contributed to the model (B = 0.484, SE = 0.136, Waid = 12.706, p = 0.000). The estimated odds ratio favored an increase of 62.3% [Exp(B) = 1.623, 95% CI (1.244, 2.118)] for participant attention for every unit increase in experience. None of the other variables were found to be statistically significant (Table 5).

Table 4. Nonparametric correlations.

Factor	AG	AT	EX	GE	GR	HC	IN	LC	MS	RA	SC	TC
Age (AG)	1	0.034	0.029	0.133**	0.030	0.036	0.017	0.035	0.266**	0.098*	0.032	0.111**
Attention (AT)	0.034	1	0.220**	0.013	0.044	0.935**	0.045	0.504**	0.023	0.059	0.016	0.114**
Experience (EX)	0.029	0.220**	1	0.000	0.044	0.213**	0.049	0.192**	0.020	0.027	0.001	0.032
Gender (GE)	0.133**	0.013	0.000	1	0.033	0.021	0.087*	0.030	0.318**	0.016	0.017	0.013
Group (GR)	0.030	0.044	0.044	0.033	1	0.058	0.005	0.002	0.048	0.075	0.030	0.033
Honesty check (HC)	0.036	0.935**	0.213**	0.021	0.058	1	0.050	0.310**	0.048	0.083*	0.007	0.132**
Income (IN)	0.017	0.045	0.049	0.087*	0.005	0.050	1	0.001	0.376**	0.072	0.294**	0.009
Logic Check (LC)	0.035	0.504**	0.192**	0.030	0.002	0.310**	0.001	1	0.009	0.031	0.022	0.139**
Marital Status (MS)	0.266**	0.023	0.020	0.318**	0.048	0.048	0.376**	0.009	1	0.137**	0.033	0.004
Race (RA)	0.098*	0.059	0.027	0.016	0.075	0.083*	0.072	0.031	0.137**	1	0.214**	0.013
Schooling SC	0.032	0.016	0.001	0.017	0.030	0.007	0.294**	0.022	0.033	0.214**	1	0.004
Time Check (TC)	0.111**	0.114**	0.032	0.013	0.033	0.132**	0.009	0.139**	0.004	0.013	0.004	1

* Correlation is significant at the .05 level (2-tailed).
** Correlation is significant at the .01 level (2-tailed).

Table 5. Logistic regression results.

Factor	B	S.E	Waid	df	Sig.	Exp(B)	95% C.I. for EXP(B)	
							Lower	Upper
Age	0.129	0.195	0.438	1	0.508	1.138	0.776	1.669
Experience	0.484	0.136	12.706	1	0.000	1.623	1.244	2.118
Gender (1)	0.047	0.237	0.040	1	0.842	1.048	0.659	1.667
Income	0.088	0.046	3.719	1	0.054	1.092	0.999	1.195
Marital Status			1.414	5	0.923			
Marital Status (1)	0.507	1.318	0.148	1	0.700	1.661	0.126	21.969
Marital Status (2)	0.785	1.380	0.324	1	0.569	2.192	0.147	32.765
Marital Status (3)	0.288	1.326	0.047	1	0.828	1.333	0.099	17.947
Marital Status (4)	20.157	15112.925	0.000	1	0.999	567589914.606	0.000	
Marital Status (5)	20.017	22927.392	0.000	1	0.999	493655138.500	0.000	
Race			1.599	3	0.660			
Race (1)	0.351	0.512	0.470	1	0.493	1.421	0.521	3.876
Race (2)	0.096	0.708	0.018	1	0.892	1.101	0.275	4.411
Race (3)	0.130	0.648	0.040	1	0.841	0.878	0.246	3.128
Schooling	0.044	0.090	0.235	1	0.628	0.957	0.803	1.142
Constant	3.250	1.717	3.582	1	0.058	0.039		

5 Discussion

Litman et al. [25] describe MTurk as "a constantly evolving marketplace where multiple factors can contribute to data quality". In this work, the attentiveness exhibited by an elite segment of the MTurk worker community was investigated. Specifically, workers holding the coveted 'Master' qualification with an 'Approval Rate' of 98% or more (i.e., completed at least 98% of the tasks they had committed to completing) and had a 'Number of HITS approved' value of 1,000 or more (i.e., the number of these activities exceeded 999). It was conjectured that these characteristics would ensure that this group of workers would be free of behavior reflecting inattentiveness and that this higher level of attentiveness would justify the additional cost attached to using workers holding the 'Master' qualification.

To confirm this hypothesis, an experimental approach was adopted in which participants were asked to complete a simple task involving reading about a hypothetical product and then answering questions on their perceptions of the product. Participant attentiveness was ascertained by using a series of questions originally proposed by Abbey and Meloy [1] and evaluating the amount of time spent on the survey. Surprisingly, the results revealed that over a fifth (22.34%) of the participants were not paying attention, having failed one of the three categories of attention checks. This result could be explained by the work of Chandler et al. [9], who examined the attentiveness of workers with an 'Approval Rate' exceeding 95% and discovered that participants were not always entirely devoted to the current task and were multitasking. In particular, 27% of participants in their sample disclosed that they were with other people while completing the study, 18% were watching television, 14% were listening to music, and 6% were chatting online [9]. This would explain the lack of attention being paid.

We contrast our findings with the work of Peer, Vosgerau, and Acquisti [32], who investigated how attention differs through two experiments. The first experiment compared workers with 'low reputation' (i.e., an 'Approval Rate' below 90%) and 'high reputation' (i.e., an 'Approval Rate' exceeding 95%). The second experiment compared what the authors describe as workers with 'low productivity' (i.e., their 'Number of HITS approved' was less than 100) and 'high productivity' (i.e., their 'Number of HITS approved' was more than 500). At least one attention check question was failed by 33.9% of the 'low reputation' workers, 2.6% of the 'high reputation' workers, 29% of the 'low productivity' workers, and 16.7% of the 'high productivity' workers. Given that we took an even more selective approach than Peer, Vosgerau, and Acquisti [32], our findings are concerning. Our failure rate of 22.34% is closer to what they classify as 'low reputation' (33.9%), and between the 'low productivity' (29%) and 'high productivity' (16.7%) workers. Three possibilities could explain this difference. First, the attention checks we used did not work as expected; second, the high level of compensation was miscommunicated; third, a seasonal influence on attentiveness exists. We explore each of these possibilities in greater detail in the following subsections.

5.1 Suitability of the Attention Checks

Abbey and Meloy [1] describe the process of forming attention check questions as a delicate task [1], the concern residing with the nature of these constructs. They give as an example the logic check which comprises questions that "require comprehension of logical relationships", such as "preferring to eat fruit over paper" [1]. The danger with such questions is that "the more subtle the statement, the less objective these checks can become" [1]. Thus, both the participants' and the researcher's responses to a question can be tainted by their interpretations. In the honor check, participants are asked to "reveal their own perceptions of their effort and data validity for the study" [1]. Effectively, the honesty check asks "a respondent to [self] identify the level of effort and attention to detail they [perceive that they] gave to the study and if that effort warrants use of their data" [1]. The weakness of this form of attention check is that respondents may have been paying adequate attention but were overly critical of themselves when submitting their responses. Consequently, they did not respond by selecting the 'strongly agree' option. An alternative approach might be to use objective tasks to gauge participant attention, as demonstrated by Aker et al. [3]. However, even these questions have issues. For example, if presented with a mathematical problem, the participant must have the skill to solve the question.

Although our study is not the first to use attention checks with survey research to identify careless (i.e., inattentive) participants [22], the use of attention checks has been questioned by some researchers, as it is believed that this can negatively interact with the survey response quality [22, 44]. Some researchers argue that attention checks should not be used at all [22, 44]. Abbey and Meloy [1] warn that the process to exclude respondents who are not paying attention "can become subjective" in cases where the study is not "largely a replication of known results with expected times, scales, or constructs". Our attention checks may have been too sensitive. If the criteria for rejection were to fail more than one attention check, the inattentiveness rate would drop to 4.61% (26/564). This rate is closer to what has been found in other studies. More specifically, it is similar to the rate of 4.17% reported by Paolacci, Chandler, and Ipeirotis [31] and closer to the finding of 2.6% by Peer, Vosgerau, and Acquisti [32] (for what they classified as 'high reputation' workers with an 'Approval Rate' exceeding 95%).

5.2 The Effect of Compensation

Another aspect to consider is compensation and its effect on participant attention. In our study, despite offering an extremely high hourly wage to our participants (above $22/hour), we found substantial evidence of inattentiveness, as the high wage did not eliminate the problem of lack of attention exhibited by participants. The magnitude of the compensation we offered can be better understood when it is compared to the median wage for MTurk workers in the United States, which in 2019 was said to be $3.01/hour [16], and the current federal minimum wage of $7.25/hour [43]. Correspondingly, our participants should have been well enticed, and no evidence of inattentiveness should have been discovered. Thus, a high wage does not eliminate the possibility of having inattentive participants whose work must be discarded. An explanation for this finding might be that participants do not consider the hourly wage but rather the total compensation offered. For example, one may prefer a reward of $0.50/hour if the total compensation of a task were $10 rather than a reward of $20/hour when the total compensation from the task was $1. A tradeoff appears to exist where, as per Aker et al. [3], increasing compensation leads to improved data quality. However, our research suggests that the ability of money to improve attention is limited after a certain point. Additional research is needed to create a better understanding of the marginal effects of wages on participants' attention and identify an optimal point that maximizes attention vis-à-vis compensation.

5.3 The Explanation of Seasonality

An alternative explanation of the varying range of inattentiveness exhibited by participants in the studies mentioned above may be found in the work of Chmielewski and Kucker [11], who replicated an experiment four times: the first between December 2015 and January 2016; the second between March 2017 and May 2017; the third between July 2018 and September 2018; and the fourth in April 2019. In their work, the percentage of participants who failed at least one attention check (which they called a "validity indicator") slowly increased from 10.4% to 13.8%, then jumped to 62%, and finally dropped to 38.2%. Given that we collected our data eight months after the conclusion of their final data collection, our inattentiveness rate of 22.34% is not only similar but might indicate a downward trend and possibly a cyclical pattern.

5.4 The Irrelevance of User Characteristics

We also attempted to ascertain whether a pattern exists that could help to predict which participants would fail our attention checks. The characteristics of age, gender, income, marital status, race, and schooling were examined, but no relationship was found concerning participant attention. It appears that the lack of attention does not reside with any specific demographic group. Instead, everyone has an equal chance of being inattentive. This outcome is slightly puzzling, as specific demographics have already been linked with participant attention, such as age [26] and culture [27]. The analysis did identify that participants' prior experience of using the technology applied in the study influenced their attention, with those having the most prior experience with the technology exhibiting the greatest attention.

5.5 Implications

This work has several noteworthy implications. The first concerns the discovery that participant inattentiveness persists within the population we investigated. This group consisted of MTurk workers with the 'Master' qualification, an 'Approval Rate' of 98% or more, and a 'Number of HITS approved' value of 1,000 or more. Coupled with the high compensation to ensure participants were highly motivated, it is evident that no 'silver bullet' exists that can reliably eliminate the manifestation of participant inattentiveness. Thus, there appears to be no justification in undertaking the additional expense associated with recruiting only participants with the 'Master' qualification. If inattentiveness can be observed under these 'optimal' conditions, this concern cannot be discounted. The fact that there is no one characteristic (i.e., age, education, gender, income, or marital status) that can be used to explain the phenomenon offers minimal hope of an informed intervention. Instead, researchers must vigilantly review participants for inattentiveness and not presume that certain criteria will ensure participants pay attention. Ultimately, the finding highlights the importance of using attention checks to identify inattentive participants and implementing a process to address these occurrences. Specifically, with an inattentiveness rate as high as 22.34%, such a practice would demand "researcher time, funds, and other resources" [11].

A tactic to mitigate the additional cost might be to refuse to compensate any participant who fails to satisfy one or a combination of attention checks. However, this involves challenges. Participants who are refused compensation may object and thus require additional (potentially costly) resources to be invested by the researcher to address those concerns. Participants who have earnestly participated as best they can but failed to produce results that pass the attention check(s) would be unfairly denied compensation. An alternative strategy to withholding payment might be to offer a low rate for participation in studies but offer a bonus for submissions matching a particular pattern. The problem with this approach is that participants may not focus on the research but on producing the illusion that they paid attention. Moreover, this may introduce biases in the responses, as participants may not respond honestly and authentically but rather as they believe the researchers want them to respond.

No simple solution exists. Consequently, to address participant inattentiveness, researchers should consider adjusting their proposals to account for the effort and costs required to identify participants who do not pay attention, address problems arising when addressing their poor performance, and recruit additional participants to replace submissions that must be disregarded.

References

1. Abbey, J., Meloy, M.: Attention by design: using attention checks to detect inattentive respondents and improve data quality. J. Oper. Manag. **53–56**, 63–70 (2017). https://doi.org/10.1016/j.jom.2017.06.001
2. Aguinis, H., et al.: MTurk research: review and recommendations. J. Manag. **47**(4), 823–837 (2021). https://doi.org/10.1177/0149206320969787
3. Aker, A., et al.: Assessing crowdsourcing quality through objective tasks. In: Proceedings of the Eighth International Conference on Language Resources and Evaluation (LREC 2012), Istanbul, Turkey, pp. 1456–1461. European Language Resources Association (ELRA) (2012)
4. Andersen, D., Lau, R.: Pay rates and subject performance in social science experiments using crowdsourced online samples. J. Exp. Polit. Sci. **5**(3), 217–229 (2018). https://doi.org/10.1017/XPS.2018.7
5. Barends, A.J., de Vries, R.E.: Noncompliant responding: comparing exclusion criteria in MTurk personality research to improve data quality. Pers. Individ. Differ. **143**, 84–89 (2019). https://doi.org/10.1016/j.paid.2019.02.015
6. Berry, D.T.R., et al.: MMPI-2 random responding indices: validation using a self-report methodology. Psychol. Assess. **4**(3), 340–345 (1992). https://doi.org/10.1037/1040-3590.4.3.340
7. Buhrmester, M., et al.: Amazon's Mechanical Turk: a new source of inexpensive, yet high-quality, data? Perspect. Psychol. Sci. **6**(1), 3–5 (2011). https://doi.org/10.1177/1745691610393980
8. Casler, K., et al.: Separate but equal? A comparison of participants and data gathered via Amazon's MTurk, social media, and face-to-face behavioral testing. Comput. Hum. Behav. **29**(6), 2156–2160 (2013). https://doi.org/10.1016/j.chb.2013.05.009
9. Chandler, J., Mueller, P., Paolacci, G.: Nonnaïveté among Amazon Mechanical Turk workers: consequences and solutions for behavioral researchers. Behav. Res. Methods **46**(1), 112–130 (2013). https://doi.org/10.3758/s13428-013-0365-7
10. Chen, J.J., et al.: Opportunities for crowdsourcing research on Amazon Mechanical Turk. Presented at the CHI 2011 Workshop on Crowdsourcing and Human Computation. https://www.humancomputation.com/crowdcamp/chi2011/papers/chen-jenny.pdf. Accessed 9 June 2021
11. Chmielewski, M., Kucker, S.C.: An MTurk crisis? Shifts in data quality and the impact on study results. Soc. Psychol. Pers. Sci. **11**(4), 464–473 (2019). https://doi.org/10.1177/1948550619875149
12. Crump, M.J.C., et al.: Evaluating Amazon's Mechanical Turk as a tool for experimental behavioral research. PLoS ONE **8**(3), 1–18 (2013). https://doi.org/10.1371/journal.pone.0057410
13. Difallah, D., et al.: Demographics and dynamics of Mechanical Turk workers. In: Proceedings of the Eleventh ACM International Conference on Web Search and Data Mining, New York, NY, USA, pp. 135–143. Association for Computing Machinery (2018). https://doi.org/10.1145/3159652.3159661
14. Fort, K., et al.: Amazon Mechanical Turk: gold mine or coal mine? Comput. Linguist. **37**(2), 413–420 (2011). https://doi.org/10.1162/COLI_a_00057

15. Goodman, J.K., et al.: Data collection in a flat world: the strengths and weaknesses of Mechanical Turk samples. J. Behav. Decis. Mak. **26**(3), 213–224 (2013). https://doi.org/10.1002/bdm.1753
16. Hara, K., et al.: Worker demographics and earnings on Amazon Mechanical Turk: an exploratory analysis. In: Extended Abstracts of the 2019 CHI Conference on Human Factors in Computing Systems, New York, NY, USA, pp. 1–6. ACM Inc. (2019). https://doi.org/10.1145/3290607.3312970
17. Hauser, D.J., Schwarz, N.: Attentive Turkers: MTurk participants perform better on online attention checks than do subject pool participants. Behav. Res. Methods **48**(1), 400–407 (2015). https://doi.org/10.3758/s13428-015-0578-z
18. Holland, A.: How estimated reading times increase engagement with content. https://marketingland.com/estimated-reading-times-increase-engagement-79830. Accessed 9 June 2021
19. Horton, J.J., Chilton, L.B.: The labor economics of paid crowdsourcing. In: Proceedings of the 11th ACM Conference on Electronic Commerce, Cambridge, Massachusetts, USA, pp. 209–218. ACM Inc. (2010). https://doi.org/10.1145/1807342.1807376
20. Hydock, C.: Assessing and overcoming participant dishonesty in online data collection. Behav. Res. Methods **50**(4), 1563–1567 (2017). https://doi.org/10.3758/s13428-017-0984-5
21. Johnson, J.A.: Ascertaining the validity of individual protocols from Web-based personality inventories. J. Res. Pers. **39**(1), 103–129 (2005). https://doi.org/10.1016/j.jrp.2004.09.009
22. Kung, F.Y.H., et al.: Are attention check questions a threat to scale validity? Applied Psychology: An International Review. **67**(2), 264–283 (2018). https://doi.org/10.1111/apps.12108
23. Laerd Statistics: Binomial Logistic Regression using SPSS Statistics, https://statistics.laerd.com/spss-tutorials/binomial-logistic-regression-using-spss-statistics.php#procedure, last accessed 2020/11/29.
24. Levay, K.E., et al.: The demographic and political composition of Mechanical Turk samples. SAGE Open **6**, 1 (2016). https://doi.org/10.1177/2158244016636433
25. Litman, L., Robinson, J., Rosenzweig, C.: The relationship between motivation, monetary compensation, and data quality among US- and India-based workers on Mechanical Turk. Behav. Res. Methods **47**(2), 519–528 (2014). https://doi.org/10.3758/s13428-014-0483-x
26. Lufi, D., Haimov, I.: Effects of age on attention level: Changes in performance between the ages of 12 and 90. Aging Neuropsychol. Cogn. **26**(6), 904–919 (2019). https://doi.org/10.1080/13825585.2018.1546820
27. Masuda, T.: Culture and attention: recent empirical findings and new directions in cultural psychology. Soc. Pers. Psychol. Compass **11**(12), e12363 (2017). https://doi.org/10.1111/spc3.12363
28. Meade, A.W., Craig, S.B.: Identifying careless responses in survey data. Psychol. Methods **17**(3), 437–455 (2012). https://doi.org/10.1037/a0028085
29. Okumus, B., et al.: Psychological factors influencing customers' acceptance of smartphone diet apps when ordering food at restaurants. Int. J. Hosp. Manag. **72**, 67–77 (2018). https://doi.org/10.1016/j.ijhm.2018.01.001
30. Palan, S., Schitter, C.: Prolific.ac—a subject pool for online experiments. J. Behav. Exp. Financ. **17**, 22–27 (2018). https://doi.org/10.1016/j.jbef.2017.12.004
31. Paolacci, G., et al.: Running experiments on Amazon Mechanical Turk. Judgm. Decis. Mak. **5**(5), 411–419 (2010)
32. Peer, E., Vosgerau, J., Acquisti, A.: Reputation as a sufficient condition for data quality on Amazon Mechanical Turk. Behav. Res. Methods **46**(4), 1023–1031 (2013). https://doi.org/10.3758/s13428-013-0434-y
33. Pittman, M., Sheehan, K.: Amazon's Mechanical Turk a digital sweatshop? Transparency and accountability in crowdsourced online research. J. Media Ethics **31**(4), 260–262 (2016). https://doi.org/10.1080/23736992.2016.1228811

34. Rand, D.G.: The promise of Mechanical Turk: how online labor markets can help theorists run behavioral experiments. J. Theor. Biol. **299**, 172–179 (2012). https://doi.org/10.1016/j.jtbi.2011.03.004
35. Rondan-Cataluña, F.J., et al.: A comparison of the different versions of popular technology acceptance models: a non-linear perspective. Kybernetes **44**(5), 788–805 (2015). https://doi.org/10.1108/K-09-2014-0184
36. Ross, J., et al.: Who are the crowdworkers? Shifting demographics in Mechanical Turk. In: CHI 2010 Extended Abstracts on Human Factors in Computing Systems, New York, NY, USA, pp. 2863–2872. Association for Computing Machinery (2010). https://doi.org/10.1145/1753846.1753873
37. Rouse, S.V.: A reliability analysis of Mechanical Turk data. Comput. Hum. Behav. **43**, 304–307 (2015). https://doi.org/10.1016/j.chb.2014.11.004
38. Salinas-Segura, A., Thiesse, F.: Extending UTAUT2 to explore pervasive information systems. In: Proceedings of the 23rd European Conference on Information Systems, Münster, DE, pp. 1–17. Association for Information Systems (2015). https://doi.org/10.18151/7217456
39. Schmidt, G.B., Jettinghoff, W.M.: Using Amazon Mechanical Turk and other compensated crowdsourcing sites. Bus. Horiz. **59**(4), 391–400 (2016). https://doi.org/10.1016/j.bushor.2016.02.004
40. Stewart, N., et al.: The average laboratory samples a population of 7,300 Amazon Mechanical Turk workers. Judgm. Decis. Mak. **10**(5), 479–491 (2015)
41. Stieninger, M., et al.: Factors influencing the organizational adoption of cloud computing: a survey among cloud workers. Int. J. Inf. Syst. Proj. Manag. **6**(1), 5–23 (2018)
42. Stone, A.A., et al.: MTurk participants have substantially lower evaluative subjective well-being than other survey participants. Comput. Hum. Behav. **94**, 1–8 (2019). https://doi.org/10.1016/j.chb.2018.12.042
43. U.S. Department of Labor: Minimum Wage. https://www.dol.gov/general/topic/wages/minimumwage. Accessed 25 Nov 2020
44. Vannette, D.: Using attention checks in your surveys may harm data quality. https://www.qualtrics.com/blog/using-attention-checks-in-your-surveys-may-harm-data-quality/. Accessed 07 Jan 2021
45. Venkatesh, V., et al.: Consumer acceptance and use of information technology: extending the unified theory of acceptance and use of technology. MIS Q. **36**(1), 157–178 (2012). https://doi.org/10.2307/41410412
46. Venkatesh, V., Bala, H.: Technology acceptance model 3 and a research agenda on interventions. Decis. Sci. **39**(2), 273–315 (2008). https://doi.org/10.1111/j.1540-5915.2008.00192.x
47. Versta Research: How to Estimate the Length of a Survey. https://verstaresearch.com/newsletters/how-to-estimate-the-length-of-a-survey/. Accessed 10 Apr 2020
48. Yang, H.C., Wang, Y.: Social sharing of online videos: examining American consumers' video sharing attitudes, intent, and behavior. Psychol. Mark. **32**(9), 907–919 (2015). https://doi.org/10.1002/mar.20826
49. Yoo, W., et al.: Drone delivery: factors affecting the public's attitude and intention to adopt. Telematics Inform. **35**(6), 1687–1700 (2018). https://doi.org/10.1016/j.tele.2018.04.014
50. Zack, E.S., et al.: Can nonprobability samples be used for social science research? A cautionary tale. Surv. Res. Methods **13**, 215–227 (2019)
51. Zimmerman, J., et al.: Field trial of tiramisu: crowd-sourcing bus arrival times to spur co-design. In: Proceedings of the SIGCHI Conference on Human Factors in Computing Systems, New York, NY, USA, pp. 1677–1686. Association for Computing Machinery (2011). https://doi.org/10.1145/1978942.1979187

Research on Service Experience Design Framework Based on Semantics to Improve the Enterprise Service Capability

Kun Zhou[1], Xi Zhang[2], and Yuanlong Gui[1(✉)]

[1] Guangdong Industry Polytechnic, Guangzhou, China
1994105012@gdip.edu.cn
[2] Guangdong University of Technology, Guangzhou, China

Abstract. In order to improve the effectiveness of meaning transmission in service design, according to the characteristics of service design, the role of semantics in service design is discussed. Based on previous research on product semantics, a semantic-based service design framework is proposed which includes four stages: extracting service propositions, semantics and touch points mapping, transforming into service capabilities, and adjusting organizational structure. Based on this framework, the enterprise's service propositions are transformed into the service capabilities that each touch point should achieve, and help enterprises adjust the organizational structure to make it more efficient and agile. Actual cases have verified the feasibility and effectiveness of the framework, which provides an effective research framework in the field of service design should summarize the contents of the paper in short terms.

Keywords: Semantic · Service design · Service proposition · Service capability

1 Introduction

In consumption society, "meaning" has become an important consumption object. When we buy a product or be keen on a brand, we are more impressed by the meaning it conveys. When functional consumption is transformed into experiential consumption, "meaning" will become an important way for enterprises to seek innovation in the next few years. The meaning needs to be expressed through a certain vector in the process of transmission. In product design, users rely on product appearance to perceive meaning. However, with the expansion of design connotation and extension, in the digital age, products have expanded into intangible services.

As there are more and more kinds of information medias, more and more conflicts between market changes and organizational structures, and stakeholder networks become more and more complex, collaboration and interaction modes become diversified, semantics would easily become vague, different, or even dissolve in such a changing environment. We need to respond to this change and rethink how semantics are constructed and expressed in service design. This paper applies semantics to service design

C. Stephanidis et al. (Eds.): HCII 2021, LNCS 13094, pp. 165–175, 2021.
https://doi.org/10.1007/978-3-030-90238-4_13

through literature research and case study and proposes a service design framework that is verified of the effectiveness by an actual project case.

This paper applies semantics to service design through the methods of literature research and case study and proposes a service design framework to verify its effectiveness through an actual project case.

2 Product Semantics

Product semantics comes from semantics and is often applied in product design. Product semantics was originally defined as "the study of the symbol quality of man-made forms in the context of the use form, and the application of this knowledge to industrial design" [1]. Krippendorf [2] later argued that product semantics are used to address the "cognitive and social environment in which they are used" and described the concern for the quality of symbols in design. Product semantics is no longer a breakthrough in function, but expanded to the field of meaning [3]. This is a paradigm shift from "design for function" to "design for meaning". Meaning becomes a bridge between users, products and environment. In product design, even a highly complex product still has a functional "form". The "form" here contains two levels of meaning. The first layer means the external shape of an object. Every object is composed of some basic shapes, such as circles, squares or triangles. The second layer is the "spirit" contained in the shape of the object. "Form" refers to the combination of the "shape" and "spirit" of an object [4]. And users can form personalized explanations based on these forms. Through the definition and research of product semantics, we know that semantics can help users understand the important information in the product.

Product semantic research is to further shape the information conveyed by the product in the meaning. That is to construct the information conveyed by the product from the perspective of design (Product-objective features), information transmission (use, operation, function-description) and meaning (user understanding and perception-construction of emotionalization). This enables the products to have characteristic, descriptive and emotional expression, which let users complete the construction of meaning and achieve the "communication" with the products.

3 Service Design and Elements of Service Design

Service is defined as "an intangible economic activity. It is the interactive behavior of value transfer between service providers and service recipients (users) which based on meeting the needs of users, creating service value as the goal [5]. The term "service design" was proposed by Bill Hollins in the field of design management in 1991 [6]. Service design is a design activity that takes users as the main perspective, and realizes the system innovation of service provision, service process, and service touch points through value co-creation among stakeholders, thereby enhancing service experience, service quality and service value. Service design is systematic, intangible, shared and collaborative [7]. It delivers value through the interaction of service providers and recipients. In the past, users rely on the color, shape, and material of the product to understand the product's function, usage, and the meaning that the brand wants to convey. But in service design, users can rely on the interaction on the touch point to understand and accept the service.

The elements of service design constitute a complete service design framework which provides ways for service design. The service propositions is the in-depth insight and elaboration of the corporate vision and user needs. It includes long-term goals, product positioning, user emotions, user experience, and common value services. The touch point is the key objective unit, showing the interaction between stakeholders. The service function includes the service process, service content and resources to meet user needs and realize semantic transmission. Service propositions, service touch points, and service functions have great meaning to service design (see Fig. 1), because they define the key information and main content of the service, and ultimately determine the quality of the service. Therefore, in service design, if users can clearly understand and use products, and perceive the value clearly and accurately, it needs to start with the service value proposition, service touch points and service functions to plan the assignment and transformation of meaning to help users obtain a better experience.

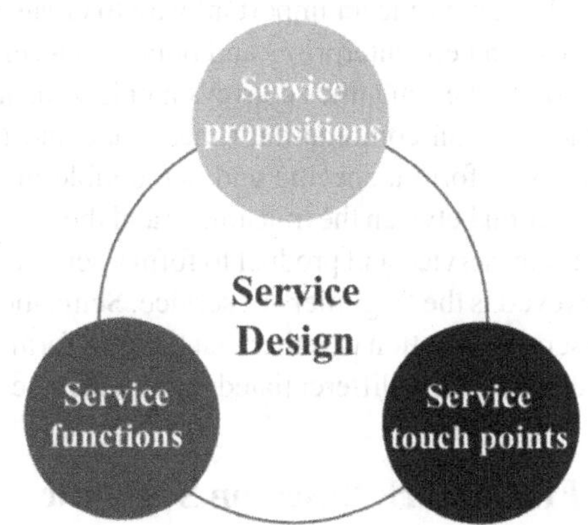

Fig. 1. Elements of service design (made by the author)

4 Semantics in Service Design

The construction of product semantics leads to the tangible elements in the product, so that the construction of meaning has concrete characteristics. Product semantics is a comprehensive system based on communication and semiotic linguistics, including product symbols, symbol meanings and users of symbols [8]. For intangible services, semantics can still help designers complete object construction, information transmission and meaning assignment.

First of all, the meaning assignment in product semantics and the service value proposition in service design are both to convey and create core emotions and opinions to users. Service proposition is the symbol of service, the "referred" in semantics. It is the symbolic value of psychology, sociality and culture conveyed in the context of use.

From a semantic point of view, creating the emotion, experience and value in a service is a kind of meaning giving and creation, which is to clearly allow users to perceive, acquire and understand the "meaning" provided by the service.

Second, service touch points involve service recipients and service providers, that is, users of symbols. In product semantic analysis, the product interface is the touch point between the enterprise and the user, and the user shape their understandings based on their own experience and knowledge. In service design, service touch points are the most basic unit to convey meaning. Users and services interact on touch points, accept and perceive meaning through virtual, physical or interpersonal touch points to form their own perception and experience. The touch points have become points of meaning transfer.

Finally, the "signifier" in product semantics has the same effect as the service function in service design. In product design, the language of product form embodies the signifier of product extension semantics [9]. At the service level, the service con-tent and interactive methods provided by the service become the "signifier" in the service. Service function is the carrier of meaning transfer, reflecting the user's motivation, needs and expectations, and at the same time an important way to create them.

In summary, semantics can help enterprises and brands to form a clear service value proposition. The "referred" in the semantics conveys a unique meaning that is different from other brands. Semantics can combine the time, space and interpersonal characteristics of service contacts to form a specific and perceptible meaning transfer point, which is the node of interaction between the transmitter and the receiver in the semantics. The combination of meaning, service and product to form a service function with unique meaning that can be conveyed is the "signifier" of service. Semantics links service value, service touch point and service function together that helps to form and to deliver unique meaning, and helps enterprises form differentiated system services.

5 Service Design Framework Based on Semantic

The semantic-based service design framework takes the formation and transmission of semantics as the basis in service design. The framework starts with the semantic "referred", and re-plans and designs the "signifier" through the "referred", which means that the service proposition must be determined first. The overall process is divided into four steps (see Fig. 2).

5.1 Service Propositions Extraction

First, service propositions are obtained through corpus analysis. The semantics conveyed by brands and products depend on carriers such as products, services, and advertisements. These carriers will show certain losses and deviations in the process of delivery, resulting in the semantic meaning received by the user is different from the semantic meaning originally intended to be conveyed. In order to obtain accurate semantics before designing, it is necessary to collect semantics from both inside and outside the enterprise. And after comprehensive analysis, the value proposition which is suitable for current brand development and user needs then can be deter-mined.

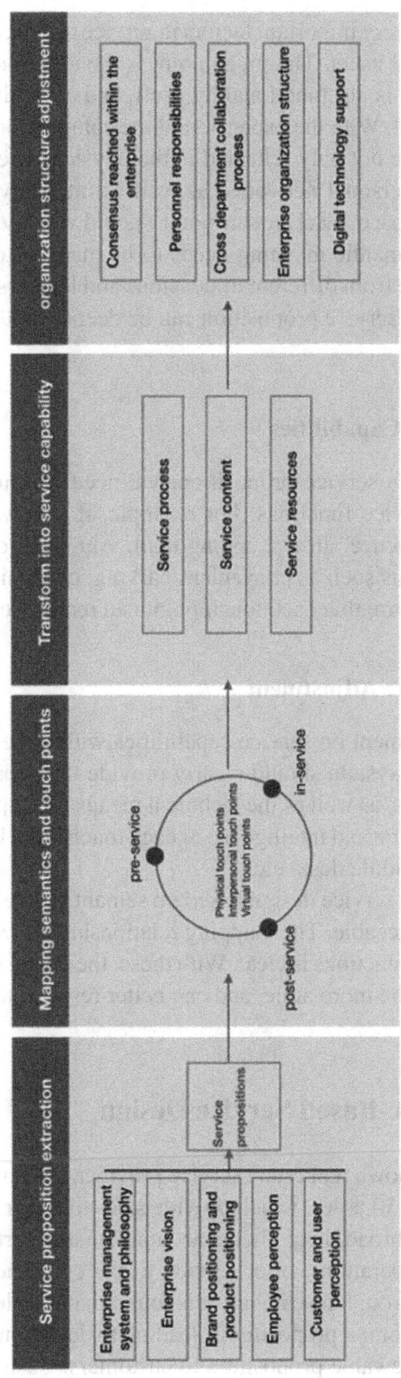

Fig. 2. Service design framework based on semantic (made by the author)

5.2 Mapping Semantics and Touch Points

The "touch point" is one of the very important factors in service design, and it is the points between products, services and users. The touch point is the entry point in the service research and design process. It is not functional by itself, and can only be effective if it is given its function and content. With the expansion of the connotation and extension of service design, touch points are not only reflected in the physical level, but also appear in virtual networks and interpersonal networks, as well as interactive touch points in the dimensions of time and space. Each touch point should convey a clear semantic meaning, and the extracted semantic meaning needs to be mapped to the touch point. Only by delivering brand value from different dimensions and levels, strengthening it in different spaces and times, the service proposition can be deeply rooted in the hearts of the people.

5.3 Transform into Service Capabilities

When we have determined the service proposition, we need to transform these key semantics into designable service functions. For example, if you want to realize the semantic meaning of "convenience" in a shopping mall, you need to transform "convenience" into service functions such as convenient parking, convenient payment, and convenient search, so as to ensure that each touch point can realize "convenience".

5.4 Organizational Structure Adjustment

The last step is to reach agreement on service capabilities within the organization and form a top-down system. The system should clearly provide the departments and personnel with various capabilities, as well as the technical means for implementation. The enterprise realizes the digitization and intelligence of each touch point by providing data, application data, front desk, middle desk, etc.

After completing the entire service design based on semantics, the path from goal to implementation is clear and traceable. The mapping relationship between various touch points, semantics, and service functions is clear. With these, the organization-al structure and cooperation methods will be more agile, and can better respond to various changes in the market.

6 The Case of Semantic-Based Service Design

The case comes from a well-known real estate service provider in China. Their products and services cover more than 150 cities, which provide sales tools for real estate sellers and sales staff domestic and abroad, help clients accumulate and sort out various user data to carry out secondary operations, offer service to the consumers. But they still face very serious problems. Through researches, it is found that people in the enterprise have a vague understanding of brand propositions, leading to deviations in publicity and sales, and unable to pass on the value proposition to customers. Customers also cannot clearly understand the core value of the product when using them, and the entire product information is overloaded, causing a serious user experience issue. The enterprises hope to re-organize the brand value to obtain a clear brand awareness. Therefore, it is necessary to improve value perception through global service touch points based on service propositions. So constructing and conveying semantics has become a new proposition.

In order to help enterprises define and convey the core value of the brand, we re-design the service based on semantics, so that customers can clearly understand the function and value of the product.

6.1 Service Proposition Extraction

Semantics need to be extracted from multiple stakeholders, both inside and outside the enterprise. In the enterprise, the vision of enterprise development is obtained by interpreting the corporate culture, corporate slogans, regulations and managers interviews; interviewing with middle-level and basic-level employees through their work experience and corporate culture perception. Outside the enterprise, the users', dealers' and other stakeholders' perception and experience with the products are collected through interviews which focus on the process of contact with the product.

These keywords need to be agreed internally and externally. In order to obtain consistency, the collected keywords were screened and clustered. Through two rounds of workshops covering all employees, 9 semantic keywords could convey the meaning of the brand, products and services were determined (see Fig. 3).

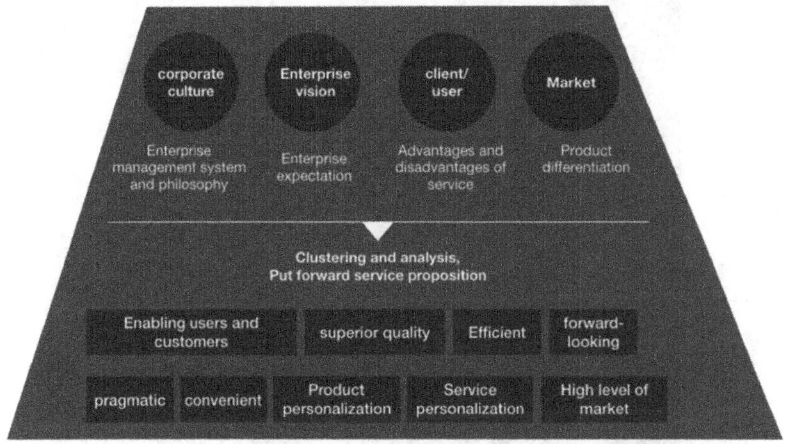

Fig. 3. Extract the service proposition (made by the author)

6.2 Mapping Semantics and Touch Points

After obtaining the service value proposition, the key semantics needs to be mapped to the various stages of the service life cycle, and the core semantics conveyed by each touch point should be clear. The service life cycle is divided into three stages:

	Pre-service		In-service						Post-service	
touch points	plan	extension	Demand docking	Requirement confirmation	Product development	Product delivery	Guidance for use	feedback	iteration	New demands
forward-looking	forward-looking idea; forward looking technology	New promotion methods; Flexible cooperation mode	forward-looking idea; forward-looking product;	Demand insight	The function of foresight Beautiful interface Leading technology				Analysis with global view	Demand insight
pragmatic	Identify the characteristics of the industry; Clear market demand; Clear competitive advantage;	Make full use of existing channels; Discover new channels; low cost;	Tap potential demand; Judge the feasibility of demand; Cost assessment;	Clear requirements	low cost	low cost	solve the problem	solve the problem	solve the problem Fast iteration	Document potential needs
Efficient			Timely service; High coordination; have good understanding; Predetermination boundary;	High coordination; Predetermination boundary	Short cycle; High coordination; Accurate function development;	Short cycle; High coordination; Accurate function development;	Timely service and training; High coordination;	Timely response; High coordination;	Timely response; High coordination; Short development cycle;	Timely service; High coordination degree; Strong understanding; Boundary prediction

Fig. 4. Mapping semantics and touch points (partial pictures, made by the author)

pre-service, in-service and post-service (see Fig. 4). Pre-service contains two touch points: planning and promotion. In-service contains six touch points: demand docking, demand confirmation, product development, product delivery, product use guidance and feedback. Post-service includes two main touch points: iteration and new demand docking. The service proposition is different in each stage. For example, in the planning stage, "forward-looking" needs to be transformed into "ideological forward-looking" and "technical forward-looking". In the iterative stage, "forward-looking" needs to be transformed into "global analysis insight."

6.3 Transform into Service Capability

After clarifying the semantic meaning of each touch point, it is necessary to configure the service content and resources in each touch point. Semantic keywords are transformed into service capabilities. In the process of transformation, each touch point adds a lot of new service capabilities based on its semantics. In the touch point "Product Usage Guidance", the service capability of "Operation Training" has been added. Different guidance programs were designed for different types of customers. For example, for corporation, they generally hand over products to sales and use them in different sales areas. Therefore, during the training, the staff of the group will generally be trained face-to-face, and then the group staff will train the sales staff in other districts. For other types of customers, the corresponding online training or written training methods will be selected according to their needs and business volume. In the "demand docking" touch point, it needs to know in advance the customer's operational needs, whether there is a special operation model or promotional activities. This requires adding activity modules to the product system to help customers customize the types of data that need to be tested, so that product can better meet the customers' needs.

After clarifying the new service capabilities of each touch point, a new service process corresponding to it has also been generated. Then the mapping from service value to service function is completed, so that the products and services can deliver value more accurately. Through products and services at different levels and dimensions, users can enjoy different experiences and form a three-dimensional and vivid perception.

6.4 Organization Structure Adjustment

After establishing the new service process, the new service process was discussed and improved through two co-creation workshops, and the new collaboration process within the enterprise was determined according to the service process. Based on the new collaboration process, the enterprise added two new departments, namely the customer management center and the strategic consulting group. The customer management center is set up to better realize the personalization of products and services. The customer management center mainly connects different types of users, divides customers and customer needs into different management types. And the entire designs including different kinds of designs are set as whole service to meet different types of customer needs, which greatly shortens the time for demand communication and confirmation. The strategic consulting group is composed of high-level managers, mainly for evaluating the value stage of the project in the early stage of the project and determining the positioning of the product. In addition, through the process of combing, the enterprise has also increased the positions of service designers. At the end of the project, the enterprise hired a service designer to undertake the iterative work of internal service processes.

After the entire project is completed, the satisfaction questionnaire is used to verify the proposed service process through corporate WeChat, email, etc. The service instruction manuals and tool manuals designed in this research help for enterprise employees familiarize with the new process and clarify personnel responsibilities through a top-down consistent method. In terms of organizational structure, by adding positions and departments to cooperate with the new product delivery process, this greatly improves product quality and efficiency and shortens the entire delivery cycle.

7 Discussion

This paper puts forward the research path to improve the service capability of enterprises through semantic analysis, which is not only the goal of this study, but also the innovation of this paper. In consumption society, "meaning" has become an important consumption object. The service capability promotion of enterprise traditionally does not pay attention to the value of "meaning" that could promote the enterprise service capability, and the enterprise mostly choose to improve the service capability in one single way, such as increasing a single service content, adjusting the organizational structure, upgrading the service information system and organizing staff raining, which is a lack of systematic and overall strategy. Therefore, this paper proposes a semantic based service design system framework to help enterprises establish value proposition expression system from the inside to outside.

Because this study only carried out a real case verification, to a certain extent, it cannot fully show its application value. In addition, there is a lack of clear and fixed methods in the process of semantic transformation. How to extract and transform semantics more accurately is still an aspect that needs continuous research and discussion in the future. Therefore, we need more practice in the further study.

Now service design are used in industries generally, which is include the real estate industry, retail, transportation, medical and other industries. Semantic based service design framework can be further applied to different industries, which will help the researcher to modify the framework and make it more practical value.

8 Conclusion

This paper extracts the service value proposition of the enterprise based on semantics, and uses it to plan the service capability and service process of the enterprise, and adjusts the organization structure of the enterprise to better respond to the new service capability. The semantic-based service design framework covers the entire link of the service, so that the service proposition is passed to each touch point. The framework solves the problem of inconsistent understanding of internal and external value propositions that existed before, and makes the service proposition perceived by customers or users more distinct and three-dimensional.

Analyzing service design from the perspective of semantics is an innovative re-search perspective of service design. When enterprises face fierce market competition, product functions or performance become homogeneous. Enterprises often need to reorganize their value propositions and give touch points more meaning. This will be the future for many enterprises to enhance user perception and an important way to enter new markets. In the future, this framework can be applied to more service products, helping enterprises to achieve business value through semantics.

References

1. Krippendorff, K., Butter, R.: Product semantics: exploring the symbolic qualities of form. Innov. J. IDSA **3**(2), 4–9 (1984)
2. Krippendorff, K.: On the essential contexts of artifacts or on the proposition that "design is making sense (of things)". Des. Issues **5**(2), 9–38 (1989)
3. Hou, R.J.: Research on the application of product semantics in the design of home furnishing products in my country. Qingdao Technological University, Qingdao (2015)
4. Jiang, X.P.: Research on the application of semantics in mechanical product design. Mach. Des. Manuf. **9**(9), 252–254 (2009)
5. Hu, F., Li, W.Q.: Define "service design". Packag. Eng. **40**(10), 37–51 (2019)
6. Hollins, G., Hollins, B.: Total Design: Managing the Design Process in the Service Sector, 2nd edn. Pitman, London (1991)
7. Gao, Y.: Research on service design innovation based on the dimension of experience value. China Academy of Art, Hangzhou (2017)
8. Lu, Z.L., Fritz, F.: Research on new energy vehicle modeling design based on product semantic analysis. Mach. Des. **34**(3), 111–116 (2017)
9. Du, H.: Research on form bionic design method based on product semantics. Packag. Eng. **15**(10), 60–63 (2015)

Design, User Experience and Human Behavior Studies

Expectation, Perception, and Accuracy in News Recommender Systems: Understanding the Relationships of User Evaluation Criteria Using Direct Feedback

Poornima Belavadi, Laura Burbach, Stefan Ahlers, Martina Ziefle, and André Calero Valdez(✉)

Human-Computer Interaction Center, RWTH Aachen University, Campus Boulevard 57, 52076 Aachen, Germany
{belavadi,burbach,ahlers,ziefle,calero-valdez}@comm.rwth-aachen.de

Abstract. Typically, a user-focused approach of evaluation of recommender systems requires the users to recollect their experiences, exposing study results to memory biases. In this paper, we describe a study conducted to test a framework, that allows recommender systems to be used and evaluated simultaneously. In this study, we asked 140 participants about their expected, perceived, and actual quality of the recommendations. We compare the performance of two recommender systems. The singular value decomposition recommendation system was able to correctly predict more than half of all evaluations and performed better than participants expected. However, users were more satisfied with the suggestions of the user-based collaborative filtering recommendation system. Our approach allows to compare actual item ratings, expected quality, and perceived quality of recommendations. Serendipity was found to be an important influencing factor for better item ratings by users. Participants rated both recommendation systems better when they perceived higher quality.

Keywords: Recommender systems · Live evaluation · User studies · Recommendation accuracy · News recommendation · Qualtrics

1 Introduction

Recommender systems use different algorithms to predict what a user likes. By utilizing user data, recommender systems draw conclusions about the preferences and interests of the user. With multinational companies using recommender systems (e.g., Amazon, Netflix etc.) to build revenue, research on improving recommender systems has been gaining interest. A good portion of which focuses on evaluation of recommender systems. The evaluation is done either by mathematically determining the accuracy of the algorithm or by conducting user

C. Stephanidis et al. (Eds.): HCII 2021, LNCS 13094, pp. 179–197, 2021.
https://doi.org/10.1007/978-3-030-90238-4_14

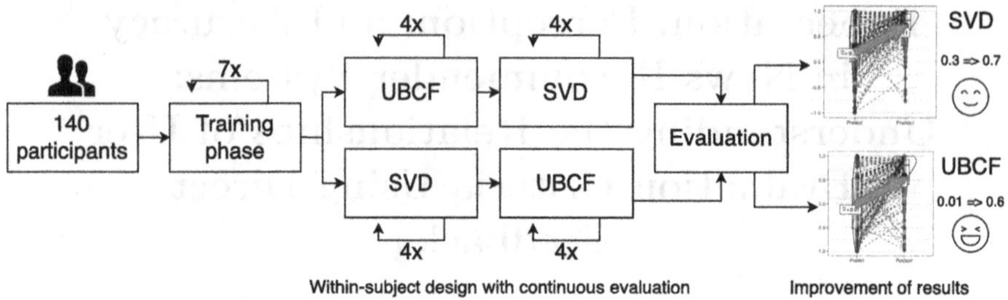

Fig. 1. User study showing the experimental design and results.

studies. Even in user studies, the recommender systems were mostly analyzed offline according to statistical criteria, such as run-time efficiency of the algorithms used or the accuracy of the recommendations. Only some studies focus on the attitudes of users of recommender systems and carried out online evaluations [18,34]. In these studies, participants were asked about their experiences with a recommender system retrospectively, but the studies did not check whether the reported experiences and opinions corresponded with actual recommendation quality.

In this paper, we describe a user study to test a framework that allows the users to use and evaluate two recommender systems (see Fig. 1). We asked the users to give feedback about the quality of the recommendations directly after exposure. The recommender system used this feedback and the data was used to re-train the algorithm instantly. We tested the framework on the evaluation of two *recommender systems for news articles*. The collected data enabled us to evaluate the recommender systems according to both statistical and user-centric criteria. We examine and compare the *quality of the recommended news items expected by the respondents*, the *perceived quality of the recommendations*, and the *actual item rating* while re-training the recommender system during use.

2 Related Work

Today, various recommender systems are being developed and used in many different application contexts such as *entertainment* [22,39], *online shopping* [7,41], *e-health* [9,12,47] and *social networks* [13,49], creating a need for better recommender systems. This has motivated research on the evaluation of recommender systems. Evaluating and developing recommender systems poses challenges. Following, we discuss some of the different types of recommender systems and challenges faced during their development.

2.1 Types of Recommender Systems

Most commonly, recommender systems are divided into three classes: *content-based recommender systems*, *collaborative filtering* and *hybrid recommender systems* [1,46]. Most recommender systems combine sub-types from different classes to compensate for the weaknesses of the individual systems and benefit from the respective strengths [45].

Content-Based Recommender Systems. Content-based recommender systems give recommendations based on the attributes of items that have been evaluated in the past. This involves comparing attributes of the evaluated items with the attributes of items that have not yet been evaluated. If the attributes of a new item correspond to the attributes of items that the user rated well, the new item is recommended [45].

Collaborative Filtering. The most widespread recommender technique being used is collaborative filtering [2,8]. Here, the ratings of users are sampled to create a user-item rating matrix. In this matrix, each column corresponds to an item that has been rated and each row corresponds to a user who assigned the rating. The value at the intersection of row and column is the rating the user assigned to the corresponding item [38]. The absence of a value represents that the user has not yet given a rating for this item [38]. The recommender system predicts these values, to identify items that could be of interest to a user by comparing their past ratings with the ratings of other users [40,48]. Collaborative filtering can be divided into two classes: *memory-based* and *model-based filtering* [8,23,42].

Memory-Based Collaborative Filtering. Memory-based collaborative filtering provides recommendations by generating either *user-based* or *item-based predictions* for items by searching for similarities in the user-item rating matrix [23]: *User-based collaborative filtering (UBCF)* compares the similarity between users. Two users are similar if the ratings of the items they both rated are similar. To generate recommendations for all the items that have not yet been rated by the user, the ratings of these items by similar users are compared. The predictions of the items correspond to the weighted average ratings of the other users. The greater the similarity between the users, the more weight is given to the rating [2,23]. *Item-based collaborative filtering* compares the similarity between items and not between users. The system calculates the similarity between all item pairs based on their ratings. To generate new recommendations for a user, the system compares the similarity between their rated items and the items that have not yet been rated. The items that have not yet been rated and show the greatest similarities are recommended to the user [23].

Model-Based Collaborative Filtering. Model-based collaborative filtering uses various techniques to calculate how a user might rate an item. This is done using machine learning, but also statistical methods such as *Bayes networks*, *cluster analysis*, *matrix factorization* [40] or *Singular Value Decomposition (SVD)*, which is a method from linear algebra that is commonly used as a method for reducing dimensionality [50]. Compared to memory-based collaborative filtering, model-based collaborative filtering scales better and takes less time to calculate the recommendation, but the development of the underlying model is more costly [2].

For our study, we use UBCF as the representative for memory-based and we use SVD as the representative for model-based techniques.

2.2 Challenges in Developing Recommender Systems

Developers of recommender system face a multitude of challenges: As recommender systems operate on large pools of data, an overlap between two users may be very small or non-existent. In addition, the distribution of ratings among the individual items can be extremely unequal. A recommender system must take this lack of data (*data sparsity*) into account [21,30]. The problem of data sparsity is especially relevant for the so-called *cold-start problem.* To classify users and provide them with appropriate recommendations, recommender systems need information on them such as their ratings. For new users who have not yet submitted ratings, this information is missing (*cold-start problem - new user)*, hence, the recommender system cannot make accurate recommendations to these users [20,43,45]. New items cause problems in a similar way (*cold-start problem - new item).* If recommender systems only consider the users' ratings and not the attributes of the items, they will never recommend items that have not yet been rated [20,43,45].

Another problem faced especially by content-based recommender systems is the problem of *overspecialization.* As they are solely based on the users' previous ratings, they only provide recommendations for new items that are similar to the ones the user has previously rated [46]. Some of these might be inappropriate to recommend (e.g., recommending a washing machine after the user has already purchased a different one).

A recommender system should also make unexpected recommendations to users that may prove to be appropriate (*serendipity*). To make frequent suitable recommendations, the system usually recommends items that are currently very popular with other users or are often rated highly. This is problematic because these items are often found by users without the help of a recommender system. Hence, a good list of recommendations should also contain less obvious items that the user would probably not find without the recommendations of the system. Balancing the accuracy and variety of recommendations is a central challenge for recommender systems [30,31,46].

Recommending News Articles. Recommending news articles poses further challenges [24,29]. Recommender systems must analyze and classify a large number of articles in a very short frame of time. This is further complicated by variations in the structure of different article types [29,51]. For no other item *topicality* plays such a decisive role. Articles that are interesting today can be uninteresting tomorrow. One requirement for recommender systems is therefore not to recommend articles that are no longer up-to-date [29,51].

To recommend suitable news articles, detailed *user profiles* must be created. The recommender system should automatically record the articles that the user has read, while preserving the privacy of the users [29].

There has been only sparse research on recommender systems for news articles (e.g., [6,11,17,26]). Beyond recommending news articles in real time [3,4, 25,32], we connect real-time recommendation with a simultaneous evaluation of the recommender system.

2.3 Offline- vs. Online Evaluation

The evaluation of recommender systems is carried out in either of two ways: online or offline. For offline evaluation, data is first collected or simulated and the recommender system is then tested in a system-centered manner. By contrast, in online evaluation, real test users evaluate a prototype or a productively used recommender system according to user-centered criteria [28].

Offline Evaluation. In offline evaluation, recommender systems are subject to the quality criteria *accuracy*, *robustness and stability*, *coverage*, and *diversity*. To measure the *accuracy*, the predictions of the system are compared with the actual assessments of the users and the number of predictions matching the assessments are recorded. For predictions that do not match, the extent to which the values correlate with each other and the extent to which they differ is examined. The more often the predictions meet (or at least correlate with) the actual user evaluations, the higher is the *accuracy* of the recommender system. To measure *robustness and stability*, the performance of the recommender system is compared before and after a deliberate manipulation of the evaluations of individual items ("shilling attack"). If the predictions about the evaluation after the attack do not deviate strongly, the recommender system is considered *robust*. *Coverage* is high if the recommender system can access all or a very large part of the item pool for the recommendations. By contrast, systems with low *coverage* are often faced with the cold-start problem for new items. Lastly, *diversity* denotes the variability between the recommended items, i.e., how strongly the recommendations differ from each other.

To perform an offline evaluation of recommender systems, a certain number of test user ratings are removed from an existing data set. The remaining data sets are used to train the recommender system. Based on this training data, the recommender system creates recommendations for the test users or predicts ratings for the items whose ratings have been removed. The recommendations generated this way are then compared with the actual recommendations and evaluated according to the criteria presented above [28]. Offline evaluations allow for the objective evaluation of recommender systems and their underlying algorithms according to statistical criteria. Compared to online evaluations, offline evaluations are simpler and more cost-effective and they require less resources. However, as offline evaluations do not measure user satisfaction with a recommender system, some studies also include online evaluations [10,14,28]. In most cases, the recommender systems that scored best in the offline evaluation are tested by users in an online evaluation in a second step. This enables to collect the users' opinions and still save resources [10,14,28].

Online Evaluation. Online evaluations aim at obtaining a sophisticated picture of the users' attitudes. Typical concepts that are often surveyed in online evaluations include *perceived accuracy*, *perceived diversity*, *novelty*, *serendipity*, *satisfaction*, *trust*, and *data privacy concerns*. *Perceived accuracy* refers to the degree to which the users feel that the recommendations match their interests. It measures the overall assessment of the perceived quality of the recommendations [33]. *Perceived diversity* denotes how much variety the users perceive in their recommendations. When users receive the same or very similar recommendations, they may be disappointed by the recommendations and their confidence in the recommender system might decrease [33]. Most users expect a recommender system to suggest items that match their interests and preferences. If users receive item recommendations that they consider new and unexpected, and this recommendation turns out to be relevant to them, they might be positively surprised. This is called *serendipity* [28]. Closely related to this is the factor *novelty* which is about whether users perceive the recommendations of a system as new [28]. User *satisfaction* is another important dimension in evaluating recommender systems. It refers to the users' thoughts and feelings during the use of the recommender system [10,33]. User *trust* in recommender systems is influenced by both the accuracy and the transparency of the recommender system. *Trust* determines whether users rely on the recommendations or not [27,44]. Lastly, *data privacy concerns* regarding the handling of user data by recommender systems influence the willingness to release data to a recommender system. The type of data required by the recommender system and the transparency of the system influence *data privacy concerns* [27,34].

2.4 Our Research Question

As we have seen, different metrics are used in either online or offline evaluation of recommender systems. A possible downside in online evaluation is that users are asked about their perceptions after having used the recommender system. As previously mentioned this may introduce memory biases. In our approach we ask participants about their evaluation of each item directly and utilize this feedback to retrain the recommender system online. Using this approach we ask the following research question.

RQ: How do users expectations and evaluations of recommender systems depend on the accuracy of recommendations, when the recommender is trained on live feedback?

3 Method

To ensure that the participants are able to test different recommender systems "live" during the online survey, we designed a framework to establish a connection between the recommender systems and the online survey. Through this connection the participants' evaluations are forwarded directly from the survey software to the recommender system and the items to be evaluated are dynamically provided by the recommender system. To offer suitable recommendations to the participants during the online survey, the recommender system creates a profile of each user. In this study, we tested the framework designed for evaluating recommender systems using an online survey. Following, we briefly describe the framework and the data preparation and finally describe the survey.

Framework. We designed the framework as a client-server architecture. We stored all the news articles to be recommended in a database which was connected to the server. The server hosted the recommender systems which were developed using the R package recommenderlab [19]. The client received the items from the server and showed it to the users via the Qualtrics survey that was connected to the client.

Data Preparation. As the data basis, we used the Million Post Corpus provided by Schabus, Skowron, and Trapp [37]. The database contains 12,087 articles with over one million comments published on the website of the Austrian daily newspaper *"Der Standard*[1]*"* from 1st of June 2015 to 31st of May 2016. To make the data usable for the recommender systems, we prepared the articles and the comments. We removed articles that were not news and therefore not suitable for the study (i.e., advertisements). We reduced the total number of articles from 12,087 to 10,309. For good readability, we limited the length of the articles to a maximum of two sections.

To ensure that no cold start problem occurs (see Sect. 2.2), artificial evaluations used to train the recommender systems were generated from the comments on the articles that contained user IDs and textual data. The comment text on the articles were converted into ratings by conducting a Sentiment Analysis using the *German sentiment vocabulary SentiWS V2.0* by Remus, Quasthoff, and Heyer [35]. If the sum of word-by-word sentiment was negative we rated the item as 1 otherwise as 5. We assigned this value as a *rating* for the article and saved it with the corresponding *Article ID* and *User ID* in a rating matrix, which we used to generate recommendations.

Online Survey. The survey consists of three parts: We first asked participants about demographic factors and attitudes. Secondly, the recommender systems were re-trained. In the third part, the users tested and evaluated two recommender systems (*UBCF* and *SVD*).

We measured all items on a six-point-Likert scale (1 - disagree very much, 2 - disagree, 3 - rather disagree, 4 - rather agree, 5 - agree, 6 - agree very much).

Demography and Attitudes. As *demographic data*, we measured *gender*, *age* and *education level* of the participants. Additionally, we measured the *computer self-efficacy (CSE)* using 8 items by Beier [5]. Moreover, we asked participants what

[1] https://www.derstandard.at/.

type of news they are interested in (*politics, sports, economics, culture, lifestyle, computer and technology, network politics* or *science*).

Expected Accuracy of Recommender Systems. We further asked the participants how accurate they expect the recommendations of certain recommender systems would be (*random, non-personalized, user-based collaborative, article-based collaborative, hybrid* and *content based*). In addition to the name of the recommender system, we described to the participants what type of recommendation the system returns and what data it requires.

Knowledge About Daily Events in Austria. Before the recommender systems are re-trained, we informed the participants that the articles shown were from the Austrian daily newspaper "*Der Standard*". We asked the participants on a six-point Likert scale (*not at all, not, somewhat not, somewhat, very* and *extremely*), how familiar they are with the daily events in Austria, as this might be a confounding variable in recommendation accuracy.

Re-training of Recommender Systems. Re-training the recommender systems is necessary to overcome the cold start problems of the UBCF and the SVD system for the participants. To re-train the recommender systems we asked participants to evaluate seven items. To select these items we need to identify informative items for evaluation. To achieve this, two other recommender systems (*RANDOM and POPULAR*) that do not have the cold start issue were used to select *five out of the seven articles.* We selected three articles *randomly* and the two *most popular.* Two further articles were selected from a set of pre-selected articles. These were chosen by three researchers by manual coding all articles based on the topics that participants indicated as favorite news topics in the survey.

Rating of Training Articles. After the participants rated these initial articles, we asked for the participant's *perception of quality, diversity, novelty, serendipity* and *relevance* of the recommendations, using seven statements (see Table 1). We

Table 1. Scale used for the evaluation of articles

Scale item (*perceived article quality*)	Construct
Recommended items were well chosen [a]	Quality
Recommendations differ significantly from each other [b]	Diversity
Recommendations provided new information [b]	Novelty
Recommendations surprised me [b]	Serendipity
I liked the items recommended [a]	Quality
Recommendations were relevant [b]	Relevance
Items recommended matched my interest [c]	Quality

[a]Source: Knijnenburg et al. [27], [b]Source: Fazeli et al. [16], [c]Source: Pu, Chen, and Hu [34]

used a subset of validated scales, as these questions would have to be evaluated often by the users. The new scale *perceived article quality* still shows a good internal reliability of Cronbach's $\alpha = .87$. However, we also intend to analyze on individual item levels.

Evaluation of Two Recommender Systems. Lastly, we asked the participants to also evaluate the overall performance of the *UBCF* and *SVD* recommender system. The two systems were tested one after the other. To reduce a possible sequence effect, the order in which the systems were tested was randomized. Every participant evaluated four items. Then, we asked the participants to rate the performance of the recommender systems using the scale as before (see Table 1). The scale showed a good internal reliability of Cronbach's $\alpha = .92$ for both recommender systems. We used five further statements by Knijnenburg et al. [27] (see Table 2) to ask the participants how they *assess the recommender systems.* The scale showed a good internal reliability for both systems (Cronbach's $\alpha = .88$ UBCF, $\alpha = .89$ SVD).

Table 2. List of statements used for evaluation of recommender systems

Scale item (*assessment recommender system*)
The system is useless
I would recommend the system to others
I liked the items recommended by the system
The system recommended too many bad items
I can find better items without the help of the system

Collection of Data. Participants were acquired between October and November 2019 using snowball-sampling by sending the survey via WhatsApp, Signal, Slack, and posting it on Facebook groups. We note that this yields a high social media usage bias, which we integrate when analyzing our findings.

Statistical Methods. We checked the internal reliability of the scales using the *R-package psych* [36] by calculating the Cronbach's α. We used parametric tests to check whether the results are significant. In addition, an α error of 5% ($\alpha = .05$) and a β error of 20% ($\beta = .2$) is permitted. With a sample size of $N = 140$, this means that correlations could be detected with an effect strength of $|\varrho| \geq .21$ [15].

4 Results

All procedures and statistical evaluations are available in our supplementary material in an OSF repository[2] We used R Version 3.4.1. and RMarkdown to analyze the data. After a presentation of our sample, we report our findings.

[2] https://osf.io/qn4as/.

4.1 Description of the Sample

The online survey was completed by $N = 140$ German Internet users. 64% of the users are female and the average age is 41 ($SD = 16.21$). The age distribution of the sample is bimodal. Users between 29 and 47 years are underrepresented. 41% of the users have a university degree. The users' *knowledge about the daily events in Austria* is limited ($M = 1.60$; $SD = .84$).

4.2 User Ratings of the Recommendations

Expected Quality of the Recommender Systems. Participants believe that recommender systems that *randomly* select an article recommend a suitable article with an mean accuracy of 23% ($SD = 19.24$; see Fig. 2, mean and standard error shown in red). *Popular recommender systems* are presumed to give good recommendations with an accuracy of 36% ($SD = 23.50$). An accuracy of slightly more than 50% is expected from *collaborative recommender systems.* Participants rated the *IBCF recommender system* slightly better with an accuracy of 52% ($SD = 20{,}854$) than *UBCF* with 51% ($SD = 19.34$). They rated *hybrid recommender systems* best. These should deliver good recommendations with an accuracy of 60% ($SD = 22.46$).

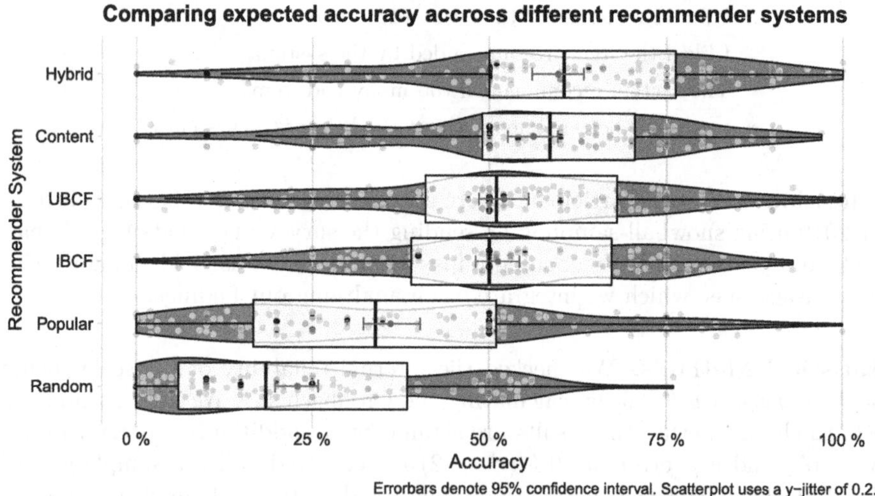

Fig. 2. How do users expect different Recommender Systems to perform?

Evaluation of the Pre-selected Articles. During the *re-training phase* (see Fig. 3), the users rated the recommendations slightly negative ($M = 3.15$; $SD = .89$). After the *re-training phase*, the users described the recommended articles as *very diverse* ($M = 4.62$; $SD = .78$). The recommended articles offered users *new information* (M = 3.79; SD = 1.08), but they were not always *relevant* to them (M = 3.45; SD = 1.23). Furthermore, the users were *surprised* by the recommendations (M = 3.69; SD = 1.26). However, when users were asked again about their ratings in retrospect, users rated the *perceived quality* of the recommended articles slightly negative (M = 3.32; SD = .95).

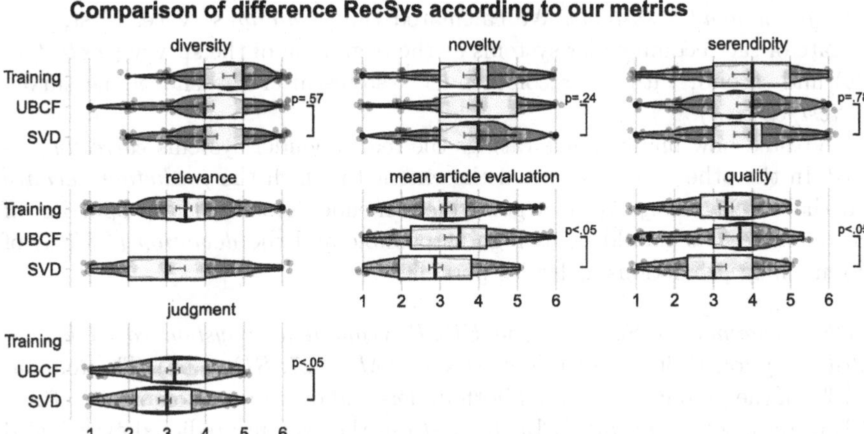

Fig. 3. Results of evaluation: comparison of different recommender systems according to our metrics

Evaluating the UBCF Recommender System. Participants rated the recommendations given by the *user-based collaborative filtering recommender system (UBCF)* slightly negative (M = 3.31; SD = 1.17; see Fig. 3). In the following overall evaluation, the users perceived the articles as *very diverse* (M = 4.05; SD = 1.06). For many users, *new information* was offered by the recommended articles (M = 3.91; SD = 1.04). The users were *surprised* by the recommendations (M = 3.76; SD = 1.06) and rated the *perceived quality* average (M = 3.43; SD = 1.15). The *overall rating* for the UBCF recommender system was slightly negative (M = 3.17; SD = 1.01) (see Fig. 3).

Evaluating the SVD Recommender System. Most of the articles recommended by the recommender system, which uses *singular value decomposition (SVD)*, were rated negative (M = 2.87; SD = 1.05). In the following *overall evaluation*, the users perceived the articles as *very diverse* (M = 4.15; SD = .96). The recommended articles provided the users with *new information* (M = 3.79; SD = 1.01), which were only of *moderate relevance* (M = 3.28; SD = 1.25). However, the users were *surprised* by the recommended articles (M = 3.71; SD = 1.16). All in all, the *perceived quality* of the recommended items (M =3.15;

$SD = 1.01$) was better than the individual ratings. Although, the *overall rating* of the recommender system was negative ($M = 2.94$; $SD = .96$) (see Fig. 3).

4.3 Progress of the Recommender Systems

After finishing the survey, we evaluated the improvement of the *two recommender systems*. For this, we compared the *actual article ratings* with the *predictions of the recommender systems*. We calculated the *predictions* twice. First, with the state of the recommender systems at the beginning of the survey (*pretest/ex ante*) and second, with the recommender systems after the end of the survey (*posttest/ex post*).

We tested how many article ratings the recommender systems *correctly predicted*. In the other case, we tested the extent to which the *prediction deviated* from the *actual rating*. With a good recommender system, the proportion of correct *predictions* should be as high as possible and the *deviation (RMSE)* of the remaining predictions as low as possible.

UBCF Recommender System. The *UBCF recommender system correctly predicted* every fourth item in the *pretest state* ($M = .27$; $SD = .07$). The remaining 73% of the predictions showed both under- and overestimation of up to three levels in terms of *actual rating* In the *posttest*, the system predicted every third rating correctly ($M = .33$; $SD = .05$). The over- and underestimates were still up to three levels but the *deviation* had slightly decreased. Although the distance between *prediction* and *actual evaluation* has changed in only 87 of 560 of the evaluated articles, there is a small, significant difference between *pretest* and *posttest* of the *UBCF recommender system* ($t(139.0) = 10.54, p < .001$).

The correlation between the *prediction* and the *actual ratings* improved significantly between the *pretest* and the *posttest* of the *UBCF recommender system* ($t(123) = 8.68, p < .001$). As shown in Fig. 4(a), in the *pretest* we found almost as many negative as positive correlations ($\mu = .01$) between the *predicted* and *actual ratings*. In the *posttest* we found mainly strong positive correlations ($\mu = .6$) and only sporadically strong negative correlations.

SVD Recommender System. The *pretest SVD recommender system* correctly predicted every third article rating ($M = .35$; $SD = .03$). For the remaining 65% of recommendations, there was both over- and underestimation of up to three levels between *predictions* and *actual ratings* The *posttest SVD recommender system* correctly predicted more than half of all article ratings ($M = .57$; $SD = .03$). The *deviation* between the *prediction* and the *actual evaluation* has improved for the remaining 43%. Thus, the over- and underestimation of the ratings was reduced to at most two levels. When looking at the *absolute distances*, it is evident that the difference between the *pretest* and the *posttest recommender system* is significant ($t(139.0) = 10.15, p < .001$).

The correlation between the *prediction* and the *actual ratings* improved significantly between the *pretest* and *posttest SVD recommender system* ($t(129) = 8.45, p < .001$). Figure 4(b) shows that the *pretest SVD recommender system*

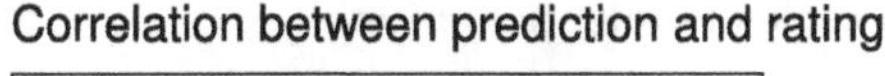

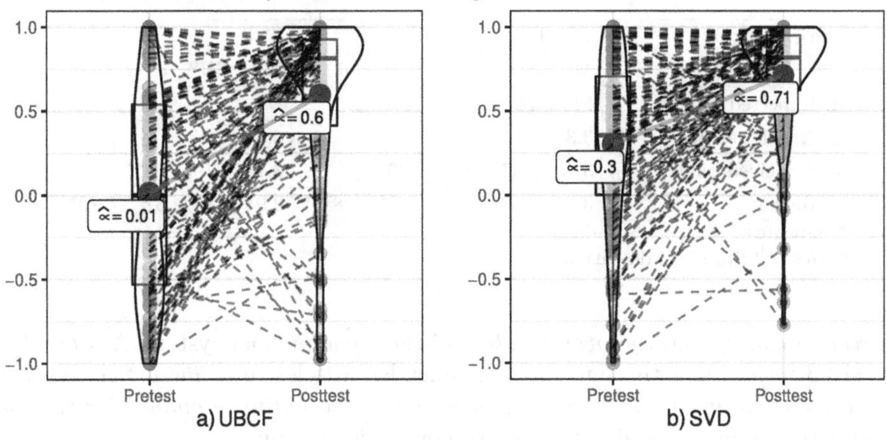

Fig. 4. Every point shown is the correlation coefficient between actual and predicted rating for a single user. Lines connect the users between pre- and post-test. Correlations above 0 indicate good predictions; correlations below 0 indicate very bad predictions. Both figures a) and b) indicate improvement, as the lines tend to move upwards. $\hat{\alpha}$ is the mean correlation coefficent for all users. The red line indicated the amount of improvement.

had more positive than negative correlations (μ = .3) between the *predicted* and *actual ratings*. The *posttest SVD recommender system* shows mostly strong positive correlations (μ = .71), whereas negative correlations occur only sporadically.

4.4 Relationships Between Evaluation Criteria

We analyzed how the *article ratings*, the *user-centered rating criteria*, the *expected article quality*, and the *overall rating* of the recommender system are correlated. As the functionality of a *SVD recommender system* is quite complex, we did not ask the participants which *quality* they *expected* from the *SVD recommender system*. Therefore, we evaluated the *expected quality* of the recommendations only for the *UBCF recommender system*.

Article Rating, Perceived Quality and Overall Rating. Table 3 shows that for both *recommender systems*, the *article ratings* have a significantly strong positive correlation with both the *perceived quality* and the *rating of the recommender system*. The better the users *rated the articles*, the better they *perceived the quality* of the recommendations and the better the *rating of the recommender system*. In the case of the *UBCF recommender system*, the *expected article quality* correlated slightly positively with the *actual article rating*.

User Centric Factors. For the *UBCF* and *SVD recommender system* (see Table 4) the *perceived novelty* correlates positively with the *article ratings*, the

Table 3. Correlation table of evaluation criteria

Variables	UBCF					SVD			
	M[a]	SD[b]	1	2	3	M[a]	SD[b]	1	2
1. Expected quality	51	21							
2. Article ratings UBCF	3.31	1.17	.27**			2.87	1.05		
3. Perceived quality	3.43	1.15	.16	.76**		3.15	1.01	.68**	
4. Rating	3.17	1.01	.19	.62**	.81**	2.94	.96	.57**	.80**

* indicates $p < .05$; ** indicates $p < .01$
[a] Mean, [b] Standard deviation

perceived quality and the *overall rating* of the recommender system. A user who received *novel* items through the recommended articles *rated the articles* themselves, the *quality of the recommendations* and the *entire recommender system* better than a user who did not receive novel information.

Table 4. Correlation table of user-centered criteria of the *User-based collaborative filtering (UBCF) recommender system* and *Singular value decomposition (SVD) recommender system*

Variables	UBCF					SVD					
	M[a]	SD[b]	1	2	3	M[a]	SD[b]	1	2	3	4
1. Perceived novelty	3.91	1.04				3.79	1.01				
2. Perceived diversity	4.05	1.06	.01			4.05	.96	.02			
3. Perceived serendipity	3.76	1.06	.05	.18*		3.71	1.16	.12	.24**		
4. Perceived relevance						3.28	1.25	.30**	.04	−.02	
5. Expected quality	51	21	.09	−.07	.04						
6. Article ratings	3.31	1.17	.45**	.05	−.15	2.87	1.05	.29**	.05	−.08	.41**
7. Perceived quality	3.43	1.15	.54**	.04	−.17	3.15	1.01	.45**	−.01	−.03	.54**
8. Rating	3.17	1.01	.44**	−.08	−.12	2.94	.96	.37**	−.03	−.02	.49**

* indicates $p < .05$; ** indicates $p < .01$
[a] Mean, [b] Standard deviation

Furthermore, in the *SVD recommender system*, the *perceived relevance* of the recommended articles correlates significantly and positively with the *article rating*, the *perceived quality* and the *overall rating*. The more *relevant* the recommended articles were for the users, the better the recommender systems were *rated*. This in turn positively influenced the *perceived quality* and the *overall rating of the recommender system*.

5 Discussion

In our study, we compared two recommender systems before (pretest) and after (posttest) an online study. Both recommender systems improved their prediction accuracy. Also, the correlation between the predictions and the actual ratings increased between the pre- and post-test.

Comparing the two recommender systems, the SVD recommender system scored better on all statistical measures (accuracy, deviation of predictions, and correlation between predictions and assessments) than UBCF. This suggests that users would also rate the SVD recommender system as better than the UBCF system. However, users preferred the articles suggested to them by the UBCF recommender system and perceived the quality as higher, thus giving the UBCF a better overall rating. This agrees with finding from McNee, Riedl, and Konstan [31], who found that the evaluation of a recommender system does not only depend on the accuracy of the recommendations.

We also looked at whether there is a correlation between the expected quality of the recommendations, the actual article ratings, and the subjective ratings of the users. The participants expected an accuracy of about 50% from the collaborative filtering recommender systems (UBCF). Our UBCF recommender system did not meet the expectations of the participants with an accuracy of only 33%. In contrast, the SVD recommender system exceeded expectations with 57%. The overall evaluation the recommender system shows that the perceived quality of a recommender system is correlated with the individual ratings of the recommended items.

If the participants experience the recommended items as relevant and novel, they rated the quality of the recommendations as higher and also rated these articles better. This shows that a framework like ours can help to investigate differences and relationships in algorithmic and other user-centric evaluations in online studies.

In our study, the participants had to rate each article individually, which takes a lot of time. In a next study, we want to compile the article recommendations of different recommender systems into a single generated news page after the system has been re-trained. Thus, we could compare a larger number of recommender systems. However, the simultaneous evaluation of several recommendations could also lead to problems in the comparability of the results. As news articles on news pages are often implicitly consumed, users judgments on whole pages would be influenced by many factors. Making isolation of factors harder to achieve.

In the future, the framework could be tested in other test scenarios, for example with recommender systems that recommend other products. Here, it must be considered that other types of items have different "shelf-lives" than News, therefore a drift in user preferences would have to be accounted for differently.

6 Conclusion and Outlook

In this study, we analyzed recommender systems not separately according to statistical measures, such as accuracy, or according to user-centric criteria, but to carry out both types of evaluation combined. Most interestingly, users evaluate recommendations differently than accuracy metrics, revealing the importance of studying recommender systems from a users perspective. Better accuracy does imply better user evaluation, but not solely so. Users are particularly bad at predicting the performance of algorithmic recommendations, stressing the importance of features like explanations and visualizations of recommendations.

A balance between accuracy and user-centric criteria is nevertheless important. Our framework allows to better explore this balance and provides a starting point for further research.

Acknowledgements. This research was supported by the Digital Society research program funded by the Ministry of Culture and Science of the German State of North Rhine-Westphalia.

References

1. Adomavicius, G., Tuzhilin, A.: Toward the next generation of recommender systems: a survey of the state-of-the-art and possible extensions. IEEE Trans. Knowl. Data Eng. **17**(6), 734–749 (2005). ISSN 1041-4347
2. Alyari, F., Navimipour, N.J.: Recommender systems: a systematic review of the state of the art literature and suggestions for future research. Kybernetes **47**(5), 985–1017 (2018)
3. Atoum, J. O., Yakti, I.M.: A framework for real time news recommendations. In: Proceedings - International Conference on New Trends in Computing Sciences, ICTCS 2017, NJ 08854, USA, vol. 2018-Janua, pp. 89–93. Institute of Electrical and Electronics Engineers (IEEE) (2017)
4. Beck, P., et al.: A system for online news recommendations in real-time with apache mahout. In: Working Notes of the 8th International Conference of the CLEF Initiative, vol. 1866. CEUR Workshop Proceedings (2017)
5. Beier, G.: Kontrollüberzeugungen im umgang mit technik. Rep. Psychol. **9**, 684–693 (1999)
6. Bogers, T., van den Bosch, A.: Comparing and evaluating information retrieval algorithms for news recommendation. In: Proceedings of the 2007 ACM Conference on Recommender Systems, RecSys 2007, pp. 141–144. Association for Computing Machinery (2007). ISBN 9781595937308
7. Burbach, L., et al.: User preferences in recommendation algorithms: the influence of user diversity, trust, and product category on privacy perceptions in recommender algorithms. In: Proceedings of the 12th ACM Conference on Recommender Systems, RecSys 2018, pp. 306–310. Association for Computing Machinery (2018). ISBN 9781450359016
8. Burke, R.: Hybrid recommender systems: survey and experiments. User Model. User-Adapt. Interact. **12**(4), 331–370 (2002). ISSN 0924-1868
9. Valdez, A.C., Ziefle, M., Verbert, K.: HCI for recommender systems: the past, the present and the future. In: Proceedings of the 10th ACM Conference on Recommender Systems, RecSys 2016, pp. 123–126. Association for Computing Machinery (2016). ISBN 9781450340359
10. Cremonesi, P., Garzotto, F., Turrin, R.: User-centric vs. system-centric evaluation of recommender systems. In: Kotzé, P., Marsden, G., Lindgaard, G., Wesson, J., Winckler, M. (eds.) INTERACT 2013. LNCS, vol. 8119, pp. 334–351. Springer, Heidelberg (2013). https://doi.org/10.1007/978-3-642-40477-1_21
11. De Pessemier, T., et al.: A user-centric evaluation of context-aware recommendations for a mobile news service. Multimed. Tools Appl. **75**(6), 3323–3351 (2015)

12. Duan, L., Street, N., Xu, E.: Healthcare information systems: data mining methods in the creation of a clinical recommender system. Enterp. Inf. Syst. **5**, 169–181 (2011)
13. Eirinaki, M., et al.: Recommender systems for large-scale social networks: a review of challenges and solutions. Future Gener. Comput. Syst. **78**, 413–418 (2018)
14. Ekstrand, M.D., Riedl, J.T., Konstan, J.A.: Collaborative filtering recommender systems. Found. Trends Hum.-Comput. Interact. **4**(2), 81–173 (2011). ISSN 1551-3955
15. Faul, F., et al.: Statistical power analyses using g*power 3.1: tests for correlation and regression analyses. Behav. Res. Methods **41**, 1149–60 (2009)
16. Fazeli, S., et al.: User-centric evaluation of recommender systems in social learning platforms: accuracy is just the tip of the iceberg. IEEE Trans. Learn. Technol. **PP**, 1 (2017)
17. Garcin, F., et al.: Offline and online evaluation of news recommender systems at swissinfo.ch. In: Proceedings of the 8th ACM Conference on Recommender Systems, RecSys 2014, pp. 169–176. Association for Computing Machinery (2014). ISBN 9781450326681
18. Ge, M., Delgado-Battenfeld, C., Jannach, D.: Beyond accuracy: evaluating recommender systems by coverage and serendipity. In: Proceedings of the Fourth ACM Conference on Recommender Systems, RecSys 2010, pp. 257–260. Association for Computing Machinery (2010). ISBN 9781605589060
19. Hahsler, M., Vereet, B., Hahsler, M.M.: Package 'recommenderlab' (2019)
20. Hanafi, M., Suryana, N., Basari, A.S.: An understanding and approach solution for cold start problem associated with recommender system: a literature review. J. Theor. Appl. Inf. Technol. **96**(9), 2677–2695 (2018)
21. Huang, Z., Chen, H., Zeng, D.: Applying associative retrieval techniques to alleviate the sparsity problem in collaborative filtering. ACM Trans. Inf. Syst. **22**(1), 116–142 (2004). ISSN 1046-8188
22. Ishida, Y., Uchiya, T., Takumi, I.: Design and evaluation of a movie recommendation system showing a review for evoking interested. Int. J. Web Inf. Syst. **13**, 72–84 (2017). https://doi.org/10.1108/IJWIS-12-2016-0073
23. Isinkaye, F.O., Folajimi, Y., Ojokoh, B.A.: Recommendation systems: principles, methods and evaluation. Egypt. Inf. J. **16**, 261–273 (2015)
24. Karimi, M., Jannach, D., Jugovac, M.: News recommender systems - survey and roads ahead. Inf. Process. Manag. **54**(6), 1203–1227 (2018). ISSN 0306-4573. https://doi.org/10.1016/j.ipm.2018.04.008. http://www.sciencedirect.com/science/article/pii/S030645731730153X
25. Lommatzsch, A.: Real-time news recommendation using context-aware ensembles. In: de Rijke, M., et al. (eds.) ECIR 2014. LNCS, vol. 8416, pp. 51–62. Springer, Cham (2014). https://doi.org/10.1007/978-3-319-06028-6_5
26. Kirshenbaum, E., Forman, G., Dugan, M.: A live comparison of methods for personalized article recommendation at Forbes.com. In: Flach, P.A., De Bie, T., Cristianini, N. (eds.) ECML PKDD 2012. LNCS (LNAI), vol. 7524, pp. 51–66. Springer, Heidelberg (2012). https://doi.org/10.1007/978-3-642-33486-3_4
27. Knijnenburg, B.P., Willemsen, M.C., Gantner, Z., Soncu, H., Newell, C.: Explaining the user experience of recommender systems. User Model. User-Adapt. Interact. **22**(4–5), 441–504 (2012). https://doi.org/10.1007/s11257-011-9118-4. ISSN 0924-1868
28. Kotkov, D., Wang, S., Veijalainen, J.: A survey of serendipity in recommender systems. Knowl.-Based Syst. **111**, 08 (2016). https://doi.org/10.1016/j.knosys.2016.08.014

29. Li, L., et al.: Personalized news recommendation: land an experimental investigation. J. Comput. Sci. Technol. **26**, 754–766 (2011). https://doi.org/10.1007/s11390-011-0175-2
30. Lü, L., et al.: Recommender systems. Phys. Rep. **519**(1), 1–49 (2012). ISSN 0370-1573
31. McNee, S.M., Riedl, J., Konstan, J.A.: Being accurate is not enough: How accuracy metrics have hurt recommender systems. In CHI 2006 Extended Abstracts on Human Factors in Computing Systems, CHI EA 2006, pp. 1097–1101. Association for Computing Machinery (2006). ISBN 1595932984. https://doi.org/10.1145/1125451.1125659
32. Phelan, O., McCarthy, K., Smyth, B.: Using Twitter to recommend real-time topical news. In: Proceedings of the Third ACM Conference on Recommender Systems, RecSys 2009, pp. 385–388. Association for Computing Machinery (2009). ISBN 9781605584355
33. Pu, P., Chen, L., Hu, R.: A user-centric evaluation framework for recommender systems. In: Proceedings of the Fifth ACM Conference on Recommender Systems, RecSys 2011, pp. 157–164. Association for Computing Machinery (2011). ISBN 9781450306836. https://doi.org/10.1145/2043932.2043962
34. Pu, P., Chen, L., Hu, R.: Evaluating recommender systems from the user's perspective: Survey of the state of the art. User Model. User-Adapt. Interact. **22**(4–5), 317–355 (2012). https://doi.org/10.1007/s11257-011-9115-7. ISSN 0924-1868
35. Remus, R., Quasthoff, U., Heyer, G.: Sentiws - a publicly available German-language resource for sentiment analysis. In: Proceedings of the 7th International Language Resources and Evaluation (LREC 2010), pp. 1168–1171 (2010)
36. Revelle, W.R.: Psych: procedures for personality and psychological research (2017)
37. Schabus, D., Skowron, M., Trapp, M.: One million posts: a data set of German online discussions. In: Proceedings of the 40th International ACM SIGIR Conference on Research and Development in Information Retrieval, SIGIR 2017, pp. 1241–1244. Association for Computing Machinery (2017). ISBN 9781450350228. https://doi.org/10.1145/3077136.3080711
38. Schafer, J.B., Frankowski, D., Herlocker, J., Sen, S.: Collaborative filtering recommender systems. In: Brusilovsky, P., Kobsa, A., Nejdl, W. (eds.) The Adaptive Web. LNCS, vol. 4321, pp. 291–324. Springer, Heidelberg (2007). https://doi.org/10.1007/978-3-540-72079-9_9. ISBN 9783540720782
39. Schedl, M., et al.: Current challenges and visions in music recommender systems research. Int. J. Multimed. Inf. Retrieval **7**, 95–116 (2018)
40. Shi, Y., Larson, M., Hanjalic, A.: Collaborative filtering beyond the user-item matrix: a survey of the state of the art and future challenges. ACM Comput. Surv. **47**(1), 1–45 (2014). https://doi.org/10.1145/2556270. ISSN 0360-0300
41. Smith, B., Linden, G.: Two decades of recommender systems at amazon.com. IEEE Internet Comput. **21**(3), 12–18 (2017). ISSN 1089-7801
42. Sohail, S.S., Siddiqui, J., Ali, R.: Classifications of recommender systems: a review. J. Eng. Sci. Technol. Rev. **10**(4), 132–153 (2017)
43. Son, L.H.: Dealing with the new user cold-start problem in recommender systems: a comparative review. Inf. Syst. **58**, 87–104 (2016). http://dblp.uni-trier.de/db/journals/is/is58.html#Son16
44. Svrcek, M., Kompan, M., Bielikova, M.: Towards understandable personalized recommendations: Hybrid explanations. Comput. Sci. Inf. Syst. **16**, 179–203 (2019). https://doi.org/10.2298/CSIS171217012S
45. Taghavi, M., et al.: New insights towards developing recommender systems. Comput. J. **61**, 319–348 (2018). https://doi.org/10.1093/comjnl/bxx056

46. Taneja, A., Arora, A.: Recommendation research trends: review, approaches and open issues. Int. J. Web Eng. Technol. **13**(2), 123–186 (2018)
47. Valdez, A.C., Ziefle, M.: The users' perspective on the privacy-utility trade-offs in health recommender systems. Int. J. Hum.-Comput. Stud. **121**, 108–121 (2019). ISSN 1071-5819. Advances in Computer-Human Interaction for Recommender Systems
48. Wang, J., de Vries, A.P., Reinders, M.J.T.: Unifying user-based and item-based collaborative filtering approaches by similarity fusion. In: Proceedings of the 29th Annual International ACM SIGIR Conference on Research and Development in Information Retrieval, SIGIR 2006, pp. 501–508. Association for Computing Machinery (2006). ISBN 1595933697
49. Zhou, X., et al.: The state-of-the-art in personalized recommender systems for social networking. Artif. Intell. Rev. **37**(2), 119–132 (2012). ISSN 0269-2821
50. Zhou, Xun, et al.: SVD-based incremental approaches for recommender systems. J. Comput. Syst. Sci. **81**(4), 717–733 (2015). https://doi.org/10.1016/j.jcss.2014.11.016
51. Özgöbek, O., Gulla, J., Erdur, C.: A survey on challenges and methods in news recommendation. In: WEBIST 2014 - Proceedings of the 10th International Conference on Web Information Systems and Technologies, vol. 2, pp. 278–285, January 2014

Partial Consent: A Study on User Preference for Informed Consent

Sven Bock[1(✉)], Ashraf Ferdouse Chowdhury[2], and Nurul Momen[3]

[1] Technische Universität Berlin, Berlin, Germany
sven.bock@mms.tu-berlin.de

[2] Stockholm University, Stockholm, Sweden

[3] Blekinge Institute of Technology, Karlshamn, Sweden
nurul.momen@bth.se

Abstract. This study evaluates the use, acceptance and desirability of consenting with partial consent. For this purpose, an app was developed that requested access permissions to various data for different tasks. The user had the means to limit the exposure of personal data by opting an additional *Maybe-* button other than the usual binary *(allow/deny)* options. Upon expiration of time-bound partial consent, the user could potentially reassess the trade-off between service and privacy consequences. Partial consent was only used by one fifth of the participants, whereby just under half of the participants stated after completing the questionnaire that they would like to use it on their private devices.

Keywords: Partial consent · Informed consent · Data protection · User study

1 Introduction

Nowadays, consenting to grant access rights to personal data has become a part and parcel of almost every step that we take in this vast digital sphere. Currently, only a binary selection is available to the users when granting rights to use an application (accept and reject). In some cases, users are conditioned in such a way that they would grant access to personal information, even if a feeling of unease overcomes them [8,11]. Furthermore, revoking the consent to granted privilege requires some efforts, technical understanding, and awareness of the consequence [12,15,18,20]. The complex legal language in which the terms and conditions are written, as well as the liberal privacy preferences make it difficult for users to prevent the disclosure of their personal data [13,14]. The question arises as to the extent to which smartphone users have sovereignty over their personal data. However, there is no incentive to review or revoke this access authorization after the initial usage [19]. In order to address such dilemma, an additional option could be added when requesting rights, so that the user does not have only a binary selection available, where she either releases her rights indefinitely or cannot use the app or service at all. This additional option

C. Stephanidis et al. (Eds.): HCII 2021, LNCS 13094, pp. 198–216, 2021.
https://doi.org/10.1007/978-3-030-90238-4_15

could allow users to consent temporarily and proceed to use the desired app or service. Upon expiration of the conditional or temporary consent, the user could evaluate the consequence of her decision made earlier and then change the validity of consent accordingly.

Here, it would be questionable whether this conditional or temporary access authorization is going to be accepted by the user, and how user-friendly this additional option could be designed for everyday usage. Furthermore, it should be determined whether there is a general need for temporary access authorization. Moreover, the question arises whether a recurring request for access authorization disturbs or even annoys the user. Thus, it would still be relevant to determine whether the user is more interested in the control of her data with repeated requests for access authorization, or in the convenience with permanent access granting scheme. Furthermore, it should be determined which data types and life contexts the temporary access authorization can best be applied. For example, the implementation of conditional or temporary access authorization in the location data can be helpful because the transmission of sensitive information is interrupted and the user can thus protect her location privacy. On the other hand, it is debatable whether the temporary access authorization can add value to the access of contacts, because if the privilege is granted once, an instance of the data set becomes accessible. Hence, the repeated access only provides insight into the same or slightly extended data set. These circumstances lead us to the following research questions:

1. Is the conditional or temporary access authorization applicable to the real-life scenarios?
2. Is the conditional or temporary access authorization accepted by the user?
3. Is the conditional temporary access authorization desired by the user?
4. What is the the user perception regarding the frequency of access authorization requests?
5. Would it be feasible to find a measurable threshold for identifying annoyance associated with recurring request from access authorization scheme with conditional or temporary consent?
6. Could a conditional or temporary consenting scheme play an important role for the user to have control over their data privacy, or does the convenience take precedence?

Furthermore, we would like to explore and correlate with the real-life contexts and data types that can be used to realize the potential of conditional/temporary/partial access authorization to protect data privacy.

In order to address the aforementioned questions, we introduced an additional option on the interface—a *Maybe* button, by which we envisioned to bring the partial/temporary/conditional consent for granting access into a practical context [17]. Due to larger user base, project feasibility, and open-source nature of the platform, our research is concentrated on the access control model of Android. Moreover, it is the most prevalent mobile operating system and hence, results would have greater impact. In this study, our goal is to investigate the usability and acceptance of the *Maybe* button, in other words—

partial/temporary/conditional consent for access authorization. Except for the latest changes in Android 11.0[1], it has been only possible to grant the rights requested by the apps and services for an indefinite period of time. Thus, the user has the choice of either granting the requested rights, or not using the app. In this study, we show that an alternative way, a third option could be implemented in the form of a *Maybe* button. The *Maybe* button was only used by one-fifth of the participants, although just under half of them stated after completing the questionnaire that they wanted to use it on their private device. This paper is organized as following: in Sect. 2, we present a brief overview of related work along with corresponding definitions. The study design and methodology are discussed in Sect. 3. Section 4 elaborates on the collected data and results. Upon discussing the findings of this study in Sect. 5, we conclude this paper in Sect. 6.

2 Background

In this section, we discuss the relevant background in the context of this article. Our discussion includes the definition of partial consent and a brief discussion on the state of the art literature.

2.1 Definition

The definition for consent can be found in several disciplines and contexts; e.g., health profession, law, the social and behavioral sciences, and moral philosophy [10]. However, our research is focused within the realm of information privacy, hence the definition from GDPR (Article 4(11)) can be deemed as relevant and prudent consideration.

The General Data Protection Regulation—GDPR provides following definition: *'consent' of the data subject means any freely given, specific, informed and unambiguous indication of the data subject's wishes by which he or she, by a statement or by a clear affirmative action, signifies agreement to the processing of personal data relating to him or her* [7].

We define partial consent as following: *it is data subject's freely given, specific, informed and unambiguous indication of a clear affirmative action, that signifies agreement to the processing of personal data relating to him or her with a condition—that allows the data subject to impose validity on the given consent by means of time or access count.*

In particular, we aim at the challenges associated with achieving the properties of consent: (a) freely given, (b) specific, (c) informed, and unambiguous. Traditionally, a binary choice is presented through the interface to acquire consent decisions from the user. Previous research had shown that users find such interfaces cumbersome to consent with the aforementioned attributes [8,15].

[1] https://www.android.com/android-11/.

Numerous services, products, blurry representation about consequences, instant psychological gratification, leveraging privacy dark patterns are held responsible for causing dilemma and consequently surrendering to the obvious outcome—consenting for life [17]. We see the potential to address this problem with partial consent, which introduced an indecisive state for the users to contemplate on the probable consequence and thus potentially allowing them reassessing the decisions taken earlier [16,17].

2.2 Related Work and Hypotheses

Expiry date for consent, conditional commitment, partial or temporary consent were introduced and discussed in the research arena in the recent past [9,11,17]. In [9], the significance of the user-consent's validity was highlighted. In [11], an alternative solution—*try before you buy* concept was discussed with consideration from various use cases. In [17], the discussion around partial consent got the momentum from different aspects; e.g., legal, social, technical feasibility, vendor's perspectives, and usability.

As a continuation of our prior research efforts, the aim of this study is therefore to determine whether the partial consent—in the form of a *Maybe* button, which entail a renewed request, is desired and accepted by the user. Based on the results of the previous studies, which confirmed that the user attach a high value to personal data, it can be assumed that most users of mobile devices will positively evaluate another function for the protection of their personal data [4,5]. Also the privacy paradox was confirmed, which states that users would always use an app that has a more user-friendly privacy policy if no compromise in functionality had to be made [3]. Brown and Norberg coined the term privacy paradox as a dichotomy of information privacy attitude and actual behavior [6]. These dichotomies do not imply irrationality or reckless behavior. Individuals make privacy-related decisions based on multiple factors, including (but not limited to) knowledge, concerns, and assessed consequences of their actions. Although respondents in the study of Acquisti and Grossklags demonstrate sophisticated attitudes toward privacy and some privacy-compliant behavior, their decision-making process appears to be influenced by incomplete information, bounded rationality, and systematic psychological deviations from rationality [1]. Now, it is questionable whether a recurring request for consent to the user will result in a reduction of functionality. If so, it must be expected that the user will not use the partial consent to protect her personal data. However, the results of another study on the influence of privacy indicator on selection behavior show that the indicator was indeed used and contributed to privacy-preserving behavior [4]. For this reason, it is assumed that the participants in this study will also use this function to set a validity for consent and thus take preventive measures. The following hypotheses can therefore be drawn:

- Hypothesis 1: The Maybe button (partial consent) is applied by the users (recording of usage)
- Hypothesis 2: The Maybe button (partial consent) is positively evaluated by the users regarding its functionality (questionnaire)

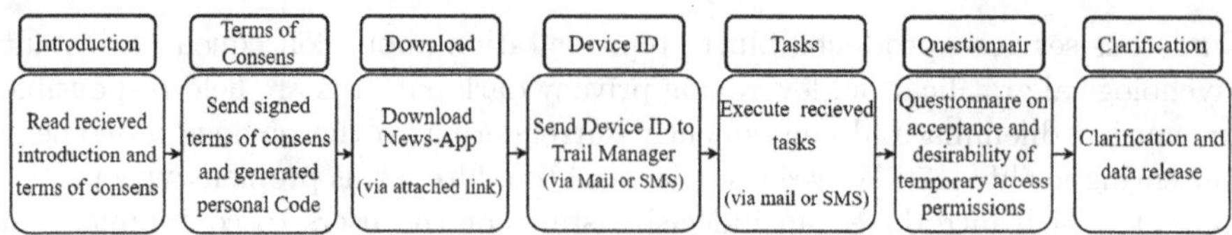

Fig. 1. Workflow of the study.

- Hypothesis 3: The function of Maybe button (partial consent) is desired by the users in their everyday life (questionnaire).
- Hypothesis 4: The users feel annoyed by the recurring request of Maybe button—partial consent (questionnaire).
- Hypothesis 5: Users prefer to control their data despite the recurring request from Maybe button—partial consent (questionnaire).

3 Study Design and Methodology

The subjects for this study were recruited via an online portal of Technical University of Berlin. Within the framework of the acquisition of the participant, the aim was to achieve a sample for the experiment that was as heterogeneous as possible so that the demographic characteristics balanced each other out and had no effect on the results. The participants were not informed about the purpose of the study while they were deceived into thinking that the study was an evaluation of a news app. The participants were offered a remuneration of €10 for the complete participation in the study. This empirical experiment does not omit an independent variable (IV), it was decided to do so because it should only be tested whether the function of the *Maybe* button—partial consent is used and accepted. Since the option of the usual accepting and rejecting is still available, it can be observed in the same sample, which options the participants prefer without changing the possible selection. The dependent variable (DV) consists of the use of the *Maybe* button. Attention is paid to whether the participants use this temporary function to perform their daily tasks or whether they opt for the previous options for accepting or rejecting.

The study was conducted as a 10-day online experiment. As illustrated by the flowchart in Fig. 1, the participants were asked to send the consent form and a self-generated personal code by e-mail to the experiment leader. In order to speed up the process, we decided to send the tasks everyday once through both SMS and e-mail. Thus, the participants received the task as a push message directly to their smartphone on which they were supposed to solve the tasks within the prototype app. They were also asked to download the prototype news app and install it on their mobile device. They were then requested to provide their "Device ID" (automatically generated) to confirm functionality and uniqueness. After receiving the "Device ID", the participants were given several tasks on a daily basis, which were associated with consenting to the access permissions.

These include, for example, sharing news and granting access to location data. When the participants were asked to consent to the required permission, a dialog box appeared in which they could choose from *Allow, Maybe,* and *Deny.* Since the experiment was also planned to be performed other countries, the user interface was designed in English, hence the partial consent scheme was implemented as the *Maybe* button.

As illustrated in Fig. 2, with the first option—"Allow" the participants consent to grant access to their data without any restrictions. If the "Maybe" option (partial consent) is selected, a further dialog box opens in which the participants can choose the validity for their consent and this process is repeated upon expiration. The option "Deny" prevents the app to access any data. The prototype app keeps the record of operating behavior (decision behavior) of the participants. After the ten-day experiment, the participants responded to an online questionnaire about user-friendliness and hypothetical everyday use of *Maybe* button/partial consent. Only the data/permission type (memory, microphone, etc.), the selected option *(Allow, Maybe, Deny)* and the time stamp (which records when the decision was made), are stored.

In the following days, participants had to perform several tasks that included e.g., receiving local news, sharing the news with other peers, saving the news content on the device, etc. In addition, the participants were asked to activate a push notification. The dialogue window described above with the three options *(Allow, Maybe, Deny)* was always presented requesting the permission for the first time. After the last task was completed, the participant was requested to respond to an online questionnaire. At first, they were asked to enter their personal code to link their participatory information with the data sent from the app. Furthermore, they were asked about individual preferences for the *Maybe* button, for how long, and for what period they would grant it. The answer options ranged from "less than 30 min" to "forever". This question was also asked in order to determine whether the temporary access authorization is desired by the participants.

The survey also consisted of a series of questions which could be answered using a scale from 1 to 10, where 1 is no consent and 10 is full consent. We also recorded whether or not the user has noticed the option of *Maybe* button and whether the participants would use it on their private devices. With a follow-up question, we tried to determine whether the recurring request to consent had caused annoyance to them. Then the participants were asked to evaluate how important it is to have control over their personal data, and what they thought about the usability of *Maybe* button. The aim was to determine whether the respondents prioritize usability over privacy protection. Lastly, the participant was asked about their personal preference and opinion about the *Maybe* button.

Subsequently, possible improvements of the *Maybe* button were suggested in free text fields. In addition, the participants were asked to indicate in which situation they would use the *Maybe* button and what their first thought was when they saw the Maybe option. They could also indicate if and why they used the *Maybe* button. Moreover, the adequacy of the given time periods should be

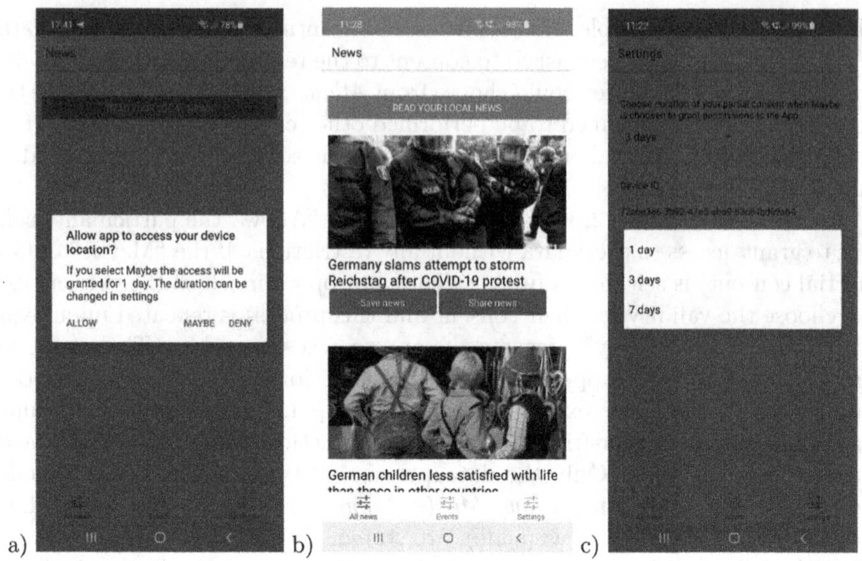

Fig. 2. Interface of the prototype app.

determined. Finally, the participants were asked about their demographic data and the highest educational degree, the technical affinity, the daily smartphone use, the importance of privacy protection and the precautions they would prefer to have. After the successful completion of the questionnaire, the information about the study was sent to all participants by e-mail. In the clarification it was revealed that the focus of the study was on *Maybe* button/partial consent. In addition, the function of the *Maybe* button was briefly explained.

4 Results

In this section, we present a summary of results from the collected data. The data collection campaign was carried out in two phases: *(a)* activity log from the prototype app and *(b)* participant responses from a questionnaire. A total of 31 German-speaking individuals (11 male, 20 female) participated in this study. The age range in the sample was between 20 and 64 years with an average of M = 29.68 and a standard deviation of SD = 9.18. Their educational backgrounds varied from high school diploma to higher academic degree from university. Among the participants, 1 person spent more than 30 min, 7 more than 1 h, 14 more than 2 h and 9 more than 5 h on their smartphones. We used Likert scale to determine participants' preference about privacy protection [2], and their responses indicated high priority—M = 5.97 (SD = 0.91). Also, 97% of the respondents stated to be active in protecting their data.

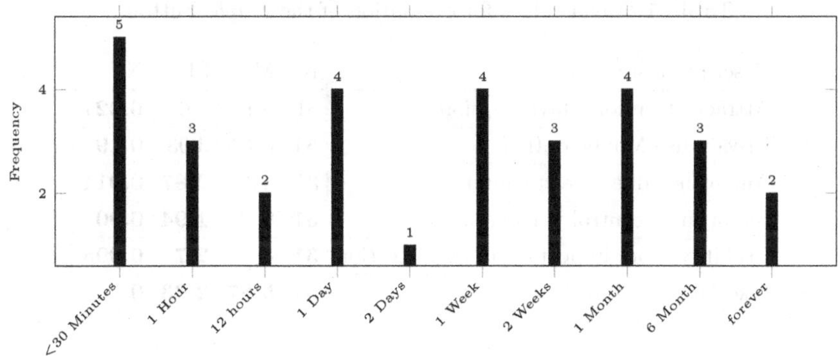

Fig. 3. Distribution of desired duration of partial consent.

4.1 Analysis of Questionnaire Response

The questionnaire focusing on the acceptance of *Maybe* button was consisted of ten questions, in which the respondents were asked to indicate whether they agree or disagree with the statements given on a scale of 1 (disagree) to 10 (fully agree). Regarding the statement (1) that the option of the *Maybe* button was conspicuous, an average value of $M = 5.1$ ($SD = 3.6$) could be calculated.

After dividing the sample into two groups, 20 participants (64.52%) disagreed with the statement (with a value of ≤ 5) and 11 participants (35.48%) agreed with the statement (with a value of ≥ 6). The statement (2) that the participants would use the *Maybe* button on their private device was calculated with an average value of 5.35 ($SD = 3.03$). After dividing the sample into two groups, 17 participants (54.9%) disagreed with the statement and 14 participants (45.1%) agreed with the statement. The statement (3) that the recurring request to provide consent causes annoyance obtained a mean value of 4.32 ($SD = 2.87$). After dividing the sample into two groups, 21 participants (67.8%) disagreed with the statement and 10 participants (32.4%) agreed with the statement The statement (4) that control of personal data is important achieved the highest mean of 8.19 ($SD = 1.94$). After dividing the sample into two groups, four participants (12.9%) disagreed with the statement and 27 participants (87.2%) agreed with the statement. Regarding the statement (5) that user-friendliness is more important than the protection of personal data, the mean value is 4 ($SD = 2.7$). After dividing the sample into two groups, 22 participants (71.1%) disagreed with the statement and nine participants (29%) agreed with the statement With regard to the statement (6) whether the option of *Maybe* button is liked, a mean value of 5.87 ($SD = 2.33$) was recorded. After dividing the sample into two groups, 16 participants (51.6%) disagreed with the statement and 15 participants (48.4%) agreed with the statement. For a better overview Table 1 is summarizing the descriptive results.

As illustrated in Fig. 3, it can be seen that the participants selected the extremes and the middle range of the scale frequently. The second question,

Table 1. Mean values for evaluation of the *Maybe* button.

Descriptive statistics	N	M	SD	NV
Attract attention Maybe option	31	5.1	3.6	**0.021**
Private use Maybe option	**31**	**5.35**	3.03	**0.19**
Annoyance due to recurring request	31	4.32	**2.87**	**0.014**
Importance control personal data	**31**	8.19	**1.94**	**0.001**
Usability more important than protection	**31**	**4**	**2.7**	0.005
Like Maybe option	**31**	**5.87**	**2.33**	**0**

which should raise how long an app is allowed to access a permission, was answered by five participants (16.1%) with 'less than 30 min'. Four participants (12.9%) stated that a time span of '1 day', '1 week' and '1 month' would be appropriate, respectively. Three participants (9.7%) stated that a duration of '1 h', '2 weeks' and '6 months' would be adequate, respectively. Two participants stated that a duration of '12 h' and permanent privilege to access permission would be appropriate. Only one person responded that a period of 'two days' would be sufficient.

The third question was asked to identify which permission is considered for partial consent, and 25 participants (80.6%) responded positively to use this feature for the location. For all other permissions, the majority of the participants indicated against the usage of partial consent (*Maybe* button): Messages (SMS)—41.9%, Calendar—41.9%, Phone memory—38.7%, Contacts—35.5%, Messages—29%, Microphone—29%. For a better overview Table 2 summarizing the results.

In the fourth question, participants were able to provide suggestions for improvement regarding the privilege granting scheme with partial consent. Nine participants (29.03%) offered no suggestion for improvement. Five participants indicated that they would like to have the option of any time for which permissions are granted with partial consent. Four of the respondents (12.9%) wanted to see improvements in the visibility of the *Maybe* button. Three participants wanted longer intervals for partial consent and improvements to the design to make it more memorable. In addition, there were individual suggestions for improvement (3.23%) regarding longer time intervals for partial consent, an overview interface that shows 'who' can access 'which' data and a timer that illustrates how long an app has been granted a certain privilege. Furthermore, it was mentioned as an improvement suggestion to grant privilege only if the app is actively used and not running in the background. Another suggestion was to grant access authorization only for a certain number of accesses and thus exclude the temporary aspect. Furthermore, an explanation, for what the requested access permissions are needed, was claimed. Another request was the translation of the interface into the mother tongue of the end user. Another possibility for improvement was the request for an explanation of the function, e.g. in the form of an info box. For the detailed overview of the results Table 3 is implemented.

In the fifth question, we wanted to identify user preference for everyday situations where the *Maybe* button could be used. Resulted data shows that the partial consent is particularly desirable when the user uses location-oriented service, which was indicated by nine respondents (29.03%). Three respondents (9.68%) indicated that they would use the feature on applications that were only briefly installed on the device for a specific task and then deleted. Three other respondents would use the feature on applications that have access to personal data, such as messages and pictures. Two of the participants (6.45%) stated that they would never use the function. In addition, four other participants each cited an everyday situation, including installing new applications, staying at work, rarely using installed applications, and scheduling. The sixth question was aimed at documenting the first impression of the individuals, and eight participants (25.81%) mentioned that they noticed the *Maybe* button, but did not give it any further thought. Five participants (16.13%) made positive comments about the *Maybe* button. The function was described as "good" and "interesting". Four participants (12.90%) each stated that they did not notice the button or did not understand the function. Only two participants (6.45%) expressed a negative opinion about the function and criticized the lack of explanation of the function. The seventh question was only intended to record whether the *Maybe* button was used by the participants, and the data showed that a total of 32.3% of the participants used it. The eighth question was intended to ascertain whether the participants had correctly understood the function of the *Maybe* button. It should be mentioned that the question was only displayed if the use of the Maybe function was confirmed in the previous question. Of these ten participants who confirmed the use of the *Maybe* button, seven (70%) were able to provide the correct explanation about the purpose of *Maybe* button. One respondent stated that he understood the function, but did not explain it. Another participant gave an incorrect explanation of the function, while one respondent gave an answer that answered the sixth question of the first impression rather than the understanding of the function. The ninth question was aimed at capturing the motives of the participants that led them to use the *Maybe* button. Eight participants (25.81%) stated that they did not use the *Maybe* button. Table 4 offers an overview regarding further detail about participants' motives behind the usage of *Maybe* button.

Upon analyzing participants' response for the tenth question, it was confirmed that the given selection of the duration of one, three and seven days for the *Maybe* button was appropriate for 19 participants (61.3%). Four participants 12.9 % stated that the intervals were too short and that there were too few options regarding the interval length. Two participants (6.5%) indicated that the intervals were too long. In total, five participants gave other reasons. Four participants gave the reason that they had not noticed the *Maybe* button or could not remember how long the intervals were. One person suggested that an option should be added in front of the *Maybe* button that would allow the app to access the permission only while the application is open and in the foreground. While correlating the given answers with each other a significant result could be found between the conspicuousness of the *Maybe* button and the control of the

Table 2. Frequencies of the permissions where the temporary access permission would be used.

		Yes	No
Location temporary	Frequency	25	6
	Percentage	80.6	19.4
Phone temporary memory	Frequency	11	20
	Percentage	35.5	64.5
SDcard memory temporary	Frequency	6	25
	Percentage	19.4	80.6
Contacts temporary	Frequency	11	20
	Percentage	35.5	64.5
Messeges Messenger temporary	Frequency	9	22
	Percentage	29	71
Messeges SMS temporary	Frequency	13	18
	Percentage	41.9	58.1
Calender temporary	Frequency	13	18
	Percentage	41.9	58.1
Microfon temporary	Frequency	9	22
	Percentage	29	71

personal data (question 1, statement 4). A Spearman-Rho test was performed, which showed a significant negative correlation of mean effect strength with $r = -.36$ and $p < .042$. Next, we examined whether the importance of privacy protection correlates with the statements from the first question. or the level of acceptance, the value of 0.05 was chosen. A Spearman-Rho test was also performed because the data set on the importance of privacy did not show a normal distribution. Only a positive ($r = 0.72$) significant ($p < .000$) correlation with a high effect size between the importance of privacy control (question 1, statement 4) was detected. All other correlations were not significant. In further analysis, it was checked whether the answers to the second question, which was to record the preferred duration of the participants for *Maybe* button, correlated with the statements of the first question. Only a significant ($p < .043$) negative correlation ($r = -0.37$) of a mean effect strength could be found with the importance of personal data control. Furthermore, with the help of another Spearman correlation, it could be determined that people with a higher technical affinity are more likely to feel annoyed by the repetitive requests than people with a lower technical affinity ($r = 0.38$, $p < .035$).

4.2 Analysis of App's Activity-Log

After preparing the application data, it was possible to see which options were selected for the individual access authorizations.

Table 3. Frequencies of the suggestions for improvement for the temporary access authorization.

Not specified	Frequency	9
	Percentage	29.03
Access permission shorter intervals	Frequency	3
	Percentage	9.68
Access permission longer intervals	Frequency	1
	Percentage	3.23
Improvements Maybe Option visibility	Frequency	4
	Percentage	12.90
Data retrieval	Frequency	1
	Percentage	3.23
Explanation Function	Frequency	1
	Percentage	3.23
At will time	Frequency	5
	Percentage	16.13
Timer	Frequency	1
	Percentage	3.23
Authorization only current use	Frequency	1
	Percentage	3.23
Limit number of accesses	Frequency	1
	Percentage	3.23
Design	Frequency	3
	Percentage	9.68
Language Customize	Frequency	1
	Percentage	3.23
Explanation data usage	Frequency	1
	Percentage	3.23

Over a period of ten days and the provision of a total of 24 tasks, four tasks were sent regarding the activation of push notification. Five participants (16.1%) did not grant this permission, while 26 persons (83.9%) granted it once with *Allow* button. The access authorization was rejected ("Deny") not once by 30 persons (96.8%), but twice by one person (3.2%). The *Maybe* button was used by two participants (6.5%) three times, two others (6.5%) twice and a single person (3.2%) once. Thus, a total of 26 participants (83.9%) did not select the option with the *Maybe* button. A total of 26 times (66.7%) the *Allow* button was selected, twice (5.1%) the *Deny* button and eleven times (28.2%) the *Maybe* button was selected. A total of ten tasks were set regarding the release of the position data (Location). Two participants (6.5%) did not even accept the access

Table 4. Frequencies of the reasons for using the *Maybe* button.

Control		Not used		No reason		Data protection		Temporary use		Mistrust app	
Frequency	%	Frequency	%	Frequency	%	Frequency	%	Frequency	%	Frequency	%
3	9.68	8	25.81	6	19.35	3	9.68	5	16.13	2	6.45

authorization, while 29 persons (93.5%) granted the authorization once (Allow). The access authorization was not even accepted by 27 persons (87.1%), but was rejected once by three persons (9.7%) and twice by one person (3.2%) (Deny). The temporary access authorization was chosen six times by two participants (6.5%). Thus a total of 29 participants (93.5%) did not choose the option with this access authorization. A total of 29 participants (74.4%) selected the *Allow* option 29 times, five times (12.8%) the *Deny* option and twelve times (30.8%) the temporary access authorization *Maybe.* In total, five tasks were set regarding the release of contact data (Contacts). Six participants (19.4%) did not even accept the access authorization, while 25 persons (80.6%) granted the authorization once (Allow). Access authorization was not even accepted once by 28 persons (90.3%), but rejected once by two persons (6.5%) and twice by one person (3.2%) (Deny). The temporary access authorization was chosen four times by two participants (6.5%), three times by two other participants (6.5%) and twice by one person (3.2%). Thus a total of 26 participants (83.9%) did not choose the option with this access authorization. A total of 25 times (64.1%) the option *Allow* was selected, four times (10.3%) the option *Deny* and 16 times (41%) the temporary access authorization *Maybe.* A total of four tasks were set regarding the release of the device storage. Four participants (12.9%) did not even accept the access authorization, while 27 persons (87.1%) granted the authorization once (Allow). The access authorization was not even accepted by 30 persons (96.8%), but one person (3.2%) rejected it three times (Deny). The temporary access authorization was chosen four times by three participants (9.7%), three times by one participant (3.2%) and twice by another participant (3.2%). This means that a total of 26 participants (83.9%) did not select the option with this access authorization. A total of 27 times (69.2%) the option *Allow* was selected, three times (7.7%) the option *Deny* and 17 times (43.6%) the temporary access authorization *Maybe* was selected. In total, two tasks were set regarding the release of the calendar (Calender). Six participants (19.4%) did not even accept the access authorization, while 25 persons (80.6%) granted the authorization once (Allow). The access authorization was not once rejected (Deny) by 30 persons (96.8%), but twice by one person (3.2%). Temporary access authorization was chosen twice by four participants (12.9%) and once by one participant (3.2%). Thus a total of 26 participants (83.9%) did not choose the option with this access authorization. A total of 25 participants (64.1%) chose the option *Allow*, two participants (5.1%) chose the option *Deny* and nine participants (23.1%) chose the temporary access authorization *Maybe.*

Regardless of the different tasks, not even one requested access authorization was accepted by two persons (6.5%), while two persons (6.5%) accepted once,

Table 5. Summary of app-activity-log from the prototype where $F = frequency, \% = percentage$ & $N =$ number of participants.

Permissions	Notification			Location			Contacts			Storage			Calendar			Total		
	F	%	N	F	%	N	F	%	N	F	%	N	F	%	N	F	%	N
Accept	26	66.67	26	29	74.36	29	25	64.10	25	27	69.23	27	25	64.10	25	132	61.97	29
Maybe	11	28.21	5	12	30.77	2	16	41.03	5	17	43.59	5	9	23.08	5	65	30.52	6
Decline	2	5.13	1	5	12.82	4	4	10.26	3	3	7.69	1	2	5.13	1	16	7.51	6

one person (3.2%) twice, two other persons (6.5%) four times and 24 persons (77.4%) five times.

In total, 25 participants (80.6%) did not refuse to access once, while four persons (12.9%) refused access once, one person (3.2%) refused it three times and another person (3.2%) refused it nine times.

The temporary access authorization was not even chosen once by 25 participants (80.6%) and once, twice, four and five times by one person (3.2%). Two other persons (6.5%) selected the temporary access authorization 17 times. From this it could be calculated that of the total of 213 requested access authorizations, 132 (62%) were accepted and only 16 (7.5%) were rejected. The remaining 65 authorizations (30.5%) were granted temporarily by selecting the *Maybe* option. We analyzed the machine log generated by the prototype app over a period of ten days and the provision of a total of 24 tasks. As summarized in Table 5, a total of 29 (93.5%) participants used the *Allow* button and six participants (19.4%) each used the *Maybe* and *Deny* buttons. We also found that only one participant chose the *Allow* button as the final consent, after using the *Maybe* button. Five other participants used the *Maybe* button, and did not use *Allow* button in the end.

After evaluating the duration chosen when using the *Maybe* button, two participants (6.5%) chose one day as the duration twice, and one participant (3.2%) chose one day seven times, twelve times, 13 times and 16 times respectively. Only one person chose a duration of three days and continued selecting it ten times in a row. Another person chose a duration of seven days and continued selecting it three times in a row. Thus, the option of 'one day' was chosen 52 times (80%), 'three days' ten times (15.38%), and 'seven days' three times (4.62%).

The shortest duration for the partial consent could be found in the push notifications with $M = 3.5$ ($SD = 1.73$). In ascending order, the duration of *Maybe* function for the calendar had an average value of $M = 3.6$ ($SD = 2.61$), the device memory of $M = 4.2$ ($SD = 2.78$), the contacts of $M = 5.2$ ($SD = 2.17$) and the location data of $M = 6.33$ ($SD = 5.51$). Overall, the average duration of granted *Maybe* button was 4.57 days.

The next step was to compare the behavior of the group that claimed to have used the *Maybe* button with the group that claimed not to have used it. The first group selected the *Allow* button on an average of 2.9 times ($SD = 2.28$), 6.3 times ($SD = 7.42$) the *Maybe* button, and 1.2 times ($SD = 2.90$) the *Deny* button. The second group selected the *Allow* button an average of 4.9 times ($SD = 0.3$), 0.1 times ($SD = 0.44$) the *Maybe* button, and 0.19 times ($SD = 0.40$)

the *Deny* button. All data records showed the absence of a normal distribution. For this reason, it was decided to perform the Man-Whtney-U-test, which gave a significant result between the groups regarding the *Allow* and *Maybe* button.

5 Discussion

In this section, we would like to revisit the hypotheses that are discussed in Sect. 2.2, and reflect on the corresponding observations of this study. We also would like to elaborate on the correlation drawn form the collected data, and highlight upon the corresponding interpretations.

5.1 Hypothesis #1

When comparing the overall frequency of access granting decisions (choosing from the three options—*Allow/Maybe /Deny*), 30% of permission requests were granted with *Maybe* button, i.e., partial consent. In addition, the *Maybe* button was used extensively by a total of six out of 31 people (19.4%), although a total of ten people (32.3%) stated that they clicked on the *Maybe* option. Here, moral pressure may also have had an influence on the information provided in the questionnaire, with participants stating that they had used the temporary access granting function, which is the focus of this lab study, even though they had not.

The hypothesis can thus only be partially confirmed because 80.6% of the participants did not use the *Maybe* button. Furthermore, of these ten people, nine participants stated that they had correctly understood the function of the *Maybe* button. Seven people gave a correct explanation of the *Maybe* button. Thus, it can be assumed that the majority of the participants who used the *Maybe* button understood its function. However, the question was only asked to those participants who also confirmed that they had clicked on the *Maybe* button. In retrospect, it would also have been interesting to see whether the participants who decided against using the *Maybe* button correctly understood how it worked.

The partial consent was predominantly (80%) assigned for one day. However, this phenomenon could occur due to the default setting of the *Maybe* button, which was set to one day, on the other hand, this behavior could also reflect on the participants' desire to release the permission only for a short period of time. While comparing the individual types of permission request with each other, it was noticeable that the *Maybe* button was the most frequently used for the memory and contacts. Therefore, it can be assumed that the *Maybe* button could be applied especially in these two categories. However, these results partially contradict the statement that the functionality of the *Maybe* button is the most desired for the location data. This behavior can be explained with the fact that the participants were subjected to a certain moral pressure to fulfill the set of tasks and therefore used *Maybe* button for access requests within this experiment that they would probably not grant in everyday life. Furthermore, it can be assumed that the participants were aware of the problem that a one-time usage of the *Maybe* button (to e.g., contacts or the memory) already allows the complete disclosure of the data, while a time limit on the release of location data only provides a subset of the data for a limited duration.

5.2 Hypothesis #2

On an average, the *Maybe* button was rather liked with a value of 5.87 on the scale. However, dividing the sample into two groups resulted in a slight majority disliking the *Maybe* button. It should be noted here, however, that over a third of the sample rated the statement with a score of 5.00. Only four participants disagreed with this statement with a lower value on the scale. Due to the large proportion of participants who rated the statement with a 5, i.e., were just on the threshold of agreeing, it would be worth considering here whether this slight displeasure could be counteracted by a prior explanation. The questionnaire also revealed that just under two-third of the participants considered the specified duration of *Maybe* button to be adequate. The desire for longer and shorter time periods, which was mentioned by the other participants, could be addressed with the help of an adjustable time for the *Maybe* button. Based on the presented findings, it is difficult to judge in this case whether the hypothesis could still be upheld. The information provided by the participants did not suggest a clear and conclusive consent to the *Maybe* button. However, there is a possibility that if this feature finds favor with half of the sample, the use of *Maybe* button could be increased from about 50% to 90% after a prior explanation.

5.3 Hypothesis #3

Four-fifth of the participants used the *Maybe* button for location data, while the use of it tended to be rejected for all other permissions. This could again be related to the problem of handing over the data once, which was already discussed in Sect. 5.2. Repeated access to memory and contact data seldom results in gaining new data after a short period of time. This is different for location data, where permanent access can create a movement profile and there are times when it is not wanted that third parties can locate the current position of the user. This could also be the reason why participants were found to be unwilling to apply the *Maybe* button for all other permissions. The high percentages for the use of *Maybe* button can be found in the SMS/messages and the calendar, following the location data. The reason for this could be that if the messages or the calendar were accessed, a third party could track when the end users were performing which activity.

Furthermore, the questionnaire reveals that over 45% of the participants would use the *Maybe* button on their private device. Overall, however, an average of 5.35 on a scale of 1 to 10 can be observed, so that a slight agreement regarding private use can be concluded. Comparing the use of *Maybe* button by 19.4% of the participants with the 45% who would use partial consent scheme on their private device after the experiment, it is noticeable that highlighting the function could potentially double the number of users. When looking at the results, this hypothesis can also only be partially upheld, because only just less than half of the participants stated that they intended to use the *Maybe* button, i.e., partial consent on their private device in everyday life.

5.4 Hypothesis #4

The results of the questionnaire show that over two-thirds of the participants did not find the recurring request for partial consent annoying. Also, on average, a mean score of 4.30 is obtained on a scale of 1 to 10, which confirms that the majority of participants disagreed with the statement that the recurring request for consent annoyed them. Another indication that granting access with partial consent is not perceived as annoying as the repeated use of the option by almost all participants who used *Maybe* button. As already mentioned in relation to hypothesis #1, only one person subsequently granted the permission permanently after using the *Maybe* button. Thus, it can be assumed that at least for the participants who used the *Maybe* button, the repeated appearance of the consent request did not cause any annoyance. For this reason, the fourth hypothesis cannot be upheld.

5.5 Hypothesis #5

Nine-tenth of the participants considered the control of their data important. Overall, a mean value of 8.19 (SD = 1.94) was achieved. Thus, it can be concluded that the control of their data was important to the participants. If this finding is now combined with the observations from Sect. 5.4, that the recurring requests for temporary access permission do not cause annoyance, and Sect. 5.3, that the *Maybe* button is desired, it could be concluded that the users prefer control over their data despite the recurring request for access permission. Furthermore, it can be determined that privacy protection scored a mean of 5.97 (SD = 0.91) on a scale of 1 to 7. Thus, it can be assumed that privacy protection was highly valued by the participants.

In addition, more than two-thirds of the participants disagreed with the statement that user-friendliness was more important than control of their data. This was confirmed by the mean value, which is only 4.0 (SD = 2.70) on a scale from 1 to 10 for this statement. Thus, it can be assumed that the participants preferred control over their data despite the recurring request for partial consent, so that the fifth hypothesis can be upheld.

6 Conclusion and Future Work

The goal of this study was to explore the acceptance and use of partial consent. For this purpose, a prototype with *Maybe* button was implemented on an emulated interface of Android's access control model. This allowed users to temporarily grant permissions for 1, 3 or 7 days, depending on the preferred duration.

The participants were given two to three tasks daily over a period of ten days. While performing the tasks, the participants were encountered with permission requests, e.g., access to the memory. However, these permission requests were simulated on the interface only, so that the prototype did not access any permission. The app merely transmitted information (only a few bytes of data) that included: *(i)* when (time stamp), *(ii)* which authorization, and *(iii)* which option *(Allow/Maybe/Deny)* was selected.

In a nutshell, the main findings from the collected data and statistical analysis of this study are: *(a)* the *Maybe* button was only used by one fifth of the participants, *(b)* almost all participants who used the *Maybe* button also correctly understood its function, *(c)* when the *Maybe* button was used, it was repeatedly selected after the selected period of time had elapsed, so that in most cases there was no final consent to the corresponding permission, *(d)* the *Maybe* button was mostly granted for a period of one day, *(e)* the *Maybe* button was most often used for access to the device memory and contacts, *(f)* the questionnaires show that *Maybe* button is highly desired for the location data, *(g)* the majority of participants stated that they liked the *Maybe* button, *(h)* most of the participants were satisfied with the validity period of the *Maybe* button, only sporadically a variable with a self-defined time span was desired, *(i)* after filling out the questionnaire, almost half of the respondents stated that they want to use the *Maybe* button on their private device. *(j)* the recurring request of partial consent did not cause annoyance to most participants. *(k)* most of the participants preferred to control their data despite the recurring requests for partial consent. *(l)* the participants who indicated that the control of their personal data was important to them, chose a shorter duration when granting access with partial consent, i.e., *Maybe* button.

For the future research, it is planned to further develop this prototype and visualize the requested permissions in order to offer a comparison between different apps and their behavior regarding privacy. This feature should enable the user to grant a temporary permission with the *Maybe* button in order to assess the trade-off between data disclosure and privacy concerns. In future, we also would like to explore the requirements and technical feasibility of *Maybe* button in different platforms, e.g., consenting to cookie notification on web interface. In case of the mobile app's interface, we would like to carry out user studies in a lab environment, and identify the challenges associated with human factors.

References

1. Acquisti, A., Grossklags, J.: Privacy and rationality in individual decision making. IEEE Secur. Priv. **3**(1), 26–33 (2005)
2. Albaum, G.: The Likert scale revisited. Mark. Res. Soc. J. **39**(2), 1–21 (1997)
3. Bock, S.: My data is mine - users' handling of personal data in everyday life. In: SICHERHEIT 2018, pp. 261–266. Gesellschaft für Informatik e.V., Bonn (2018)
4. Bock, S., Momen, N.: Nudging the user with privacy indicator: a study on the app selection behavior of the user. In: Proceedings of the 11th Nordic Conference on Human-Computer Interaction: Shaping Experiences, Shaping Society. ACM (2020)

5. Bock, S., Momen, N.: A study on user preference: influencing app selection decision with privacy indicator. In: Stephanidis, C., Marcus, A., Rosenzweig, E., Rau, P.-L.P., Moallem, A., Rauterberg, M. (eds.) HCII 2020. LNCS, vol. 12423, pp. 579–599. Springer, Cham (2020). https://doi.org/10.1007/978-3-030-60114-0_39
6. Brown, B.: Studying the internet experience. HP laboratories technical report HPL 49 (2001)
7. European Commission: Regulation (EU) 2016/679 of the European Parliament and of the Council of 27 April 2016 on the protection of natural persons with regard to the processing of personal data and on the free movement of such data (General Data Protection Regulation). Off. J. Eur. Union L119 (2016)
8. Norwegian Consumer Council: Deceived by design, how tech companies use dark patterns to discourage us from exercising our rights to privacy. Norwegian Consumer Council Report (2018)
9. Custers, B.: Click here to consent forever: expiry dates for informed consent. Big Data Soc. **3**(1) (2016)
10. Faden, R.R., Beauchamp, T.L.: A history and Theory of Informed Consent. Oxford University Press, Oxford (1986)
11. Fritsch, L.: Partial commitment – "try before you buy" and "buyer's remorse" for personal data in big data & machine learning. In: Steghöfer, J.-P., Esfandiari, B. (eds.) IFIPTM 2017. IAICT, vol. 505, pp. 3–11. Springer, Cham (2017). https://doi.org/10.1007/978-3-319-59171-1_1
12. Fritsch, L., Momen, N.: Derived partial identities generated from app permissions. In: Fritsch, L., Roßnagel, H., Hühnlein, D. (eds.) Open Identity Summit 2017, pp. 117–130. Gesellschaft für Informatik, Bonn (2017)
13. Fuchs, C.: Facebook, web 2.0 und ökonomische überwachung. Datenschutz Datensicherheit-DuD **34**(7), 453–458 (2010)
14. Hatamian, M., Wairimu, S., Momen, N., Fritsch, L.: A privacy and security analysis of early-deployed COVID-19 contact tracing android apps. Empir. Softw. Eng. **26**(3), 1–51 (2021)
15. Momen, N.: Measuring apps' privacy-friendliness: introducing transparency to apps' data access behavior. Ph.D. thesis, Karlstad University, Department of Mathematics and Computer Science (2020)
16. Momen, N., Bock, S.: Neither do i want to accept, nor decline; is there an alternative? In: Stephanidis, C., Antona, M. (eds.) HCII 2020. CCIS, vol. 1226, pp. 573–580. Springer, Cham (2020). https://doi.org/10.1007/978-3-030-50732-9_74
17. Momen, N., Bock, S., Fritsch, L.: Accept - maybe - decline: introducing partial consent for the permission-based access control model of Android. In: Proceedings of the 25th ACM Symposium on Access Control Models and Technologies, SACMAT 2020, pp. 71–80. ACM (2020)
18. Momen, N., Fritsch, L.: App-generated digital identities extracted through android permission-based data access - a survey of app privacy. In: SICHERHEIT 2020, pp. 15–28. Gesellschaft für Informatik e.V., Bonn (2020)
19. Momen, N., Hatamian, M., Fritsch, L.: Did app privacy improve after the GDPR? IEEE Secur. Priv. **17**(6), 10–20 (2019)
20. Momen, N., Pulls, T., Fritsch, L., Lindskog, S.: How much privilege does an app need? Investigating resource usage of android apps (short paper). In: 2017 15th Annual Conference on Privacy, Security and Trust (PST), pp. 268–2685 (2017). https://doi.org/10.1109/PST.2017.00039

UI Development of Hardcore Battle Royale Game for Novice Users

Woo Jin Choi and Chang Joo Lim(✉)

Korea Polytechnic University, 237, Sangidaehak-ro, Siheung-si, Republic of Korea
{Chldnwls94,scjlim}@kpu.ac.kr

Abstract. Recently, while the number of users who enjoy online games has increased rapidly, the number of users of battle royale games has continued to decline. This study aims to solve novice users' departure problem in battle royale genre through improving the user interface (UI).

First, the study conducted a research on UI trends of battle royale games that are currently in service and carried out ideation of UI design for novice users. Second, as an experimental game, Battleground was selected and built as a simulator that can compare and evaluate the original UI and newly proposed UI based on trend research. Finally, usability test was carried out on 50 participants, the results were statistically analyzed and compared.

As a result, it is confirmed that the proposed UI has achieved high usability for both novice and hardcore users. This suggests that the UI design proposed in this study can maintain the hardcore contents of battle royale genre and also lower the barriers to entry for novice users.

Keywords: Digital games · Online games · Hardcore games · Battle royale · User Interface

1 Introduction

1.1 Research Background

Recently, as the inflow phenomenon of new online game user is active, the game user base is changing from hardcore users to inexperienced and novice users [1]. Therefore, the importance of attracting and maintaining users is being emphasized, in a situation where various genres of games are globally being released on various platforms simultaneously [2]. On the other hand, battle royale, which have become a popular genre starting with Player Unknown's Battleground (PUBG, Battleground) in 2017, are showing a continuous decline in the number of game users. Battlegrounds, which once had average of 1 million concurrent users currently maintains 300,000, while Apex Legends and Ring of Elysium, which have released consecutively maintains average of 200,000, and 250,000 [3]. Battle royale games are failing to maintain novice users due to genre characteristics which has high barriers to entry and gives significant impact on stress levels [4].

C. Stephanidis et al. (Eds.): HCII 2021, LNCS 13094, pp. 217–228, 2021.
https://doi.org/10.1007/978-3-030-90238-4_16

1.2 Research Methods

This study aims to solve the question, "Can the departure problem of novice users in the battle royale genre be resolved through user interface (UI) improvement?" and approaches consists of three stages: UI trend research, prototyping and evaluation [5]. First, in UI trend research, three different shooting-based hardcore battle royale games that are currently in service were selected. The UI design features of games were classified according to the heuristic evaluation by Jacob Nielsen [6]. Based on the classified characteristics, the ideation of UI design that can enhance the learnability and efficiency of novice users was carried out. Second, as an experiment game, Battleground was selected, and built as a simulator that core experience of game is playable [7]. Also it can compare and evaluate the original UI and newly proposed UI. Finally, based on experimental results and usability test, UI improvement ideas of hardcore battle royale game for novice users were proposed.

2 Hardcore Battle Royale Game

2.1 Usability Heuristics

Shooting-based battle royale games usually includes in-game weapon customization systems. It runs through the Inventory UI and help users to upgrade their weapons in a unique way. However, this system requires user to recognize and be familiar with information of variety attachable parts for each weapons, which makes battle royale as a hardcore genre.

Therefore, this study highlights the features captured from the UI trend research and provides insight into some shortcoming features [8]. UI trend research was focused on identifying the novice users' pain point when executing inventory UI that are provided in hardcore battle royal games. First, three different battle royale games (Battleground, Apex Legend, and Ring of Elysium) were selected, which have high concurrent users and are registered in the popular category in Steam: Global online game digital distribution service [9]. The classified UI features were defined in Table 1, according to the five factors of heuristic evaluation.

Table 1. Usability heuristics of hardcore battle royale games

Usability heuristics		Battleground	Apex legend	Ring of elysium
1	Visibility of system status	Blur background & 80% screen usage	Blur background & 30% screen usage	Opacity background & 50% screen usage
2	Consistency and standards	Monochrome flat rectangle	Black flat rectangle	White flat rectangle
3	Recognition rather than recall	Content highlight	Content highlight & color rating	Content highlight

(*continued*)

Table 1. (*continued*)

Usability heuristics		Battleground	Apex legend	Ring of elysium
4	Flexibility and efficiency of use	Drag and drop customization	Wheel and click customization	Drag and drop customization
5	Aesthetic and minimalistic design	Simplified item information	Simplified item information	Simplified item information

2.2 UI Design Trends

Battleground. Battleground's inventory UI is designed with flat rectangle of solid color, reduces users' confusion with strong consistency. It also provides (1) content highlight effect of compatible weapon slot which allows users to perform many tasks quick in urgent situations.

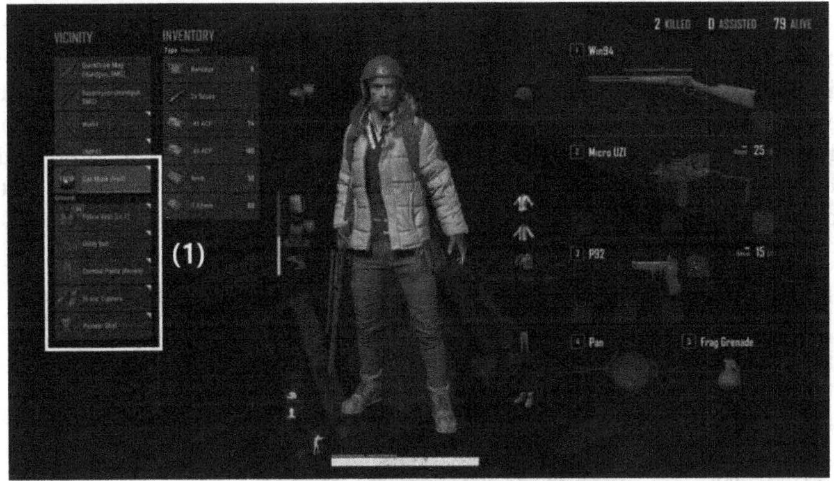

Fig. 1. Battleground inventory UI

Apex Legend. Apex Legend's Inventory UI is designed to help users easily recognize the value of parts by dividing the performance of items and weapon parts with (1) Color rating. It also provides automatic replacement of weapon parts in a situation of acquiring high performance parts. This feature shortens the time of early farming and leads to a fast pace of the game overall (Fig. 2).

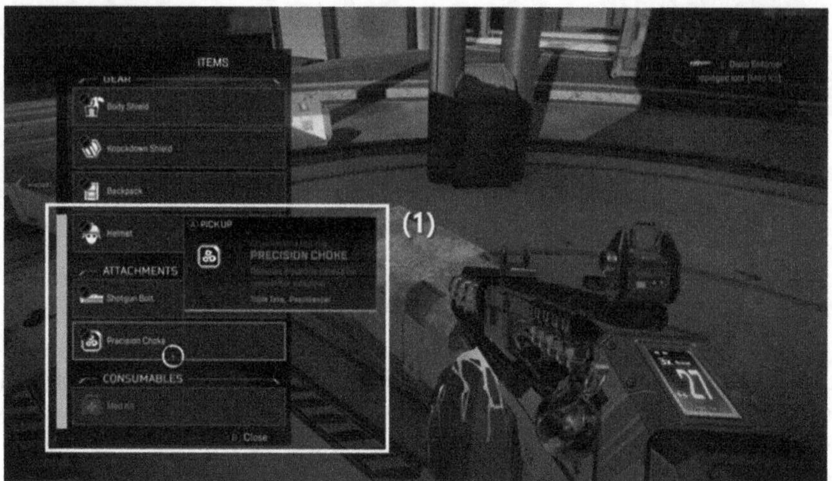

Fig. 2. Apex Legend inventory UI

Ring of Elysium. Ring of Elysium's inventory UI is designed with flat rectangle with white opacity, and (1) automatically classifies items and weapon parts when acquired. The weapon customization slots intuitively deliver the type of parts that can be attached and whether it is attachable or not through (2) saturation and shape. It also provides automatic attachment feature for the acquired parts, according to the priority of the weapons set by the user (Fig. 3).

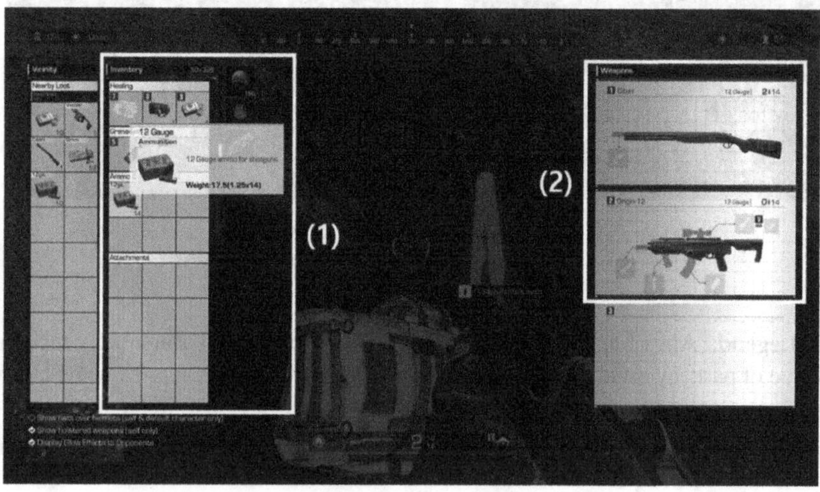

Fig. 3. Ring of Elysium inventory UI

2.3 Ideations for Novice Users: Battleground

As a result of UI trend research, inventory UI of the selected games had three common features. First, it is viewed with an opaque background that does not completely block gameplay, allowing users to control information while continuously recognize the situation of the battlefield. Second, inventory items induce interaction with users through highlighting contents. Third, it first delivers simplified information and then materialize it according to users' behavior. In addition, the features such as automatic classification, simplified behavior, and color rating shown in two games (Apex Legend, Ring of Elysium) are considered to be efficient for novice users.

In this study, Battleground which has the largest user pool in battle royal genre was selected as an experimental game. Battleground's inventory UI can be positive for experienced hardcore users to come up with creative gameplay ideas, however it is rather an entry barrier for inexperienced novice users due to difficulty in recognizing and learning. Furthermore, unlike hardcore users, novice users are more occasionally or periodically exposed to the online game, which means more guidance of learning is needed. Therefore, this study has proposed a new UI design for novice user that could enhance the selective attention and simplified behavior of the original UI.

3 Materials and Methods

3.1 Game Simulator

System Configuration. The overall system configuration of the game simulator is shown through diagram (Fig. 1) below. The simulator consists of a client and a server, built with Unity 3D: Cross-platform game engine. The client includes two games: Game A and Game B. All contents of each game are the same except UI. According to the system behavior, it manages and stores user info, gameplay data, and usability test results. The stored data is sent to the server via the User Data Protocol (UDP) method, and received data packets were saved as an excel file format (Fig. 4).

Game Scenario. The game consists of three levels (Level 1, Level 2, Level 3) and one level contains three tasks. Tasks are divided into three parts: (1) weapon parts collection, (2) weapon customization, and (3) enemy annihilation. When all three tasks are completed, system automatically proceeds to the next level (Fig. 5).

Data Collection. The games included in the simulator measure three quantitative data to analyze the result in more detail [11]. The game system records users' weapon customization time, level complete time, and number of level failures. These three metrics are intended to measure the learnability and efficiency of battle royale games inventory UI. First, the weapon customization time is the time it takes for a user to open a customizing panel and modify the part in the same way as the modification method suggested in the task. Second, the level complete time is the elapsed time of a user to complete all three tasks in one level. Finally, the number of level failures is the number of times a user is eliminated for failing the task at a given time. In the event of users' level failure, the level will be restarted and complete time will be accumulated.

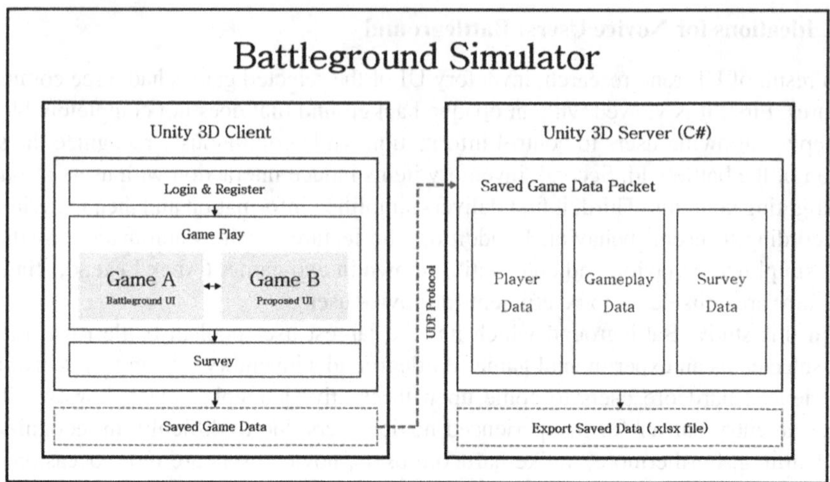

Fig. 4. System configuration diagram

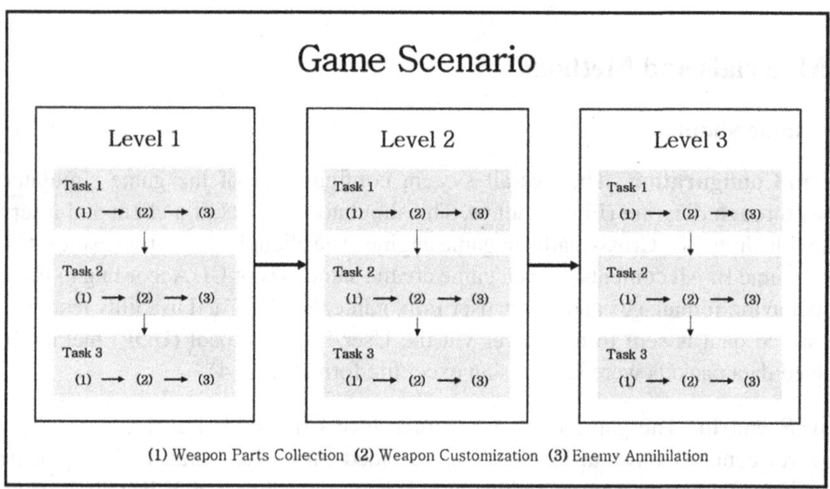

Fig. 5. Game scenario diagram

3.2 Game UI Implementation

Game A: Original UI. Game A (Fig. 6) includes an original inventory UI which imitated the Battleground's UI layout, visual concept, and customization system. It is displayed on the entire screen, including the inventory UI on the left side and weapon status UI on the right side. The collected weapon parts can be attach with mouse click or drag-and-drop methods, and (2) highlight and (3) customized effects is expressed in the same way as original.

Game B: Proposed UI. Game B (Fig. 7) includes newly designed UI which was proposed in this study. It is displayed on the right side of the screen, including the (1)

Fig. 6. Game A: Battleground user interface

inventory UI and weapon status UI. It also (2) automatically classifies the collected weapon parts in slot. The weapon parts can be attached with a mouse click and also provide highlight effect and specific information by mouse hover.

Fig. 7. Game B: Proposed user interface

3.3 Experiment

Participants. The experiment was conducted by recruiting a total of 50 participants (male: 42; female: 8; average age 24). All participants except two had experience in battle royale games. The group distribution was divided into a novice and expert group, based on self-evaluation of participants' user experience, game play time and proficiency before the experiment (Table 2).

Table 2. Demographic characteristic of participants

Variable		Number (N)	%
Gender	Male	42	84
	Female	8	16
Experience of hardcore FPS game	Yes	48	96
	No	2	4
User skill	**Novice**	**28**	**56**
	Expert	**22**	**44**

Procedure. Participants sat in designated seats and conducted an experiment with a brief overview of the simulator. When participants completed both games, the experiment was ended and usability test was carried out (Fig. 8).

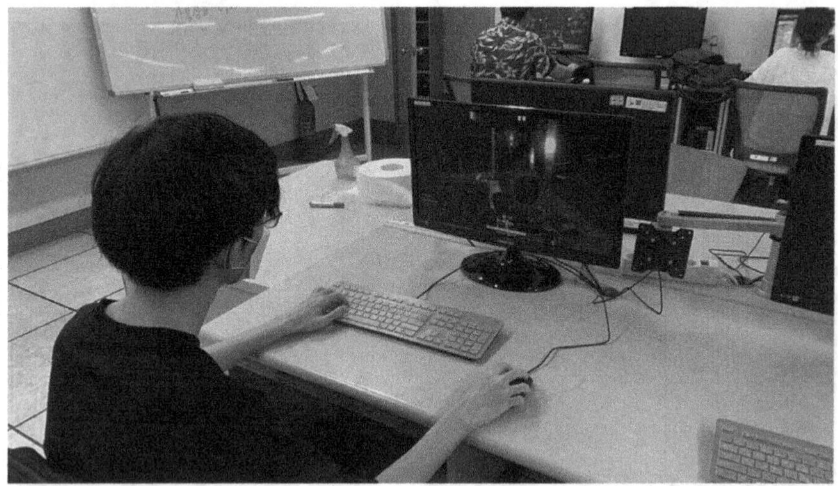

Fig. 8. Experiment environment

4 Results

This study used a method to statistically analyze and compare results from the experiment to understand magnitude of the effect the proposed UI has on the novice and expert users [10]. In Table 3, the customization time is the average time taken by participants to perform weapon customization among the three tasks presented for each level, the play time is the average time taken to complete the level, and the level failure is the average number of times that user failed to complete the level.

The novice group showed significant results in Game B, which were measured relatively low for all collected data compared to Game A. In particular, the average customization time was reduced compared to the expert group by 10.43 s at Level1 (L1), 28.01 s at Level2 (L2), and 17.39 s at Level3 (L3). The expert group showed similar aspects to the novice group, except that L2 and L3 average play time in Game B was higher than in Game A.

Table 3. Experimental result

Game	User skill	Variable	L1	L2	L3
Game A: Battleground UI	Novice	Customization time(s)	32.04	61.21	96.61
		Play time(s)	457.68	515.32	807.02
		Level failure	0	0.36	6.75
	Expert	Customization time(s)	13.68	33.51	70.60
		Play time(s)	332.84	337.13	559.91
		Level failure	0	0.32	3.24
Game B: Proposed UI	Novice	Customization time(s)	**15.29**	**54.01**	**72.77**
		Play time(s)	**328.95**	**504.36**	**655.83**
		Level Failure	0	**0.18**	**5.43**
	Expert	Customization time(s)	**7.36**	**54.32**	**64.15**
		Play time(s)	**223.11**	531.24	610.93
		Level failure	0	**0.14**	**1.05**

Both groups showed the longest average play time and the highest number of level failure in L3 of each game, which is found to be the most difficult and requires a long time and iterate to complete. Therefore, this study collected the change in weapon customization time of users that have level failures more than three times in L3 to measure the learnability and efficiency of both games. The result is showed through graph (Fig. 9) below.

As a result, both groups achieved a small time reduction of 1.15 s in Game A, while a large time reduction of 13.66 s in Game B. In addition, customization time reduction in Game B was large until the third failure, and gradually narrowed down after the fourth failure. This result confirms that Game B is more efficient and ease to learn than Game A, and gives positive impact on both groups.

The usability test result of both groups in Game B: Proposed UI is shown below in Fig. 10. The assessment was conducted on a five-point scale for the five factors of heuristic evaluation used in trend research, and average value of each factor was drawn through a radial graph.

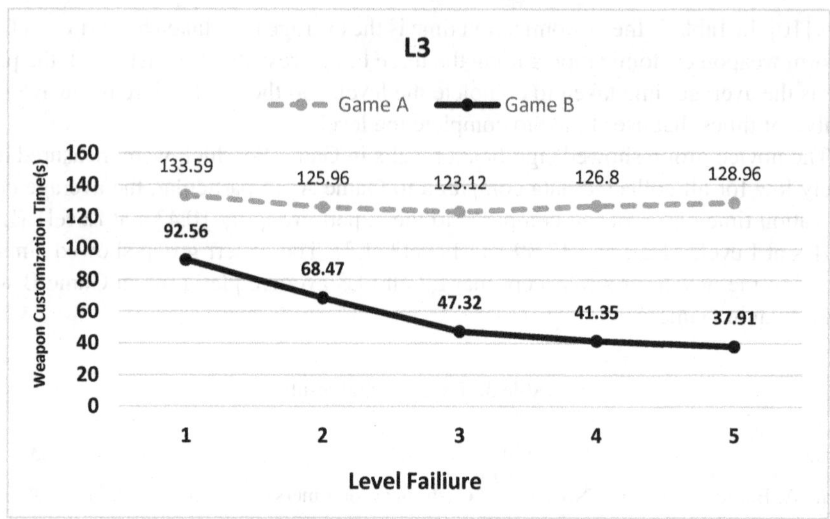

Fig. 9. Participants learnability and efficiency in L3

Both groups scored similar for all factors, indicating that the satisfaction of Game B: proposed UI is almost the same for novice and expert users. Efficiency and visibility were scored higher than average value of factors in each group (Novice: 3.70, Expert: 3.89). Efficiency (Novice: 4.24, Expert: 4.45) scored the highest among all factors, which indicates Game B: Proposed UI becomes easier with iteration. Visibility (Novice: 4.03, Expert: 3.97) also scored higher than the average, which indicates Game B: Proposed UI helped users to easily understand the UI status.

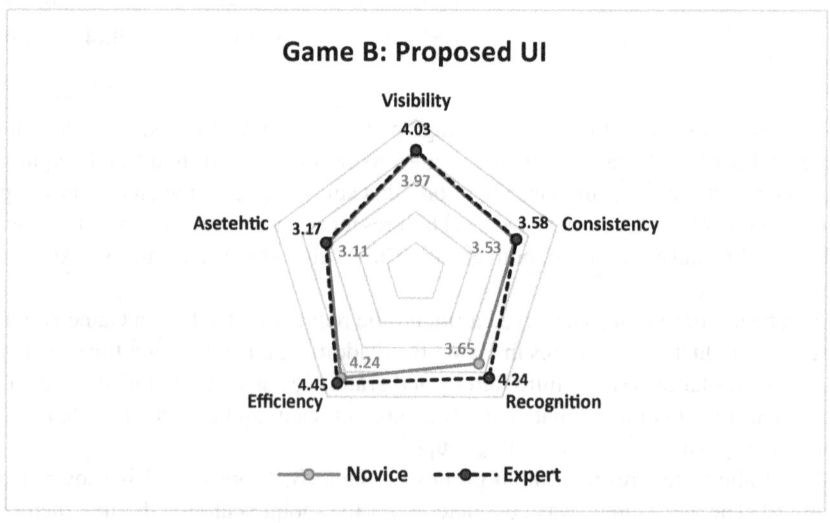

Fig. 10. Usability test diagram of proposed UI

5 Discussion

In the novice group, it is confirmed that learnability was relatively low in Game A, due to inexperience, lack of affordance, and distraction. On the other hand, in Game B, learnability and efficiency appeared to be high due to classified information architecture, and simplified behaviors that users can take. This can be viewed that novice users prefer a layout group with similar attributes of information and an intuitive UI design that intends simplified behavior.

In the Expert group, users have showed better results in Game B, which means it is also more efficient and easier for skilled users. All measurements in the expert group were relatively higher than novice group, which can also be viewed as skilled users with user experience can learn quick enough with time investments and iterations, even if the UI is not intuitive and complex to use.

The result indicates that proposed UI has improved usability for both novice and hardcore users, and the need for well-explained layout design, smooth gaze flow and affordance.

6 Conclusion

This study proposed a solution to prevent departure of novice users in the hardcore battle royal games through UI development. Battle royale games have made various attempts such as updating game contents to maintain novice users, however this study has focused on UI improvement in order of UI trend research, game simulator development, and usability test.

As a result, it is confirmed that the proposed UI has achieved high usability for both novice and expert users. This suggests that UI improvement can keep hardcore contents of battle royale games and also lower the barriers to entry for novice users. If a game company that currently serving battle royale game is planning to design or improve UI for novice users, the proposed UI design and positive results based on this study is expect to be helpful. Furthermore, the research should be continued to refine, expand, and improve the unique UI design characteristics of battle royal genres and its accuracy and convenience.

Acknowledgement. This paper was supported by Korea Institute for Advancement of Technology (KIAT) grant funded by the Korea Government (MOTIE) (P0012725, HRD Program for Industrial Innovation).

References

1. Desurvire, H., Wiberg, C.: User experience design for inexpereienced gamers: GAP—game approachability principles. In: Bernhaupt, R. (eds.) Game User Experience Evaluation. Human-Computer Interaction Series, pp. 169–186. Springer, Cham (2015). https://doi.org/10.1007/978-3-319-15985-0_8

2. Brown, E.: The life and tools of a games designer. In: Bernhaupt, R. (eds.) Evaluating User Experience in Games. Human-Computer Interaction Series, pp. 73–87. Springer, London (2010). https://doi.org/10.1007/978-1-84882-963-3_5
3. Steam & Game Stats. https://store.steampowered.com/stats/?l=koreana. Accessed 01 June 2021
4. HaekaI, M.F., Moch, B.N., Muslim, E.: Stress emotion evaluation in battle royale game by using electroencephalogram. In: ICONESTI (5), pp. 1–5 (2020)
5. Koeffel, C., Hochleitner, W., Leitner, J., Haller, M., Geven, A., Tscheligi, M.: Using heuristics to evaluate the overall user experience of video games and advanced interaction games. In: Bernhaupt, R. (eds.) Evaluating User Experience in Games. Springer, London, pp. 169–186 (2010). https://doi.org/10.1007/978-1-84882-963-3_13
6. Nielsen, J.: Ten usability heuristics. http://www.useit.com/papers/heuristic/heuristic_list. Accessed 11 June 2021
7. McAllister, G., White, G.: Video game development and user experience. In: Bernhaupt, R. (ed.) Game User Experience Evaluation. Human–Computer Interaction Series, pp. 11–35. Springer, Cham (2015). https://doi.org/10.1007/978-3-319-15985-0_2
8. Mueller, F., Bianchi-Berthouze, N.: Evaluating exertion games. In: Bernhaupt, R. (ed.) Game User Experience Evaluation. Human–Computer Interaction Series, pp. 239–262. Springer, Cham (2015). https://doi.org/10.1007/978-3-319-15985-0_11
9. Steam: Browsing Battle Royale. https://store.steampowered.com/tags/en/battleroyale/. Accessed 01 June 2021
10. Brown, M., Kehoe, A., Kirakowski, J., Pitt, I.: Beyond the gamepad: HCI and game controller design and evaluation. In: Bernhaupt, R. (ed.) Game User Experience Evaluation. Human–Computer Interaction Series, pp. 263–285. Springer, Cham (2015). https://doi.org/10.1007/978-3-319-15985-0_12
11. Lemay, P., Maheux-Lessard, M.: Investigating experiences and attitudes toward videogames using a semantic differential methodology. In: Bernhaupt, R. (ed.) Evaluating User Experience in Games, pp. 89–105. Springer, London (2010). https://doi.org/10.1007/978-1-84882-963-3_6

The Reaches of Crowdsourcing: A Systematic Literature Review

Samantha Dishman and Vincent G. Duffy(✉)

Purdue University, West Lafayette, IN 47906, USA
{sdishman,duffy}@purdue.edu

Abstract. With the continued increased use of the Internet and mobile devices, the study of how people interact with each other is more relevant than ever. This development in communication has introduced a new question of how this can lead to a collective effort controlled by the masses. This collectivist system has been coined crowdsourcing. This study is a systematic literature review that aims to find emerging trends within the topic of crowdsourcing. This review was done using tools such as Harzing's Publish or Perish, Scopus, MAXQDA, Vicinitas, and VosViewer, and CiteSpace. Bibliometric analyses were done using these tools to show the emergence of crowdsourcing and in what ways researchers and the public are interested in the topic. These results are visualized using maps and word clouds. The results of the analysis showed an interest in potentially using crowdsourcing in the healthcare and transportation industries. Also, these results showed the emergence of new subtopics such as disaster preparedness and job recruitment techniques. It is imperative to learn how these potential avenues for crowdsourcing can be utilized and how it will uniquely affect each industry and its current job design. These findings are highlighted in the literature review and subsequent discussion.

Keywords: Crowdsourcing · Platform · User · Crowd · Data collection

1 Introduction and Background

Advancements in technology have allowed for communication between individuals to be easier than ever. This has led to the development of a new model of gathering and utilizing data, in which a large number of people participate in an activity in order to reach a collective goal. This phenomenon is referred to as crowdsourcing. Since this is a relatively new concept that has emerged within the past two decades, the definitions can vary between sources. In his overview of the intricacies of crowdsourcing, researcher Jakob Pohlisch cites several definitions of crowdsourcing that experts have developed before giving the following comprehensive definition:

Crowdsourcing is a type of participative online activity in which an individual, an institution, a non-profit organization, or company proposes to a group of individuals of varying knowledge, heterogeneity, and number, via a flexible open call, the voluntary undertaking of a task. The undertaking of the task, of variable complexity and modularity,

C. Stephanidis et al. (Eds.): HCII 2021, LNCS 13094, pp. 229–248, 2021.
https://doi.org/10.1007/978-3-030-90238-4_17

and in which the crowd should participate bringing their work, money, knowledge and/or experience, always entails mutual benefit. The user will receive the satisfaction of a given type of need, be it economic, social recognition, self-esteem, or the development of individual skills, while the crowdsourcer will obtain and utilize to their advantage what the user has brought to the venture, whose form will depend on the type of activity undertaken (Pohlisch et al. 2021).

Due to the adaptable nature of crowdsourcing for many different goals, several different methodologies for differentiating the types of crowdsourcing have emerged. According to Pohlisch, there are four types of crowdsourcing which are distinguished by the *differentiating* value between contributions and the *deriving* value from contributions. The first type is called a crowd rating, which is based upon many homogeneous contributions whose value is derived from their aggregate value (Pohlisch et al. 2021). The next type results from the aggregate of heterogeneous contributions referred to as crowd creation (Pohlisch et al. 2021). Next is crowd processing, which relies on a large number of individual homogeneous contributions (Pohlisch et al. 2021). Finally, crowd solving relies on individual heterogeneous contributions meant to solve a singular problem (Pohlisch et al. 2021). These types of crowdsourcing Pohlisch outlined can be visualized in Fig. 1.

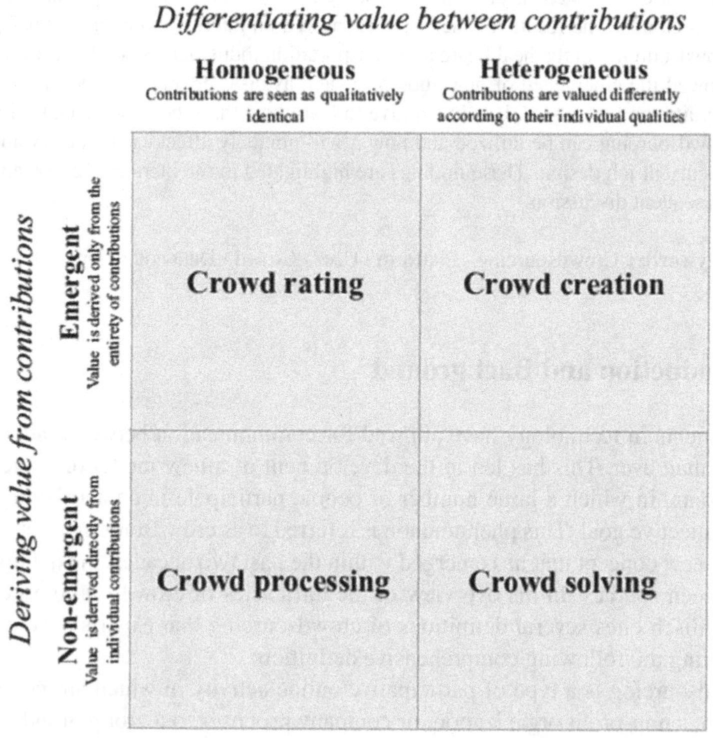

Differentiating value between contributions

Deriving value from contributions	**Homogeneous** Contributions are seen as qualitatively identical	**Heterogeneous** Contributions are valued differently according to their individual qualities
Emergent Value is derived only from the entirety of contributions	**Crowd rating**	**Crowd creation**
Non-emergent Value is derived directly from individual contributions	**Crowd processing**	**Crowd solving**

Fig. 1. Types of crowdsourcing (Pohlisch et al. 2021).

Regardless of the type of crowdsourcing, there are two main roles within the process: the crowdsourcer and the crowd. The crowdsourcer is the content owner who enables the crowd to come together in order to solve a problem or generate ideas. The second and arguably most important role within a crowdsourcing platform is the role of the crowd. While the role of the user is important for any system, the role of the crowd is unique for crowdsourcing because the system cannot even begin without it. There are other roles an organization may choose to utilize, such as a third-party IT platform. However, since this is not a requirement for organizations, this literature review will focus solely on the crowdsourcer and the crowd.

While crowdsourcing is a relatively new concept, the act of online communities coming together is not new. In order to better understand the development of crowdsourcing, one first must understand the history of socially centered work systems and online communities. Socially centered work systems, or "participatory ergonomics" is seen as a byproduct of sociotechnical work system theory developed in the mid-twentieth century by Emery and Trist (Taveira and Smith 2012). However, the term "participatory ergonomics" was not developed until the 1980s by Noro, Kogi, and Imada when discussing ergonomics in the context of the degree to which individuals were involved in its practice (Taveira and Smith 2012). At this time, participatory ergonomics solely referred to workers having autonomy in controlling their work activities and outcomes in order to achieve desirable goals (Taveira and Smith 2012). The study of participatory ergonomics is the first instance of the worker, or the masses, contributing something-in this case labor-that benefits both the worker and the organization to reach a goal.

The next portion of crowdsourcing is the role of the online community. In the *Handbook of Human Factors and Ergonomics*, Ozok and Zaphiris define online communities as "social aggregations that emerge from the Net when enough people carry on those public discussions long enough, with sufficient human feeling, to form webs of personal relationships in cyberspace" (Zaphiris and Ozok 2012). This definition is important for the purpose of analyzing crowdsourcing because many driving forces behind why crowdsourcing works can be seen in it. First, the concept of "human feeling" driving online communities can be seen in many crowdsourcing platforms. For instance, the platform GoFundMe is a crowdfunding website in which people can set up a donation webpage in the hopes the public will take interest and help their fundraising efforts. This crowdsourcing platform relies solely on human feeling driving people to contribute and to share their stories with the world. Without instigating this feeling, the site would gain no traction. The goal of crowdsourcing is to initiate participation by providing an online community for this human feeling to flourish, thus driving users to participate with no reward other than the desire to help improve the system and help reach a collective goal.

2 Purpose of Study

The purpose of this study is to do a systematic literature review of the articles available on crowdsourcing. These findings can be useful for future researchers to study the intricacies of crowdsourcing and what circumstances need to be true for it to be successful. It is also important for organizations interested in possibly utilizing this method of data collection in order to see how exactly this change would affect their work environment and potential changes they must make within their organizational design. The goal of this literature review is to outline the growth of the topic in academic and non-academic settings and to highlight some industries that have successfully utilized this method. The tools used to collect data for the review are CiteSpace, Google Scholar, Harzing's Publish or Perish, MAXQDA, Scopus, VosViewer, Vicinitas, and Web of Science.

3 Data Collection

3.1 Database Searches

This literature review entailed gathering data from multiple online databases. Every database search used "crowdsourcing" as the search term. While other literature review topics may call for varying the keyword search in order to gather a broad range of subtopics, the term "crowdsourcing" encompasses all subtopics such as "crowdfunding" or "crowdsourcing platform," which are discussed later in the discussion section. This is the motivation for not varying the search term.

Table 1 shows the databases used, the term searched, and the number of results. The only notable exceptions are Harzing's Publish or Perish and Vicinitas. Harzing's Publish or Perish was intentionally stopped at 200 articles but can allow up to 1,000 articles. In the case of Vicinitas, the platform has a maximum number of 2,000 Tweets that can be analyzed.

Table 1. Database search table with key search terms and results

Database	Search term	Results
Scopus	"crowdsourcing"	16,020 articles
Web of Science	"crowdsourcing"	9,717 articles
Google Scholar	"crowdsourcing"	291,000 articles
Harzing's Publish or Perish	"crowdsourcing"	Stopped at 200 articles
Vicinitas	"crowdsourcing"	Maximum 2,000 Tweets

3.2 Trend Analysis

A trend analysis of articles published related to crowdsourcing was done using Scopus and Web of Science. This trend analysis shows the exponential jump in interest in the topic within the past decade, and where in the world this interest is taking place.

Figure 2 shows the number of documents published per year on Scopus. The first year displayed on the x-axis represents two documents published in 1984. The emergence of the topic remains low for more than two decades until approximately 2008, which is the first year more than ten articles are published related to crowdsourcing. This year is significant to the emergence of crowdsourcing because, after this time, one can see exponential growth in interest in the topic. At its peak in 2019, the number of documents found on crowdsourcing on Scopus exceeded 2,000. At the time of the search, 471 documents had been published thus far in 2021. If that trend were to continue throughout 2021 then over 1,700 documents on crowdsourcing will be found on Scopus at the end of the year.

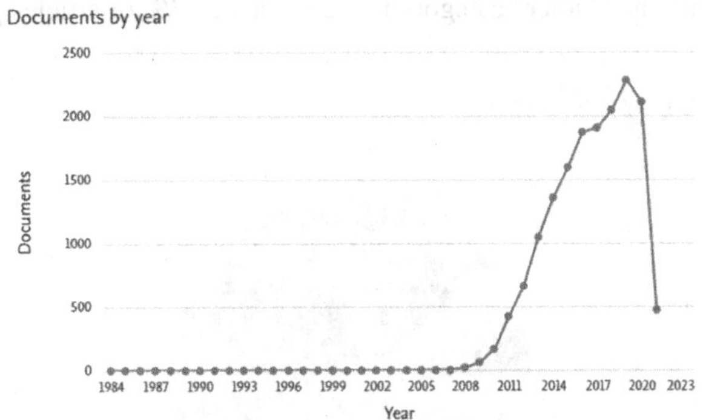

Fig. 2. Trend analysis of articles per year related to crowdsourcing. (*Scopus* n.d.)

Figure 3 shows where the interest in crowdsourcing is concentrated when the search was done in 2021 on Web of Science. The majority of interest comes from the United States, followed by China and England. Multiple significant observations that can be pulled from this data. One of which is that despite crowdsourcing's collectivist nature, the country that shows the most interest in it is famously individualistic. Additional observations about the United States' interest in crowdsourcing will be discussed in discussion and future work sections.

Fig. 3. Treemap from Web of Science of countries publishing articles related to crowdsourcing. (*Web of Science* n.d.)

3.3 Emergence Indicator

The next part of the literature review analyzes the emergence of trends and what areas of study have the highest interest in crowdsourcing. Figure 4 shows the percentage of articles found on Scopus by subject area. The area with the highest interest is Computer Sciences with 38.1% of articles published. This is not surprising since crowdsourcing is impossible without gathering large amounts of data from many sources. It is also significant that the topic reaches such a wide range of areas such as Environmental Sciences, Business, Medicine, and Social Sciences. A total of eleven subject areas are covered, including the "Other" category that contains 12.3% of articles published.

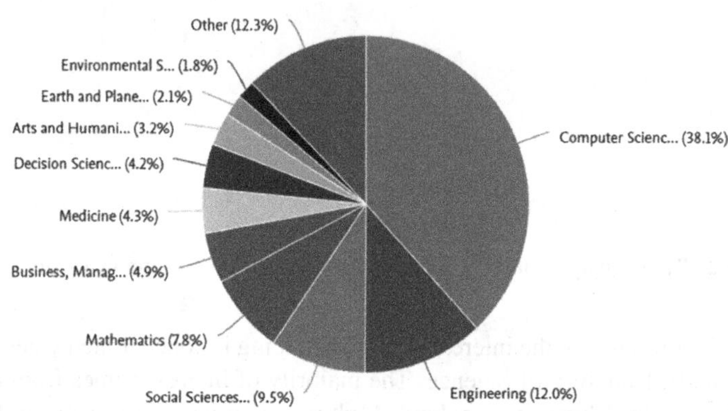

Fig. 4. Pie chart distinguishing areas of study that have research related to crowdsourcing. (*Scopus* n.d.)

3.4 Engagement Measures

After analyzing the emergence of the topic, the next step in the literature review is to analyze the engagement and social media interaction. This was done using the tool Vicinitas, which analyzes engagement of topics on the social media platform Twitter. The search resulted in a maximum of 2,000 Tweets by 1,800 users, with an influence of 18.1 million. Influence is defined by Vicinitas as the sum of the number of followers from each user that posted a Tweet related to the topic of interest. These findings show the incredible popularity of the subject in non-academic circles.

Figure 5 shows the engagement timeline and post timeline from Vicinitas when the search was done in April 2021. These graphs show a steady and significant trend in the topic on Twitter. This is equally as significant as similar data found on Scopus and Web of Science because it shows an interest outside of the realm of academia. Vicinitas found that engagement among Twitter users reached 4,500. Engagement is defined by Vicinitas as the sum of the likes and retweets a post on the subject receives. "Likes" are a way for a user to show interest but not post the Tweet to their page, while "retweeting" posts the Tweet to their timeline and thus exposes all of their followers to the content.

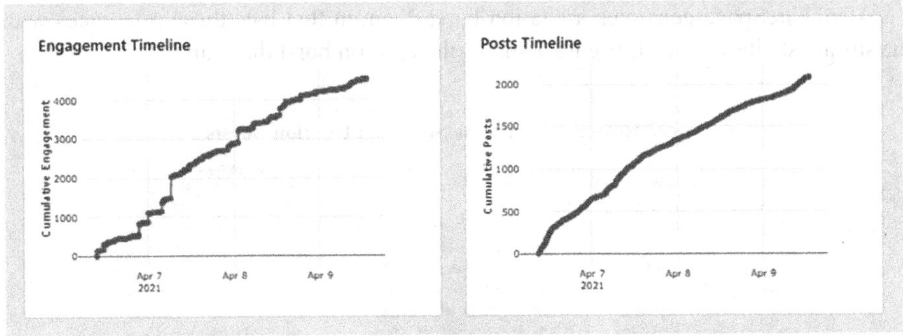

Fig. 5. Engagement and post timelines from Vicinitas search. (*Vicinitas* n.d.)

4 Results

4.1 Co-citation Analyses

The next step in the literature review is to perform co-citation analyses. This is done in order to see the connection between highly cited articles on the subject and the breadth of their reach. The first co-citation analysis was done with CiteSpace using data from Web of Science. CiteSpace has a feature that allows the user to showcase how these highly cited articles are connected by doing a cluster analysis. In this analysis, the clusters were crowdsourcing optimization, incentive mechanism, mobile edge computing, and crowdfunding. This feature can be seen in the citation web in Fig. 6.

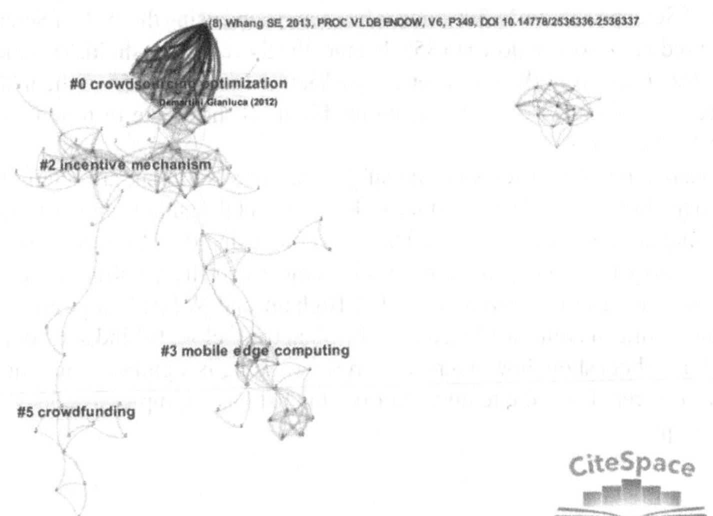

Fig. 6. Co-citation map from CiteSpace using Web of Science data. (*CiteSpace*, *Web of Science* n.d.)

Also, CiteSpace generates a citation burst diagram that lists those references with the strongest citation burst. Figure 7 shows the citation burst diagram.

Top 25 References with the Strongest Citation Bursts

References	Year	Strength	Begin	End	2016 - 2019
Zhao YX, 2014, INFORM SYST FRONT, V16, P417, DOI 10.1007/s10796-012-9350-4, DOI	2014	2.93	2018	2019	
Kryvasheyeu Y, 2016, SCI ADV, V2, P0, DOI 10.1126/sciadv.1500779, DOI	2016	2.44	2018	2019	
de Albuquerque JP, 2015, INT J GEOGR INF SCI, V29, P667, DOI 10.1080/13658816.2014.996567, DOI	2015	2.44	2018	2019	
To H, 2014, PROC VLDB ENDOW, V7, P919, DOI 10.14778/2732951.2732966, DOI	2014	2.25	2018	2019	
Tong YX, 2016, PROC INT CONF DATA, V0, P49, DOI 10.1109/ICDE.2016.7498228, DOI	2016	2.23	2018	2019	
Garcia-Molina H, 2016, IEEE T KNOWL DATA EN, V28, P901, DOI 10.1109/TKDE.2016.2518669, DOI	2016	1.95	2018	2019	
Singer Y, 2013, P 22 INT C WORLD WID, V0, PP1157, DOI 10.1145/2488388.2488489], DOI	2013	1.82	2018	2019	
Kong XJ, 2017, IEEE T IND INFORM, V13, P1202, DOI 10.1109/TII.2017.2684163, DOI	2017	1.67	2018	2019	
Chittilappilly AI, 2016, IEEE T KNOWL DATA EN, V28, P2246, DOI 10.1109/TKDE.2016.2555805, DOI	2016	1.67	2018	2019	
Cohen B, 2014, ORGAN ENVIRON, V27, P279, DOI 10.1177/1086026614546199, DOI	2014	1.67	2018	2019	
Hildebrand M, 2013, REPROD HEALTH MATTER, V21, P57, DOI 10.1016/S0968-8080(13)41687-7, DOI	2013	1.67	2018	2019	
Cheng P, 2016, IEEE T KNOWL DATA EN, V28, P2201, DOI 10.1109/TKDE.2016.2550041, DOI	2016	1.67	2018	2019	
Chandler J, 2016, ANNU REV CLIN PSYCHO, V12, P53, DOI 10.1146/annurev-clinpsy-021815-093623, DOI	2016	3.2	2017	2017	
Shapiro DN, 2013, CLIN PSYCHOL SCI, V1, P213, DOI 10.1177/2167702612469015, DOI	2013	2.84	2017	2017	
Buhrmester M, 2011, PERSPECT PSYCHOL SCI, V6, P3, DOI 10.1177/1745691610393980, DOI	2011	2.73	2016	2016	
Peer E, 2014, BEHAV RES METHODS, V46, P1023, DOI 10.3758/s13428-013-0434-y, DOI	2014	2.53	2017	2017	
Doan A, 2011, COMMUN ACM, V54, P86, DOI 10.1145/1924421.1924442, DOI	2011	2.48	2016	2016	
Berinsky AJ, 2012, POLIT ANAL, V20, P351, DOI 10.1093/pan/mpr057, DOI	2012	2.21	2017	2017	
Zheng HC, 2011, INT J ELECTRON COMM, V15, P57, DOI 10.2753/JEC1086-4415150402, DOI	2011	1.98	2016	2016	
Hauser DJ, 2016, BEHAV RES METHODS, V48, P400, DOI 10.3758/s13428-015-0578-z, DOI	2016	1.77	2017	2017	
Stewart N, 2015, JUDGM DECIS MAK, V10, P479	2015	1.77	2017	2017	
Chandler J, 2015, PSYCHOL SCI, V26, P1131, DOI 10.1177/0956797615585115, DOI	2015	1.77	2017	2017	
Johnson DR, 2012, TEACH PSYCHOL, V39, P245, DOI 10.1177/0098628312456615, DOI	2012	1.77	2017	2017	
Litman L, 2015, BEHAV RES METHODS, V47, P519, DOI 10.3758/s13428-014-0483-x, DOI	2015	1.77	2017	2017	
Ranard BL, 2014, J GEN INTERN MED, V29, P187, DOI 10.1007/s11606-013-2536-8, DOI	2014	1.73	2016	2016	

Fig. 7. Citation burst diagram from CiteSpace co-citation analysis. (*CiteSpace* n.d.)

Another co-citation analysis was done using the tool VosViewer and gathering data from Web of Science. Of the 24,396 cited references found using the Web of Science data, VosViewer reduced those to down to 355 that met the five cites threshold. This means that of the articles pulled from Web of Science, VosViewer found that 355 of them have been cited at least five times in other publications. Figure 8 shows the resulting co-citation map from VosViewer.

After analyzing how articles on the subject of crowdsourcing are intertwined, the next step in the literature review is to take a closer look at the authors and institutions that are publishing articles on the subject. This was done using data from Scopus. Figure 9 shows a graph of the leading authors on the subject and the number of their articles found on Scopus. The top two authors J.P. Bigham and W.S. Lasecki are tied at 62 publications, while the third leading author L. Chen trails close behind at 61 documents. These high numbers show how interest in crowdsourcing is significant and thus makes it an emerging area. Even the tenth author on this list has an impressive 36 documents found on Scopus.

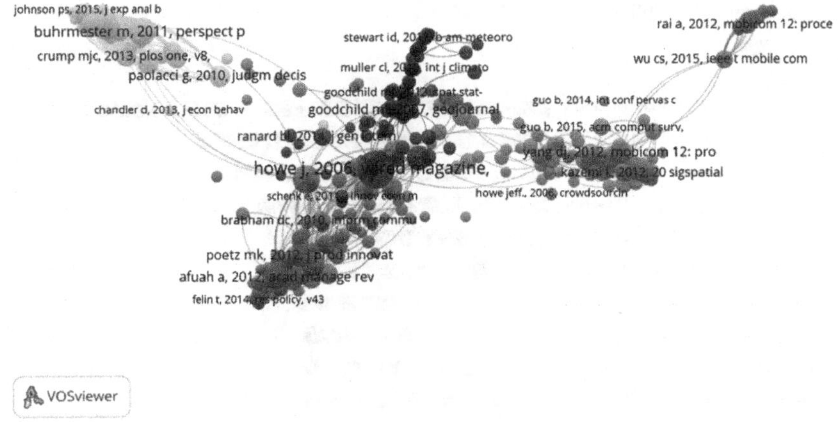

Fig. 8. VosViewer map of co-citation analysis using Web of Science data. (*VosViewer*, *Web of Science* n.d.)

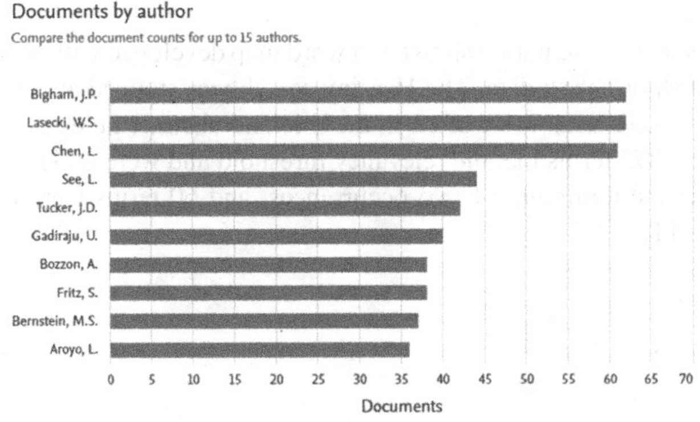

Fig. 9. Leading Authors from Scopus search (*Scopus* n.d.)

Figure 10 shows the Scopus data of leading institutions dedicating research to crowdsourcing. This graph shows the institution along with the number of documents found on Scopus. The top institution is Carnegie Mellon University with 277 documents found in Scopus. Next are two more American Universities Stanford University and University of Washington with 227 and 184 documents, respectively. To get a better idea of which nations are dedicating research to the subject, see Fig. 3.

4.2 Content Analyses

The next step in the literature review is to analyze and map the content of the articles published. This was done using VosViewer, Vicinitas, and MAXQDA to visually represent topics discussed in articles. This is key for crowdsourcing since it is such a broad subject that can be applied to many different scenarios.

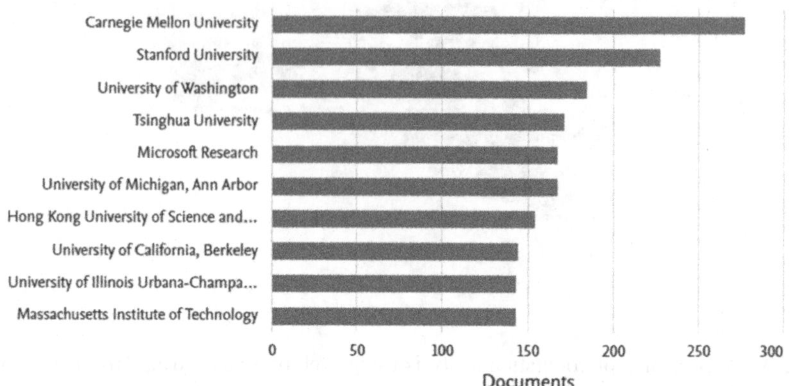

Fig. 10. Leading institutions from Scopus search (*Scopus* n.d.)

The first visual representation shown is a word map developed with VosViewer using Harzing's Publish or Perish data. The Harzing search was stopped at 200 articles. This map shows the words used most often in the title and abstract fields of articles. Of all the terms pulled, 373 terms met the relevancy threshold and were pulled by VosViewer, 61 of those met the minimum of two occurrences, and 60 terms were connected and mapped in Fig. 11.

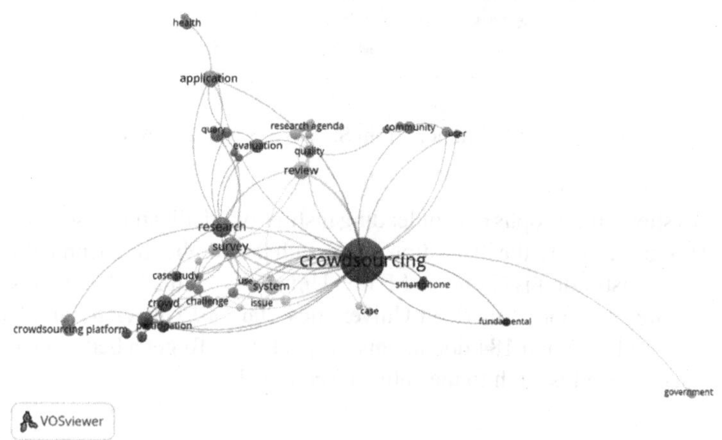

Fig. 11. Text data analysis from VosViewer using Harzing data. (*VosViewer*, *Harzing's Publish or Perish* n.d.)

The following table, Table 2, is a cluster analysis of the keyword map from Fig. 11. It can be seen that VosViewer categorizes and distinguishes similar terms by color. This text analysis found there were many different subtopics that all use similar language surrounding crowdsourcing. VosViewer found eight different clusters each comprised of approximately 4–7 nodes. These clusters are application, participation, platform, research, review, system, use, and user. One could argue that these terms could be clumped further into three categories. The first of these categories contain the first three terms application, participation, and platform. These terms are related because they all pertain to the logistics of crowdsourcing and how it is done using crowd participation. Examples of crowdsourcing platforms are the travel application Waze and the vacation rental website Airbnb. The next three clusters research, review, and system are all related to the study of crowdsourcing and its intricacies as a system model. This would be where the leading institutions and leading authors come into play when looking at remerging areas of study on crowdsourcing. The last two clusters are pertinent to crowdsourcing: use and user. The user is essential to crowdsourcing because the entire model is built upon the need for user interaction. These two clusters can be connected to every subtopic in crowdsourcing.

Table 2. Cluster analysis of VosViewer text map results (*VosViewer* n.d.).

Cluster	Number of nodes
Application	4
Participation	6
Platform	4
Research	7
Review	6
System	4
Use	7
User	7

The next visual representation in Fig. 12 is a word cloud generated by Vicinitas. Recall that Vicinitas gathers data from Tweets on the social media platform Twitter. This word cloud displays the most common words used in Tweets about crowdsourcing, with the largest words being the most common. This keyword visualization is important because it highlights issues and subjects that people potentially not involved in academia are interested in regarding crowdsourcing. Some common words highlighted by Vicinitas are "medical," "government," and "workers."

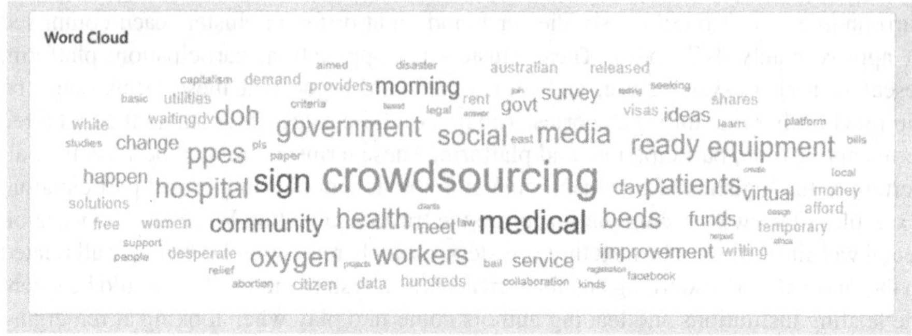

Fig. 12. Word cloud from Vicinitas search. (*Vicinitas* n.d.)

The next figure, Fig. 13, shows another word cloud generated by MAXQDA after gathering articles from varying sources such as Google Scholar, Scopus, and Web of Science. These documents were compared to find the most commonly used terms similar to Vicinitas. However, it should be noted that this word cloud was edited to remove superfluous things such as prepositions, numbers, and other unnecessary terms. This word cloud differs from the Vicinitas word cloud because it represents commonly used words in research areas on crowdsourcing. This can be seen in the word cloud as there are research-related terms present such as "task," "model," and "performance."

Fig. 13. Word cloud generated by MAXQDA. (*MAXQDA* n.d.)

5 Discussion

In order to fully understand crowdsourcing and its implications on job design, the next part of the literature review will analyze specific case studies on crowdsourcing as it is applied to various industries.

Healthcare. The emergence of crowdsourcing in the healthcare industry began as a way to advance health information technology (Desai et al. 2020). Within a decade, this strategy became more commonplace across subfields of the healthcare industry as a method of gathering information used to treat patients (Desai et al. 2020). This has drastically changed the healthcare industry on various levels from research to patient care.

An example of an application of this technique can be seen as a result of the COVID-19 pandemic. In an article entitled "Crowdsourcing a crisis response for COVID-19 in oncology" published in *Nature Cancer* magazine, Aakash Desai and colleagues outline the usefulness of such a strategy in the field of oncology. The article states that 7% of all crowdsourcing efforts map to oncology (Desai et al. 2020). This technique is especially helpful for cancer research because of its relatively quick turnaround time for collecting data and comparatively low costs (Desai et al. 2020).

At the time of publication in May of 2020, there was little known about how COVID-19 uniquely affected patients with a comorbidity like cancer. Given that patients with cancer are often elderly and immunocompromised, there was a pressing need for oncologists to understand how the virus affected their patients and what side effects they may be left with long-term. The article outlines over ten initiatives that were started in the first months of the virus's emergence, many of which reached international participation. One of these initiatives was the COVID-19 and Cancer Consortium (CCC19), which originated on Twitter, and grew to comprise over 90 cancer centers over the world (Desai et al. 2020). Consortiums like this involve a registration survey that allows healthcare professionals to provide information on COVID-19 patients with cancer (Desai et al. 2020). The information available to healthcare professionals ranges from demographic information to cancer diagnoses and treatments, and course of illness and outcomes (Desai et al. 2020).

Despite the COVID-19 crisis being an exceptional event, it is evident that crowdsourcing can effectively be utilized in the healthcare industry moving forward. This can change the way medicine is practiced and how healthcare professionals gather data. Some possible limitations of the method would be the willingness of international participation. The COVID-19 pandemic was an incentive for countries to jointly approach a common concern, however, that willingness might change if there is no incentive.

Disaster Management. Similar to the response of the healthcare industry to the COVID-19 pandemic, crowdsourcing is often used in response to other types of crises, like natural disasters. In a study done at Carnegie Mellon University, researchers looked at the social media platform Twitter as a potential source for disaster preparedness and response in Padang, Indonesia.

The advantages of using social media are its efficiency and ease of access from a mobile device, and its disadvantages are that data might be biased and information can quickly become out of date (Carley et al. 2016). To evaluate these advantages and disadvantages, researchers analyzed how Twitter is used during times of non-crisis. Researchers unsurprisingly found that Twitter usage is drastically different in cities versus rural areas, and disaster management would be affected by those differences (Carley et al. 2016). Additionally, the population of users in cities has sub-populations with informal leaders that could potentially be recruited to spread important information to their followers in a time of crisis (Carley et al. 2016). However, it was found that many of these informal leaders are teenage girls- who hold little authority outside of social media (Carley et al. 2016).

While it is possible for government agencies to potentially utilize Twitter for disaster preparedness, this would require a willingness from informal community leaders to assist in spreading pertinent information. Despite these areas for improvement, researchers found that social media has many potential benefits regarding disaster management. Twitter is projected to continue to have high usage (Carley et al. 2016). This, in conjunction with prevalent geo-tagging information, can be useful in the aftermath of a disaster (Carley et al. 2016). The social media platform Facebook has already implemented such a feature by allowing users to mark themselves as "safe" after a disaster.

Researchers concluded that this research should serve as a baseline for future work on crowdsourcing disaster management, but it shows promise that government officials and disaster response groups support such efforts (Carley et al. 2016).

Atmospheric Sciences. Another industry that can be improved with the use of crowdsourcing techniques is the atmospheric sciences. In practice, this would consist of gathering weather information around the world from amateur weather stations, smart devices, or social media. Potential applications of this type of system are farming, risk assessment, modeling, and forecasting (Muller et al. 2015). This method of data collection could be helpful in both densely populated areas and areas in which meteorological technology is unavailable (Muller et al. 2015). According to Muller and colleagues, "the most important value of such information may be in what it tells us about local activities in various geographic locations that go unnoticed by the world's media" (Muller et al. 2015).

The aim of research in this area is to assess the quality control issues of crowdsourced meteorological information and to explore future potential applications. There are currently dozens of crowdsourcing initiatives within meteorology, one of which is Citizen Science. Citizen Science is a collaborative research network involving the public (Muller et al. 2015). Participants can use technology to gather data, or they can also be a "virtual sensor" themselves and interpret their surroundings (Muller et al. 2015).

While some aspects of crowdsourcing data look promising to meteorologists, the issue of quality assurance is always a factor. It has been shown that there is a positive relationship between the accuracy of the information and the number of contributors (Muller et al. 2015). However, there has also been evidence to suggest that accuracy of information declines if there are too many contributors (Muller et al. 2015). Therefore, it is essential to rate the quality of information that is contributed by the public. With the decline of traditional meteorological networks and the increase in demand for real-time information, there is an obvious use for crowdsourcing in atmospheric sciences (Muller et al. 2015). The utilization of such data could ignite a change in the field of how data is collected entirely. Similar to the COVID-19 crisis in the healthcare field, climate change affects every country and could be an incentive for a collectivist approach to gathering and using information.

Transportation. The previously mentioned uses for crowdsourced data have all been in the context of a crisis. However, that is not the only instance in which crowdsourced data is becoming more and more prevalent. The transportation industry is another field that has successfully researched and utilized crowdsourced data. An example of this is the mobile phone application Waze. Waze is a map application that gives directions like other travel applications but is set apart because it provides real-time data on road conditions that are entirely provided by Waze users.

In an article published in 2020 by Nelson et al., researchers analyzed data gathered from the fitness application Strava, which allows users to track their bike route and post it as a form of social media. The goal of this research is to provide a generalized strategy for utilizing this type of data (Nelson et al. 2021). The obstacles to overcome in pursuit of this goal are like those in any case of crowdsourced data, which is the potential bias of information and a potentially unrepresentative sample set (Nelson et al. 2021). Using Strava data, researchers gathered geographic data from users in five cities in order to identify variables important for predicting ridership (Nelson et al. 2021). They concluded that important variables necessary for utilizing such data include the number of riders, percentage of trips categorized as commuting, the safety precautions taken, and income (Nelson et al. 2021).

By creating a generalized model for gathering crowdsourced information, researchers and developers can reduce the cost of gathering this data themselves. Also, government officials can use this data to support public transit proposals without having to survey the public, thereby reducing costs for the taxpayer.

6 Conclusion

It is evident that when implemented, crowdsourcing can be an invaluable method for data collection that offers unique advantages and disadvantages for various industries. As outlined in the findings of the literature review, interest in this topic has exponentially grown within the past decade. However, not all literature on the topic concludes that crowdsourcing is a viable solution for industries. This section aims to outline the different perspectives on the value and potential of crowdsourcing.

In the 2021 article entitled "A Recipe for Success: Crowdsourcing, Online Social Networks, and Their Impact on Organizational Performance," authors Daniel Palacios-Marqués et al. claim that the use of social media and crowdsourcing have a positive impact on organizational performance (Palacios-Marqués et al. 2021). They attribute this to the concept of "openness" (Palacios-Marqués et al. 2021). In this instance, openness refers to a company's ability to learn by gathering and transforming knowledge and enabling inter- and intra-organizational communication (Palacios-Marqués et al. 2021). By enabling this open line of communication, a company, in turn, develops an image of transparency to the public, which makes human resources more likely to participate (Palacios-Marqués et al. 2021). This trust and openness are the core building blocks that drive crowdsourcing to be successful. This is because, according to Palacios-Marqués et al.,

> Open culture is related to open innovation because it promotes creative thinking and the development of different points of view. Using crowdsourcing, companies obtain solutions to both present and latent problems thanks to the contribution of the crowd. The search for alternative solutions through crowdsourcing makes sense in companies with an organizational culture that is open to new perspectives. (Palacios-Marqués et al. 2021)

On the other hand, some researchers have found flaws in the crowdsourcing model. For crowdsourcing to be successful it must rely on the minds within the crowd, which is inherently risky for a business. That doubt in the capability of the crowd is what researchers Jie Ren et al. call "the boundary of creativity." In their research, they found that the crowd is only useful for general tasks and tend to fall short in executing specialist tasks (Ren et al. 2021). More specifically, the creativity of the crowd is hindered by their lack of contextual knowledge, and the diversity of backgrounds negatively affects their level of practicality (Ren et al. 2021). However, this study also found that this lack of practicality can be remedied by allowing participants to view each other's solutions (Ren et al. 2021). This caveat shows promise because the current method of crowdsourcing can easily be reworked to account for this decline in practicality.

Another study by Linus Dahlander and Henning Piezunka supports this theory that crowdsourcing has potential but is currently being incorrectly leveraged. They outline four pitfalls of crowdsourcing that companies typically fall prey to. They are improper planning, not building the crowd, having narrow attention, and lacking accountability (Dahlander and Piezunka 2020). The first pitfall refers to companies that fail to recognize the critical steps that are involved in gathering crowdsourced data and become overwhelmed by the task (Dahlander and Piezunka 2020). Next, not building the crowd is a pitfall in which a company is not active enough in the early stages of advertising and fails to draw enough attention to the problem, thereby not receiving a large response (Dahlander and Piezunka 2020). The goal of crowdsourcing is to gather a wide range of ideas that the company may not have thought of on its own. The next pitfall of having narrow attention is when a company correctly advertises for ideas, but they only show interest in those that they would have generated internally (Dahlander and Piezunka 2020). Finally, lack of accountability refers to a company not responding to submissions or asking for crowd input and not following through with their wishes, thereby losing support (Dahlander and Piezunka 2020). Crowdsourcing utilizing the masses by making them a decision-maker, and this pitfall is analogous to asking the crowd to sit at the table but not allowing them to speak.

Though there are possible disadvantages to implementing a crowdsourcing model in an industry, these disadvantages are not irreparable. By identifying possible challenges, researchers are continuing to perfect the generalized model for crowdsourcing that can be applied to any industry.

7 Future Work

The crowdsourcing applications are numerous, and many studies are currently underway to investigate how to implement such a task and the impact it has on a company. One of these studies, funded by the National Science Foundation, has produced an article by Dong Wei and colleagues entitled "Recommending Deployment Strategies for Collaborative Tasks."

In this article, Dong et al. give a generalized model for developing crowdsourced tasks. They include necessary parameters such as a lower bound on the quality of submissions, an upper bound on the time to task completion, and an upper bound on costs (Wei et al. 2020). They have found that the key to building a collaborative task is to define three dimensions: structure, organization, and style (Wei et al. 2020). Structure is defined as the decision to solicit work sequentially or simultaneously (Wei et al. 2020). Organization is whether to organize the deployment collaboratively or independently (Wei et al. 2020). Finally, style is the decision to rely only on the crowd or to involve machine-learning algorithms as well (Wei et al. 2020).

Concerning job design, crowdsourced information is a growing factor in the job recruitment process. In an article entitled "Investigating Web-Based Recruitment Sources: Employee testimonials vs word-of-mouse," authors Greet Van Hoye and Filip Lievens claim that crowdsourced employment recruitment sites such as Glassdoor.com carry more credibility to prospective employees when investigating a potential employer (Hoye and Lievens 2007). They coined this phenomenon as "word-of-mouse" testimonials (Hoye and Lievens 2007). Their research showed that prospective employees found these word-of-mouse testimonials as more credible when judging organizational attractiveness as opposed to employee testimonials (Hoye and Lievens 2007). Additionally, they found that employee testimonials were only favored higher than word-of-mouse testimonials when they focused on individual employees rather than the organization as a whole. This information is incredibly used for industries when designing their recruitment plan. The influence of crowdsourced job information websites has on organizational perception is too great for companies to ignore.

Crowdsourcing would not be possible without the field of human-computer interaction (HCI). Authors Thomas Malone et al. state that the goal of HCI is "to understand and to design interactions between people and machines" (Malone et al. 2015). Crowdsourcing has formed from masses of people participating in this type of human-machine interaction to gather and coordinate a collective effort. Researchers Brandtner et al. akin crowdsourcing to an open-innovation process that utilizes web processes and Web 2.0 tools to integrate outside capabilities into a large and diverse population (Brandtner et al. 2014). Furthermore, they focus on the HCI side of this open-innovation process, stating, "an important but often overlooked aspect of these platforms is the human computer interaction, and in particular the motivation and willingness of individuals to contribute to those platforms" (Brandtner et al. 2014). They found five factors that successful open-innovation crowdsourcing platforms possess, the most influential being social interaction, which is a direct result of HCI (Brandtner et al. 2014). Social interaction through the Internet is greater than ever and will continue to grow and become more entwined in everyday life. It is up to the research and development of HCI and social computing to keep up with the open-innovation process in order to successfully utilize the strength of the masses.

Another topic deeply related to the HCI aspect of crowdsourcing is the idea of collective intelligence. The MIT Center for Collective Intelligence defines collective intelligence as "groups of individuals doing things collectively that seem intelligent" (MIT). Their mission statement is to answer the question, "what if the goal were to create combined human/machine systems that were more intelligent than either people or machines could be alone?" (MIT). Collective intelligence and crowdsourcing are congruent in the core of their missions: to leverage the masses by utilizing technology to gain what could not be possible without a mass collective effort. Utilizing the crowd would not be possible without the growing ease of the HCI. It has created the means for various industries to participate more easily and to reach a wider audience, thereby making the case for crowdsourcing much simpler.

In the article entitled, "Collective Intelligence," author Dr. Jan Marco Leimeister outlines five success factors that determine a quality collective intelligence application. The first is control, which refers to the loss of control a company must face after opening previously closed structures to utilize the crowd (Leimeister 2010). He goes on to state that a company must decide how much they want to open up the environment to the crowd, which can vary and is crucial for success (Leimeister 2010). The next factor is diversity vs. in-depth expertise, meaning that one must ascertain the ratio of crowd-collected information versus the amount of highly trained individuals they will utilize (Leimeister 2010). Crowdsourced intelligence on its own is not comprehensive, and without the expertise to translate the mass amounts of data, it will be useless. As stated previously, this is one of the core concerns of crowdsourcing and is a problem unique to HCI. The next indicator is engagement (Leimeister 2010). Engagement can is influenced by a wider variety of factors including both motivation and ease of access. The ease of access issue has become almost obsolete with the help of HCI, however, there is still the question of sample size biases that can arise from disclosing those still without internet access. Next is the question of policing, which may initially be perceived as in direct contradiction with the premise of crowdsourcing, however, Leimeister states policing of participants is necessary (Leimeister 2010). This is because of the higher rate of respondents, then the higher likelihood of inappropriate or malicious conduct (Leimeister 2010). It is important to note that policing in the form of punishment may be effective, however, it may affect the level of engagement from participants. The last factor that contributes to the success of collective intelligence efforts is that of intellectual property, which refers to the decision of how to acquire intellectual property that is acquired using a collective intelligence model (Leimeister 2010).

The potential areas for the application of collective intelligence have grown recently as a result of the development of applications and user-generated content (Leimeister 2010). This could not have been possible without the integration of HCI in everyday life. By having a portable computer in the form of smartphones, users can interact more directly than ever before. Furthermore, the combined effort of many users has enabled the masses to obtain greater influence, thus creating a collective intelligence network. An example of such a collective intelligence network is Wikipedia. Wikipedia provides more than four million English language articles and is seen as the largest English language encyclopedia in the world (Leimeister 2010). It is an example of successful collective intelligence because of its motivation through social computing. Articles may be edited; however, its policing system provides a safeguard from false information. Other platforms have followed the Wikipedia model, such as Intellipedia of the CIA and the project network Amazee (Leimeister 2010).

It is evident that crowdsourcing is becoming a more sought-after tool for industries to change the way they gather data and utilize communication technology. Some industries may be more apt than others, however, the more commonplace crowdsourcing becomes the more generalizable it will become. The future of crowdsourcing looks promising and will undoubtedly change numerous industries. As authors Palacios- Marqués et al. state, "the limits of crowdsourcing lie within the imagination" (Palacios-Marqués et al. 2021).

References

Brandtner, P., Auinger, A., Helfert, M.: Principles of human computer interaction in crowdsourcing to foster motivation in the context of open innovation. In: Nah, F.-H. (ed.) HCIB 2014. LNCS, vol. 8527, pp. 585–596. Springer, Cham (2014). https://doi.org/10.1007/978-3-319-07293-7_57

Carley, K.M., Malik, M., Landwehr, P.M., Pfeffer, J., Kowalchuck, M.: Crowd sourcing disaster management: the complex nature of twitter usage in Padang Indonesia. Saf. Sci. **90**, 48–61 (2016). https://doi.org/10.1016/j.ssci.2016.04.002

CiteSpace (n.d). http://cluster.cis.drexel.edu/~cchen/citespace/. Accessed 05 Apr 2021

Dahlander, L., Piezunka, H.: Why crowdsourcing fails. J. Organ. Des. **9**(1), 1–9 (2020). https://doi.org/10.1186/s41469-020-00088-7

Desai, A., et al.: Crowdsourcing a crisis response for COVID-19 in Oncology. Nat. Cancer **1**(5), 473–476 (2020). https://doi.org/10.1038/s43018-020-0065-z

Harzing's Publish or Perish (n.d.). https://harzing.com/resources/publish-or-perish. Accessed 09 Apr 2021

Hoye, G.V., Lievens, F.: Investigating web-based recruitment sources: employee testimonials vs word-of-mouse. Int. J. Sel. Assess. **15**(4), 372–382 (2007). https://doi.org/10.1111/j.1468-2389.2007.00396.x

Leimeister, J.M.: Collective intelligence. Bus. Inf. Syst. Eng. **2**(4), 245–248 (2010). https://aisel.aisnet.org/bise/vol2/iss4/6

Malone, T., Bigham, J., Bernstein, M., Adar, E.: Human-computer interaction. In: Handbook of Collective Intelligence, pp. 57–84. Essay, MIT Press (2015)

MAXQDA (n.d.). https://www.maxqda.com/. Accessed 28 Apr 2021

MIT: MIT Center for Collective Intelligence. https://cci.mit.edu/about

Muller, C.L., et al.: Crowdsourcing for climate and atmospheric sciences: current status and future potential. Int. J. Climatol. **35**(11), 3185–3203 (2015). https://doi.org/10.1002/joc.4210

Nelson, T., et al.: Generalized model for mapping bicycle ridership with crowdsourced data. Transp. Res. Part C Emerg. Technol. **125**(2020), 102981 (2021). https://doi.org/10.1016/j.trc.2021.102981

Palacios-Marqués, D., Gallego-Nicholls, J.F., Guijarro-García, M.: A recipe for success: crowdsourcing, online social networks, and their impact on organizational performance. Technol. Forecast. Soc. Change **165**, June 2020. https://doi.org/10.1016/j.techfore.2020.120566

Pohlisch, J.: An introduction to internal crowdsourcing. In: Ulbrich, H., Wedel, M., Dienel, H.-L. (eds.) Internal Crowdsourcing in Companies. CMS, pp. 15–26. Springer, Cham (2021). https://doi.org/10.1007/978-3-030-52881-2_2

Ren, J., Han, Y., Genc, Y., Yeoh, W., Popovič, A.: The boundary of crowdsourcing in the domain of creativity. Technol. Forecast. Soc. Change **165** (2021). https://doi.org/10.1016/j.techfore.2020.120530

Scopus (n.d.). https://www.scopus.com/search/form.uri?dis-play=basic&zone=header&origin=. Accessed 09 Apr 2021

Taveira, A.D., Smith, M.J.: Social and organizational foundations of ergonomics. In: Handbook of Human Factors and Ergonomics (2012). https://doi.org/10.1002/9781118131350.ch9

Vicinitas (n.d.). https://www.vicinitas.io/. Accessed 09 Apr 2021

VOSviewer (n.d.). https://www.vosviewer.com/. Accessed 23 Mar 2021

Web of Science (n.d.). https://apps-webofknowledgecom.ezproxy.lib.pur-due.edu/WOS_GeneralSearch_input.do?product=WOS&searchmode=General-Search&SID=7EKUw7yEVcCGVrUwOlu&preferencesSaved=. Accessed 09 Apr 2021

Wei, D., Basu Roy, S., Amer-Yahia, S.: Recommending deployment strategies for collaborative tasks (2020). https://doi.org/10.1145/3318464.3389719

Zaphiris, P., Ozok, A.A.: Human factors in online communities and social computing. In: Salvendy, G. (ed.) Handbook of Human Factors and Ergonomics (2012). https://doi.org/10.1002/9781118131350.ch44

A Bibliometric Analysis on Cybercrime in Nigeria

Monica Okwuchkwu Enebechi[1](✉), Chidubem Nuela Enebechi[1,2](✉), and Vincent G. Duffy[2](✉)

[1] Security and Strategic Studies, Nasarawa State University, Keffi, Nigeria
[2] School of Industrial Engineering, Purdue University, West Lafayette, IN 47907, USA
{cenebech,duffy}@purdue.edu

Abstract. The development and improvement of the internet with new distinctive features and abilities has widened the access to computer technology and created new opportunities for work and business activities. It has also posed threats to people using it either for business or leisure. According to Brenner (2007), the emergence of technology and electronic communication has brought a far-reaching increase in the incidence of criminal activities. As technology evolves, so does the nature of the crimes committed through the internet. There are numerous crimes of this nature committed daily on the internet. Criminal activity on the Internet is wide-ranging and can include an assortment of offenses such as illegal interception, copyright violation, stalking, money laundering, extortion, fraud, and resource theft involving the illegal use of computers (Broadhurst and Choo 2011). Internet-based cybercrime has become a huge menace threatening the socio-economic and technological advancement of Nigeria. Since the advent of the internet, cybercrime has become a recurring decimal in Nigeria. The main objective of this study is to create more awareness of cybercrime in Nigeria through a bibliometric analysis while enlightening the public on the existence of cybercrimes. The analysis was carried out through software like Harzing, VOSviewer, maxQDA, bibExcel, Mendeley, and Citespace. The software is used to establish rising scope in the field of cybersecurity.

Keywords: Cybercrime · Nigeria · Cybersecurity · Bibliometric analysis

1 Introduction and Background

1.1 Problem Statement

There is no denying the fact that the internet has changed the way we do things. It has become a driving force in every aspect of human endeavor. Cyberspace is constantly evolving, so too is the threat of cybercrime on national security, prosperity, and quality of life of the citizenry and the world as a global village.

C. Stephanidis et al. (Eds.): HCII 2021, LNCS 13094, pp. 249–269, 2021.
https://doi.org/10.1007/978-3-030-90238-4_18

Cybercrime may be referred to as any form of misconduct in cyberspace. It is simply defined as the criminal use of the Internet. Cybercrime is believed to have started in the 1960s in the form of hacking. This was followed by privacy violations, telephone tapping, trespassing, and distribution of illegal materials in the 1970s. The 1980s witnessed the introduction of viruses (Loader 2000; Ali et al. 2014). The fast pace of development of ICT from the 1990s till today has added to the list of criminal exploits in cyberspace (Gercke 2006). Today, the Internet is used for espionage and as a medium to commit terrorism and transnational crimes. With e-banking gaining ground in Nigeria and other parts of the world, customers and online buyers are facing great risk of unknowingly passing on their information to fraudsters (Atili 2011). Hackers get information about those who have made purchases through websites and then make fake cards, which they use with less detection.

The contribution of the internet to the development of the nation has been marred by the evolution of new waves of crime. The internet has become an environment where the most lucrative and safest crime thrives. Cybercrime has become a global threat from Europe to America, Africa to Asia, and the other parts of the world. Cybercrime has come as a surprise and a strange phenomenon that for now lives with us in Nigeria. With each passing day, we witness more and more alarming cases of cybercrimes in Nigeria, with each new case more shocking than the one before. The problem confronting the society, Nigeria inclusive is how to create effective and efficient awareness of cybercrime with a proper enlightenment campaign.

Table 1. Top repeated terms derived from Harzing database, the keywords searched were "Cybersecurity", "Cybercrime in Nigeria".

Term	Occurrences	Relevance
Punishment	15	4.39
Cryptocurrency	5	1.84
Criminology	9	1.71
Prevention	40	1.73
Law	78	0.22

A search was done in the Harzing database to derive the top reoccurring terms in the field of "cybersecurity" as seen in Table 1 above. The term with the highest relevance is "Punishment" with a relevance score of 4.39 and a maximum occurrence of 15. While the term with the lowest relevance as it pertains to the search is the term "Law" with a relevance score of 0.22 and an occurrence of 78 across all the papers used to carry out the bibliometric analysis. The research methodology section of this paper dives into how Harzing was incorporated in the bibliometric analysis and how the terms are intertwined. As seen below, Fig. 1 shows the terms repeated throughout the various re resources used to carry out the analysis. Later sections of the paper shed more light on the resources and tools that were used to carry out the analysis and how the terms various terms originated.

Since it was difficult to find a variety of literature using the search terms "Cybercrime in Nigeria", "Cybersecurity" was used as part of the search terms for deriving metadata. The metadata was then fed to the various analysis tools like Vosviewer and Citespace to carry out the bibliometric analysis.

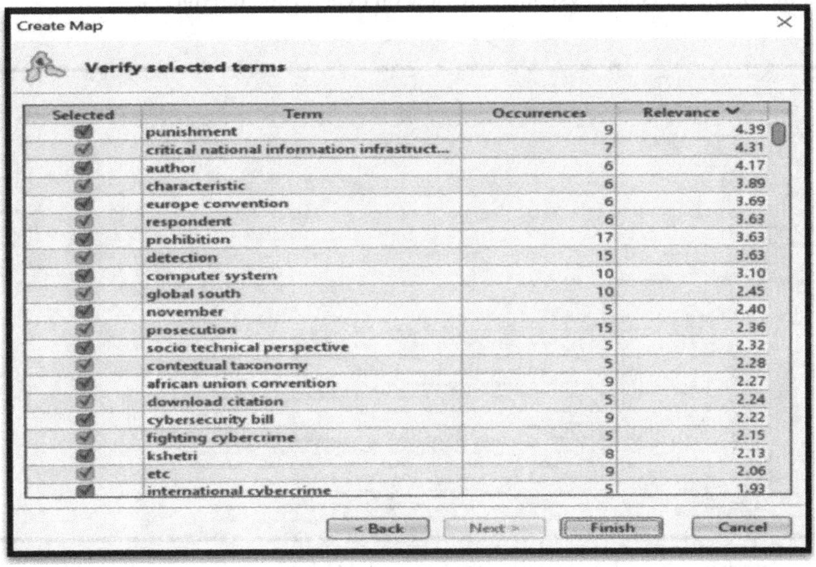

Create Map

Verify selected terms

Selected	Term	Occurrences	Relevance
✓	punishment	9	4.39
✓	critical national information infrastruct...	7	4.31
✓	author	6	4.17
✓	characteristic	6	3.89
✓	europe convention	6	3.69
✓	respondent	6	3.63
✓	prohibition	17	3.63
✓	detection	15	3.63
✓	computer system	10	3.10
✓	global south	10	2.45
✓	november	5	2.40
✓	prosecution	15	2.36
✓	socio technical perspective	5	2.32
✓	contextual taxonomy	5	2.28
✓	african union convention	9	2.27
✓	download citation	5	2.24
✓	cybersecurity bill	9	2.22
✓	fighting cybercrime	5	2.15
✓	kshetri	8	2.13
✓	etc	9	2.06
✓	international cybercrime	5	1.93

< Back | Next > | Finish | Cancel

Fig. 1. The top terms in the table above were originated from metadata gotten through a search in the Harzing database.

2 Research Methodology

2.1 Harzing

A search was done on Harzing using the google scholar option. Harzing is a software tool used to retrieve and analyze academic citations from various data sources like Google Scholar, Microsoft Academic Search, Scopus, Web of science, Crossref, etc. (https://harzing.com/resources/publish-or-perish). The keywords used for this search were "Cybersecurity" and "Cybercrime in Nigeria". The results produced from the keywords were derived with the publications of about 45 citation years, the citations were published between the year 1970 till present. A total of 820 papers were yielded from the result. The papers had 7,114 cites and about 139,49 citations per year were also produced as indicated in Fig. 2 below.

Most of the steps implemented in the methodology section of this paper were derived from Fahimnia et al.'s "Green supply chain management: A review and bibliometric analysis. The paper gives formidable insights on how to appropriately carry out a systematic review and bibliometric analysis. In this paper, Fahimnia et al.'s paper was adapted to examine the issues regarding cybercrime in Nigeria and also to conduct a suitable bibliometric analysis and systematic review on cybercrime in Nigeria.

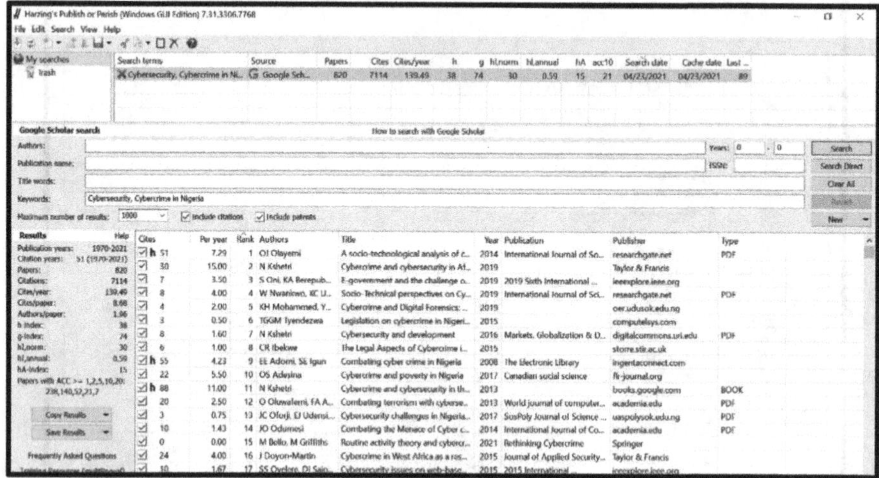

Fig. 2. Search is done in Harzing using the keywords "Cybercrime in Nigeria" and "Cybersecurity" (https://harzing.com/resources/publish-or-perish)

2.2 VOS Viewer

VOS Viewer was also one of the core tools used to create a visualization piece and network graph of all the relevant terms repeated in the various papers.

VOS viewer is a visualization tool for creating visualized scientific landscapes, the software is a very important tool for constructing and visualizing bibliometric networks that show how terms are connected (https://www.vosviewer.com/).

The metadata derived from Harzing is fed to the VOS viewer to create the list of relevant terms with multiple occurrences as shown in Fig. 1 above.

After putting the metadata, the next step is to create a network map based on text data using the co-occurrence terms. The data source selected for reading the data was the "read data from bibliographic database files." In most cases, the bibliographic database files can be derived from Scopus, Web of Science, Dimensions, PubMed. VOS viewer also supports other file types like reference managers like RIS, EndNote, and RefWorks, the option to be selected will be "read data from reference manager files." Harzing has an option to save the metadata files as a RIS version extension. This allows users to create their visualized bibliometric network graph on a VOS viewer.

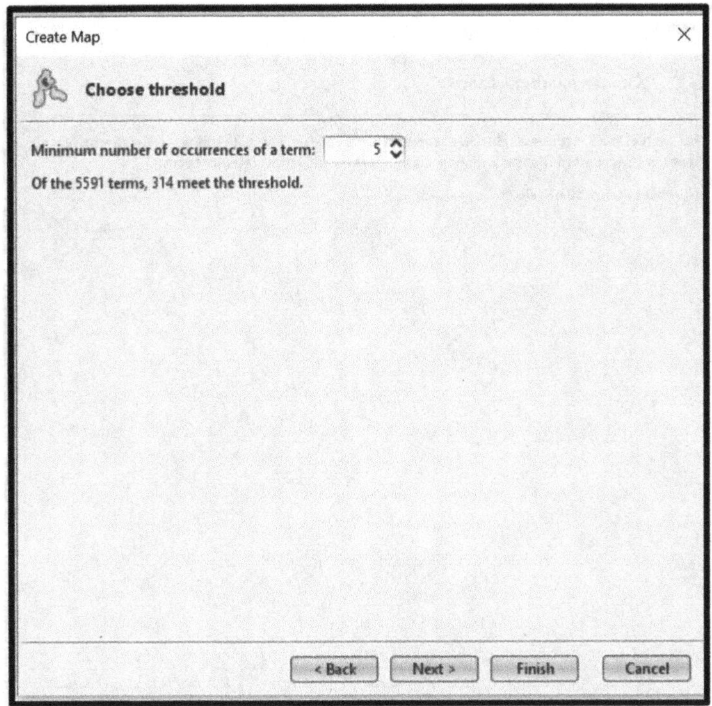

Fig. 3. Minimum occurrence and threshold on VOSviewer. (https://www.vosviewer.com/)

As indicated in Fig. 3 above a minimum of 5 occurrences were used for this search, to obtain efficient results. Figure 3 also shows that of the 5591 terms from the data 314 of the terms were able to meet up with the threshold. The values used for the occurrence and threshold along with the metadata derived from the search on the Harzing software can be used to replicate the VOSviewer visualization piece indicated in Fig. 5 below.

Figure 4 shows that a total of 300 terms were selected to create the visualized bibliometric network graph shown in Fig. 5. The default number of relevant terms selected by the VOS viewer is usually 60% but to create the most efficient result showing a well robust bibliometric network map 300 terms were used to carry out the bibliometric analysis in this paper.

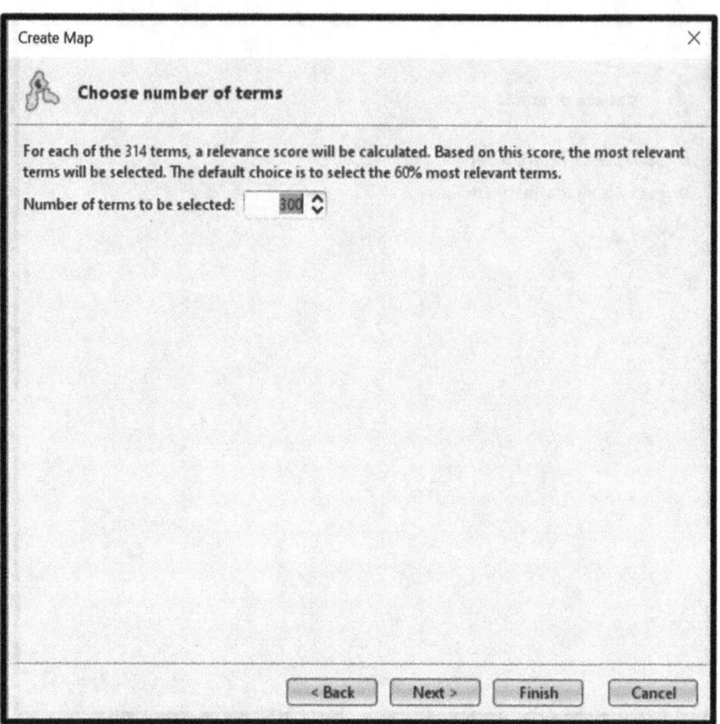

Fig. 4. 300 terms were selected as the most relevant terms on VOSviewer. (https://www.vosviewer.com/)

Cybercrime is one of the fastest, growing criminal activities on the planet. Organized crimes are using cyberspace more frequently to target credit card information and personal and financial details for internet fraud. In Nigeria and other parts of Africa, the perpetrators of this illegal act have even upgraded their nefarious activities from the physical to the mystical, in what is currently called *Yahoo Yahoo+* after the *yahoo yahoo* (Nigeria Prince). While the former involves using occult powers to target individuals for their scams, the latter is where the victims are told stories of the perpetrators claiming to be a Prince with wealth that may need processes. This, undoubtedly, has increased Nigeria's notoriety in the world rating on cyber-related offenses (Atili 2011).

Cyber-attacks now can greatly harm society in new and critical ways. Online fraud and cyber-attacks are just a few examples of computer-related crimes that are committed on an extremely large scale every day (Gercke 2006). For a long time, cybercrime has tarnished Nigerian international reputation and greatly discouraged the investment of foreign investors. The phenomenal rise of mobile communication and the drive from the Central Bank of Nigeria towards a cashless economy has contributed greatly to the growth of cybercrime hindering the socio-economic development of the country. This is as a result of engendering lack of trust and confidence in profitable transactions, promotes denial of opportunities to innocent Nigerians abroad, and causes loss of employment and revenue loss. Cybercrime has aided other illicit activities in Nigeria such as intellectual plagiarism, disruption of public services, drug trafficking, terrorism, and or social media abuse.

Although electronic scams or spam emails are generally believed to be linked to Nigeria, the scam is now prevalent in many other African countries and the targets are usually innocent individuals who could be in any part of the world.

Therefore, it is important to carry out a bibliometric analysis on this topic to gain a better understanding of how cybercrime contributed negatively to the growth of not only Nigeria but also other communities around the world. Figure 5 below shows terms such as "Cybercriminal", "policy", "law", "legislation", "cyberspace", "cybercrime legislation", "phishing", "technology", "fraud", "cybercrime law", "prevention", "society", "and "awareness". As previously stated, the terms were derived from a search done on Harzing. The search terms were "cybercrime in Nigeria" and "cybersecurity". The search yielded a metadata result with was used in the VOS viewer to construct and visualize the bibliometric network map as shown below in Fig. 5.

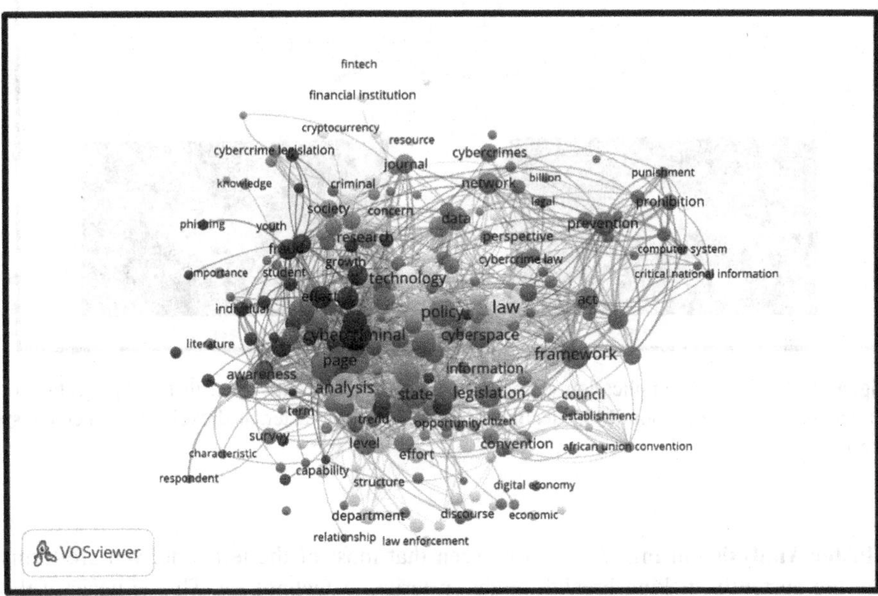

Fig. 5. VOSviewer visualization network graph created with metadata from the Harzing search above. (https://www.vosviewer.com/)

2.3 CiteSpace

CiteSpace is a very useful citation tool used to create trends and patterns that visualize scientific literature. In this paper, a citation cluster analysis and a citation burst were created using CiteSpace (http://cluster.cis.drexel.edu/~cchen/citespace/). A citation burst can be used to conduct a visual analysis of articles in a database as seen in Fig. 8, the cluster analysis done through CiteSpace can also be seen in Fig. 7 (Fig. 6).

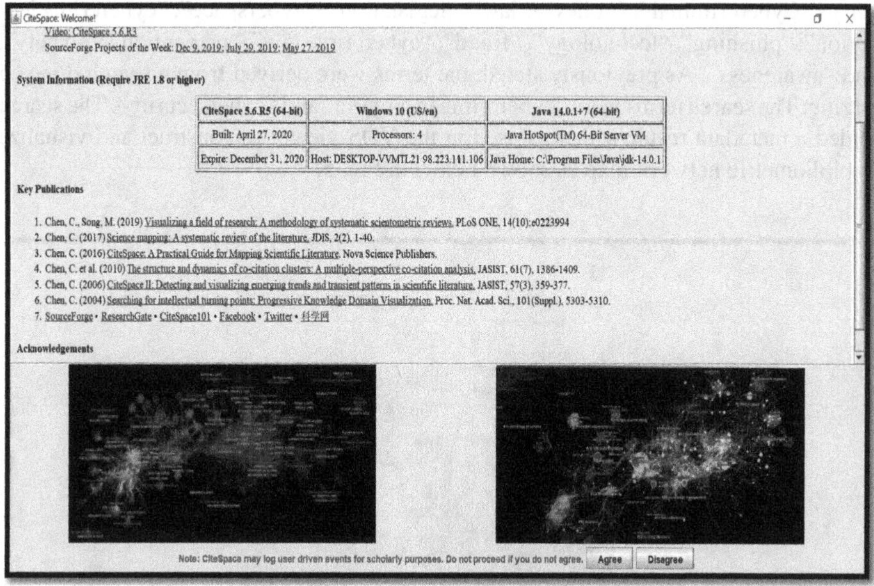

Fig. 6. CiteSpace java application analyzing and visualizing data trend in the Application of Human Factors and Ergonomics in Healthcare System (http://cluster.cis.drexel.edu/~cchen/citespace/)

Cluster Analysis. In Fig. 7, it can be seen that most of the terms derived are terms that are currently making breakthroughs in terms of technology. This is because the papers used for the analysis in CiteSpace were more recent articles. The search term "Cybercrime in Nigeria" did not yield sufficient results on the Web of Science database. Hence a search on "Cybersecurity" was done instead as seen above some of the prevalent terms that were repeated frequently across all the articles used for the analysis are terms like "deep learning", "security", "autonomous vehicles", "judgment analysis", "informational security and "adversarial machine learning". These terms are all linked to the world of cybercrime and cybersecurity in one way or the other. The cluster analysis as seen below in Fig. 7 was extracted based on keywords feature in the CiteSpace software.

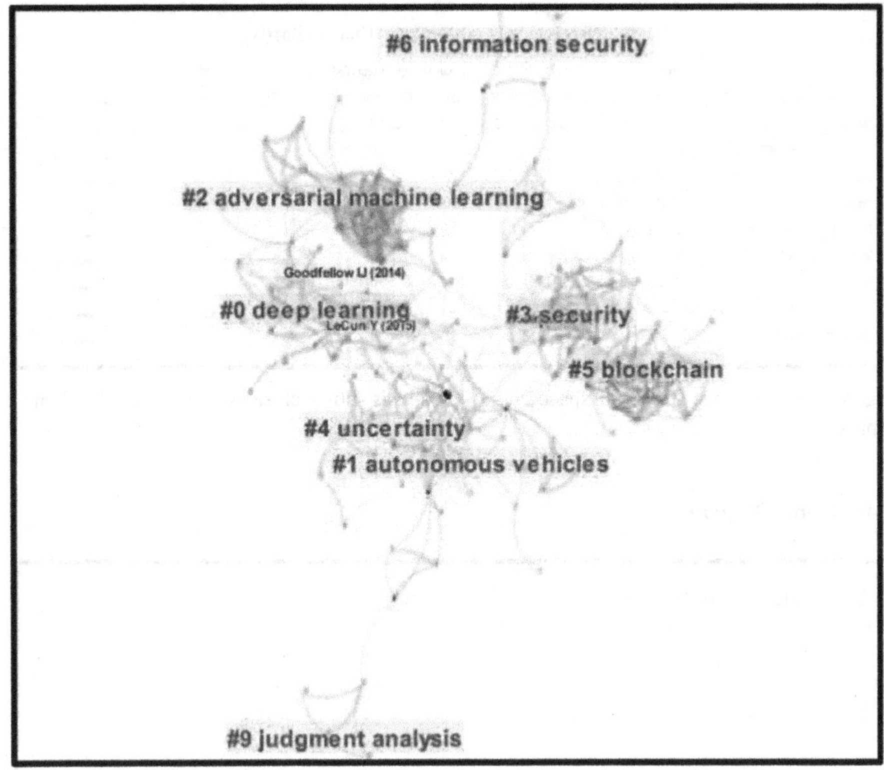

Fig. 7. Cluster Analysis for Citespace (http://cluster.cis.drexel.edu/~cchen/citespace/)

Citation Burst. A citation burst is a very important and useful tool in CiteSpace because it points to the top authors and references with the top citations in a specific field. Figure 8 shows the top 10 authors and references with the strongest citation burst. The results are shown in Fig. 8 below, from the year 2012–2016. Although the initial search was for the year 1995–2021. The results below show that there were stronger citations in the field of cybersecurity between 2018–2019.

It makes a lot of sense that as the year advances so does the citation strength of the field of cybercrime grows. The internet creates unlimited profit-oriented opportunities for commercial, social, and other human activities. The advent and introduction of cybercrime has put the users of the Internet at a higher risk. The global village currently records an increasing criminal behavior. News of cybercriminal activities continues to fill the pages of the newspaper; it is central to world news and has become a global problem. There is hardly a place where computers and internet facilities are found that cases of crime are not recorded.

Top 10 References with the Strongest Citation Bursts

References	Year	Strength	Begin	End	1995 - 2021
Sridhar S, 2012, P IEEE, V100, P210, DOI 10.1109/JPROC.2011.2165269, DOI	2012	5.37	**2016**	2017	
Hahn A, 2013, IEEE T SMART GRID, V4, P847, DOI 10.1109/TSG.2012.2226919, DOI	2013	3.88	**2017**	2018	
NIST, 2014, FRAM IMPR CRIT INFR, V0, P0	2014	3.32	**2017**	2018	
Sridhar S, 2014, IEEE T SMART GRID, V5, P580, DOI 10.1109/TSG.2014.2298195, DOI	2014	3.7	**2018**	2019	
Christidis K, 2016, IEEE ACCESS, V4, P2292, DOI 10.1109/ACCESS.2016.2566339, DOI	2016	3.39	**2018**	2019	
Ransford B, 2017, PACE, V40, P913, DOI 10.1111/pace.13102, DOI	2017	3.28	**2018**	2019	
Mitchell R, 2014, ACM COMPUT SURV, V46, P0, DOI 10.1145/2542049, DOI	2014	2.87	**2018**	2019	
Mejri MN, 2014, VEH COMMUN, V1, P53, DOI 10.1016/j.vehcom.2014.05.001, DOI	2014	2.87	**2018**	2019	
Biggio B, 2014, IEEE T KNOWL DATA EN, V26, P984, DOI 10.1109/TKDE.2013.57, DOI	2014	2.87	**2018**	2019	
Ganesan R, 2016, ACM T INTEL SYST TEC, V8, P0, DOI 10.1145/2882969, DOI	2016	2.87	**2018**	2019	

Fig. 8. Citation Burst from CiteSpace citation Analysis (http://cluster.cis.drexel.edu/~cchen/citespace/)

2.4 Trend Diagram

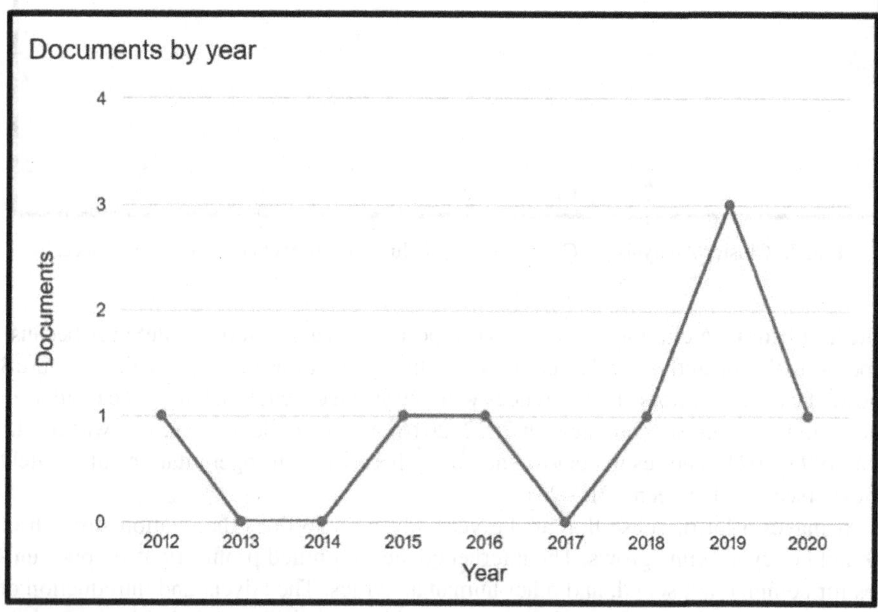

Fig. 9. Trend graph data of keywords "Cybercrime in Nigeria OR "Cybersecurity" between the year 2012–2020.

Figure 9 illustrates a trend graph that was obtained through publication data from the Scopus database. The trend graph reveals information about data on "Cybercrime in Nigeria" from the year 2012 up until the present. There was a limited amount of data for the search "Cybercrime in Nigeria", this can be tied to the reason that awareness about cybercrime in Nigeria was not made until recent years. The trend graph as seen in

Fig. 9 also shows that there is a gradual increase in easily accessible data on cybercrime in Nigeria due to the spike between 2017–2019 (Figs. 10 and 11).

2.5 Engagement Measure Vicinitas

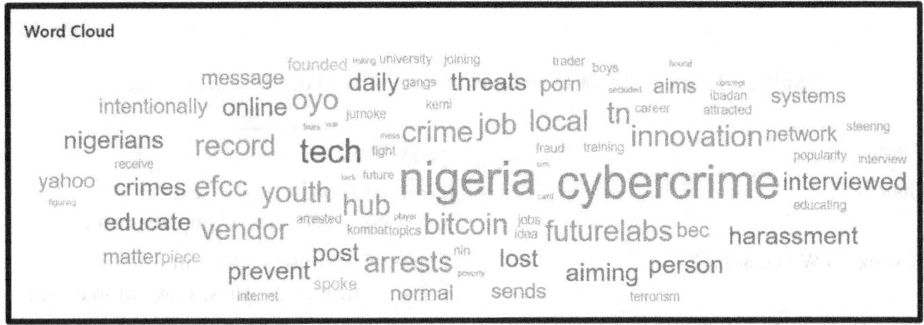

Fig. 10. Wordcloud derived from Vicintas Twitter analytics tool, the search term used was "cybercrime in Nigeria" (https://www.vicinitas.io/)

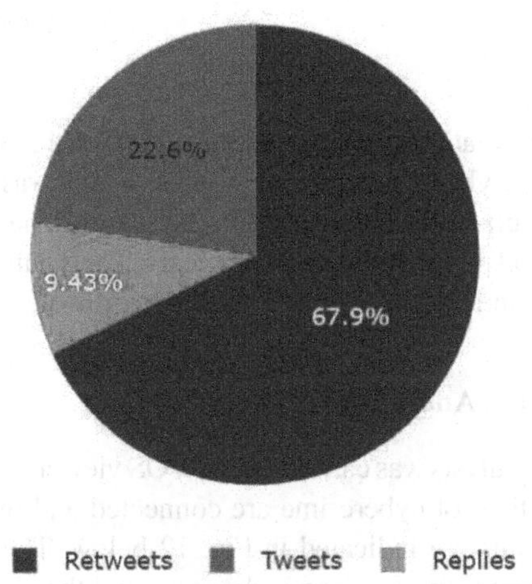

Fig. 11. Pie chart showing user engagement on twitter for the topic "cybercrime in Nigeria." (https://www.vicinitas.io/)

Vincintas is an analysis tool that helps to track and analyze real0time data using tweets of social media campaigns, hashtags, and brands on Twitter. The analysis on Vincintas was done to gauge and measure the engagement of individuals on Twitter in terms of the topic of cybercrime in Nigeria.

2.6 Author Relationship

Table 2. Leading author table derived from Scopus database (reference)

Author	Keywords
Blair R	National security; Cybersecurity; Cybercrime; Nigeria; Computer crime
Nwankwo W., Ukaoha K.C	Crime; Cyberlaw; Cybercrime; Cybersecurity; Forensics; Law enforcement
Okon P., Kayode-Adedeji T., Afolayan T.-A., Iruonagbe C	Awareness; Cybercrime; Cybersecurity; Nigeria; Policy; Rural communities; Sango Ota
Onwuka E.N., Afolayan D.O., Abubakar W., Ibrahim J.I	Cybercrime; Cybersecurity; Internet; Nigeria; Online risk; Telecommunication
Doyon-Martin J	Combatting cybercrime; cybercrime in West Africa; Electronic waste

The author relationship table displayed in Table 2, was extracted through a search of the keywords "Cybercrime in Nigeria". The search was done using the Scopus database, a total of 8 results were obtained from the search. The results in Table 2 show the top 5 authors who had publications in recent years. Their publications dive into various aspects of cybercrime in Nigeria as seen in the keywords of Table 2 above.

2.7 Co-citation Analysis

A co-citation analysis was carried out on VOS viewers to show how authors and their lead papers in the field of cybercrime are connected and intertwined. The results produced from the analysis are indicated in Fig. 12 below. The metadata used for this analysis was also derived from the Scopus database and then fed into VOS viewer to create the network map of authors.

The results reveal that the authors who fall into the same nodes or same color have either work that is closely related. Table 3 shows the leading authors in the field of "Cybercrime in Nigeria." The result indicates that OJ Olayemi is one of the authors with the highest 7.29 citations per year on the paper "*A socio-technological analysis of cybercrime and cybersecurity in Nigeria.*" Several other authors like Soni, KA Berepubo et al. also had impactful papers like "*E-government and the challenge of cybercrime in Nigeria*", with a total of 3.5 citations per year.

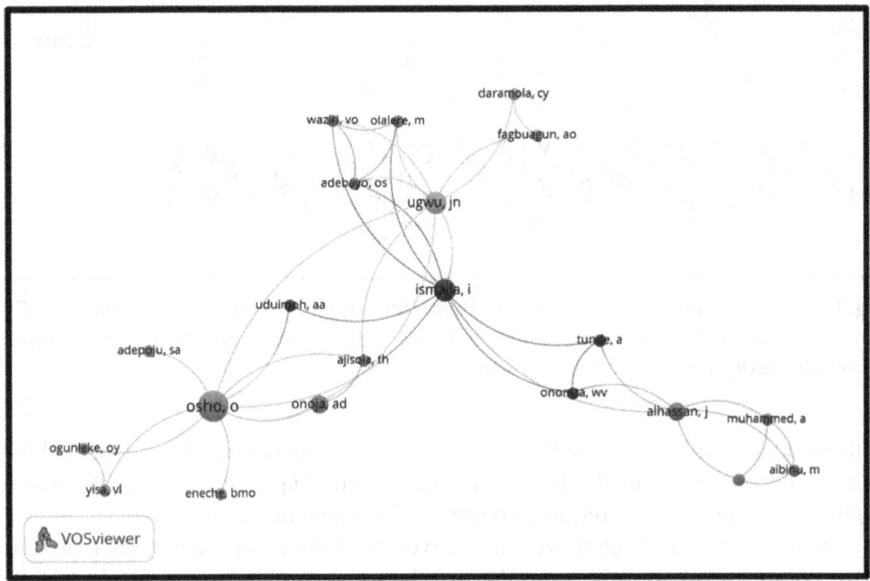

Fig. 12. Co-citation analysis of the terms "Cybercrime in Nigeria" OR "Cybersecurity" (https://www.vosviewer.com/)

Table 3. Leading Authors derived from Harzing using the Web of Science database

Author	Publisher	Citation per year
OJ Olayemi	researchgate.net	7.29
S Oni, KA Berepubo et al	ieeexplore.ieee.org	3.5
CR Ibekwe	storre.stir.ac.uk	1
JC Oforji, EJ Udensi, KC Ibegbu	uaspolysok.edu.ng	0.75
AN Ayofe, O Oluwaseyifunmitan	researchgate.net	1.33

2.8 Pivot Chat

The pivot chat was created through bibExcel, which is another useful software for accomplishing analysis of bibliographic data. Through bibExcel data was generated

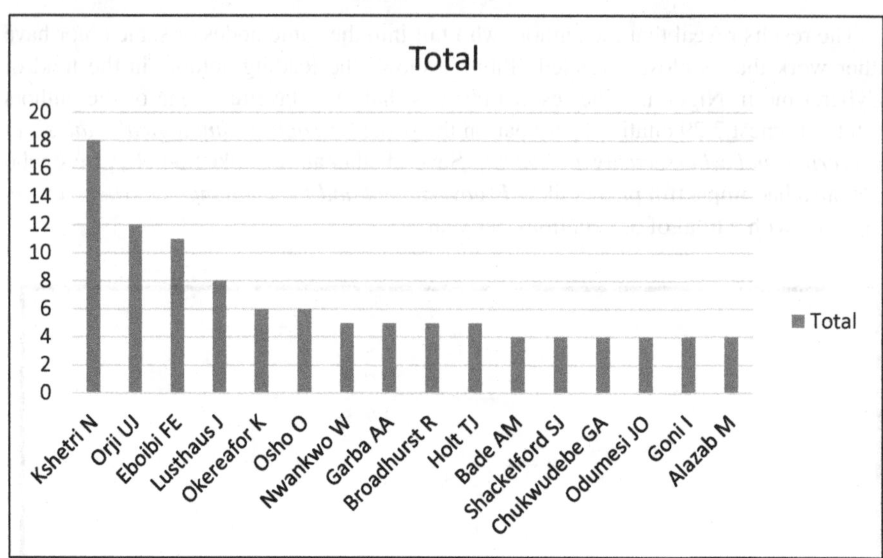

Fig. 13. Pivot chat of the top Authors with Papers that dive into the major components of "Cybercrime in Nigeria", this was derived from bibExcel. The pivot chat was plotted on excel (https://homepage.univie.ac.at/juan.gorraiz/bibexcel/)

with metadata derived from the Scopus database. The data generated from biExcel was then imported to excel for further processing and building of the pivot chat with top authors in the field of Cybercrime in Nigeria as indicated in Fig. 13 above.

The terms italicized below were all derived from the citespace metadata analysis, these terms were a few of the terms that stood out and constantly repeated through the amalgamation of data from Scopus, web of science, and other mediums of gathering data used in this paper. The terms are also well aligned with the information from the top authors in Fig. 13 indicating the effectiveness of this bibliometric analysis.

Information Security: According to a platform for computer scientists called geek for geeks, "Information Security is the practice of preventing unauthorized access, use, disclosure, disruption, modification, inspection, recording or destruction of information". This definition is in line with the definition by "A review of the legal and regulatory frameworks of Nigerian Cybercrimes Act 2015" published by Felix E. Eboibi as indicated in Pivot chart shown above Fig. 13.

Deep Learning: Deep learning is a subset of Artificial Intelligence (AI). Investopedia defines Artificial intelligence (AI) as "the simulation of human intelligence in machines that are programmed to think like humans and mimic their actions." When it comes to cybersecurity, deep learning can be very impactful because AI systems get more advanced the more data they are exposed to data. They not only analyze data but also use their analysis to become more learned through identifying trends and patterns and thus becoming immensely autonomous in the realm of cybersecurity. Figure 13 above indicates that Kshetri has the highest amount of relevant information on cybercrime in Africa. In 2003 Kshetri stated that "Cybersecurity is considered to be as a luxury and not a necessity." This is one of the main reasons why cybercrime continues to increase in emerging African economies (Ksheri 2019). According to Duffy and Duffy (2020) cybersecurity is an evolving field in the area of Human-Computer-Interaction (HCI)", human factors also happens to be an important facet of HCI that is intwined with cybercrime and cybersecurity. Therefore, it is apparent that implementing advanced technological techniques from deep learning will help African countries mitigate the constant damage inflicted by cyber attackers thereby improving the Human-Computer Interaction experiences of users in the African nation and worldwide.

Block Chain: According to Forbes, Blockchain is a very effecting tool that can contribute to the eradication of cybercrime through "boosting cybersecurity by preventing denial of service and other attacks in air service traffic". The main concept behind blockchain is decentralization, when data is spread over to different locations, it makes it harder for attackers to find its origination and maliciously manipulate it leading to less exposure of malleability in the cyber world. Orji UI happens to be among the top authors with relevant information about cybercrime in Nigeria as indicated in Fig. 13. "Cybercrime targeted to electronic banking generally reduce customer trust in the electronic transaction and impedes the adoption of electronic banking services as well as e-commerce" (Orji 2019). Nigeria will greatly benefit from implementing high-level blockchain to decentralize data thereby leaving attackers fruitless in their degenerated activities.

2.9 Mendeley

Figure 14 shows the capture of some of the papers used to carry out the bibliometric analysis in this paper. Some of the papers used for the bibliometric analysis in this paper were obtained for the 3rd edition of Safety and Health for Engineers. Most of the other articles used for the bibliometric analysis were acquired from a variety of databases as indicated like Scopus, Web of Science, PubMed, etc.

The two main chapters used to produce the bibliometric analysis used in this paper originated from the textbook "Safety and Health for Engineers" by Roger L. Brauer. The chapters are "*Fundamentals of Safety Management*", chapter 34, and chapter 36 "*System Safety*". When it comes to safety, most people generally think of physical safety or being protected from hazardous materials and dangerous situations. But safety encompasses internet safety and protection from risky situations online like cybercrime. Cybercrimes are broadly categorized into three categories, namely crime against individuals, property, and government (Gordon and Sarah 2006). Each category can use a variety of methods and the methods used vary from one criminal to another. Below describes each of the three as:

1. Individual: This type of cybercrime can be in the form of cyberstalking, distributing pornography, trafficking, and "grooming".
2. Property: Susan and Brenner (2010) assert that just like in the real world where a criminal can steal and rob, even in the cyber world criminals resort to stealing and

robbing. In this case, they can steal a person's bank details and siphon off money; misuse the credit card to make numerous purchases online; run a scam to get naïve people to part with their hard-earned money; use malicious software to gain access to an organization's website or disrupt the systems of the organization.

3. Government: Although not as common as the other two categories, crimes against a government is referred to as cyber terrorism. If successful, this category can wreak havoc and cause panic amongst the civilian population.

Authors	Title	Year	Published In
Yang, Samuel C.; Wen, Bo	Toward a cybersecurity curriculum model for undergraduate business schools: A survey of AACSB-accredited institutions i...	2017	Journal of Education for Business
Brauer, Roger L.	Safety and Health for Engineers: Third Edition	2016	Wiley
Oni, Samuel; Araife Berepubo, Karina; Atinuke Oni, Aderonk...	E-government and the challenge of cybercrime in Nigeria	2019	2019 6th International Conference on eDemocracy and eGovernment, ICEDEG 2019
Doyon-Martin, Jacquelynn	Cybercrime in West Africa as a Result of Transboundary E-Waste	2015	Journal of Applied Security Research
Oluyinka, Solomon; Shamsuddin, Alina; Wahab, E...	A study of electronic commerce adoption factors in Nigeria	2013	International Journal of Information Systems and Change Management
Okon, Patrick; Kayode-Adedeji, Tolulope; Afolayan, Tayo Adi...	Cybersecurity awareness among rural communities in Sango Ota, Ogun State, Nigeria	2019	14th International Conference on Cyber Warfare and Security, ICCWS 2019
Ibrahim, Suleman	Social and contextual taxonomy of cybercrime: Socioeconomic theory of Nigerian cybercriminals	2016	International Journal of Law, Crime and Justice
Longe, Olumide B.; Chiemeke, S. C.	Cyber crime and criminality in Nigeria - What roles are internet access points in playing?	2008	European Journal of Social Sciences
Veresha, Roman V.	Preventive measures against computer related crimes: Approaching an individual	2018	Informatologia
Duffy, Brendan M.; Duffy, Vincent G.	Data mining methodology in support of a systematic review of human aspects of cybersecurity	2020	Lecture Notes in Computer Science (including subseries Lecture Notes in Artificial Intelligence...
Onwudebelu, U.; Ugwoke, U.C.; Igbinosa, G.O.	E-Governance initiatives in Nigeria	2012	Proceedings of the 2012 IEEE 4th International Conference on Adaptive Science and Technolog...
Okon, P.; Kayode-Adedeji, T.; Afolayan, T.-A.; Iruonagbe, C.	Cybersecurity awareness among rural communities in Sango Ota, Ogun State, Nigeria	2019	14th International Conference on Cyber Warfare and Security, ICCWS 2019
Nwankwo, W.; Ukaoha, K.C.	Socio-technical perspectives on cybersecurity: Nigeria's cybercrime legislation in review	2019	International Journal of Scientific and Technology Research

Fig. 14. Some of the papers and references used for this bibliometric organized in the Mendeley Software. (https://www.mendeley.com/?interaction_required=true)

2.10 maxQDA

An analysis was done using the 2020 version of MaxQDA. The analysis produced a word cloud image with several terms that are connected to the application of the realm of "Cybercrime in Nigeria".industry. The word cloud image is displayed in Fig. 15 below. Some of the terms that were derived from the software include "hazardous", "safety", "design", "cybercrime", "standard management". These terms are constantly repeated and resurface with the different tools used to carry out this bibliometric analysis.

Fig. 15. Bibliometric Analysis with MaxQDA (https://www.maxqda.com/qualitative-analysissoftware)

3 Results

3.1 List of 10 Ways Insights

The list of 10 ways was meticulously considered for the most vital reference pieces used in the bibliometric analysis of this paper. According to Duffy's *Improving efficiencies and patient safety in healthcare through human factors and ergonomics*, a List of 10 ways helps to develop one's ability to evaluate research and to systematically view research papers in various streamlined ways (Duffy 2009).

The list of 10 ways, played an important role in understanding why "Cybercrime in Nigeria" is an important topic that deserves attention. It also helped to streamline the important pieces of literature that contribute significantly to this topic.

According to Okeshola (2013), cybercrime is the most complicated problem in cyberspace and many nations are battling to protect their cyberspace from criminals, for national security and integration. Cybercrime refers to unlawful practices carried out using computers, electronic and ancillary devices. It involves disruption of network traffic, email bombing, distribution of viruses, identity theft, cyberstalk, and cybersquatting (Fanawopo 2004).

Maitanmi (2013) asserts that cybercrime is a type of crime committed by criminals who make use of a computer as a tool and the internet as a connection to reach a variety of objectives such as illegal downloading of music files and films, piracy, spam mailing, and the likes. Cyber-crime evolves from the wrong application or abuse of internet services. Since the advent of the internet, cybercrime has become a recurring decimal in Nigeria.

Ifukor (2006) stated that cybercrime includes all forms of crime committed through the use of the internet. They are referred to as internet fraud. This means using one or more components of the internet such as chat rooms and emails among others, to present fraudulent solicitation to prospective victims or to defraud individuals or financial institutions. According to Laver (2005) drug cartels, organized crime, international money launderers, and computer hackers are unleashing themselves on the information highways and they are becoming even more successful.

4 Future Work and Discussion

The cost of cybercrime to any nation is enormous and can completely ruin the country's economy if the proper security strategies are not put in place. Foreign Investments into that economy can begin to dwindle. Several governments of the world are continuously carrying out research to improve their cybercrime attack countermeasures.

Apart from local economic sabotage and fraud, a large number of Nigerians particularly the youths are also engaged in perpetrating internet fraud by duping individuals and corporate institutions abroad through spurious economic deals. Fraudulent practices through cybercrime have thrown up emergency millionaires, even billionaires in the economic system which is injurious to the country's economic growth as most of such funds acquired illegally are not used productively to promote the economy. Such funds could not be easily traced by law enforcement agencies. Indeed, economic sabotage resulting from cybercrime cannot be over-emphasized.

Nevertheless, cybercrime requires the knowledge of expertise/specialists in computer technology and internet protocols for proof. Therefore, further research needs to be done for more progress to be made in the realm of cybersecurity.

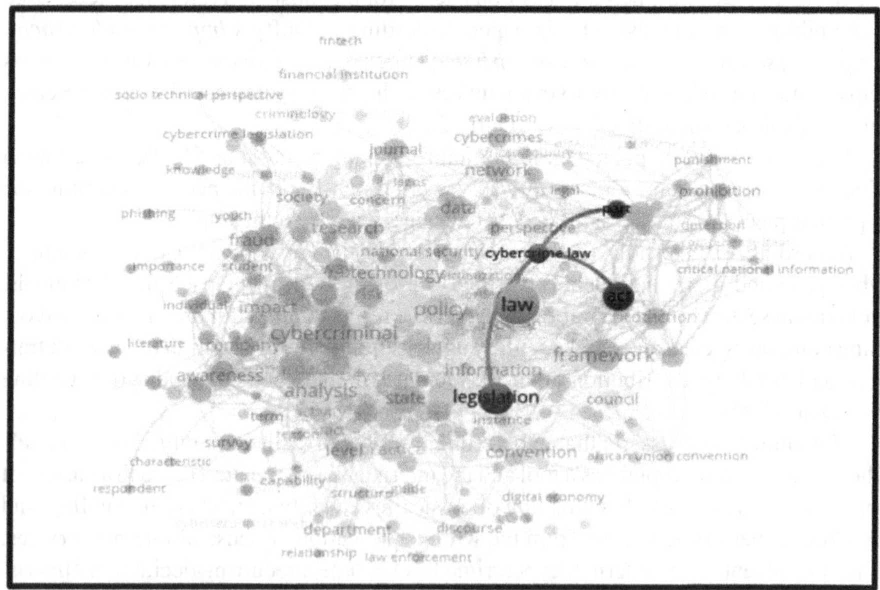

Fig. 16. VOSviewer visualization network graph, the term focus in this cluster is "Law" (https://www.vosviewer.com/)

For example, The United States government has an independent agency called The National Science Foundation (NSF). The NSF is in charge of advancing the progress of science, a mission accomplished by funding proposals for research and education made by scientists, engineers. In Fig. 17 below some awards and proposals.

An example of some of the awards include "curriculum to Broaden Participation in Cybersecurity for Middle School Teachers and students (CyberMiSTS), this award aims at exposing students at an early age to cybercrimes and gives middle school teachers the opportunities to educate their students about the consequences of cybercrime and how to protect themselves from being exposed to cyber threats. The NSF also has other awards geared towards cybersecurity in Education, Healthcare, Network Security, and various STEM fields. Nigerian and many more countries should emulate the mission of NSF and fund research inclined to promote a safe and conducive environment online for all its citizens (Fig. 16).

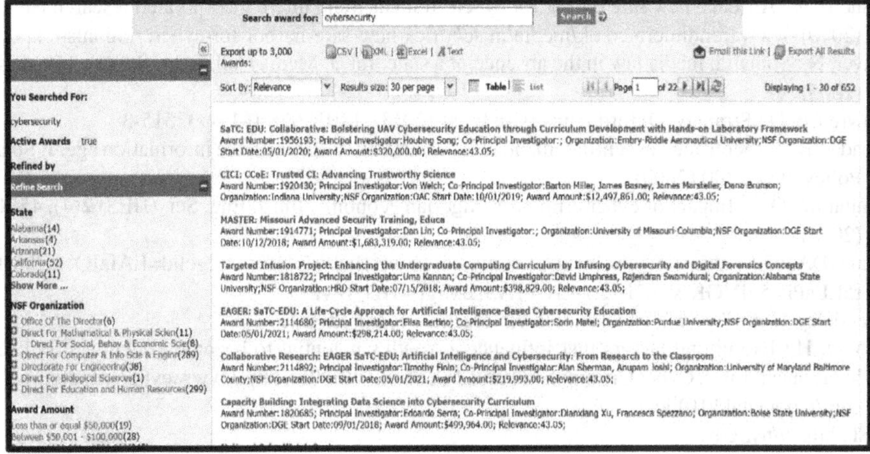

Fig. 17. NSF award search for the term "Cybersecurity" (https://www.nsf.gov/awardsearch/)

References

Aina, L.O.: Research in Information Sciences: An African Perspective, pp. 1–31. Stirling-Horden, Ibadan (2002)

Atili, A.: Want! Law for tackling cybercrime. The Nations Newspaper, p. 41 (2011)

bibExcel. https://homepage.univie.ac.at/juan.gorraiz/bibexcel/

Brenner, S.: Law in an Era of Smart Technology, p. 374. Oxford University Press, Oxford (2007)

CiteSpace. http://cluster.cis.drexel.edu/~cchen/citespace/

Dasuki, E.A.: Why EFCC is losing war on web scam. Vanguard Newspaper, p. 34 (2011)

Doyon-Martin, J.: Cybercrime in West Africa as a result of transboundary e-waste. J. Appl. Secur. Res. **10**(2), 207–220 (2015). https://doi.org/10.1080/19361610.2015.1004511

Duffy, B., Duffy, V.: Data mining methodology in support of a systematic review of human aspects of cybersecurity. In: Duffy, V.G. (ed.) HCII 2020. LNCS, vol. 12199, pp. 242–253. Springer, Cham (2020). https://doi.org/10.1007/978-3-030-49907-5_17

Fahimnia, B., Sarkis, J., Davarzani, H.: Green supply chain management: a review and bibliometric analysis. Int. J. Prod. Econ. **162**(C), 101–114 (2015). https://doi.org/10.1016/j.ijpe.2015.01.003
Fanawopo, S.: FG moves to enforce cyber crime laws (2004)
Forbes. https://www.forbes.com/?sh=179fbb342254
Gercke, M.: The slow wake of a global approach against cybercrime. Comput. Law Rev. Int. **2**(2), 141 (2006)
Geeks for Geeks. https://www.geeksforgeeks.org/
Greenwald, S.E.: Combating cybercrime in Nigeria. Electron. Libr. **26**(5), 716–725 (2014)
Harzing. https://harzing.com/resources/publish-or-perish
Howell, H., Lind, U.: Customs arrest suspected hacker. The Vanguard Newspaper, p. 53 (2009)
Ifukor, M.O.: Cybercrime: a challenge to information and communication technology (ICT). Commun. J. Libr. Inf. Sci. **8**(2), 38–49 (2006)
Investopedia. https://www.investopedia.com/
Kifordu, A., Nwankwo, W., Ukpere, W.: The role of public-private partnership on the implementation of national cybersecurity policies: a case of Nigeria. J. Adv. Res. Dyn. Control Syst. **11**(8 Special), 1386–1392 (2019)
Kshetri, R.: Africa: a new safe harbor for cybercriminals trend micro incorporated research paper (2010). www.trendmicro.co.uk/media/misc/africa-new-safe-harbor-for-cybercriminals-en.pdf
Laver, N.: Somalia: media law in the absence of a state. Int. J. Media Cult. Polit. **8**(2–3), 159–174 (2005)
Lawrence, D.: Strategy. Oxford University Press (2013). ISBN 978-0-19-932515-3
Loader, B.: Cybercrime: law enforcement, security and surveillance in the information age. J. Soc. Policy **30**(1), 300 (2000)
Maitanmi, O.S.: Impact of cyber crimes on Nigerian economy. Int. J. Eng. Sci. (IJES) **2**(4), 4551 (2013)
MaxQDA2020. https://www.maxqda.com/qualitative-analysissoftware?gclid=EAIaIQobChMIr5LLsPiw5gIVGKrsCh2IBglDEAAYASAAEgLy9fD_BwE
Mendeley. https://www.mendeley.com/?interaction_required=true
Moore, H.: Examining factors that influence a Youth's potential to become a victim of online harassment. Int. J. Cyber Criminol. **4**(1 and 2), 685–698 (2005). http://www.cybercrimejournal.com/mooreetal2010ijcc
NSF. https://nsf.gov/
Nwankwo, W., Ukaoha, K.C.: Socio-technical perspectives on cybersecurity: Nigeria's cybercrime legislation in review. Int. J. Sci. Technol. Res. **8**(10), 47–58 (2019)
Odunfa, A.: Nigeria: report on cyber threat calls for quick passage of bill (2014). http://www.allafrica.com/stories/201405080279.Html
Okeshola, F.B.: The nature, causes and consequences of cyber crime in tertiary institutions in Zaria-Kaduna State, Nigeria. Am. Int. J. Contemp. Res. **3**(9), 98–114 (2013)
Okon, P., Kayode-Adedeji, T., Afolayan, T.-A., Iruonagbe, C.: Cybersecurity awareness among rural communities in Sango Ota, Ogun State, Nigeria. In: 14th International Conference on Cyber Warfare and Security, ICCWS 2019, pp. 294–303 (2019)
Onwuka, E.N., Afolayan, D.O., Abubakar, W., Ibrahim, J.I.: Survey of on-line risks faced by internet users in the Nigerian telecommunication space. In: CEUR Workshop Proceedings, vol. 1830, pp. 28–33 (2016)
Roger, L.B.: Second Edition Safety and Health (2006)
Scopus. https://www.scopus.com/search/form.uri?display=basic&zone=header&origin=#basic
Steffani, M.: Cyber security analysis of Turkey. Int. J. Inf. Secur. Sci. **1**(4), 112–125 (2006)
Stremlau, G., Osman, W.: The privacy privilege: law enforcement, technology and the constitution. J. Technol. Law Policy **7**, 123 (2015)
Veresha, R.V.: Preventive measures against computer related crimes: approaching an individual. Informatologia **51**(3–4), 189–199 (2018). https://doi.org/10.32914/i.51.3-4.7

Vincintas. https://www.vicinitas.io/
VOSviewer. https://www.vosviewer.com/
Wada, T., Longe D.A., Paul, S., Danquah, O.R.: Action speaks louder than words – understanding cyber criminal behavior using criminological theories. J. Internet Bank. Commer. **17**(1), 1 (2012)
Waziri, F.: Antigraft campaign: the war, the worries. The Punch, 1st March 2009, p. 1 (2009)
Weaver, C.: Principles of Cybercrime, 2nd edn. Cambridge University Press, Cambridge (1998)
Web of Science. http://login.webofknowledge.com

Differences in Product Selection Depend on Situations: Using Eyeglasses as an Example

Yuri Hamada[1(✉)], Atsuya Nagata[2], Naoki Takahashi[2], and Hiroko Shoji[2]

[1] Aoyama Gakuin University, 5-10-1 Fuchinobe, Chuo-ku, Sagamihara, Kanagawa 252-5258, Japan
hamada@ise.aoyama.ac.jp

[2] Chuo University, 1-13-27 Kasuga, Bunkyo-ku, Tokyo 112-8551, Japan

Abstract. Conventional research on decision support focuses on rational decision-making. However, consumer decision-making is considered to be ambiguous and varies depending on the situation and context. Therefore, the difference in the selection results depending on the usage of products was investigated. Specifically, the variation in the choice of eyeglasses according to the situation of use was examined. First, the variability of product selection based on the number of attribute types of the selected product that vary with the purpose of the purchase was quantified. This revealed that product selection depends on the situation. Furthermore, a cluster analysis was used to group the participants to show the differences in the structures from the decision tree analysis. This may lead to decision support for product selection.

Keywords: Decision making · Product selection · Decision tree

1 Introduction

Since the 1990s, the widespread use of the Internet and cell phones has increased the access to information. In addition, the spread of e-commerce sites has made it possible for people to purchase products online without having to go to the actual stores. In this way, we are surrounded by a vast array of products and services, and the size of the market is increasing. Therefore, there is a need for a method to support consumers in their selection of products that match their preferences.

Conventional research on decision support assumes that decision-makers have clear goals and criteria and make rational decisions based on them. However, in actual decision-making, the criteria often change dynamically depending on the situation and context, and people are required to make decisions based on their own sensitivity. However, few studies have considered decision-making as a dynamic process and clarified its characteristics.

To clarify the characteristics of the dynamically changing decision-making process, the authors investigated the influence of interactions with product information on the selection result [1, 2]. Specifically, we investigated how the order in which product attributes were presented changes the outcome in the selection of a watch. We found

C. Stephanidis et al. (Eds.): HCII 2021, LNCS 13094, pp. 270–279, 2021.
https://doi.org/10.1007/978-3-030-90238-4_19

that most participants chose different products depending on their interactions with the product information. In this way, the authors clarified the variations in the selection results depending on the order of presentation of product attributes in previous studies. Although these studies did not consider the usage of the product, consumers' choices are thought to vary depending on the usage of the product in an actual purchase.

In this study, the change in selection results depending on the use of the product was investigated. The differences in the choice of eyeglasses depending on the situation in which the product is used were considered.

2 Related Research

For studies on decision-making, researchers in decision science have used a model of rational judgement in which people choose alternatives based on the notion that a certain alternative has the highest expected utility [3]. The Analytic Hierarchy Process (AHP) was proposed by Saaty for rational decision-making [4]. The AHP is a method that shows rationality and guides decision-making by quantitatively measuring the importance of evaluation criteria. The method has been applied to many decision support systems [5]. In this way, conventional decision-making research does not assume decision-making in which the criteria change depending on the situation.

However, Simon argues that human decision-making is not completely rational and has limitations, what he calls 'limited rationality [6]'. Many studies, including behavioural economics, claim that human decision-making is influenced by emotional factors. Motterlini points out that the decision-maker's criteria are dynamic and change depending on the situation and that actual decisions are made emotionally [7]. Underhill states that consumers' buying behaviour 'is becoming less and less likely to be influenced by the situation outside the store [8]'. This means that consumers do not come to a store with a clear idea of what they want to buy. Rather, they come with ambiguous desires and are influenced by the impressions and information that they receive in the store. Thus, people's preferences are now considered to be formed during the decision-making process instead of being predetermined [9].

As mentioned above, consumer decision-making is ambiguous, and the resulting choice varies depending on the situation and context. Therefore, the authors investigated how product selection results change depending on the order in which the product attributes are presented [1, 2]. As a result, participants who had a clear image of the product they wanted chose a very similar product, while those who were not particular about the product tended to select a product with a low similarity. A comparison of the results between men and women showed that women were more influenced by the interaction with product information than were men and that the variability of the selection results was greater. In these studies, we focused on watches and did not consider the place and situation of use. However, in actual purchases, the products to be selected vary depending on the usage.

In this study, the effects of the situation on the choice of eyeglasses were examined.

3 Experimental Method

In the experiment, 207 eyeglasses were prepared. Thirty male and female participants in their teens and twenties, either graduate or undergraduate students, participated in the experiment. The following 12 attributes were extracted: name, price, shape, colour, front material, lens width, lens height, bridge width, temple length, frame type, nose pad, and country of origin. Part of the actual presentation screen is shown in Table 1.

Next, the eyeglasses were used in five situations: on the run, at home, at a wedding, in job hunting, and on a trip. The participants were asked to choose products based on the situations. Product selection was set in the order of frame colour, frame shape, and price. After all the products were selected, the participants were asked to fill out a questionnaire on whether they wore glasses, whether the glasses were suitable for each situation, and what the important points were.

In conducting this study, the necessary procedures were performed in accordance with the ethical rules for research involving humans at the Faculty of Science and Engineering, Chuo University.

Table 1. Presentation screen

No.	1	2	3
Name	A NU AN-07 DMGR	A NU AN-08 BK	A NU AN-08 BLUSS
Picture			
Price	16500	16500	16500
Shape	Boston	Boston	Boston
Colour	Gray	Black	Blue
Front Material	Metal	Metal	Metal
Lens Width (mm)	47	47	47
Lens Height (mm)	43	40.3	40.3
Bridge Width (mm)	23	22	22
Temples Length (mm)	145	145	145
Frame Type	With frame	With frame	With frame
Nose Pad	Clings	Clings	Clings
Country of Origin	China	China	China

4 Results

As a result of investigating whether or not there was a difference in the selection of products depending on the situation of use, only one participant out of 30 selected all the same glasses. To quantify the variability by situation, the types selected for each attribute were counted. The number of types of each of the three attributes (price, shape, and colour) selected in each of the five situations is shown in Table 2. The entries for each participant number indicates the number of choices for each attribute, and the maximum, minimum, mean, and variance for all participants are also shown. Table 2 shows that the order of variance is colour > price > shape. This suggests that colour is the most influential factor for the situation.

Table 2. Number of selected types per attribute

No	Price	Shape	Colour	Average
1	2	2	1	1.7
2	2	1	1	1.3
3	2	2	2	2.0
4	1	1	1	1.0
5	2	3	2	2.3
6	3	3	3	3.0
7	1	2	1	1.3
8	3	4	5	4.0
9	5	4	2	3.7
10	4	4	5	4.3
11	5	4	3	4.0
12	4	4	5	4.3
13	4	3	5	4.0
14	3	3	2	2.7
15	2	2	3	2.3
16	3	2	2	2.3
17	3	3	4	3.3
18	4	3	3	3.3
19	3	2	3	2.7
20	2	3	1	2.0
21	4	3	3	3.3
22	3	2	2	2.3

(continued)

Table 2. (*continued*)

No	Price	Shape	Colour	Average
23	3	2	2	2.3
24	4	3	3	3.3
25	3	4	3	3.3
26	3	4	4	3.7
27	3	3	3	3.0
28	4	3	3	3.3
29	3	4	4	3.7
30	4	3	4	3.7
Maximum	5	4	5	4.3
Minimum	1	1	1	1.0
Average	3.1	2.9	2.8	2.9
Variance	1.0	0.8	1.6	0.8

5 Analysis Method

The decision-making process is represented as a decision tree. A decision tree is an algorithm that performs classification by creating one rule after another, dividing the data [10]. The decision tree takes an attribute for the node and a value for the arc. A value can be a single value or a set of values, including intervals. The leaf (the bottom node) represents the choices, including multiple choices. When framed this way, one product of the selection process (i.e., the decision-making process) is presented as a decision tree.

In this study, 30 decision trees were created, one for each of the participants including their selections for the five situations. Next, to examine the differences in the structures of the decision trees, the number of nodes and the number of selected types for each attribute were compared. Specifically, the number of selected types was used to perform hierarchical clustering and classify the participants. In addition, the numbers of nodes in the decision tree for each group were compared.

6 Analysis Results and Discussion

The number of nodes and the depth of each participant's decision tree along with the maximum, minimum, mean, and variance are shown in Table 3.

Figure 1 shows the decision tree of participant 2. It can be seen from the figure that the number of nodes in the decision tree was seven and the depth was three. The average number of types of the three attributes of the participants in this experiment was approximately 1.3. Figure 2 shows the decision tree of participant 27, and it can be seen from this figure that the number of nodes in the decision tree was 15 and the depth was five. The average number of types of the three attributes of participant 27 in this experiment was approximately 3.0. Figure 3 shows the decision tree of participant 10, for whom the number of nodes in the decision tree was 25 and the depth was 7. The average number of types of the three attributes for participant 10 in this experiment was approximately 4.3. This suggests that the average number of types of the three attributes tends to be lower for participants with fewer nodes and less depth in the decision tree.

Next, to examine the differences in the structure of the decision tree, the number of nodes in the decision tree and the number of selected types for each attribute were compared. Hierarchical clustering was performed using the number of selected types, and the participants were classified into two groups. Figure 4 shows Hierarchical clustering result. The numbers of nodes in the two groups were compared, and a t-test was conducted. The differences in the number of nodes for each group are shown in Table 4. Table 4 shows that the average number of nodes in the group with less change in selection by situation was smaller than that in the group with more change in selection. The results of the t-test showed a P-value of 0.0006, which is significant at a 5% level, indicating that the participants with less change in choice due to the situation also had fewer nodes in the decision tree.

Table 3. Number of nodes and depth for decision trees

No	Nodes	Depth
1	9	4
2	7	3
3	9	4
4	3	1
5	15	5
6	15	5
7	7	3
8	14	6

(*continued*)

Table 3. (*continued*)

No	Nodes	Depth
9	21	5
10	25	7
11	19	6
12	23	6
13	19	7
14	17	6
15	13	5
16	20	7
17	15	5
18	21	7
19	9	4
20	17	6
21	23	7
22	15	4
23	19	6
24	15	5
25	17	5
26	19	5
27	17	5
28	19	6
29	15	5
30	21	7
Maximum	25	7
Minimum	3	1
Average	15.9	5.2
Variance	28.1	2.0

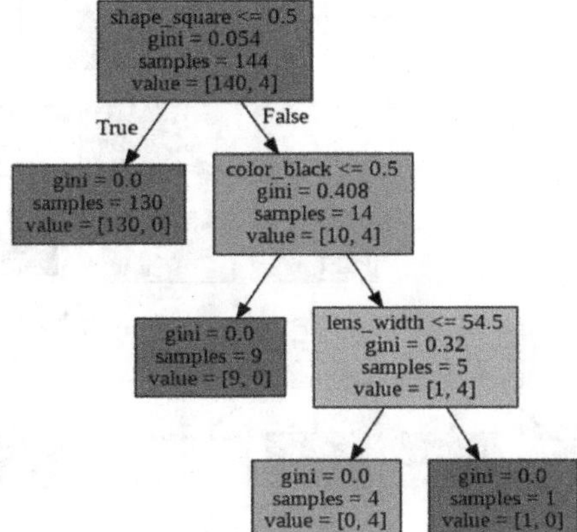

Fig. 1. Decision tree of participant 2

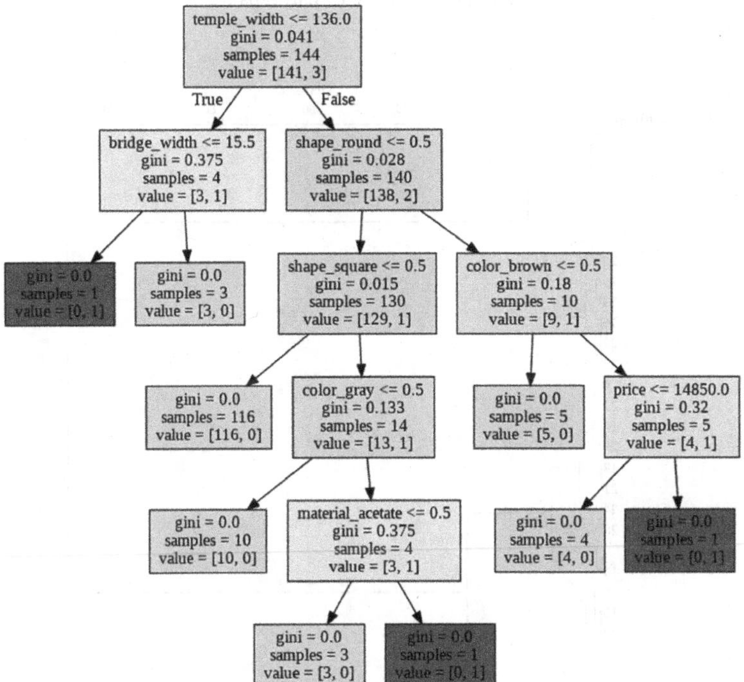

Fig. 2. Decision tree of participant 10

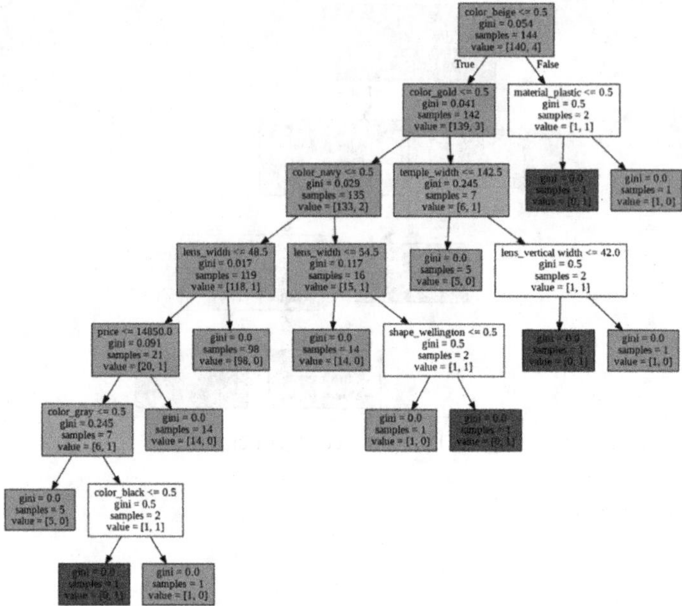

Fig. 3. Decision tree of participant 7

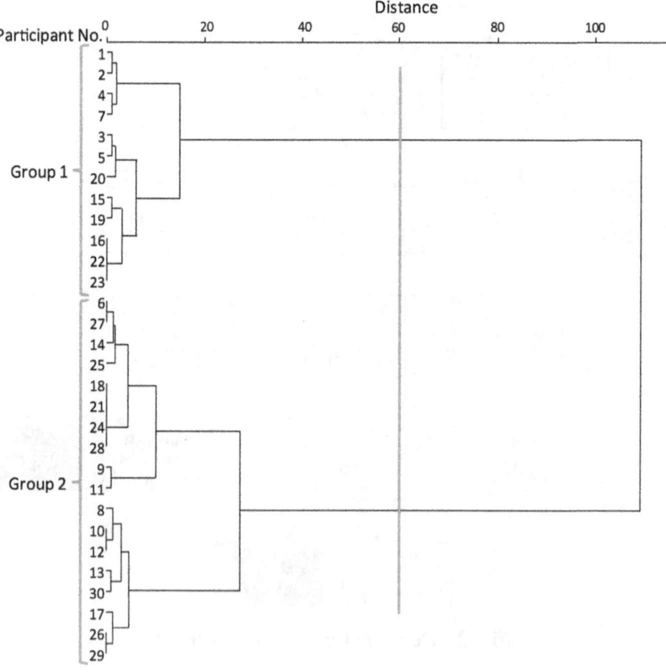

Fig. 4. Hierarchical clustering result

Table 4. Comparison of the number of nodes per group

	Group 1 with a large number of selections	Group 2 with few selections
Maximum	20	25
Minimum	3	15
Average	11.7	18.6
Variance	23.7	10.0

7 Conclusion

In this study, the variation in the selection of eyeglasses depending on the situation of use was investigated. First, the variability of the product selection based on the number of attribute types of the selected product that vary with purchase purpose was quantified. This revealed that product selection depends on the situation. Furthermore, a cluster analysis was used to group the participants to show the differences in the structure of the decision tree analysis. This may lead to a decision support for the product selection.

Future plans involve using the findings of this study to develop a method of providing product information that is appropriate for each individual. In addition, it is necessary to conduct experiments with a larger number of participants, considering the attributes of the consumers and their expertise in the products. It is also necessary to consider more products and services as target products.

Acknowledgements. The authors would like to thank Editage (www.editage.com) for English language editing.

References

1. Hamada, Y., Fukuda, K., Shoji, H.: Influence of selection process on its result in product purchase. Trans. Jpn. Soc. Kansei Eng. **18**(1), 47–53 (2019)
2. Hamada, Y., Fukuda, K., Shoji, H.: Influence of interaction with product information on selection results. Trans. Jpn. Soc. Kansei Eng. **18**(3), 209–214 (2019)
3. Stanovich, K.E.: Decision Making and Rationality in the Modern World. Oxford University Press, Oxford (2009)
4. Saaty, T.L.: The Analytic Hierarchy Process. McGraw-Hill, New York (1981)
5. Fujimoto, K.: The Science of Knowledge: Decision Support and Net Business. Ohmsha, Tokyo (2005)
6. Herbert, A.S.: Models of Man: Social and Rational. Wiley, Hoboken (1957)
7. Motterlini, M.: The Economy is Moved by Emotions: The First Behavioral Economics. Kinokuniya, Tokyo (2008)
8. Underhill, P.: Why We Buy: The Science of Shopping. Simon & Schuster, New York (2000)
9. Solvic, P.: The construction of preference. Am. Psychol. **50**, 369–371 (1995)
10. Terada, M., Tsuji, S., Suzuki, T., Fukushima, M.: A New Textbook on Data Analysis with Python. Nikkei Printing Co. (2018)

Identifying Early Opinion Leaders on COVID-19 on Twitter

Zahra Hatami(✉), Margeret Hall, and Neil Thorne

University of Nebraska at Omaha, Omaha 68182, USA
{zhatami,mahall,nthorne}@unomaha.edu

Abstract. This study aims to empirically identify opinion leaders on Twitter from the lens of Innovation Diffusion theory. We analyzed pandemic-specific tweets from casual users as well as from the US President to map their conversation for the purpose of finding opinion leaders over a three month period at the onset of the pandemic. By applying network analysis following with cluster enrichment as well as sentiment analysis, we recognize potential thought leaders, but we could not find strong evidence for opinion leaders according to the Innovation Diffusion theory. We interpret that users tweet for two different purposes - tweets to elicit agreement and tweets to elicit debate.

Keywords: Opinion leaders · Network analysis · Sentiment analysis

1 Introduction

Social media is a place that allows people share their ideas, experiences, thoughts, and opinions to broad audiences. There are many blogs, websites, and other types of platforms that people can access that allow for an easier and faster way to get one's word out on specific a matter [2]. These platforms are provided by virtual social networks, and users of these networks expand their communication to exchange information and share content collectively or individually [31]. The combination of mass media and interpersonal communication is the most effective way to reach and convince people of new ideas [28]. Rather than forming one's own opinion through research and experience, research suggests that people tend to be heavily influenced by social media content [27]. This is especially relevant to forming opinions because people are majorly influenced by each other. Social influence becomes a by-product of these exchanges. Social media will cause to individual or organizational groups share their ideas and become the opinion leaders when they influence other people's opinions or interests through the interweb of thoughts that is the online world [6]. Social media plays an important role in growing, challenging, and sometimes diverting people opinions. Interestingly, it has been pointed out that news without facts and opinions without merit spread faster on social media platforms than their factual counterparts [7]. It is important to take note of this framing when discussing the spread of opinions and information in regard to COVID-19 since the virus and pandemic can also

C. Stephanidis et al. (Eds.): HCII 2021, LNCS 13094, pp. 280–297, 2021.
https://doi.org/10.1007/978-3-030-90238-4_20

be the subject of misinformation on platforms worldwide [7,13,25]. The aim of this study is to identify the formation of opinion leader/leaders and describe the communicative behavior of common and popular users of virtual social networks during the onset of the COVID-19 pandemic. We use two data sets compared of (1) lay users and (2) the tweets of US President Donald Trump to compare between and within how the opinion formation process initiates. We choose only the US President instead of various agencies of the US government due to President Trump's outsized influence on the platform in use, Twitter. Reasonably, between the force of the office of the President and on his most-used platform he will serve as an opinion leader in a way that other officials do not.

In order to assess if what characteristics of opinion leaders exist, we perform a network analysis and using the Gephi and its layout algorithm [12] to identify potential opinion leaders to observe how information from them spreads between other users. In the next step, sentiment analysis is used to evaluate if other authors'opinion converged to these potential opinion leaders or not. We applied these analyses to the S-shape Curve theory of innovation diffusion [27] to evaluate the following research questions:

RQ1 - Do opinion leaders exist in general tweets about the pandemic and to what extent do their opinions diffuse through social media?
RQ2 - If opinion leader(s) exist, does the diffusion of opinions follow the S-Curve model?
RQ3 - To what degree do opinion communities and the US President agree?

In the next section, we continue with related work, followed by a description of the data utilized and the analytical approaches. Further, we discuss the result obtained from the network and sentiment analysis and apply the result of those analysis in S-shape Curve across the AuthorIDs as opinion leaders. The last section will be the conclusion with a discussion of limitations and future work.

2 Theoretical Background

2.1 Diffusion of Innovation Theory and S-Shape Curve

The speed of innovation and its importance in real world in terms of dissemination and acceptance of innovation has attracted the attention of researchers in different disciplines [23]. In 1912, Gabriel Trid introduced the S-shaped curve [15]. He realized the existence of a typical curve under the diffusion curve If the innovation established successfully over time. This curve showed the extent of innovation and imitation in social interaction over time. In fact, the S-shaped curve showed that at the beginning of the innovation introduction, the acceptance rate is low, then rises, and finally falls again (Fig. 1). The diffusion of innovation theory (DOI) was first proposed by Rogers. According to this theory, innovation is *an idea, behavior, or object* that is new to its audience. Innovation dissemination is the process by which innovation is transmitted to members of a social system through specific communication channels over a period of time [27].

DOI says that the first two and a half percent of the population are early innovators, which are critical to initial uptake of new ideas, behaviors, or objects. The next twelve and a half to thirteen and a half percent of the population are early adopters, thirty-four percent comprise the early majority, the next thirty-four percent is the late majority, and the last sixteen percent are laggards. The process of acceptance is comprised of all the changes that take place in people's eyes, from the first time something about a product, service or idea is introduced until they accept it or alternatives arrive. Accepting the new idea is divided into several sections over time because not everyone responds in the same way. Some accept the new idea quickly; some accept it late, and some do not like the new product, service or idea at all. In other words, DOI is the process that defines four important elements, including an innovation, communication channel, time, and social system: that is, innovation is transmitted to members of a social system through specific communication channels over a period of time. In accordance to the diffusion innovations theory, people who are exposed to a new product or idea early tend to carry and spread their thoughts more rapidly than those who have late exposure, making them the opinion leaders in various networks. [5,11]. Their influence in various communication networks makes opinion leaders attractive targets for initial contact to companies developing and integrating marketing strategies [6].

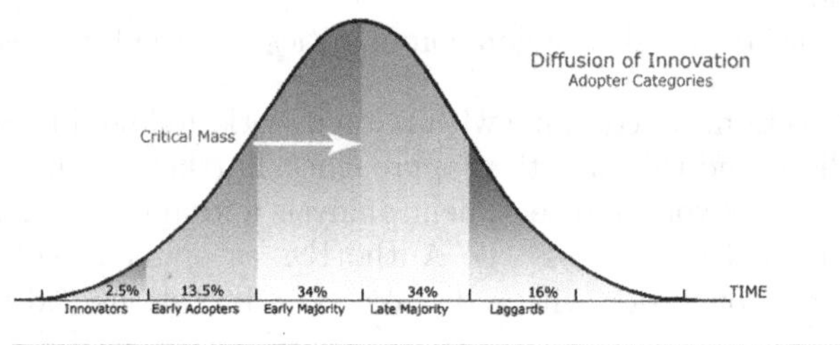

Fig. 1. S-shape curve [15]

2.2 Opinion Leadership on Social Media

We live in a world where almost anything we wish to know and anyone we wish to speak to is only one click away. While social media supports many social advancements like transparency and connectivity, it also has its drawbacks. With online options now being readily available, more people are flocking to the internet to broaden their social circles [2]. This also goes for voicing one's opinions. Rather than forming an opinion with research to back it up, people tend to be heavily influenced by social media [19,25]. There are many blogs, websites, and platforms that people can access these days that allows for an easier and faster way to get one's word out on specific matters [29].

This lack of barriers is especially relevant to forming opinions; research shows people are majorly influenced by each other, rather than regular news sources when making a judgment or opinion [19]. The most popular (or sometimes those with the largest volume) take shape and the form of opinion leaders. Standard theoretical models of communication are built on the assumption that content-based influence is binary – content is created, and content is received. Social platforms upend this assumption as individuals and aggregators are simultaneously creating, consuming, and editing information that is diffused. And due to low barriers, opinion leaders have an even easier time diffusing their opinions than they would if they were to go out into the physical world [10].

Opinion leaders are important when discussing what medium people get their news from. Strong opinion leaders are those whose words keep their audience engaged with their preferred news and wanting to know more [30]. Recently, it has been shown that people have been distancing themselves from news coming from the wide range of traditional media outlets [8]. Instead, people are turning to social media as their primary news source. People will spread information through their social media as a type of signal on their opinion or virtues, which can affect other's opinions [25]. By virtue of their recommending a media outlet, opinion leaders increase the amount of trust that their peers and followers have in that media outlets [28].

With all the people expressing their thoughts and words, social media has formed an opinion climate [16]. In opinion climates, people's opinions are influenced by disagreements, what the majority think, etc. [18]. There are notably many public figures who have made use of social media's opinionated atmosphere. In regards to politics, politicians have various mediums of communication at their disposal [4]. Over time, communication mediums can become associated to certain political leaders. Franklin D. Roosevelt is associated with the radio due to his fireside chats, as are Winston Churchill and the British Royal family. John F. Kennedy is likewise associated to television. Barack Obama was widely proclaimed as the first social media president of the United States [3]. Now it is observed that the microblog platform twitter is associated to President Donald Trump most notably due to his active usage of it to communicate with the general public [24].

2.3 Sentiment Analysis

Sentiment analysis is a subcategory of natural language processing that is concerned with identifying and measuring the positive and negative polarity associated to a given text. This is commonly done on a scale ranging from -1 to 1, where -1 is negative sentiment, 0 is neutral, and 1 is positive sentiment. Various mechanisms are considered in hopes of making the sentiment analysis most accurate, including word frequencies, parts of speech, position of terms, negation, syntax, and the types of words used. LIWC is a well-established and recognized tool for the analysis of sentiment on social media; this package analyzes sentiment across over 70 variable types [4].

Sentiment analysis presents challenges. Sarcasm is difficult to detect, and context plays a significant role in accurately rating sentiment. The word "unpredictable" for example might generally be regarded as negative, but within the context of a movie review this likely denotes positive sentiment [17]. Some of the most important mechanisms for sentiment analysis are prior polarity of words, as well as identifying the parts of speech tags for words within a given text [1]. The sentiment analysis tool VADER is shown to be extremely accurate and reliable regarding measuring sentiment within the context of microblogs [14]. So good in fact that it outperforms humans within this context.

Sentiment analysis on the topic of Coronavirus within the context of twitter has been explored [20,26]. Both studies found the general public tends to regard Coronavirus in tweets with a neutral tone. However, the World Health Organization tends to regard the virus in a positive tone which is most likely a result of an active attempt by the World Health Organization to provide hope and assurance that the state of things is well-managed or will improve [26]. This is in comparison to [13], which by employing LIWC found that Chinese users had significantly fluctuating emotions after the onset of the pandemic.

3 Data and Approaches

Network and sentiment analysis were applied in different sets of tweets collected from Twitter platform. In the following sections, for each type of analysis, different approaches for collecting, filtering, analyzing data will be explained by detail.

3.1 Data Acquisition and Filtering in Network Analysis

In order to identify opinion leader/leaders, English-speaking lay users' conversations were extracted from the Twitter platform. The data was collected from Twitter by via a keyword search for the term 'Coronavirus". The Trump tweets were extracted using the Trump Twitter Archive (http://www.trumptwitterarchive.com/). There are ten variables in the analysis:

- Tweet'ID: Unique identifier assigned per tweet.
- Conversation ID: the Tweet ID of the conversation tree's root.
- Date posted
- Author'ID: Unique identifier for users, (truncated in the analysis)
- Tweet Text
- Number of Replies: Number of replies for the ConvID or AuthorID[1].
- Number of Retweets: Number of retweets for the ConvID or AuthorID.
- Number of Favorites: Number of favorites for the ConvID or AuthorID.
- Is Reply?: Whether the tweet is reply.
- Is Retweet?: Whether the tweet is retweet. We extracted tweets from Jan/16/2020 to March/31/2020. The network analysis is only applied to the user analysis due to limitations in the archiving process of the Trump tweet data set.

[1] IDs for who was involved to the ConvID with his/her tweet.

4 Correlation Network Analysis

Correlation network analysis is applied to time series data in order to find hidden information when they cannot be found under current circumstance [21,22]. The correlation network model is (CNM) built by assuming that each conversation ID (ConvID) is a node (vertex) in the graph and two ConvIDs are connected by an undirected edge, if and only if, their correlation coefficient is 95% or more. For each ConvID, we removed duplicates and were left with 5,965 IDs. The Pearson correlation network was applied in the network since the data was normally distributed. Due to the high density between ConvIDs, first a cluster analysis was applied on the network in order to group different objects whose degree of correlation between two objects is maximal. After the cluster analysis, cluster enrichment was applied in each cluster by adding different parameters relevant to each node. Cluster enrichment brought the opportunity to identify specific ConvIDs in each cluster and their author IDs as a potential opinion leader/leaders. Furthermore, sentiment analysis was applied in the potential opinion leader/leaders' texts with the purpose of finding the content of actual opinion leader/leaders for each cluster. Cluster naming is a result of the sentiment analysis and manual inspection of the corpora. We employed the software package Gephi due to its flexibility and multi-task architecture functionality. Clusters were identified in the obtained correlation network by using Modularity on the Statistics pane in Gephi [12]. Fruchterman-Reingold layout algorithm was used in order to visualize the clusters [9]. The clusters were the sub-networks and all other parameters from data file were added to each cluster for further analysis with the purpose of doing cluster enrichment for that specific cluster. With Cluster enrichment all correspond parameters added to ConvIDs.

4.1 Experimental Results

Of the possible 6,291 ConvIDs in the data set, 5,965 ConvIDs were involved in the network based on the above referenced network criteria. As the clusters were formed with high correlations among the nodes, we could infer that the overall behavior of the nodes within each cluster is the same. The 5,965 nodes were placed in three clusters (Table 1). Hence, these three clusters were considered for further analysis.

Since a retweet is a re-posting of a tweet and helps to quickly share the tweet with different followers, we looked at potential opinion leaders if their tweets have the highest number of retweets, or the highest number of replies and favorites (Table 2). According to the number of replies, retweets, and favorites for each ConvID, we divide the data into five numeric categories. Rank one shows the highest number and rank five showed the lowest number of replies, retweets, and favorites. In each cluster, ConvIDs belong to rank one, two and three of retweet's (Table 3), and Authors' IDs belonging to the ConvIDs were selected as the potential opinion leaders.

Table 1. Three clusters and possible number of nodes

Cluster number	# of ConvID
1	3018
2	2833
3	114

Table 2. Rank: retweets, replies, favorites

Rank	#of Retweets	#of Replies	#of Favorites
1	>50000	>10000	>100000
2	49999-10000	9999-1000	99999-50000
3	9999-5000	999-500	49999-10000
4	4999-1000	499-100	9999-1000
5	<1000	<100	<1000

Cluster One: Politics. Cluster one was the biggest cluster in size which was extracted from correlation network and it is also the most dense cluster. This cluster contains 3,018 ConvIDs (Fig. 2). Cluster one leans heavily towards political debate; tweets in this cluster are broadly polarized rather than defending one or the other political stance.

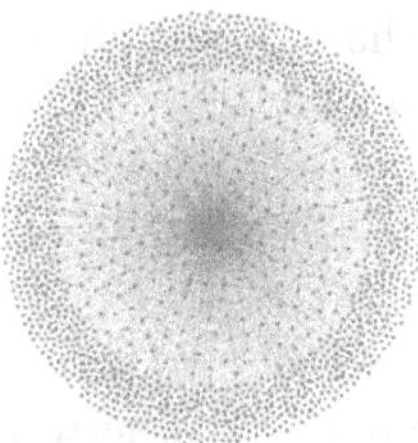

Fig. 2. Representation of cluster 1 based on 3,018 included ConvIDs

Table 3 shows the information for the top five ConvIDs in cluster one. Author_ Count stands for the number of authors involved the ConvID, followed with summation of replies (Replies_sum), retweets (Retweets_sum)and favorites (Favorites_sum) for that ConvID. The largest chain we found has fourteen unique authors involved with ConvID "ID123571". The text of the initial tweet reads: *I just landed at JFK after reporting on #coronavirus in Milan and Lombardy—the epicenter of Italy's outbreak—for @vicenews. I walked right through US customs. They didn't ask me where in Italy I went or if I came into contact with sick people. They didn't ask me anything.* There were 9,426 replies, 63,658 retweets, and 225,772 favorites for ConvID ID123571 across the fourteen authors.

Table 3. ConvID and corresponded parameters

ConvID	Author_Count	Replies	Retweets	Favorites
ID123571	14	9426	63658	225772
ID123992	12	46893	98093	450408
ID123880	2	4376	33607	347788
ID124037	3	3065	31436	112706
I D123829	1	894	27965	161995

In the next step, we assess the number of replies, retweets, and favorites for each author involved with the ConvID. 125,972 tweets have one or more replies. As mentioned earlier, we assess as potential opinion leadership as the number of retweets, replies and favorites belonging to the high rank category (Table 2). In this cluster 50 AuthorIDs satisfied the condition of being potential opinion leader since their tweets were retweeted at least 5000 times. The highest number of replies and retweets are associated with author ID "ID375687". We recognized this author ID as the likely majority opinion leader in the cluster (Table 4).

Table 4. Example-Potential opinion leader in cluster one - The colors correspond with the categories which are separated into ranks in Table 2.

ConvID	AuthorID	Replies	Retweets	Favorites
ID123571	**ID375687**	9423	63654	225758
	ID107475	0	2	8
	ID438028	0	0	0
	ID258551	0	2	0
	ID278881	0	0	2
	ID562670	1	0	1
	ID175144	0	0	1
	ID110820	0	0	0
	ID278002	0	0	0
	ID337139	0	0	0
	ID999286	0	0	0
	ID751703	1	0	1
	ID260956	1	0	1
	ID156202	0	0	0

Cluster Two: Health. There were 2,833 ConvIDs in this cluster. Following the same procedures as in cluster one, led to assessing fourteen potential opinion leaders in this cluster. Cluster 2 is mainly comprised of conversations about travel, with some discussions around health precautions (Fig. 3).

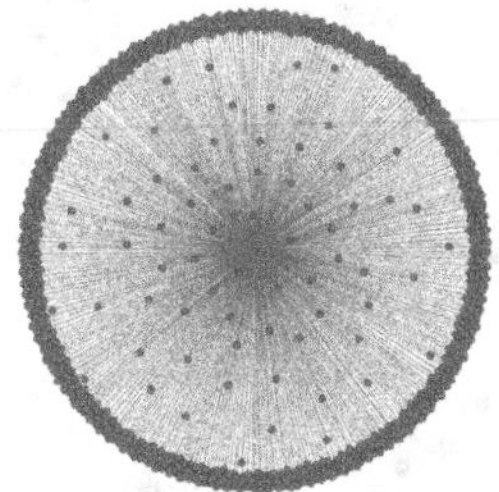

Fig. 3. Representation of Cluster 2 based on 2,833 included ConvIDs

Table 5 reveals the information for the top five ConvIDs in this cluster. The following table (Table 6) shows that four authors were involved with ConvID “ID124501”. There were total number of 795, 23,115 and 46,889 for replies, retweets and favorites for this ConvID. After applying the same procedure used in cluster, we could recognize AuthorID “ID739844192” as the potential opinion leader for this cluster. AuthorID “ID739844192” is the author of the longest chain in the cluster (Table 6).

Table 5. ConvID and corresponded parameters

ConvID	Author_Count	Replies	Retweets	Favorites
ID124501	4	795	23115	46889
ID123540	8	3369	13493	25074
ID124189	1	3369	11352	25074
ID124507	19	1545	10579	19580
ID124463	2	853	9056	11725

Table 6. Example-Potential opinion leader in cluster two- The colors correspond with the categories which are separated into ranks in Table 2.

ConvID	AuthorID	Replies	Retweets	Favorites
ID124501	**ID73984419**	795	23115	46888
	ID10682742	0	0	1
	ID10018652	0	0	0
	ID326630677	0	0	0

Cluster Three: Economics. Cluster three was the smallest cluster that extracted from the network. This cluster contained 114 ConvIDs and is largely structured around debates around re-opening the economy (Fig. 4).

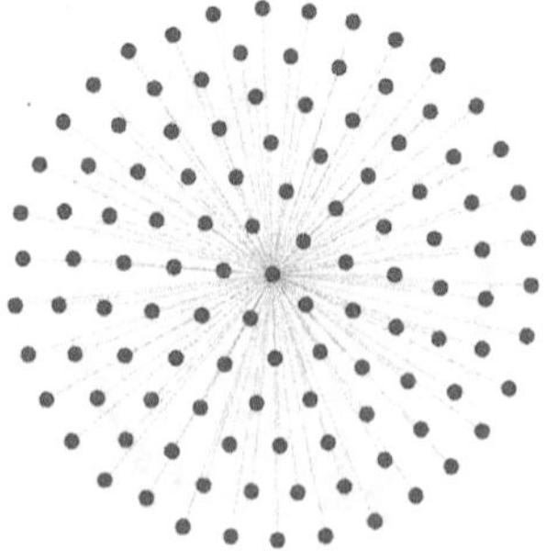

Fig. 4. Representation of cluster 3 based on 114 included ConvIDs

Only one potential opinion leader was recognized in this cluster by following cluster enrichment analysis. Table 7 shows that nine authors were involved with ConvID "ID124509". There were 16,952 replies, 6,991 retweets, and 14,584 favorites for that ConvID from those nine authors. After applying the cluster enrichment, we identified AuthorID "ID106048" as a potential opinion leader among the rest of authors. Excluding two replies and eight favorites, the total number of replies, retweets and favorites goes to AuthorID "ID106048" (Table 8).

Table 7. ConvID and corresponding attributes

ConvID	Author_Count	Replies	Retweets	Favorites
ID124509	9	16952	6991	14584

Table 8. Example-Potential opinion leader in cluster three- The colors correspond with the categories which are separated into ranks in Table 2.

ConvID	AuthorID	Replies	Retweets	Favorites
ID124509	ID106048	16950	6991	14576
	ID314113	0	0	0
	ID140960	0	0	0
	ID121713	0	0	0
	ID489604	0	0	2
	ID165602	0	0	1
	ID286792	0	0	1
	ID852179	0	0	3
	ID594091	1	0	1
	ID5940917	1	0	0

4.2 Network Analysis Result

Based on the similarities between ConvID in each clusters, the results of the first analysis indicated 64 potential opinion leaders among three clusters. In each cluster, users classified in the group of potential opinion leaders based on the number of retweets, replies and favorites on their tweets if they were belonging to the rank one, two or three. In some cases, there were authors that participated several times in the conversation but none of the tweets from those authors were heavily retweeted, replied to or favorited. This is a pattern more indicative of followership. There was a case that an opinion-leading author participated in different ConvIDs which landed in different clusters (Table 9). A visual inspection of this user suggests that the mainly tweet links to mainstream news articles, with a slight left-leaning orientation. Due to the two themes of the clusters it is reasonable to see users who cross the topics.

Table 9. A potential opinion leader in different clusters

Cluster Number	ConvID	AuthorID	Replies	Retweets	Favorites
One	ID124440	ID245953	1921	12266	28917
Two	ID124367	ID245953	959	6911	13193

After recognizing the group of authors' IDs as potential opinion leaders (those who satisfied the criteria), we apply sentiment analysis to assess the similarities and differences between the clusters.

5 Sentiment Analysis

5.1 Data Description

Our data set consists of 554,803 unique tweets from lay users and 1,004 unique tweets from President Trump, where only 20 per cent are coronavirus-related. Due to the limited sample size of Presidential tweets, we include the full 1,004 tweets in this analysis; the potential introduced bias was assessed as preferable to extremely unbalanced samples. The lay user corpus contains 16,774,014 words and the Trump corpus contains 57,273 words. 67 tweets from 57 lay users have 1,000 or more retweets, replies, or favorites; this data is not released in the Trump archive.

5.2 Polarity of the Data

Table 10 shows the average sentiment scores per tweet set as assessed by VADER. Figure 5 displays the results over time, split into five, 14-day units across the four communities. Trump's polarity remains positive across the entire observation as compared to the community clusters. Splitting Trump tweets by Coronavirus/non-Coronavirus tweets shows only a very slight difference in polarity was observed between the two datasets, which supports that Trump has maintained an overall positive tone around the pandemic (Table 10).

Table 10. Tweet sentiment averages

Trump Coronavirus-Related	0.204
Trump Non-Coronavirus	0.24
General Coronavirus-Related	−0.039

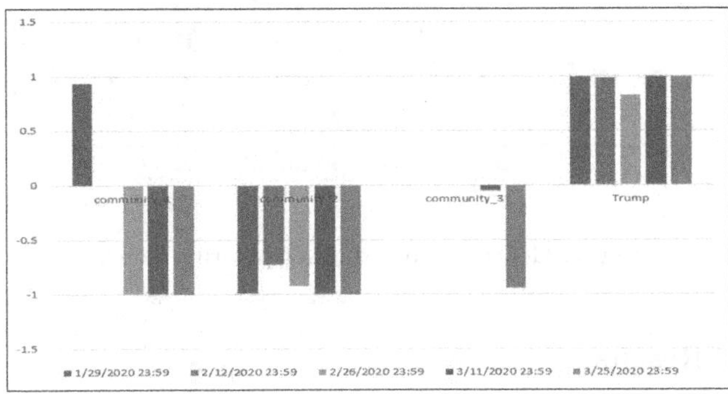

Fig. 5. VADER polarity over time

Figure 6 and Fig. 7 depict scatterplots of polarity ratings the datasets. The scatterplots showcase in a visual way that no obvious patterns are observed across the communities.

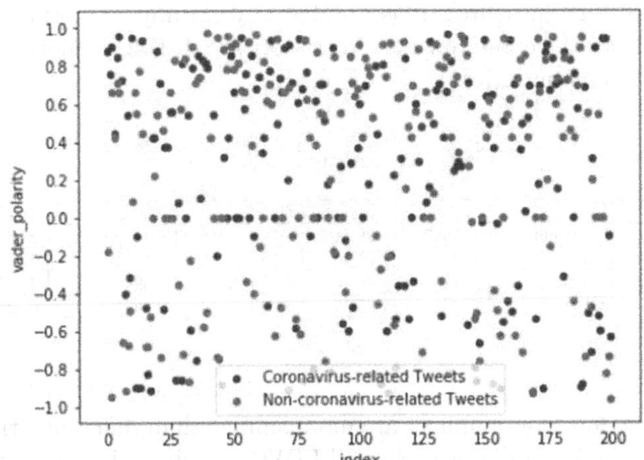

Fig. 6. Trump Dataset Polarity scores

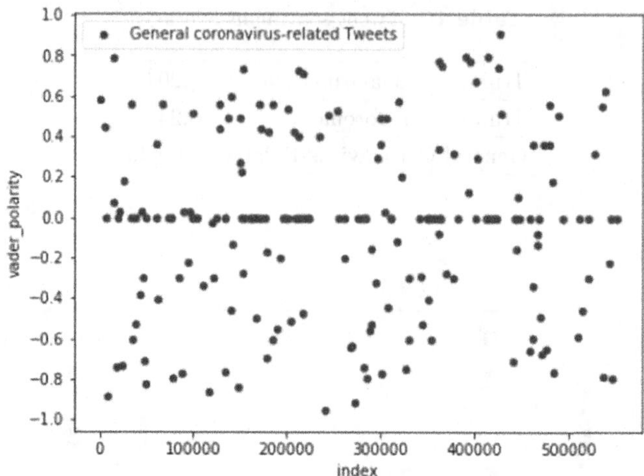

Fig. 7. General public dataset polarity scores

5.3 LIWC Results

LIWC 2015 introduced four composite variables, Analytic, Clout, Authentic, and Tone. Each is reflective of a spectrum of sentiment, where Analytic reflects to the degree that a person uses formal or informal and hierarchical or narrative thinking; Clout is a measure of expertise and confidence; Authentic reflects the degree to which the person is guarded or open; and Tone is a measure of emotional polarity. We first compared the three communities to the Trump tweets via T-Test as this test is robust to deviations from normality. Table 11 displays the LIWC values and Table 12 shows that the four communities are significantly different on all four measures. It is notable that Trump's emotional polarity is much higher than the communities - his tweets are distinctly more positive than the other three communities.

Table 11. Detected LIWC values

	WC	Analytic	Clout	Authentic	Tone
All Tweets Community 1	106547	88.71	76.06	9.13	29.91
All Tweets Community 2	113230	93.90	72.54	7.66	22.89
All Tweets Community 3	3982	90.96	72.74	5.69	25.77
President Trump Tweet	57272	85.36	81.48	8.56	62.26

Overall, the four communities are more similar than different - the results of a Pearson correlation analysis across all LIWC variables show that the communities are highly correlated (Table 13).

Table 12. T-test

One-Sample Test						
	t	df	Sig.(2-tailed)	Mean Difference 95% Confidence Interval	Lower	Upper
Analytic	49.750	3	.000	89.73250	83.9924	95.4726
Clout	36.268	3	.000	75.70500	69.0621	82.3479
Authentic	10.300	3	.002	7.76000	5.3623	10.1577
Tone	3.855	3	.031	35.20750	6.1459	64.2691

Table 13. Correlation table

Correlation Table				
	All Tweets_ Community 1	All Tweets_ Community 2	All Tweets_ Community 3	President Trump_Tweets
All Tweets_ Community 1 Pearson Correlation Sig.(2-tailed) N	1 .000 93	1.000** .000 93	0.999** .000 93	1.000** .000 93
All Tweets_ Community 2 Pearson Correlation Sig.(2-tailed) N	1.000** .000 93	1 .000 93	0.999** .000 93	1.000** .000 93
All Tweets_ Community 3 Pearson Correlation Sig.(2-tailed) N	0.999** .000 93	0.999** .000 93	1 .000 93	0.999** .000 93
President Trump_Tweets Pearson Correlation Sig.(2-tailed) N	1.000** .000 93	1.000** .000 93	0.999** .000 93	1 .000 93

Next we display world clouds comprised of all tweets across the four communities (Figs. 8, 9, 10 and 11). While an imperfect measure, word clouds support quickly assessing the most frequent terms employed in a corpus in order to assess the themes in use. When comparing the four communities it is important to remember that the Trump corpus includes all tweets and not only coronavirus tweets. Still, it is striking that in this data the terms Coronavirus or COVID-19 do not appear (Fig. 11). Community 1 is the only community where the term 'flu' appears, which reflects the politicized debate around if the pandemic was simply a bad flu (Fig. 8). Community 3's has significantly more terms than the other three communities; this is an indication that the themes of the cluster are more fluid than in the other communities (Fig. 10).

5.4 Discussion

In this study we used network analysis and text analysis to extract information from users' tweets and their conversation around the beginning of the Coronavirus pandemic. From network analysis with applying cluster enrichment, we identified three clusters (RQ1). Due to the high polarity of the discussion around novel Coronavirus it was expected that the clusters would align to political views; they rather split thematically. This was an unexpected result. We identify potential opinion leaders based on common metrics like volume of retweets, likes, and favorites. The potential opinion leaders cannot be considered as actual opinion leaders since there was no evidence of conforming general opinions to the potential leaders' opinion. Recalling the diffusion of innovation theory, while the sentiments of authors shift as time goes, we could not draw the relationship between the concept of DOI and the result of the text analysis. In the case of

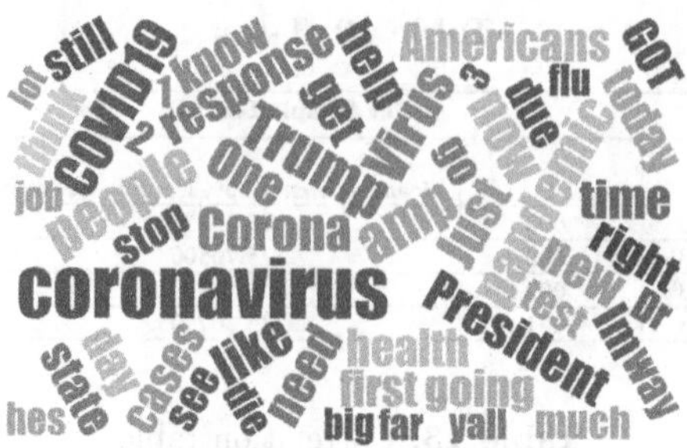

Fig. 8. Community 1's Coronavirus-related tweets

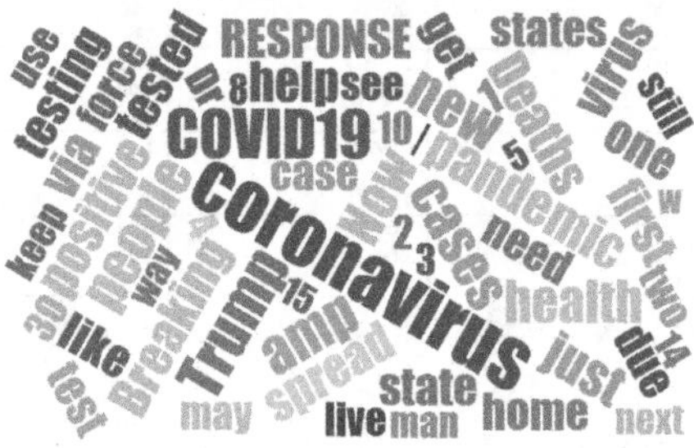

Fig. 9. Community 2's Coronavirus-related tweets

Fig. 10. Community 3's Coronavirus-related tweets

modelling the S-Curve based on DOI theory, we could not find the correspond pattern between users' percentage of their adoptions during time based on DOI. The DOI posits that the first two and a half percent of the population are innovative, the next twelve and a half or thirteen and a half percent of the population follow, thirty-four percent are the majority, the next thirty-four percent is the last majority, and the last sixteen percent are the delayed group. We could not replicate this (RQ2).

We also applied sentiment analysis using VADER and LIWC and we mapped the tweet clusters against the tweets of US President Donald Trump for the same time period. The positive and negative polarities were identified by using

Fig. 11. Trump's Coronavirus-related tweets

VADER as a whole and over time. The result showed that the sentiments of the community clusters shifted over time whereas Trump's tweet remained the same. However, from a content perspective, the clusters are all largely correlated. This is an indication that for the assessed social media platform Twitter, President Trump may serve as an overall opinion leader considering topics (RQ3).

6 Implications and Conclusion

This article attempted to validate the DOI theory in the face of a pandemic via social media opinion making and taking. The onset of the pandemic resulted in a lot of generalized confusion, and mis- and dis-information; we hypothesized that the uptake of opinions may be traceable across social media. The innovation of diffusion theory shows how and why a new idea is spread in organizations and social networks and culture. Based on this theory, the innovation does not have to be a very new idea, but it is enough to be an idea, an object, or a way that is fresh and new from the point of view of the people who accept it. As such, it would be a useful model to understanding how people set opinions with social media content.

We planned to assess opinion leadership with applying network analysis following with cluster and text analysis. While thematic clusters were established, clear opinion leadership was not. As a next step we compared the thematic clusters to the tweets of US President Donald Trump, who is a clearly established figure on Twitter as well as the presumptive leader of the pandemic response efforts. We could find that users' attitudes changed over time; however they did not follow the patterns proposed in the DOI since the dissemination is not follow the process by which an innovation is disseminated between members of a social system over time using specific channels. We also find that thematically Trump's tweets are similar to the rest of the corpus, whereas in terms of positive and negative sentiment he maintained a strikingly different tone.

One observation is worth to noting that the cluster enrichment process and opinion leadership analysis had large differences between replies, retweets, and favorites. An intuitive interpretation is there are two types of tweets: one is many people agreed on, another group is many people are debating on. This induces our further study, which clearly needed to be considered separately and differently.

7 Limitations and Future Work

We recognize that the data which describes the onset of the pandemic and tweets that have been sent out since our time frame have the potential to be different; as such we make no claims about opinion leadership on social media about the pandemic as a whole. It is unclear if a different time period would allow a different view to converge. We acknowledge the imbalance in the cluster sizes of the communities and the Trump tweets, and recognize that the difference in sample sizes may cause bias in the analysis.

In future work we intend to apply a time-lagged correlation model tied to independently-established major events in order to assess the existence of uptake patterns which may be otherwise obfuscated. We would like to extend the analysis to include other government agencies playing key roles such as the Center for Disease Control or governors in hard-hit states.

Acknowledgement. A big thanks to Rui Yang from The University of Nebraska at Omaha for providing a dataset of English-speaking lay users' conversations which were extracted from the Twitter platform. For citations of references, we prefer the use of square brackets and consecutive numbers. Citations using labels or the author/year convention are also acceptable.

References

1. Agarwal, A., Xie, B., Vovsha, I., Rambow, O., Passonneau, R.: Sentiment analysis of Twitter data. In: Proceedings of the Workshop on Languages in Social Media, LSM 2011. Citeseer (2011)
2. Amedie, J.: The impact of social media on society, p. 20
3. Birgisdóttir, L.K.: The rising influence of social media in politics: how barack Obama used social media as a successful campaign tool in the 2008 and 2012 elections. Ph.D. thesis (2014)
4. Caton, S., Hall, M., Weinhardt, C.: How do politicians use Facebook? An applied social observatory. Big Data Soc. **2**, 1–18 (2015)
5. Chan, K.K., Misra, S.: Characteristics of the opinion leader: a new dimension. J. Advert. **19**(3), 53–60 (1990)
6. Cho, Y., Hwang, J., Lee, D.: Identification of effective opinion leaders in the diffusion of technological innovation: a social network approach. Technol. Forecast. Soc. Chang. **79**(1), 97–106 (2012)
7. Cinelli, M., et al.: The COVID-19 social media infodemic. arXiv preprint arXiv:2003.05004 (2020)
8. Elisa, S.: Social media outpaces print newspapers in the U.S. as a news source (2018). https://pewrsr.ch/2rsoHtb
9. Enright, A.J., Ouzounis, C.A.: Biolayout-an automatic graph layout algorithm for similarity visualization. Bioinformatics **17**(9), 853–854 (2001)
10. Franch, F.: (Wisdom of the crowds) 2: 2010 UK election prediction with social media. J. Inf. Technol. Polit. **10**(1), 57–71 (2013)
11. Goldsmith, R.E., De Witt, T.S.: The predictive validity of an opinion leadership scale. J. Mark. Theory Pract. **11**(1), 28–35 (2003)

12. Heymann, S., Le Grand, B.: Visual analysis of complex networks for business intelligence with gephi. In: 2013 17th International Conference on Information Visualisation, pp. 307–312. IEEE (2013)
13. Hou, Z., Du, F., Jiang, H., Zhou, X., Lin, L.: Assessment of public attention, risk perception, emotional and behavioural responses to the COVID-19 outbreak: social media surveillance in china. Risk Perception, Emotional and Behavioural Responses to the COVID-19 Outbreak: Social Media Surveillance in China (3/6/2020) (2020)
14. Hutto, C.J., Gilbert, E.: VADER: a parsimonious rule-based model for sentiment analysis of social media text. In: Eighth international AAAI Conference on Weblogs and Social Media (2014)
15. Kaminski, J.: Diffusion of innovation theory. Can. J. Nurs. Inf. **6**(2), 1–6 (2011)
16. Katz, E.: The two-step flow of communication: an up-to-date report on an hypothesis. Public Opin. Q. **21**(1), 61–78 (1957)
17. Kharde, V., Sonawane, P., et al.: Sentiment analysis of twitter data: a survey of techniques. arXiv preprint arXiv:1601.06971 (2016)
18. Kramer, A.D., Guillory, J.E., Hancock, J.T.: Experimental evidence of massive-scale emotional contagion through social networks. Proc. Natl. Acad. Sci. **111**(24), 8788–8790 (2014)
19. Lee, J.K., Choi, J., Kim, C., Kim, Y.: Social media, network heterogeneity, and opinion polarization. J. Commun. **64**(4), 702–722 (2014)
20. Manguri, K.H., Ramadhan, R.N., Amin, P.R.M.: Twitter sentiment analysis on worldwide COVID-19 outbreaks. Kurdistan J. Appl. Res. 54–65 (2020)
21. Mao, S., Xiao, F.: Time series forecasting based on complex network analysis. IEEE Access **7**, 40220–40229 (2019)
22. Miśkiewicz, J.: Analysis of time series correlation. The choice of distance metrics and network structure. Acta Phys. Pol. A **121**, B-89 (2012)
23. Nooteboom, B.: Learning and Innovation in Organizations and Economies. OUP, Oxford (2000)
24. Ott, B.L.: The age of Twitter: Donald J. Trump and the politics of debasement. Crit. Stud. Media Commun. **34**(1), 59–68 (2017)
25. Pennycook, G., McPhetres, J., Zhang, Y., Lu, J.G., Rand, D.G.: Fighting COVID-19 misinformation on social media: experimental evidence for a scalable accuracy-nudge intervention. Psychol. Sci. 0956797620939054 (2020)
26. Rajput, N.K., Grover, B.A., Rathi, V.K.: Word frequency and sentiment analysis of Twitter messages during coronavirus pandemic. arXiv preprint arXiv:2004.03925 (2020)
27. Rogers, E.M.: Diffusion of Innovations. Simon and Schuster (2010)
28. Schacht, J., Hall, M., Chorley, M.: Tweet if you will: the real question is, who do you influence? In: Proceedings of the ACM Web Science Conference, pp. 1–3 (2015)
29. Siddiqui, S., Singh, T.: Social media its impact with positive and negative aspects. Int. J. Comput. Appl. Technol. Res. **5**(2), 71–75 (2016)
30. Turcotte, J., York, C., Irving, J., Scholl, R.M., Pingree, R.J.: News recommendations from social media opinion leaders: effects on media trust and information seeking. J. Comput.-Mediat. Commun. **20**(5), 520–535 (2015)
31. Wilson, S.M., Peterson, L.C.: The anthropology of online communities. Annu. Rev. Anthropol. **31**(1), 449–467 (2002)

Impact of the Cyber Hygiene Intelligence and Performance (CHIP) Interface on Cyber Situation Awareness and Cyber Hygiene

Janine D. Mator(✉) and Jeremiah D. Still

Old Dominion University, Norfolk, VA 23507, USA
{jmator,jstill}@odu.edu

Abstract. A common theme across cybersecurity solutions is a lack of transparency for end-users. Our prototype, Cyber Hygiene Intelligence & Performance (CHIP), was purposefully designed to improve end users' cyber hygiene and cyber situation awareness. We begin by addressing current cybersecurity training solutions, their inability to continuously impact user cyber hygiene and cyber situation awareness (CSA), and how end users' needs for transparency are overlooked. We then illustrate the major stages of our design process for the medium-fidelity CHIP prototype, including defining, ideation, prototyping, and analysis. For each design stage, we describe our methodologies, major decision points, design considerations, and outcomes. We then highlight our between-groups survey experiment that measured cyber hygiene and CSA for an experimental group, who received CHIP notifications while completing web-based tasks, compared to a control group, who completed the same tasks without any notifications. Our findings show promise for a web application-based solution like CHIP to increase cyber hygiene and CSA for specific cyber threats.

Keywords: Human-centered design · Cyber situation awareness · Cyber hygiene

1 Introduction

End users have been simultaneously recognized as the frontline defense and the weak point of cybersecurity systems [1, 2]. Consequently, cybersecurity training has become an essential and often mandatory task for users. However, traditional training is infrequent and only tests theoretical knowledge, which translates poorly to cyber hygiene (e.g., complying with dangerous website warnings) in the real world [3, 4]. Users typically complete cybersecurity training once a year through 30-min modules [4]. This assumes that brief, infrequent, and impersonal training is sufficient to improve user cyber hygiene.

C. Stephanidis et al. (Eds.): HCII 2021, LNCS 13094, pp. 298–309, 2021.
https://doi.org/10.1007/978-3-030-90238-4_21

Clearly, cyber hygiene remains a prominent and costly concern [5]. Approximately 90% of cyberattacks are preventable through basic cyber hygiene and best practices [6]. Common examples of poor cyber hygiene decision points include answering account security questions with information that is readily available online (e.g., a mother's maiden name) [7, 8], complying with phishing emails [9], proceeding to dangerous websites [10], and saving/auto-filling passwords in a web browser [11]. Cybersecurity training modules typically discourage these behaviors but fail to prevent them in practice [12]. Organizations may attempt to increase compliance with good cyber hygiene practices by enforcing "Big Brother" software that monitors and reports employees' online behavior. However, these programs are mainly intended for IT administrators to oversee large networks, rather than help end-users increase their awareness of cyber threats.

Even for those with good cyber hygiene, cyber situation awareness (CSA) is another concern. CSA refers to perception, comprehension, and outcome projections of a cyber threat [13]. Conventional cybersecurity tools like firewalls and anti-virus software fail to promote CSA, as they primarily operate in the background (e.g., desktop applications) and only provide information retroactively. Future solutions will need to ensure users not only possess practical knowledge of cyber threats, but how to apply their knowledge [14].

Current cybersecurity solutions lack transparency, leaving end-users out-of-the-loop. For interface designers, transparency refers to how well an interface shows users what they can do, how they can access these functions, and why the system responds as it does [15]. Therefore, transparency in a cybersecurity interface should enable users to understand the nature of a threat, the behavior that triggered it, and its intended consequences. There is a need for solutions that promote good cyber hygiene and CSA. This paper proposes a prototype interface called Cyber Hygiene Intelligence and Performance (CHIP) designed to meet these needs.

2 Overview

In the previous section, we introduced cyber hygiene, cyber situation awareness, and how both have been neglected in the implementation of current cybersecurity solutions for end-users. In the following section, we narrate the design process of our mid-fidelity CHIP prototype. The main stages in our iterative design process are summarized into the four subheadings Define, Ideate, Prototype, and Analyze. In each section, we illustrate design considerations, methodologies implemented during development, and outcomes revealed during that stage. We also discuss our between-groups survey experiment findings that assessed the impact of CHIP notifications on cyber hygiene and CSA compared to a control group. Finally, we discuss our goals, findings, and challenges while designing CHIP.

3 Design Process

3.1 Define

In the first stage of our design process, we defined our goals, stakeholders, and user requirements. Our goals were twofold: to provide users with digestible notifications that increase CSA and further their understanding of cyber hygiene. Uniquely, end-users will learn as they complete daily computing activities.

Requirements gathering from context experts narrowed our scope of cyber hygiene to four target behaviors: 1) preventing web browsers from auto-saving passwords, 2) deleting or reporting suspicious emails, 3) answering security questions with unique information, and 4) complying with dangerous website warnings. These target behaviors were selected due to their prevalent risk factors in cybersecurity. Many users store authentication credentials within web browsers, which poses the threat of unauthorized logins and malware downloads [11]. Additionally, phishing emails are often not recognized by users. As a result, millions of users have their personal information stolen each year [9]. Users also make poor decisions when creating answers to account security questions. These answers can often be found in public records; for example, a mother's maiden name [7]. Finally, users place themselves at greater risk of phishing and malware attacks when they ignore warnings about potentially dangerous websites [10].

We conducted a hierarchical task analysis (HTA) for each target behavior to break down the series of goals required at each cyber hygiene decision point (see Fig. 1 below, for example).

0: Create strong security questions
1: Read web app warning about using personal information in security questions
  1.1: Understand the cybersecurity goal and, ideally, why it is necessary
2: Select a security question
  2.1: If necessary, reread/refer to recommendations on security questions and answers (e.g., base responses on a fictional character)
3: Type response to security question based on recommendations
4: Select a second security question
  4.2: If necessary, reread/consult recommendations
5: Submit security questions and responses on the webpage
6: Read web app confirmation message and positive feedback

Fig. 1. HTA for creating strong security questions.

We next conducted a competitive analysis to better understand key stakeholders and the user requirements of related solutions. Our spreadsheet came to include categories such as the solution's company/developer, program or feature name, domain (i.e., Department of Defense or online banking), method of delivery (i.e., mobile app), and the actions available to managers and/or end-users. As we had previously seen, current solutions in the cybersecurity marketplace tended not to be targeted to end-users, but to corporations and government agencies. Not only was pricing unrealistic for individual users, but descriptions assumed more technical language (e.g., cascaded learning framework, PCAP file, and CVE matching). Clearly, such solutions are not intended to further the end-users' mental model or significantly boost CSA.

Our competitive analysis allowed us to identify a wide range of potential stakeholders, including end-users, IT and cybersecurity professionals, cybersecurity educators, and private sector stakeholders. We recognized end-users as the primary stakeholder, but we also identified the role of IT and cybersecurity professionals to maintain the web app and further assist users with questions, support tickets, etc. Cybersecurity educators comprised another class of stakeholders, as our web app could further education of sporadic training modules and inspire training certificates based on real-world user behavior. Internet service providers are one example, as they are responsible for providing a line of defense against cyberattacks [6].

During this early stage, we also summarized user requirements for the application's interface. We had previously identified the need for users to receive transparent information and timely feedback. Thus, we summarized the need for our interface to clearly communicate the nature of the cyber threat, the actions available to the user, how to carry them out, and why they are necessary. We also established the need for users to receive information that is specific to the task at hand and to access additional information and feedback as desired.

3.2 Ideate

The ideation stage primarily involved brainstorming and storyboarding to find an appropriate name, platform, and representation for our interface. In early brainstorming sessions, we tested combinations of words to find an acronym that suited our interface's overarching goals. This resulted in the name CHIP, an acronym for Cyber Hygiene Intelligence and Performance.

While brainstorming, we also considered the previously identified user requirements to establish a method of delivery. We found that the ideal platform to encourage our target behaviors was a web application, which could gather and communicate information directly within the user's browser. The development of web apps has also enabled richer and more interactive content compared to desktop applications (e.g., firewalls and antivirus software), which are considered more "cumbersome and monolithic" [16].

Next, we considered the identity of our web application. We found that the best representation to meet user needs was a helpful bot or virtual assistant, commonly portrayed as avatars using conversational language [17]. On the other hand, previous literature also informed the need to avoid qualities that are *too* human-like. Schneiderman [18] contends that designing intelligent agents to mimic humans outright is inappropriate and ineffective. Our web app's more ambiguous and informal nature was also an intentional shift away from "informant" programs whose central goal is to escalate non-compliance.

Through paper storyboarding, we visualized how CHIP might interact with end-users. The premise of each storyboard entailed a user reaching a cyber hygiene decision point, which triggered a CHIP notification. For each target cyber hygiene behavior, we sketched a series of 3–5 panels. Each panel contained a concise description of the user's actions and/or CHIP's responses in the form of a push notification from the web app. The notifications began with brief, actionable information (e.g., "Just say nope" to password-saving, or "Be original" to create account security questions), followed by additional information on the nature of the threat, such as the hacker's motivations.

See Figs. 2 and 3 for sample storyboards.

User enters new login credentials to access an online account.

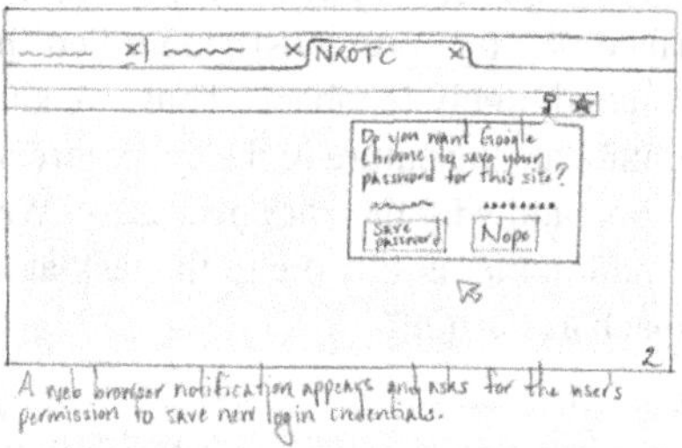

A web browser notification appears and asks for the user's permission to save new login

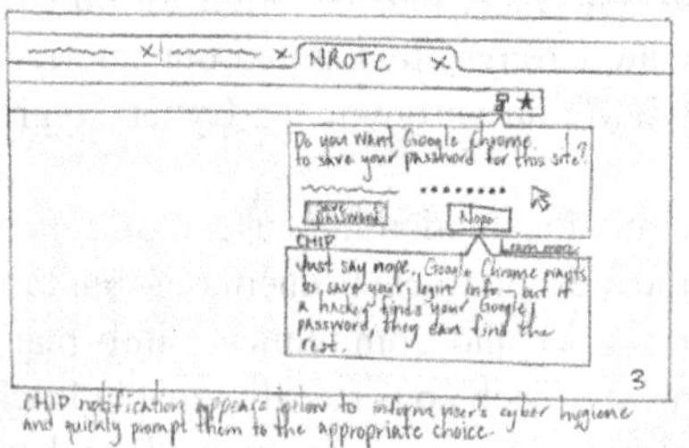

CHIP notification appears below to inform user's cyber hygiene decision and quickly prompt them to the appropriate choice.

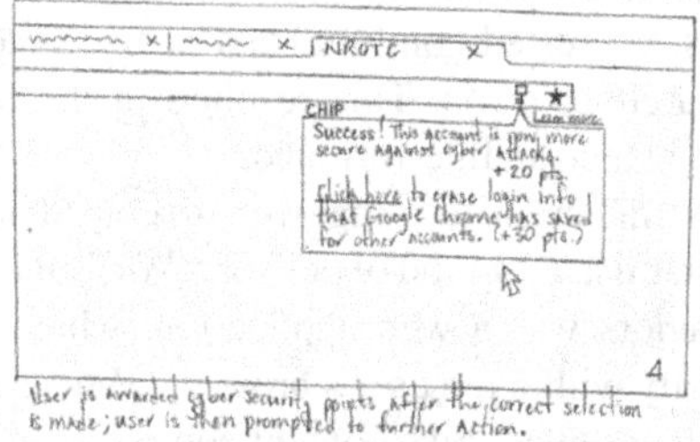

User is awarded cyber security points after the correct selection is made; user is then prompted to further action.

Fig. 2. "Denying browser password storage" storyboard.

User is creating a new account on a webpage when they reach the security questions form.

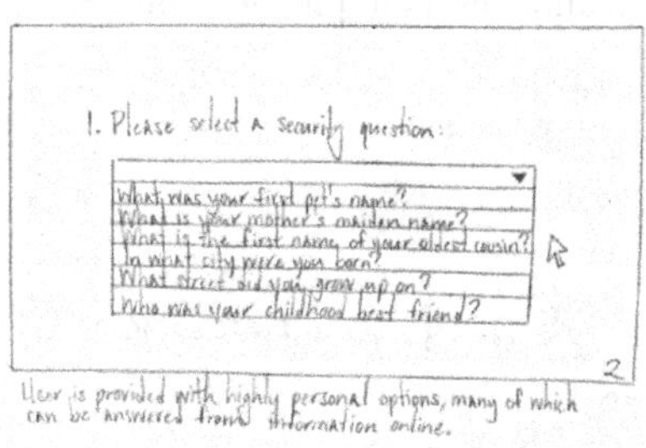

User is provided with highly personal options, many of which can be answered from information online.

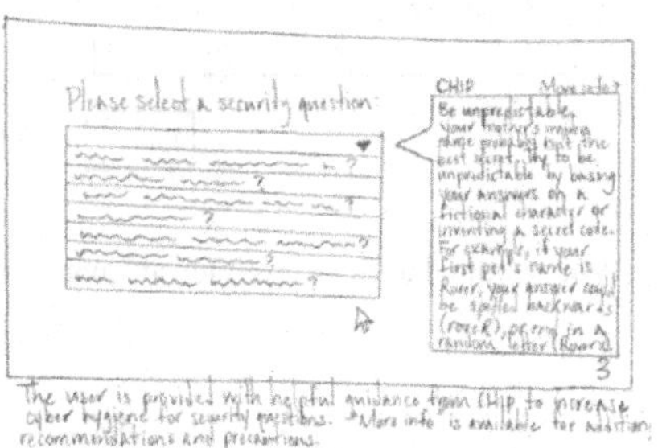

The user is provided with helpful guidance from CHIP to increase cyber hygiene for security questions. "More info" is available for additional recommendations and

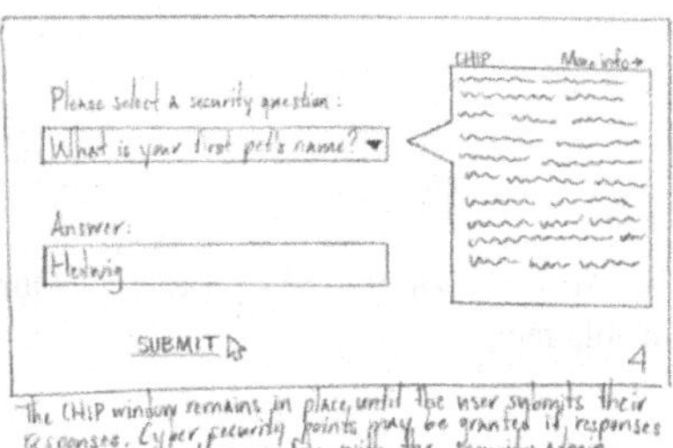

The CHIP window remains in place until the user submits their responses. Cybersecurity points may be granted if responses do not match those on file with the security admin.

Fig. 3. "Strengthening security questions" storyboard.

3.3 Prototype

Using paper sketches and storyboards from the ideation process, we developed a mid-fidelity prototype through Adobe XD. We focused on developing web app notifications. From previous literature, we had learned that users tend to relate to bots despite knowing that they rely on artificial intelligence [19]. According to Friedman [20], "users tend to be more receptive to conversational messages that feel more casual and less like system notifications." We therefore designed CHIP notifications to leverage conversational and straightforward language, omitting unnecessary technical jargon (see Fig. 4). Notifications appeared as speech bubbles emerging from the user's primary task.

In later iterations, however, we further developed the bot representation by adding a logo-like avatar (see Figs. 5 and 6). At this point, we needed to be wary of developing a relatable bot without mimicking a human entity (for further discussion, see Schneiderman, 2020). Therefore, we designed the CHIP avatar to be abstract (i.e., genderless and faceless) and removed the chat bubble response option from the user, which we ultimately found unnecessary. Instead, we included a collapsible drop-down menu that would allow users to pursue additional information and actions on their own terms.

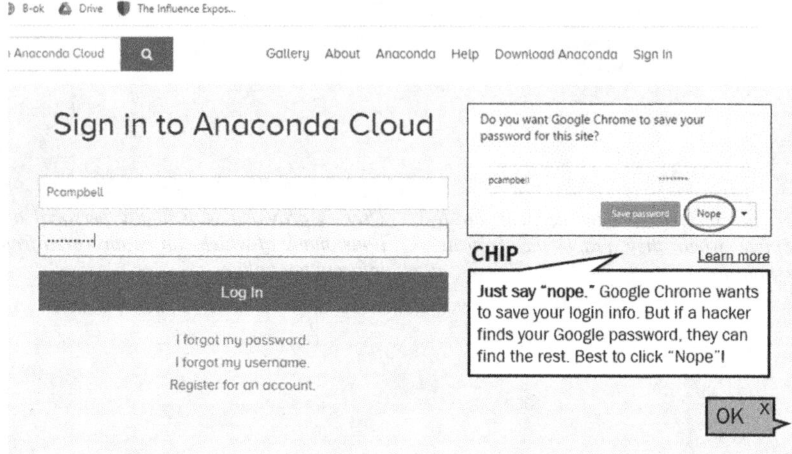

Fig. 4. An early Adobe XD prototype. Our prototype would later incorporate an avatar and collapsible menu.

Fig. 5. CHIP avatars. As users demonstrate better cyber hygiene, they may progress to higher tiers (e.g., bronze to silver, silver to gold).

Other design decisions during the prototyping stage were based on heuristic standards. For example, Fitts's Law states that the time required for a user to reach a "target" (i.e., a command button) is a function of the size and distance to the target. Our solution for CHIP as a web app allows notifications to float beside the task at hand, allowing users to quickly access information in the periphery. This solution also prevents a pop-up window from obstructing the user's primary task, which might lead users to close the notification window without reading it [21]. Once users' attention is directed toward the task, the amount of time required for them to make a decision depends on the sum and complexity of their choices. This heuristic, known as Hick's Law [22], resulted in simplifying choices to "more info" and "actions" with a handful of action choices nested within the drop-down menu, such as "contact help desk" or "report phishing email." Another heuristic known as the goal-gradient effect states that "the tendency to

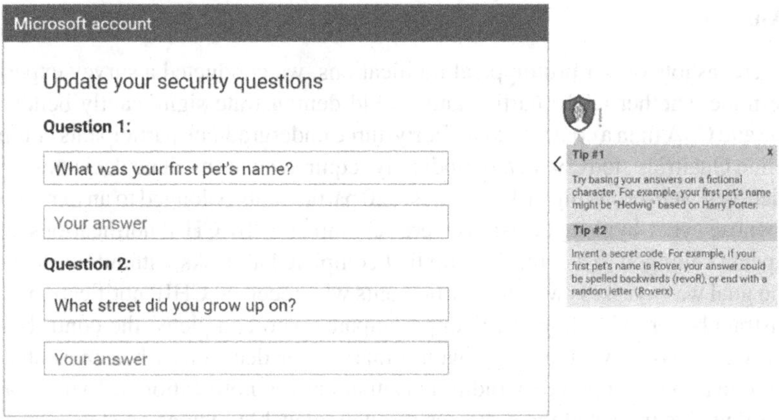

Fig. 6. Late iteration of the CHIP prototype. While creating account security questions, the user receives a push notification to access tips to create secure responses, for example, basing responses on a fictional character instead of using personally identifiable information (PII) that may be accessed online.

approach a goal increases with proximity to the goal" [23], and informed our decision to utilize progress indicators for cyber hygiene goals (see Fig. 7). In this case, the progress bars illustrate users' task completion and motivate them to reach the goal (e.g., creating a stronger password).

Fig. 7. Progress indicators are shown as users incrementally strengthen a weak password.

3.4 Analyze

Using screenshots of our prototypical notifications, we conducted a survey experiment to determine whether CHIP participants would demonstrate significantly better cyber hygiene and CSA than a control group. Forty-three undergraduate participants completed our online Qualtrics study, which periodically required cyber hygiene decisions to complete fictional web-browsing tasks. Of these, 20 participants belonged to an experimental group whose cyber hygiene decisions were accompanied by CHIP notifications, and 23 participants belonged to a control group that completed all tasks without notifications.

One goal was to assess whether participants who received CHIP notifications would demonstrate better cyber hygiene than participants who completed the control survey. That is, we assessed whether users would make safer decisions when presented with CHIP notifications compared to traditional situations (no notifications). To make a decision, participants indicated their choices through clickable regions on static screenshot images. They completed one scenario for each of the four cyber hygiene behaviors, with responses coded as either good cyber hygiene (e.g., navigating away from a dangerous website) or bad cyber hygiene (e.g., proceeding to the dangerous website). Preventing password auto-saves, deleting or reporting a phishing email, changing their answer to the account security question, and complying with an unsafe website warning (e.g., not continuing to the site) were all considered safe decisions. Saving the password, replying to or forwarding the phishing email, maintaining the same security answer, and continuing to the potentially dangerous website were considered unsafe decisions. For CHIP participants, the page margins of cyber hygiene decision points were supplemented by a notification window containing additional information about the cyber threat.

Following each cyber hygiene decision, all users completed a modified version of the 10-D SART, a valid and reliable measure of situation awareness [24] originally developed by Selcon and Taylor [25] to evaluate aircrew system design. Advantages to the 10-D SART include high ecological validity, ease of use, and the ability to be administered after various tasks due to the general nature of its questions [24, 25].

Independent-samples t-tests revealed that CHIP participants made significantly safer decisions during the password-saving ($p < .001$), phishing email ($p < .001$), and account security question scenarios ($p < .001$). CHIP users also demonstrated significantly greater CSA compared to a control group for the phishing email scenario ($p = .048$) and risky website scenario ($p = .032$), and when averaged across all four scenarios ($p = .048$) (see Fig. 8).

These findings show promise for CHIP as an interface that promotes cyber hygiene and CSA for specific behaviors. Given that user interactions with CHIP were limited to selections on simple static screenshot images, these results are especially encouraging.

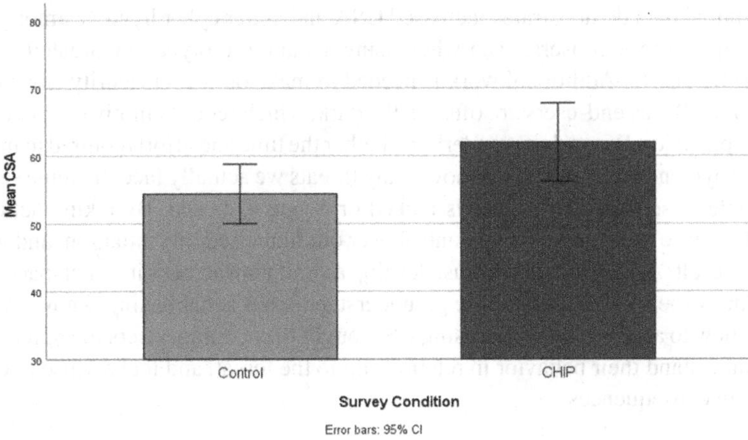

Fig. 8. Mean CSA for each survey condition across all four scenarios.

4 Conclusion

Our goal in describing CHIP's design process is not to produce a formal set of guidelines or principles for developing human-centered cybersecurity systems, but to begin a conversation in this emerging area. Too many current cybersecurity solutions are driven by the needs of large organizations rather than end users themselves. Our solution is primarily targeted toward end users, with implications for the ecosystems *within* users' work environments.

As human-centered solutions like CHIP evolve, it is essential to consider whether these systems could potentially replace traditional cybersecurity training, or simply function as a "red flag" virtual assistant. We must also consider how to respond when systems like CHIP make mistakes and how these mistakes will be tolerated. Users may prefer real-time animation that only occurs when CHIP has high confidence in a critical issue. Therefore, it may be beneficial to employ a survey to understand users' cybersecurity needs, preferences, and values. This information might positively impact end-users' experience with CHIP.

We faced many challenges in fully testing and implementing CHIP. A specific challenge included how to collect data (e.g., how to scrape this information from a variety of websites) and how to store/encrypt this information (e.g., who should maintain control/own the data). A technical question remains as to how we might collect personal information like login credentials, given a wide variety of authentication processes. It also remains to be seen how we might deliver personal feedback in a consistent manner across a variety of platforms. These challenges will need to be overcome for CHIP to be successful. Simply considering strong authentication, CHIP must access all of the user's passwords and know their preference or employer requirements for authentication. Knowing passwords can combat reuse and make end-users aware of recent breaches (i.e., requiring a password change). Online services have a wide variety of authentication security standards. However, if their employer requires very strong authentication, CHIP can help end-users maintain consistency across their online accounts.

Our initial data demonstrates increased CSA and better cyber hygiene among CHIP users compared to non-users, even when using a static prototype not situated within a real-world scenario. Additional work is needed to make our cybersecurity environment more visible. We as end-users are often in the dark, which reduces motivation to comply with best practices. We are left wondering whether the time and effort required to improve our cyber hygiene keeps us safer or how many threats we actually face. In home security, we are able to see whether a door is locked or a gate is closed. By taking action, we establish a sense of ownership and control over our home security situation, and we feel safer as a result. As designers, we must develop a similar awareness in cyberspace. Users should know their cyber situation (e.g., understand what is happening, know what can be done, how to access help). Increasing CSA ought to encourage users to appreciate the threat, understand their behavior in relationship to the threat, and feel a sense of control over future consequences.

Acknowledgments. This work was supported in part by the Commonwealth Cyber Initiative, an investment in the advancement of cyber R&D, innovation and workforce development. For more information about CCI, visit cyberinitiative.org. We thank everyone at MI Technical Solutions for their involvement in identifying cyber hygiene concerns and technical constraints.

References

1. Ng, B.Y., Xu, Y.: Studying users' computer security behavior using the Health Belief Model. In: Pacific Asia Conference on Information Systems (PACIS) 2007 Proceedings, vol. 45, pp. 423–437 (2007)
2. Assante, M.J., Tobey, D.H.: Enhancing the cybersecurity workforce. IT Prof. **13**(1), 12–15 (2011)
3. Aloul, F.A.: The need for effective information security awareness. J. Adv. Inf. Technol. **3**(3), 176–183 (2012)
4. Nagarajan, A., Allbeck, J.M., Sood, A., Janssen, T.L.: Exploring game design for cybersecurity training. In: 2012 IEEE International Conference on Cyber Technology in Automation, Control, and Intelligent Systems (CYBER), pp. 256–262 (2012)
5. Cain, A.A., Edwards, M.E., Still, J.D.: An exploratory study of cyber hygiene behaviors and knowledge. J. Inf. Secur. Appl. **42**, 36–45 (2018)
6. Carter, A.: The Department of Defense cyber strategy. The US Department of Defense, Washington, DC (2015). https://archive.defense.gov/home/features/2015/0415_cyber-strategy/final_2015_dod_cyber_strategy_for_web.pdf. Accessed 10 June 2021
7. Griffith, V., Jakobsson, M.: Messin' with Texas deriving mother's maiden names using public records. In: Ioannidis, J., Keromytis, A., Yung, M. (eds.) Applied Cryptography and Network Security, ACNS 2005, vol. 3531, pp. 91–103. Springer, Heidelberg (2005). https://doi.org/10.1007/11496137_7
8. Rabkin, A.: Personal knowledge questions for fallback authentication: security questions in the era of Facebook. In: Proceedings of the 4th Symposium on Usable Privacy and Security, pp. 13–23 (2008)
9. Sheng, S., Holbrook, M., Kumaraguru, P., Cranor, L.F., Downs, J.: Who falls for phish? A demographic analysis of phishing susceptibility and effectiveness of interventions. In: Proceedings of the SIGCHI Conference on Human Factors in Computing Systems, pp. 373–382 (2010)

10. Malkin, N., Mathur, A., Harbach, M., Egelman, S.: Personalized security messaging: nudges for compliance with browser warnings. In: 2nd European Workshop on Usable Security. Internet Society (2017)
11. Harris, M.A., Patten, K., Regan, E.: The need for BYOD mobile device security awareness and training. In: Proceedings of the Nineteenth Americas Conference on Information Systems (2013)
12. Caldwell, T.: Training–the weakest link. Comput. Fraud Secur. **9**, 8–14 (2012)
13. Franke, U., Brynielsson, J.: Cyber situational awareness – a systematic review of the literature. Comput. Secur. **46**, 18–31 (2014)
14. Hutchins, E.M., Cloppert, M.J., Amin, R.M.: Intelligence-driven computer network defense informed by analysis of adversary campaigns and intrusion kill chains. Leading Issues Inf. Warfare Secur. Res. **1**(1), 113–125 (2011)
15. Denis, C., Karsenty, L.: Inter-usability of multi-device systems: a conceptual framework. In: Multiple User Interfaces: Cross-Platform Applications and Context-Aware Interfaces, pp. 373–384 (2004)
16. Google Chrome Developers: Extensions and apps in the chrome web store. https://developer.chrome.com/docs/webstore/apps_vs_extensions/. Accessed 9 June 2021
17. Rafailidis, D., Manolopoulos, Y.: Can virtual assistants produce recommendations? In: Proceedings of the 9th International Conference on Web Intelligence, Mining and Semantics, pp. 1–6 (2019)
18. Shneiderman, B.: Human-centered artificial intelligence: three fresh ideas. AIS Trans. Hum.-Comput. Interact. **12**(3), 109–124 (2020)
19. Geiger, R.S.: Are computers merely "supporting" cooperative work?: towards an ethnography of bot development. In: Proceedings of the 2013 Conference on Computer Supported Cooperative Work Companion, pp. 51–56 (2013)
20. Friedman, V.: Privacy UX: Better notifications and permission requests. https://www.smashingmagazine.com/2019/04/privacy-better-notifications-ux-permission-requests/. Accessed 9 June 2021
21. MacKenzie, I.S.: Fitts' law as a research and design tool in human-computer interaction. Hum.-Comput. Interact. **7**(1), 91–139 (1992)
22. Proctor, R.W., Schneider, D.W.: Hick's law for choice reaction time: a review. Q. J. Exp. Psychol. **71**(6), 1281–1299 (2018)
23. Kivetz, R., Urminsky, O., Zheng, Y.: The goal-gradient hypothesis resurrected: purchase acceleration, illusionary goal progress, and customer retention. J. Mark. Res. **43**(1), 39–58 (2006)
24. Endsley, M.R., Garland, D.J. (eds.): Situation Awareness Analysis and Measurement. CRC Press (2000)
25. Selcon, S.J., Taylor, R.M.: Evaluation of the situation awareness rating technique (SART) as a tool for aircrew systems design. Paper Presented at the AGARD AMP Symposium 'Situational Awareness in Aerospace Operations', Neuilly Sur Seine, France, pp. 23–52 (1990)

The Effect of Social Media Based Electronic Word of Mouth on Propensity to Buy Wearable Devices

David Ntumba and Adheesh Budree(✉)

Department of Information Systems, The University of Cape Town, Rondebosch, Cape Town, South Africa
ntmdav001@myuct.ac.za, adheesh.budree@uct.ac.za

Abstract. There has been an increase in both wearable devices and social media usage. This study sets to describe how social media influences the buying intentions of Wearable Devices (WD). In doing so, constructs from Information Acceptance Model (IACM) had been compared to the resulting themes from 9 one-on-one interview responses and a relationship between social media and wearable devices was formed.

Objectives: Primary objectives were to determine the relationship between social media and; (i) WD sales, (ii) WD features and application information (iii) ongoing interactions on WD's.

Secondary objectives were to determine (i) if the availability of a WD affected the volume of social media interactions and (ii) if the value associated with a WD was related to the social media posts of that device.

Design, Methodology, and Approach: This study took a qualitative approach to research and used phenomenology as a stance of exploring social media's influence on purchase intentions towards the wearable device. A positivist philosophical standpoint was applied to deduce themes from the ICAM by touching on interpretivism elements in analyzing the data collected from semi-structured interviews. The data was then categorized into themes, refined into sub-themes and then assessed according to the research propositions as well as the ICAM frameworks constructs.

Findings: Social media was viewed more of as an information outlet as opposed to a driver of purchase intention. From this, it was seen that word of mouth was the biggest driver in WD awareness. Information Quality, Attitudes towards information were the main influencers that social media has towards purchase intention from the ICAM model. This was evident in the construct scores which showed that there was a relationship between the adoption of social media information and purchase intention, that there is a relationship between the features of a WD and social media usage as well as showing that there is a relationship between the users' perception of the type of social media platform and the WD presented on that platform.

Research Implications: This study explored the influence of social media on the buying intention of wearable devices as well as provided channels for further exploratory research into the human perceptions of information provided on social media.

C. Stephanidis et al. (Eds.): HCII 2021, LNCS 13094, pp. 310–325, 2021.
https://doi.org/10.1007/978-3-030-90238-4_22

Keywords: Social media · E-commerce · Influence · Electronic word of mouth · Wearable devices · Smartwatch · Information acceptance model

1 Introduction

Recent growth in social media and the usage thereof have shown a growing trend in online marketing and means of advertisements (Jashari and Rrustemi 2017) With the current online marketing trend and emergence of new technical devices in the forms of smartwatches and fitness bands, also known as wearable devices (WD), have been gained market traction seeing immense adoptions from markets for numerous reasons (Page 2015). Wearable devices have gathered interest as an emerging technology because of its visibility (Wu et al. 2016).

This study aims to describe the influence social media has on individuals' purchase intentions of wearable devices. To do this study has set out to achieve the following objectives:

- Determine the relationship between social media interactions and WD sales,
- Determine whether there is a cycle amounts social media interactions and WD features or applications,
- Determine whether there are specific social media interactions for a WD,
- Determine whether the WD availability affects the volume in social media interactions and
- Determine whether the value associated with the WD is related to social media posts.

2 Literature Review

Wearable devices are IoT artifacts and have various form factors, use specifications (Wairimu and Sun 2018). This study focuses on two WD form factors namely: smartwatches and smart bands which have distinct factors between them (Wu et al. 2016).

These devices are linked to social media applications which are also evident via the gamification of user data in mobile applications (von Entress-Fürsteneck et al. 2019). However, since social media and its influence through its interactions were at the core of this study, only social media platforms had been reviewed.

2.1 Social Media and Electronic Word of Mouth

Social media is a key element in this study and has been categorized in various forms and allowing for communications to be conducted online and the sharing of information between connected parties (Lee 2013).

In reviewing the literature on social media, influence and behavioral literature it was seen that. electronic Word of Mouth (eWOM) explains how information is shared between internet users (Erkan and Evans 2016a, b). Furthermore, eWOM is the widespread of statements and/or communications made by potential, current, former customers about the product or company over the internet (Balakrishnan et al. 2014). Internet-based communications have allowed for a greater reach and audience when compared to traditional forms of word of mouth such as television and newspapers (Chuah et al. 2016). Therefore, social media was noted as an appropriate form of eWOM after the factors of eWOM and social media were discussed (Erkan and Evans 2016a).

Social media has been closely related to web 2.0 platforms (Lee 2013). Below is an overall view of the social media components that are interconnected to form social media. The interconnection of their components has led to the phenomenon known as the socialization of information which can be conveyed to a larger global network (Lee 2013).

Figure 1 shows how the components of social interaction, communication media, and content interact with one another and merge to form social media. It further depicts how information can spread and be adopted by viewers. Therefore, the following section will cover the eWOM factors concerning social media as well as its various platforms.

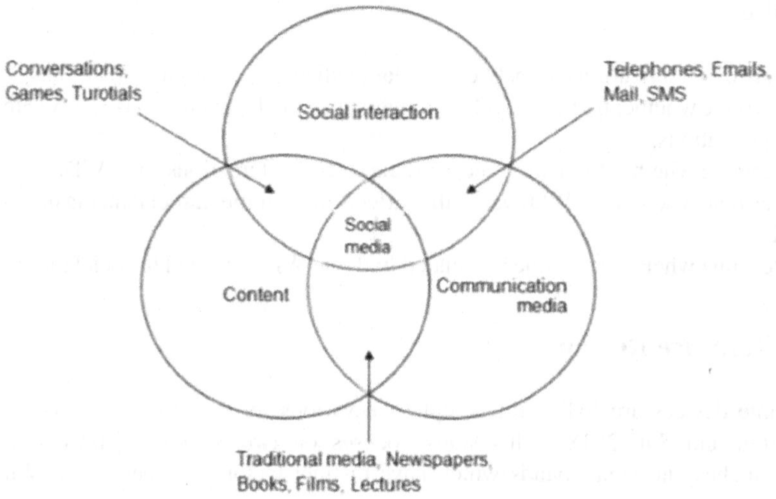

Fig. 1. Social media Components (Lee 2013, p. 24)

2.2 Types of eWOM

Electronic Word of Mouth has been noted to be a combination of customer reviews and discussion forums of which involve e-commerce sites (Erkan and Evans 2016a, b). This allows for members of those platforms to create a shared meaning of the brands or products discussed and portrayed and though those with purchase intention may not be able to physically review the device they rely highly on the information passed within eWOM platforms such as online discussions, forums, online customer review sites, blogs, social networking sites, and online brand/shopping sites (Teo et al. 2018).

2.3 Social Media Platforms

The following section explores the social media platforms contributing to eWOM as well as list the defining factors of each of the sites. Social media sites as defined by Lee (2013) may have multiple forms and this study focused on social media literature that covered factors such as influence and buying intention as they form the construct if the ICAM model which will be covered in the following sections. Literature shows that key social media sites are Facebook and Instagram and Twitter.

Facebook. Facebook, created 2004, has a substantial annual growth with registered members generating billions of views daily thus becoming an integral part of social media users' daily activities (Ellison et al. 2007). Interactions and content on Facebook are in both graphical and audio forms which include features such as friends list, walls, pokes, events and chat groups among others. The primary driving feature of Facebook is the friend's list because it allows users to publicly present their content to their connections which can be viewed, shared and further commented on (Nadkarni and Hofmann 2012).

Given the nature of eWOM platforms and Facebook's penetration and ability to generate large streams of content, this study has been directed towards a younger age group as a target population (Lee et al. 2018). Additional studies have also shown that there is a substantial Facebook sign up base within the age group of 12–17 years of age, who once members have high login rates (Pempek et al. 2009).

Firms have also used Facebook within their marketing strategy, using company Facebook pages firms can ensure brand loyalty from their customers who like and view the content on those pages (Lee et al. 2018). Therefore, the firms themselves can generate eWOM content of their products, though few users are noted to engage with them (Lee et al. 2018). Low engagement rates are since Facebook pages act as bulletin boards and users can view the content without necessarily providing any form of response (Pempek et al. 2009). This means that firms cannot rely on the number of shares and reactions on their posts as a true reflection of how many individuals have viewed.

Factors contributing to Facebook usage are due to the delivery of rich content coupled with the daily engagements from users that use their real identities thus removing anonymity and ensuring credibility (Lee et al. 2018).

Instagram. Instagram, founded in 2010, has a primary focus on sharing photographic content amongst users. The shared photos may be edited and beautified to generate a greater attraction and response to them through the online community known as "followers" who are also able to search for one another (Teo et al. 2018). However, Instagram allows for user profiles to be kept private meaning that content shared may only be viewed by authorized or approved individuals (Chen 2018) This provides a different perspective to the credibility of the posts on Instagram as the members viewing or posting the information are known to the sharer of the post to some degree.

Instagram's growth has been attributed to the improvements made in smartphones and tablet cameras to produce high-resolution images for sharing. Like Facebook, firms have also used Instagram for marketing purposes (Teo et al. 2018).

2.4 Internet of Things (IoT) Devices

IoT devices allow for interactions between people, technology and surrounding objects by connecting multiple devices on the internet primarily through cloud computing (Ray 2018) The connection of IoT devices allows for visual, audio and logical information to be collected and the actioning of jobs which includes transferring such data to other devices which is referred to as the devices "talking" to each other for users to make decisions (Al-Fuqaha et al. 2015). IoT also offers market opportunities for vendor manufactures. Although Al-Fuqaha et al. (2015) defines the market opportunities in the health care sector for mobile health, there is no definite mention that this is done using WD.

2.5 Wearable Devices

Wearable devices have been used in fitness tracking, this entails collecting and quantifying user data and offering it in a manner that is useful to the user by using graphics and metrics (Mekky 2014). User data is gathered through the combination of the accelerometer and gyroscope sensors built into the WDs (Kim et al. 2018). The data measured by WD comes in two forms namely; data from physical activities and physiological data. Physical activity data is based on what the user does such as a count of the user steps or calories. Alternatively, physiological data is the user's physiological data which is based on the changes in the users' body is which includes heart rate and temperature (Rupp et al. 2018).

Data from WDs may be used to provide the user of their health data and also drive the gamification of health applications (Lister et al. 2014). In the gamification of health applications, the use of rewards, levels, leader boards, and goal settings increases the attractiveness of respectively performing physical tasks, therefore, promoting users to become more active (Lister et al. 2014).

Wearable devices have the following characteristics:

- A portable device that does not hinder operational use.
- Allow for hands-free usage
- Make use of various sensory and measuring features (camera, GPS) to provide information about the users surrounding environment.
- Provide ambient communication to the user using notifications and alerts
- Constantly gather information about the user and surrounding environment, hence always active.

(Johnson 2014)

Using the above characteristics of WD the following section will explore the literature on smartwatches and smart bands whereby smartwatches literature focuses on purchase intention, usability, adoption, and retention. This forms a basis for the constructs which need to be considered when studying purchase intention. Smart bands have been seen in literature to be noted as fitness trackers and there has been a general health-themed trend in the literature concerning them. Also, in this section, the functions and or uses of both smart bands and smartwatches will be explored.

Smartwatch. Smartwatches, although taking the form of a traditional timepiece, have multiple functions and features that can be linked to smartphones and user accounts (Wu et al. 2016). Smartwatches can work independently of the user's smartphone device or maybe paired with such to all for cross-functionality (Gimpel et al. 2019).

Furthermore, new user experience has been created in the fashion industry trend in which users have preferences for various designs. This illustrates smartwatches, like mobile phones, to be attractive to users based on hedonic (the joy derived from using technology) or functional factors (Wairimu and Sun 2018).

Popular smartwatch vendors include Apple watches, Samsung Galaxy Gears, and Fitbits and are bought by a vast demographic group (Cho et al. 2019). However, the sales growth of smartwatches has seen a steady decline despite adoption studies report that users have a high intention to acquire smartwatches ((Chuah et al. 2016), (Wu et al. 2016)).

Drivers for Intent to Buy Smartwatches. The following are advantages of smartwatches which drive the intention to buy smartwatches:

- The smartwatches' form factor is already socially accepted, and the familiar wristwatch design allows users to feel comfortable wearing them.
- The interface of smartwatches provides a minimalistic approach to the information provided.
- Users can customize the information that is portrayed and, on the smartwatch, and this is further extended by the means of downloading applications.
  (Johnson 2014).

Limitations to Intent to Buy Smartwatches. The following are disadvantages of smartwatches which speaks to the decline in smartwatch purchase intention seen in Wairimu and Sun (2018):

- Some designs although innovative are unable to measure all the data required.
- Users may be reluctant to using smartwatches based on the uncertainty and lack of control in what is being done with their data.
- Being an IoT device, users may need to be constantly connected to an internet source for their devices and or applications to process and display their data figures.
  (Johnson 2014).

Uses and Adoption. Literature has also shown that customers benefited from the continuous use of smartwatches because they are a novelty technology. Meaning, the more they use it, the more they realize how it fits into their lives which is a gradual process (Nascimento et al. 2018) It has been noted that smartwatch purchases can derive from effective marketing from vendors, by someone else having the device hence WOM eWOM or because of them being an early adopter of smartwatches. This, however, does not guarantee usage of such (Nascimento et al. 2018). However, this study focuses on purchase intention and usage will only be considered when evaluating the eWOM from prior users of WD.

Other literature on smartwatches focused on factors surrounding purchase intention (Hsiao and Chen 2018), continuance intention (Nascimento et al. 2018), behavioral intention (Wu et al. 2016) and adoption with few studies also focusing on factors that influence purchase intention in the context of smartwatches (Hsiao and Chen 2018).

Smart Bands. Smart bands, also referred to as fitness trackers, like smartwatches are worn on the wrist and have similar features as the smartwatch. However, one distinct difference between smart bands and smartwatches is the screen size in which smart bands may have smaller or even no screen at all in the case of the Jawbone. Therefore, the Smart band form factor is that of a bracelet and the smartwatch form factor is that of a wristwatch (Lunney et al. 2016).

Smart band devices primarily track user information with step take, calorie count and work out intensity being the main measures. Research smart bands are still in early development however seeing as the main vendors in smart bands are Misfit, Jawbone, Germainly, and Fitbit with Fitbit also providing smartwatches this study has combined smartwatches and smart bands purchase intentions as WD purchase intention (Kaewkannate and Kim 2016).

3 Research Design

This study was descriptive as it examined the influence social media has on human behavior which, in addition to the (Erkans and Evans 2016) study of social media's influences buying intention through the use of eWOM, focuses on describing the phenomenon of how social media influences buying intention of WD.

A qualitative strategy had been applied within a cross-sectional timeframe whereby data was collected using 1-on-1 interviews and then responses were interpreted into themes that were then categorized and compared to the constructs of the ICAM model as a point to deduce relationships between them.

The target population forms a basis on analyzing the responses that are generated from the interviews and this study has focused on the South African 'youth' group. These individuals between the ages of 18–35 as age group (Bolton et al. 2013). Individuals of this age group, referred to as Generation Y, are highly skilled at using social media as well as ownership of mobile devices (Marketing Charts 2018). Given their high usage and ownership of both social media profiles and mobile devices, as well as being amongst the top percentile of social media users, generation Y stands as a harbinger of how future social media interactions will take place given their early exposure to the technology (Bolton et al. 2013).

In sampling the target population, it was ensured that interviewees own a social media account and are either looking to buy, own or have sold WD in connection to various social media interactions.

In gathering data ethical considerations had been made in such the interview had been performed only once consent had been given and participants were not obligated by any means to respond. Furthermore, the data recorded in the interview were solely for the use of the stated research objectives. Thus, data captured could not be represented in a manner that can identify the individual of which ensured anonymity.

3.1 Research Questions and Objectives

This study had focused on answering the following question:

How do social media influence and individuals' intention in buying a wearable device?

Following the research question was a list of sub research questions which supported the main research question as see below:

OB1. Can the acceptance of social media information be positively related to the consumers' buying intention?
OB2. How does social media introduce or increase the knowledge that users have of a WD?
OB3. Is there a relationship between social media interactions and the features or applications on WD?
OB4. What is the likelihood of users buying WD before and after being exposed to social media?
OB5. Is there a relationship between an interaction made on a given social media site and its following or credibility?

In answering the research question, primary, noted by 'OA', and secondary, noted by 'OB', objectives support the use of research tools, data collection and are listed below.

Primary Objectives

OA1. Determine the relationship between social media interactions and WD sales
OA2. Determine whether there is a cycle amounts social media interactions and WD features or applications
OA3. Determine whether there are specific social media interactions for a WD.

Secondary Objectives

OB1. Determine whether WD availability affects the volume of social media interactions.
OB2. Determine whether the value associated with the WD is related to social media posts.

3.2 Research Model and Propositions

This study had made use of the theoretical model from a study focused on social media's influence on customers' purchase intentions through their eWOM interactions known as the Information Acceptance Model (ICAM) (Erkan and Evans 2016b). Therefore Fig. 2 below illustrates the ICAM constructs of Information Quality, Information Credibility, Needs of Information, Attitude towards information, Information Usefulness, Information Adoption, and Purchase Intention.

Using the ICAM research model from Fig. 2, the following propositions are made to support the objectives:

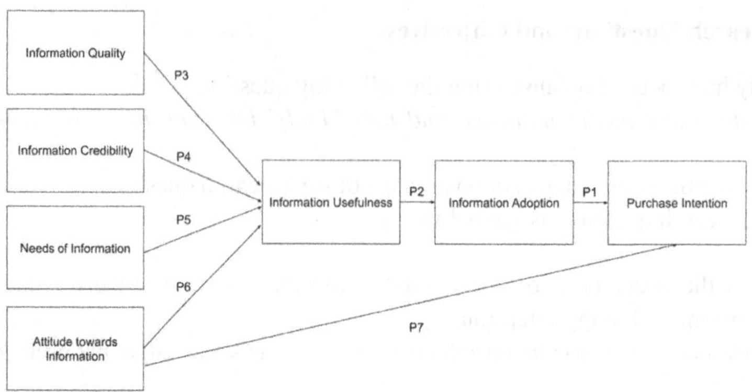

Fig. 2. Information Acceptance Model (Erkan and Evans 2016a, b)

OA1. There is a relationship between the adoption of social media information and purchase intention
OA2. There is a relationship between the usefulness of social media information and the adoption of social media information
OA3. There is a relationship between WD features and social media usage
OA4. There is a relationship between the credibility of social media information and intention to buy WD
OA5. There is a relationship between the number of WD alternatives and the views of social media information
OA6. There is a relationship between the social media platform on the user's perception of the WD presented on that platform.
OA7. There is a relationship between the attitudes of towards social media information and purchase intention

4 Research Analysis and Findings

In undertaking the data analysis this study made use of the Memoing and Axial Coding methods to organize and classify the data that had come in (Bhattacherjee 2012). In analyzing data, it was seen that responses and sentiments from owners of WDs had differed from potential customers.

4.1 Interview Response Themes

From the responses, respondents had elaborated on how various social media had influenced their interests in a WD, however, it was seen that social media was viewed to be merely a channel of information as instead of a driver of purchase intention.

This is supported by respondents expressing that they first have a set goal or reasoning into buying a WD and then proceed to rely on social media to confirm the choice that they had already made. Post validating the IACM model responses from the various interviews were used and the following table of themes was found:

Table 1. Interview themes

Theme	Brief description
Accuracy	The measure of Information found on a Particular Site versus the Actual WD
Trusted/Credible	Relative to the influence of the site/site interaction has
Hands-On Experience	Physically holding the WD
Feature	A technical feature between the WD and/or Social media
Posts	An Interaction on social media
Speed	The time is taken for information to be gathered
Usability	How easily can the WD be used
Paid	Any Monetary Transaction needed to gain information
Value for Money	The measure of a potential/existing sale versus the WD and/or its features
Not on Social media	Alternative routes of gathering information
Encourage	Positive engagement of Social media
Achievement	The Completion of a task or goal set out by Social media or WD
Competition	The Ranking of Achievements
Reasoning	The motive behind reviewing a WD
Alternative	In Place of WD or Social media
Awareness	Increase in Audience
Comments	Posts made in response
Word of Mouth	The spread of information through experience
Bias	Subjective Posts
Personal Preference	Personal Subjective Views on Social media and/or WD
Brand Loyalty	Personal Preference regarding a specific vendor(s)

From Table 1 the refined themes are as follows: Reliability, Experience, Word of Mouth, Monetary, Exclusion, Participation, Personal Preference, and Brand Loyalty had arisen.

In unpacking these themes it was seen that 'Reliability' encompassed the respondents view that WD information on social media was accurate, credible and trustworthy, had been retrievable promptly (relating to the 'Speed' theme) as well as being usable this was noted in the responses whereby it was said "Ï would first look around on Twitter to check on the device" also another respondent said, "I like to look at reviews to get an idea of the smartwatch".

Word of Mouth. Word of Mouth covered 'Posts', 'Alternative', 'Awareness', 'Comments' and 'Word of Mouth' themes related to the social media information itself and respondents' views on such. Phrases such as "I would spend more time on social media if there were posts on different devices" and "I have become more aware of devices on social media" support the Word of Mouth Theme.

Experiences. The experience was closely tied to how social media information allowed the user to know the WD without necessarily owning it. The grouped themes here were: 'Hands-on Experience' and 'Feature'. Here respondents have mentioned that the use of Youtube reviews aided them in knowing the WD before deciding.

Monetary Value. Monetary themes such as 'Paid' and 'Value for Money' indicated that further factors needed to be considered in intention to buy a WD. One respondent mentioned, "I was going to get the higher Fitbit model, but it was too expensive".

Exclusivity. Exclusion reflected cases whereby the respondent was not on a social media site and or their intention to buy a WD was not influenced by social media information. Responses such as "I will go to the store" and "I will check the company website" were used.

Participation. Participation was seen in the 'Encourage', 'Achievement' and 'Competition' themes which showed that engagement in social media information in terms of liking and reposting views affected the respondents' intentions to buy the WD. Key phrases such as "I would post my steps on Facebook" and "I have joined the group" were noted in support of the Participation theme.

Personal Preference. Personal Preference covered 'Reasoning' and 'Personal Preference' themes and related to how the user felt about a WD and the social media information from their backing and ideals. Here respondents noted, "I would like to research before making a decision".

Brand Loyalty. Brand Loyalty grouped 'Bias' and 'Brand Loyalty' themes together as respondents have shown that the choice of the WD had been impacted by the smartphone that they have been using as well as the brand that they have become accustomed to. Here the phrases "I only look at Samsung Products" as well as "This Smartwatch pairs well with Samsung Phones" illustrated how brand loyalty and bias were linked.

4.2 Summary of Themes

'Word of mouth' (WOM) was the biggest driver in WD awareness. This WOM would be based primarily on the smaller social groups be it the forums or the direct WhatsApp groups themselves. This relates to the IACM model whereby factors such as information usefulness and attitude towards usage are called. However, it was also noted through the responses that although social media is an influence to WD sales it does not represent the final sales transaction itself, rather it acts as a 'self-advertising' means of contact whereby the users can acquire as much information of the WD as they require, which

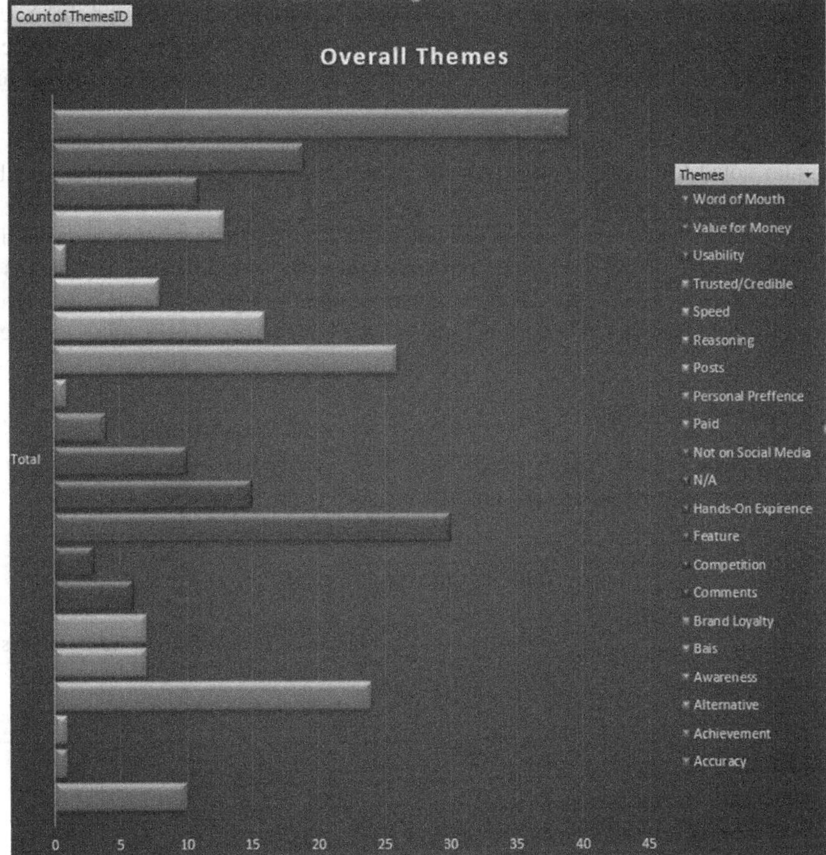

Fig. 3. Theme graph

again relates to the IACM model through needs for information. This is further supported by Fig. 3 below where word of mouth had the highest occurrence of themes.

Concerning the word of mouth, it was seen that themes relating to device features, personal preference, and awareness are closely coupled. This is because the respondents have sought information on social media after knowing or seeing the product in prior posts, therefore, social media aids in retaining or intensifying the users' awareness will and/or desire to buy the WD as well as aids in comparing various device features.

Another phenomenon noticed through the responses was that of brand loyalty. Therefore, a user had a smartphone of a brand then their searches on social media will be driven/filtered by that brand. Therefore, the influence of social interactions will be driven by the members closer to the brand. This brings about the issue of the source which drives the influence on the Social media platform. According to the responses the individual or company plays a significant role in attracting then to a specific WD, these may range to experts in the field, reviewers critiques or even the company themselves this then assures the users that the information on the Social media site is of quality and credible source.

Although many comments are seen to be biased, responses have shown that Social media can only provide direction as to which the users 'should' buy but cannot directly affect the buying decision of the users due to external factors such as price, ergonomics as well as availability.

Key social media platforms where noted as Facebook, Instagram and Youtube with Instagram noted to be the main drivers in allowing users to have an almost tangible experience with the WD.

Lastly, it was noted that social media had aided the users in viewing and discussing the usefulness of the devices by a measure of their features. Should not all the features be met, this will form a loop between the users searching through alternatives, discussing them and concluding that WD does not match their budget, brand or features required.

4.3 Summary of Findings

The primary and refined themes were linked to the ICAM constructs and summary of the primary and refined themes per construct is seen in the table below.

Table 2. Count of constructs

Construct	Number of primary themes	Number of secondary themes
Information Quality	2	2
Source Credibility	1	1
Needs for Information	4	2
Attitude towards Information	4	2
Perceived Usefulness	1	1
Information Adoption	6	2
Purchase Intention	4	2

Table 2 showed that the Information adoption construct has the highest count of themes. This supported the Erkans and Evans study where it was seen that information would more readily adopt due to viewers already deeming the source from which it originates as useful whereby "Since people usually receive the eWOM information from their friends and acquaintances in social media, they may already think that the information will be useful" (Erkan and Evans 2016a, b, p. 52).

Information quality and information usefulness were considered to have a positive impact on information usefulness (Erkan and Evans 2016a, b, p. 52). However, Unlike the (Erkan and Evans 2016a, b) study information quality and information usefulness do not have a direct impact on information usefulness rather it is seen that the needs for information has more overlapping themes. This supports the (Teo et al. 2018) study that individuals require information on the items they intend to purchase.

5 Conclusion

Initially, this study had aimed to describe how social media influences individuals to purchase intentions towards wearable devices as outlined by the (Erkan and Evans 2016a, b) literature. In the review of literature, it was seen that there has been an intent to acquire WD's (Chuah et al. 2016; Kaewkannate and Kim 2016; Wu et al. 2016; Hsiao and Chen 2018). However, there was little reference as to what role social media plays in doing so. From that gap in the literature, this study had applied the Information Acceptance Model and had deduced that social media influence is derived from its ability to produce large amounts of information.

Furthermore, from the responses, it was seen that individuals had other external influences as well as internal subjective norms for intending to buy a wearable device. These norms and influence were categorized within an array of resulting themes such as 'Value for Money' and 'Reasoning'. However, it was seen that of those influenced by social media, a significant amount has been deduced to intending to buy a wearable device due to being exposed to the electronic word of mouth found on those social media sites. Emerging themes such as 'Personal Preference' and 'Brand Loyalty' were seen to affect the type of exposure that individuals would see on social media sites whereas, on the other hand, other individuals had taken to spending more time on social media to find the best possible WD to buy.

By monitoring relevant social media platforms, it was seen that the interactions on social media had influenced the perceptions of users towards a WD and therefore their intention to buy it. These interactions were also seen to be more effective when related to social media groups which supported the (Erkan and Evans 2016a, b) literature in that information from friends on social media sites is useful and is easily adopted.

Therefore, the ICAM constructs of the relationship between social media information adoption and purchase intention, the relationship between WD features and social media usage as well as the relationship between the social media platform and the users' perception of that platform stood out as high contributors to social media influence.

This study's value was derived from providing an understanding of how vendors, organizations and individuals can market their WD products as well as how consumers who intend on buying the devices are influenced by social media. Moreover, this study served as the basis for further means of examining findings by applying various contexts and methodologies for social media influence.

Overall it was deduced from the objectives and prepositions that this study had managed to determine a link between Social media and WD buying intention though the over-exposure and advertising of such, this does not necessarily represent a definite sale.

References

Al-Fuqaha, A., Guizani, M., Mohammadi, M., Aledhari, M., Ayyash, M.: Internet of things: a survey on enabling technologies, protocols, and applications. IEEE Commun. Surv. Tutor. **17**(4), 2347–2376 (2015)

Balakrishnan, B.K., Dahnil, M.I., Yi, W.J.: The impact of social media marketing medium toward purchase intention and brand loyalty among generation Y. Procedia Soc. Behav. Sci. **148**, 177–185 (2014)

Bhattacherjee, A.: Social Science Research, 2nd edn. Anol Bhattacherjee, Tampa (2012). https://open.umn.edu/opentextbooks/BookDetail.aspx?bookId=79

Bolton, R.N., et al.: Understanding generation Y and their use of social media: a review and research agenda. J. Serv. Manag. **24**(3), 245–267 (2013)

Chen, H.: College-aged young consumers' perceptions of social media marketing: the story of Instagram. J. Curr. Issues Res. Advertising **39**(1), 22–36 (2018)

Cho, W., Lee, K.Y., Yang, S.: What makes you feel attached to smartwatches? The stimulus–organism–response (S–O–R) perspectives. Inf. Technol. People **32**(2), 319–343 (2019)

Chuah, S.H., Rauschnabel, P.A., Krey, N., Nguyen, B., Ramayah, T., Lade, S.: Wearable technologies: the role of usefulness and visibility in smartwatch adoption. Comput. Hum. Behav. **65**, 276–284 (2016)

Ellison, N.B., Steinfield, C., Lampe, C.: The benefits of Facebook "friends:" social capital and college students' use of online social network sites. J. Comput.-Mediat. Commun. **12**(4), 1143–1168 (2007)

Erkan, I., Evans, C.: The influence of eWOM in social media on consumers' purchase intentions: an extended approach to information adoption. Comput. Hum. Behav. **61**, 47–55 (2016a)

Erkan, I., Evans, C.: The influence of eWOM in social media on consumers' purchase intentions: an extended approach to information adoption. Comput. Hum. Behav. **61**, 47–55 (2016b)

Gimpel, H., Nüske, N., Rückel, T., Urbach, N., von Entreß-Fürsteneck, M.: Self-tracking and gamification: analyzing the interplay of motivations, usage and motivation fulfilment (2019)

Hsiao, K., Chen, C.: What drives smartwatch purchase intention? Perspectives from hardware, software, design, and value. Telemat. Inform. **35**(1), 103–113 (2018)

Jashari, F., Rrustemi, V.: The impact of social media on consumer behavior–case study Kosovo. J. Knowl. Manag. Econ. Inf. Technol. **7**(1), 1–21 (2017)

Johnson, K.M.: Literature review: an investigation into the usefulness of the smart watch interface for university students and the types of data they would require (2014)

Kaewkannate, K., Kim, S.: A comparison of wearable fitness devices. BMC Public Health **16**(1), 433 (2016)

Kim, S., Lee, S., Han, J.: StretchArms: promoting stretching exercise with a smartwatch. Int. J. Hum.-Comput. Interact. **34**(3), 218–225 (2018)

Lee, D., Hosanagar, K., Nair, H.S.: Advertising content and consumer engagement on social media: evidence from facebook. Manag. Sci. **64**(11), 5105–5131 (2018)

Lee, E.: Impacts of social media on consumer behavior: decision making process Turun ammattikorkeakoulu (2013). https://www.openaire.eu/search/publication?articleId=od_1319::728109eb8b479459b8eb5be8470655f4

Lister, C., West, J.H., Cannon, B., Sax, T., Brodegard, D.: Just a fad? Gamification in health and fitness apps. JMIR Serious Games **2**(2), e9 (2014)

Lunney, A., Cunningham, N.R., Eastin, M.S.: Wearable fitness technology: a structural investigation into acceptance and perceived fitness outcomes. Comput. Hum. Behav. **65**, 114–120 (2016)

Marketing Charts: Tech update: Mobile & social media usage, by generation (2018). https://www.marketingcharts.com/demographics-and-audiences-83363

Mekky, S.: Wearable computing and the hype of tracking personal activity. Paper Presented at the Royal Institute of Technology's Student Interaction Design Research Conference (SIDeR) Stockholm, Sweden (1–4) (2014)

Nadkarni, A., Hofmann, S.G.: Why do people use Facebook? Pers. Individ. Differ. **52**(3), 243–249 (2012)

Nascimento, B., Oliveira, T., Tam, C.: Wearable technology: what explains continuance intention in smartwatches? J. Retail. Consum. Serv. **43**, 157–169 (2018)

Page, T.: Barriers to the adoption of wearable technology. I-Manager's J. Inf. Technol. **4**(3), 1 (2015)

Pempek, T.A., Yermolayeva, Y.A., Calvert, S.L.: College students' social networking experiences on Facebook. J. Appl. Dev. Psychol. **30**(3), 227–238 (2009)

Ray, P.P.: A survey on internet of things architectures. J. King Saud Univ.-Comput. Inf. Sci. **30**(3), 291–319 (2018)

Rupp, M.A., Michaelis, J.R., McConnell, D.S., Smither, J.A.: The role of individual differences on perceptions of wearable fitness device trust, usability, and motivational impact. Appl. Ergon. **70**, 77–87 (2018)

Teo, L.X., Leng, H.K., Phua, Y.X.P.: Marketing on Instagram: social influence and image quality on perception of quality and purchase intention. Int. J. Sports Mark. Sponsorship (2018)

von Entress-Fürsteneck, M., Gimpel, H., Nüske, N., Rückel, T., Urbach, N.: Self-tracking and gamification: analyzing the interplay of motivations, usage and motivation fulfilment (2019)

Wairimu, J., Sun, J.: Is smartwatch really for me? An expectation-confirmation perspective (2018)

Wu, L., Wu, L., Chang, S.: Exploring consumers' intention to accept smartwatch. Comput. Hum. Behav. **64**, 383–392 (2016)

Applying Exploratory Testing and Ad-Hoc Usability Inspection to Improve the Ease of Use of a Mobile Power Consumption Registration App: An Experience Report

José Eduardo[1], Anderson Paiva[1], Victor Ferreira[1], Simara Rocha[1], Ítalo Santos[1], Luís Rivero[1,2(✉)], João Almeida[1,2], Geraldo Braz Junior[1,2], Anselmo Paiva[1,2], Aristofenes Silva[1,2], Hugo Nogueira[3], Eliana Monteiro[3], and Eduardo Fernandes[3]

[1] Núcleo de Computação Aplicada, Universidade Federal do Maranhão, São Luis, Brazil
{joseeduardo,andersonpaiva,rogeriovictor,simara,francyles, luisrivero,jdallyson,geraldo,paiva,ari}@nca.ufma.br

[2] Programa de Pós-Graduação em Ciência da Computação, Universidade Federal do Maranhão, São Luis, Brazil

[3] Equatorial Energia S/A, São Luis, Brazil
{hugo.nogueira,eliana.monteiro, eduardo.fernandes}@equatorialenergia.com.br

Abstract. As the number of mobile devices has increased, software development teams have focused on releasing mobile applications, allowing users to carry out transactions, access information and improve their lifestyle more efficiently. Nevertheless, even when providing useful means for carrying out daily tasks, users report dissatisfaction or frustration when using these applications. For energy companies, mobile applications that fail to provide both usefulness and ease of use may reduce their adoption and an increase in the company's workload, as users will require company workers to solve problems they could solve on their own. In this paper, we report how we applied exploratory testing and ad-hoc usability inspection to identify improvement opportunities during the development of a mobile application that would allow users to measure their power consumption, supporting social distancing in the context of the COVID-19 pandemic. After identified a set of functional and usability problems, the development team redesigned the application, which was perceived as both useful and easy to use from the point of view of the managers that requested it. Also, we report lessons learned that are useful for practitioners willing to replicate this experience.

Keywords: Exploratory testing · Usability inspection · Mobile application · Experience report

Supported by Equatorial Energia.

C. Stephanidis et al. (Eds.): HCII 2021, LNCS 13094, pp. 326–341, 2021.
https://doi.org/10.1007/978-3-030-90238-4_23

1 Introduction

Studies in the field of Energy Consumption show that, in 2021, almost 30% of all national energy consumption in Brazil was residential [2]. Also, as the number of households is increasing, power companies need to find new ways to manage a high number of clients [18]. This scenario causes frequent financial losses, as the lack of ability to monitor energy theft and processing energy reading errors is becoming more common.

Meter readers are workers that carry out the power consumption reading and inspection for power companies. To do so, they need to visit the consumers' homes. As this process is done manually, it became prone to errors, which in turn can increase the managerial problems of power companies, as customers may be charged less regarding their energy consumption; or customers being overly charged can file lawsuits against the company. Also, having to physically visit different locations, this activity can pose risks to the health of readers due to the exposure to changes in climate and hazards present at the visited locations. Finally, in the current scenario of the COVID-19 pandemic, direct contact with clients may also violate the social distance suggested by the world health organization [13].

Considering the above, companies in the energy sector have sought alternative means to read the energy consumption of clients with more effectiveness and efficiency. One possible solution to this problem may be self reading, as this approach could be more efficient, reduce readings costs and minimize health risks to workers. However, there is still a problem regarding the correctness of the reading. This problem motivated the Equatorial Brazilian power company that provides around 22% of the overall power in Brazil to fund the "AutoEnergyReading" project. Its goal was to develop and evaluate a mobile application embedding an Artificial Intelligence service that would allow recognizing the power consumption of a power meter. With such data, clients could perform their energy readings and generate reports regarding: (a) power consumption history; (b) power consumption comparison in different time periods; and (c) cost estimation of bills.

To improve the quality and acceptance of the application, we applied testing and usability evaluation approaches to identify improvement opportunities and redesign the user interface of the proposed mobile application. During the three-month duration of the testing stage of the AutoEnergyReading project, we applied an iterative approach through the Scrum methodology [12], each month being a sprint, to identify the systems' testing scenarios and to propose improvement opportunities. The testing team applied the following techniques: (a) Interviews, document analysis and scenarios design, to identify the power reading process and define in which contexts it could occur; (b) Exploratory Testing, to define testing scenarios for evaluating the functionalities embedded within the application; and (c) Usability ad-hoc inspection and validation, to identify improvement opportunities in the design of the application and discuss the proposals with the end users. In this paper, we describe how a software development team can apply functionality testing approached combined with usability evaluation methods to redesign an application. The lessons learned from this experience can be useful to novice software engineering teams willing to replicate the applied process.

The remainder of this paper is organized as follows. Section 2 presents concepts regarding software testing and usability evaluation, while presenting related work of how evaluation approaches have been applied in the improvement of mobile applications. Section 3 provides further information on the "AutoEnergyReading" project. Then, Sect. 4 shows the testing approaches employed for guaranteeing the quality of the proposed mobile application and its redesign. Moreover, Sect. 5 discusses the lessons learned from this experience report. Finally, Sect. 6 concludes the paper indicating future works.

2 Background

2.1 Exploratory Testing and Usability Assessment

Software testing is a crucial activity in software development, allowing to identify defects before releasing the software into the market, and reducing the costs of software maintenance [15]. Software testing involves the execution of software/system using manual or automated tools to evaluate one or more of its properties [5]. Among the properties of a software that can be evaluated, we can consider the correctness of the response of the system and the degree of usability that the application presents [16]. By improving these attributes, one can improve the usefulness and ease of use of an application, as users will be able to carry out their tasks with ease.

Several approaches have been proposed to evaluate quality attributes in software through software testing. Nevertheless, we highlight the use of exploratory testing and usability inspections. Exploratory testing is a commonly applied approach to test software without pre-designed test cases [1]. Thus, the tester has the freedom to explore the software and design new tests based on personal experience during test execution, which makes them dynamically designed, executed and modified. Applying an exploratory testing approach can be motivated by the fact that in an agile software development process, the documentation necessary for defining test cases may not be available [9]. Therefore, through experience, testers can explore the application under evaluation and identify defects. Although there is no need for prior application domain knowledge to carry out an exploratory testing, the tester needs to be knowledgeable in test techniques and able to use the accumulated knowledge about where to look for defects [1]. It is noteworthy that although exploratory testing may allow evaluating different quality attributes, the number of identified defects will depend on the degree of experience of the tester in such attribute, thus other evaluation approaches may be required to evaluate specific quality attributes such as usability [1].

Many usability evaluation methods (UEMs) have been proposed in order to improve the usability of different kinds of software systems [7]. UEMs are divided into two categories [6]: (a) user testing, in which empirical methods, observational methods and question techniques are applied when users perform tasks on the system; and (b) inspections, in which experienced inspectors review the usability aspects of the evaluated software artifacts. Due to the pandemic of the COVID-19 and the need for social distancing [13], as we could not carry out traditional software testing, we decided to focus on usability inspections. One of the main generic usability inspection methods is the Heuristic Evaluation [14]. This method, provides a set of usability guidelines which describe common properties of usable interfaces. By considering these guidelines, inspectors can verify usable properties and find usability problems when they are violated.

Combining a testing approach with a usability evaluation approach may be useful for improving the quality of an application under development, as not only functional issues may be found, but also usability problems affecting the ease of use of the application. However, as applying multiple testing approaches in different time periods may not be cost effective [1], software development team need to apply different processes to identify defects in less time. In the following subsection, we discuss how have researchers and practitioners combined these approaches to improve the effectiveness and efficiency of the evaluation process.

2.2 Related Work

Several researchers and practitioners have analyzed the advantages and disadvantages of applying different functional and usability evaluation approaches. We carried out an informal literature review trying to identify experience reports in which these evaluation approaches were applied. Below, we present our findings.

In their work, Granić and Ćukušić [6] report the results of the combination of usability evaluation approaches for assessing an e-learning platform. The evaluation methodology combined end-user assessments and expert inspections, thus providing a detailed students', teachers' and experts' feedback. The authors report how the evaluations were performed and lessons learned, such as why students participate or do not participate in the evaluations. Also, they evaluated the engagement of users with functionalities of the applications. Although usability attributes are evaluated, the authors do not focus on the proper system response regarding its functionalities.

In another work, Jewell and Franco [10] compare the advantages of different usability evaluation methods, such as Lab Testing, Pre-Session Assignments, and Online Usability Services. Using illustrations from recent studies of online shopping sites, the authors detail the advantages and limitations of each method and claim that employing them in combination could improve the quantity and quality of findings. Although the authors highlight the motivation behind the combination of different methods, they do not focus on functional testing, which could also support the identification of defects in an application under evaluation.

With regards to the evaluation of functionality, Zaharias and Poylymenakou [19] developed a usability evaluation method that extended a functional evaluation including usability measures for e-learning, focusing on a learner experience. The method evaluated positive users' affective engagement. However, the authors do not mention how functionality was assessed. Also, as the authors proposed a questionnaire that required the filling by users, it could not be applied without end users participating in the evaluation.

Although some authors focus on allowing the evaluation of both functionality and usability [3], they do not provide examples of how to carry out the evaluation in real software development scenarios. Also, we did not identify papers reporting such an evaluation with lessons learned in an agile development process for mobile applications. Thus, this experience report could be useful to software engineers willing to improve the quality of their applications in terms of functionality and usability. In the following sections, we present the development context of the application and its evaluation and redesign process.

3 Mobile Application for Self-reading of Power Consumption

3.1 Application Context

As mentioned before, in Brazil, to register how much energy a customer has consumed, it is necessary to physically travel to the location where the energy consumption meter is installed. As the power company needs to provide an employee to carry out the collection and some places are difficult to access, the registration process can be difficult or even dangerous. Considering such difficulties, an application was developed which allows the customer to record the energy consumption and send the data to the responsible company. The application allows to recognize the energy consumption of an energy meter. This data can then be sent to the company for validation and, based on these data, customers can generate reports for further analysis on their power consumption.

By placing the consumer as an integral part of the consumption reading process, self-reading promotes a closer relationship between the client and the company. Also, the company can reduce the costs related to the execution of physical readings, as well as to mitigate the occurrence of errors and fraud, mainly in areas of difficult access and inspection, like the countryside. Also, the self-reading scenario respects social distancing and other norms of public health recommended due to the pandemic of COVID-19. Considering that in Brazil alone, more than 60% of the population uses smartphones, which is one of the highest rates among emerging economies [17], the self-reading approach was developed as a mobile application.

Although there are some reports on the development process of mobile applications for power consumption reading [8], the authors of this paper did not identify a paper in such context where the evaluation process of both the functionalities of the application and its usability where described. Also, further information on how to correct identified problems is necessary to guide novice software engineers in the quality improvement of mobile applications, even if for other contexts different than the one described in this paper. In the following subsection, we present further details on how the application was developed.

3.2 Characterization of the Project

The development project was called "AutoEnergyReading" (or Auto Leitura de Consumo in Portuguese). This project had a duration of 24 months (from January 2019 until December 2020). The development team had 14 team members, including: project managers, software and database analysts, AI specialists, UI specialist, software developers and testers.

To develop the mobile application, the development team applied the Scrum agile methodology, as it is a popular agile development processes in the Brazilian development community [12]. In Scrum, the development team carries out software development activities through iteration and incrementation to manage rapidly-changing project requirements [11]. There are several roles in the Scrum methodology: (a) the project owner, responsible for being the voice of business inside a Scrum project; (b) the Scrum Team, formed by its developers, testers and other roles within the project; and (c) the Scrum Master, who is responsible to keep the team focused on the practices and values that are needed to be applied inside the project; and is also responsible to help the team whenever they face some problem during the development process. In our context, the product owner was a manager inside the Equatorial Energia power company. This manager would carry out meetings at least once a month to evaluate the progress of the project. Also, in these meetings, representatives from the Equatorial power company from different sectors (e.g. image, customer relationship, billing and others) would also attend, so that they could verify if the implementation of the system was not contradicting any company policies. Within the project, the Scrum master was a software engineer with more than 10 years of experience in project management; also there was an experienced (more than 5 years) usability and UX consultant with experience in system analysis, user interface design and usability evaluations. Finally, the Scrum team was composed of both experienced (more than 3 years of experience) and inexperienced (graduate students) software engineers.

The Scrum team implemented deliverables of the project at the end of each "Sprint", which is a period of time (in our case, 1 month) to create a usable increment of the product. During the analysis and testing stages of the development process, the team applied the following techniques interviews, exploratory testing and usability evaluation. The analysis and design activities of the project began on month 1 of the project and ended in month 6. The long duration of this stage was defined due to the need to define the requirements of the application for power consumption self-reading, while also defining the scenarios and interaction of the application. Also, we needed to analyze the Equatorial's database, as users needed to generate reports based on their power consumption history. Furthermore, during the testing activities of the project, we applied exploratory Testing, to define testing scenarios for evaluating the functionalities embedded within the application. Along with the functionality tests, we applied usability ad-hoc inspections and validations, to identify improvement opportunities in the design of the application and discuss interface design change proposals with the managers at the Equatorial company.

In the next section, we present an overview of the application, describing how it works and its evaluation. We will provide examples of employed artifacts, types of defects and how we made changes in the user interface to improve the quality of the application.

4 Evaluation and Redesign of the Mobile Application

After applying the interviews and document analysis, we identified the data needed to simulate usage scenarios of the application. At all, 5 main usage scenarios of the application were identified: (a) user and account management, (b) navigation and data access, (c) image-based, voice-based and text-based power reading, (d) report generation, and (e) client company communication. These scenarios were defined in order to allow clients from different user profiles to use the application for reading power consumption. For instance, an image-based power reading was proposed to allow low-literacy users to use the app. In the following subsections, we present the activities we carried out to evaluate these scenarios in terms of functionality and usability.

4.1 Combined Exploratory Testing and Usability Evaluation

As mentioned above, the development process applied Scrum principals due to the rapid changes in the requirements. Thus, there was a lack of update of the requirements of the system. Although some documentation was created in the initial sprints, several updates had been made by the time the mobile application was developed. As a results, functional testing could not be performed based on the available documentation. Also, automated testing was discarded as a testing option due to the difficulty in implementing the changes on time.

Considering the lack of documentation, the development team also decided to apply exploratory testing. Within this approach, the development team designed test cases considering the usage scenarios and the following techniques: (a) partitions, which is responsible for assessing the population of the data base, the number of registers and its partitions; (b) limit values, which tests the extreme limits of the input values; and (c) decision tables, in which cause-effect graphs are considered in the development of test cases. For each test case, we defined a test script to manually replicate each test whenever necessary. Each script had the following information: code, name of the scenario, summary, preconditions for executing the scenario, page for starting the interaction, set of steps in the scenario and expected results within these steps. Table 1 shows an example of a test case from the power reading scenarios category.

During the testing stage, which lasted 3 months, for each developed scenario, the test team reported wrong system behaviors to the Scrum master who would inform the development team of the problems and the expected results. After developing the functionalities, new test scenarios would be discussed and the old ones would be reviewed, updated (if necessary) and re-executed to verify the proper functioning of the mobile application.

Table 1. Test case scenario in which usability problems were identified even after successfully performing the scenario.

Test Case		
Code	TC10	
Name	Perform a power reading using the camera and processing the image obtaining a success message.	
Summary	After accessing the reading types, the user may want to carry out a reading using the camera. At this moment, the application must turn on the camera. The AI service will process the images captured by the camera, detecting the meter and searching for reading numbers of power consumption. After identifying a valid reading, the system must lock the screen to the obtained image and register the numbers, confirming the reading with the user.	
Precondition	The user must be logged in the system and permission to access the camera must have been provided.	
Page for Starting the Interaction	Main menu.	
Steps	**User Actions**	**Expected Results**
1	Tap on the "Reading" option.	
		System will present the menu for types of reading.
2	Select "Camera" option.	
		System will turn on the camera and present the image being registered. The AI service will process the images.
3	Focus camera on meter numbers.	
		System will recognize a set of numbers, lock the image corresponding to the identified numbers and present the numbers on the screen. A confirmation button will be activated to save the reading.
4	Select "Confirm" option.	
		System will send image to server and success message will be presented. Message: The reading has been stored.
Results	**Passed**	
Doubts and Suggestions	During confirmation, how can the user cancel the action or restart a reading? Further details should be provided in the success message. For example: What was the stored number? In which account? In which Date? Perhaps the success message should be: A reading number has been successfully stored. Contract Number: XXXXX Reading Date: XX/XXXX Last Reading: XXXXX Updated Reading: XXXXX Obs. See corrections based on this test case in Figure 1 (Navigation and Information Redesign)	

After all tests were performed and a complete version of the system was released, the usability and UX consultant was contacted to retest the entire system and also carry out an ad-hoc usability inspection. To do this, the consultant re-executed all test cases and during each test, he verified usability attributes. Since the consultant had more than 5 years of experience evaluating the usability of web and mobile applications, he did not need to check usability heuristics, thus characterizing the inspection as ad-hoc [4]. Consequently, during the evaluation process, if usability principals from mobile applications were not met during the interaction of the analyzed test case, a suggestion/doubt was included in the test report and it was discussed with the development team and the representatives from the Equatorial company. Table 1 shows an example of the comments made by the consultant at the end of the test case.

The functional testing through exploration allowed correcting issues with regards to wrong outputs of the system and missing information according to the requirements. Furthermore, with regards to the identified problems, 30 doubts and suggestions were indicated by the consultant. Figure 1 shows some of the screens in which usability problems have been identified. For instance, in part A (Home Screen) we identified problems regarding navigation and information. In this screen, for instance, there is no information the clients account and when (s)he needs to measure the power consumption. Also, there is a problem regarding the inability to change accounts. Although this is a necessary option, it is not visible in the main screen. Furthermore, the space was not properly employed used, as there is a blank space in the interface. Finally, regarding the fonts, they are small to be read easily.

With regards to Fig. 1 part B, we identified some problems with regards to the interaction or confusing signals. For instance, messages were provided in red, making the user think that an error had happened when, in fact, it was a message with instructions on how to use the system. Furthermore, there was no feedback on the functionalities. When pressing the record button, no information was provided by the interface, which made the functionality unclear.

With regards to interaction steps and messages, other problems were found in Fig. 1 part C. For instance, there is a lack of information on loading time. If a report has a lot of registers, the interface will only load the report, which will freeze the screen. Also, some information was not provided to the user, as units of measure of the power consumption in some reports. Also, there were no standards within the interface regarding the reports, presenting the reports with different types of widgets. After identifying the user interface problems, the design team was led by the consultant in the redesign process, which will be explained in the following subsection.

4.2 Redesigning the Application Based on the Identified Problems

The redesign process was performed by the usability and UX consultant, who made suggestions for most of the identified usability problems. First, each encountered usability problem was discussed with the Scrum master of the development team. If the scrum master agreed with the identified problems and their

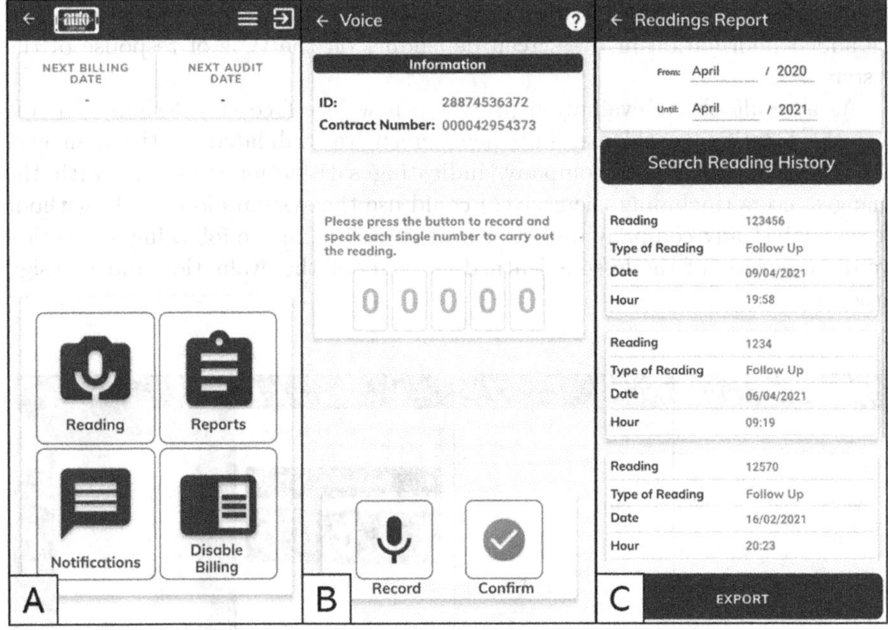

Fig. 1. Screens of the initial version of the mobile application that contain usability problems

degree of severity, the consultant would present improvement suggestions. These suggestions were discussed in meetings with the product owner and the different company representatives in order to guarantee the product changes do not go against the power company's policies. At all, around 4 meetings were performed, in which the changes in the application were discussed. If a suggestion was against the company's policy, the consultant would made changes based on the discussion of the meetings, until a new version of the refined proposal was developed. All changes were documented using a report and a high fidelity prototype.

Figures 2, 3 and 4 show examples of the new version of the mobile application. Each figure shows how we redesigned the application to correct the identified problems shown in Fig. 1. For instance, in Fig. 2 we redesigned the home screen, so users would have access to all functionalities without entering a separate menu. Additionally, in this screen the customer can view his account information and relevant information regarding measurement dates. Finally, regarding the interaction with the reading through camera option, we included further options in a minimized way, such as adding small buttons and messages through the interaction process. With regards to Fig. 3, we added visual information to represent the changes in the processing of a task. For instance, a new circle appears to indicate that the record has started. Additionally, message colors were changed and in between processing messages were added. The same happens when generating reports, as shown in Fig. 4. We made the reports screens similar, and we added information in the screen depending on the type of response of the system.

As a result of the evaluation process a new interface was designed to correct the identified problems. This new screen was validated by the managers of the Equatorial power company, indicating satisfaction and that with the changes, users (including themselves) could use the system more easily without going against any company design or process policy. In the following subsection we discuss some of the lessons learned throughout the evaluation and redesign process.

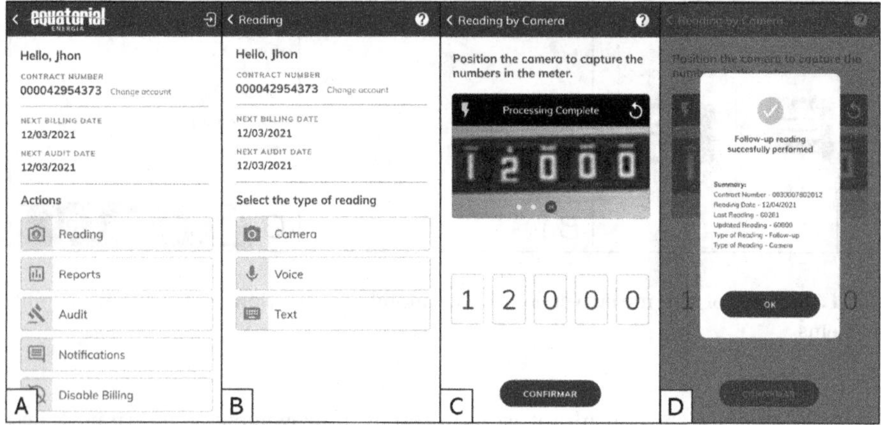

Fig. 2. Redesigned version of the application - menu and meter reading through camera

5 Discussion and Lessons Learned

Based on this experience of evaluating and redesigning the mobile application for self-power consumption reading, we were able to obtain lessons learned regarding the feasibility of combining different evaluation approaches. We will present these lessons learned reporting the decisions we made and how they were positive or negative for the project. These lessons are mainly based on the opinion of the design team as well as the managers of the project.

5.1 How the Applied Methods Contributed to the Identification of Problems

With regards to the exploratory testing, we were able to identify test cases without having access to updated documentation. The use of different testing approaches allowed us to identify problems regarding the system's response such as: (a) responses that did not match the expected processing of the system;

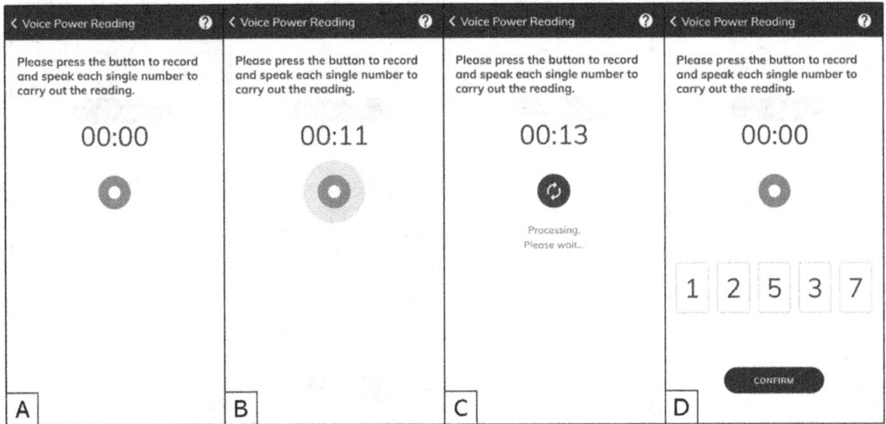

Fig. 3. Redesigned version of the application - interaction of the meter reading through voice

and (b) parts of the system that did not provide a response or froze the application. Also, the testing team was able to ask questions when the application presented results differently than how they were discussed during team meetings. This happened as there was little documentation and sometimes the developers had different interpretations of the behavior of the system. Thus, the approach allowed further validation of the application and a better understanding of the requirements from the point of view of the software team.

Through the use of the replication of the test cases along with the ad-hoc usability inspection, we were able to identify several problems as shown in Sect. 4.1. As the consultant had many years of experience on mobile applications evaluation, he knew what types of problems were more common and made several comments on the way the application worked. Also, we suggested several changes in the application to improve its quality. When combining the exploratory testing and an ad-hoc usability testing, an experienced inspector can reduce the time necessary to check for different types of defects. Also, by simulating different usage scenarios, the test team can verify to what extent these scenarios meet usability principals, mainly when making mistakes during the interaction, or changing between tasks and navigating through the application.

5.2 Difficulties in the Application of the Methods

Although the application of the methods was easy from the point of view of the testing team, some difficulties arose, which will be discussed as follows. The exploratory testing allowed identifying problems without the need of updated documentation. However, as the software evolved, some of the test cases also became obsolete. As a result, not updating these test cases usually involved consulting with the Scrum master to verify whether an identified functionality problem was really a problem. This turned the retesting process difficult for

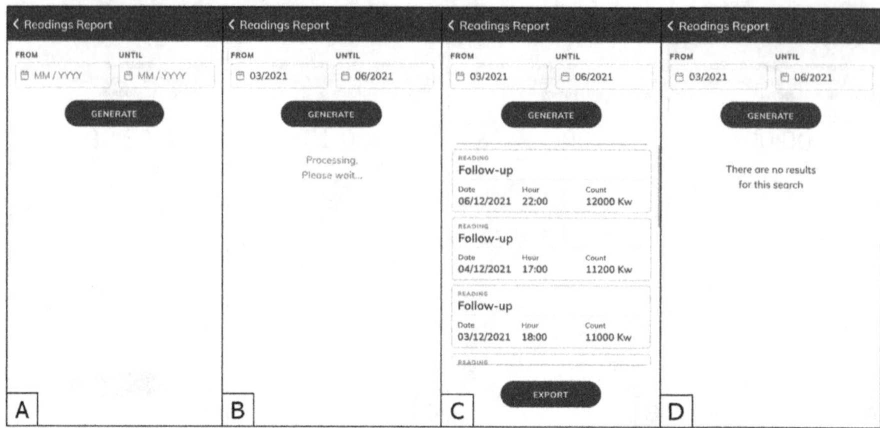

Fig. 4. Redesigned version of the application - system messages and reports

testers and the usability and UX consultant. Furthermore, part of the testing time was spent on discussing what the expected results of the application were, as well as updating the test cases.

With regards to the ad-hoc inspection, we were only able to carry out the evaluation due to the experience of the consultant. Not all development teams may have such an experienced team members. However, investing some time and providing an appropriate checklist with usability principals, an inexperienced inspector may be able to identify similar problems. Another problems was caused by the delayed of the identification and correction of the problems. When the consultant was asked to carry out the evaluation, the development team had already released a version of the application to the managers, which means that the development project was reaching its end. However, the suggestions of the consultant were necessary as, although the clients had approved a version of the application, it was still necessary to make changed to make it easier to use from the point of view of the managers at the Equatorial power company. Thus, although the changes were welcome, this changes implicated in further development time, and another round of testing and validation. In future projects, the execution of evaluation approaches for different software quality attributes should be executed together if possible, to avoid such degree of rework.

5.3 How the Identified Problems Contributed to the Identification of Improvement Opportunities

The combination of the evaluation techniques was necessary and had benefits on the overall testing process. First, by receiving a set of test cases, the consultant was guided through the exploration of the application, as the test cases forced him to check all functionalities within the application. Also, as a final version of the application was being tested by the consultant, he could focus on all problems with regards to usability issues. Furthermore, in case a functionality problem arose, the consultant could also point it out, managing to provide a single document with a report on both functionality problems and usability problems.

The identification of usability problems often came with a suggestion from the point of view of the consultant, mainly when noticing that relevant information was missing, or that there was no layout, image or font standardization. It also helped that the consultant was external to the development process, as when he executed test cases, he noticed usability problems. This happened because he was trying to understand the application following the flow of the test cases, which were organized. Additionally, being an external reviewer helped him question design decisions. Finally, as the consultant is also part of the target audience (he would also need to self-read his power consumption), the solutions were also thought from the point of view of possible end users.

The discussion of the identified problems with the Equatorial power company allowed for the refinement of the suggestions and prototypes. One of the main problems was that although some changes were accepted by the managers, they could not be approved considering that they could go against the company's policies. For instance, color had to maintain the same, as well as images and interaction processes, to maintain a standard with other mobile applications that the company provided its clients. This, however, is helping the company take notes on what changes could be necessary not only on the AutoEnergyReading app.

6 Conclusions and Future Work

In this paper, we reported the lessons learned from applying exploratory testing and ad-hoc usability inspection in one of the stages of a large development project. The "AutoEnergyReading" project was carried out by the Equatorial S/A Brazilian power company, which allowed the development and evaluation of a mobile application embedding an Artificial Intelligence service that would allow recognizing the power consumption of a power meter with audio or image. After identifying usage scenarios and a total of 99 test cases, several functionality and usability issues were corrected through the application of different evaluation techniques.

The results show that combining exploratory testing with an ad-hoc usability inspection, software development teams can be guided through the evaluation process of functionalities, while allowing the verification of usability attributes. Although in our case, the usability and UX consultant who carried out the ad-hoc inspection was very experienced, other inspection approaches could be applied, given that the evaluator will have a set of scenarios to test, in which he can check for other problems related to quality attributes such as usability. Finally, the final validation interviews with the managers of the Equatorial power company allowed us to further improve the proposed suggestions to meet the customers' expectations and company policies. As future work, we intend to continue the development of the application, however verifying to what extent the final version of the application meets the customers expectations. We intend to verify other attributes that could also affect the user experience of the users, specially those who live in the countryside, such as system response, or if the UI is appropriate for users that are digitally illiterate. Through the lessons learned within this paper and by providing details for its replication, we intend to encourage software companies to combine software testing and usability evaluation approaches for cost-effective identification of defects, especially in mobile applications with different end-user profiles.

Acknowledgments. This work was supported by AutoLeitura project funded by Equatorial Energy under the Brazilian Electricity Regulatory Agency (ANEEL) P&D Program Grant number APLPED00044_PROJETOPED_0036_S01. Additionally, this work was supported by the Foundation for the Support of Research and Scientific Development of Maranhão (FAPEMA), the Coordination for the Improvement of Higher Education Personnel (CAPES) and the National Council for Scientific and Technological Development (CNPq).

References

1. Afzal, W., Ghazi, A.N., Itkonen, J., Torkar, R., Andrews, A., Bhatti, K.: An experiment on the effectiveness and efficiency of exploratory testing. Empir. Softw. Eng. **20**(3), 844–878 (2015)
2. ANEEL: Consumption and distribution revenue reports - Brazilian national electric energy agency (ANEEL). ANEEL Oficial Website (2021). (in Portuguese)
3. Bertot, J.C., Snead, J.T., Jaeger, P.T., McClure, C.R.: Functionality, usability, and accessibility. Perform. Measure. Metrics **7**, 17–28 (2006)
4. Damian, A.L., Marques, A.B., Silva, W., Barbosa, S.D.J., Conte, T.: Checklist-based techniques with gamification and traditional approaches for inspection of interaction models. IET Softw. **14**(4), 358–368 (2020)
5. Garousi, V., Mäntylä, M.V.: A systematic literature review of literature reviews in software testing. Inf. Softw. Technol. **80**, 195–216 (2016)
6. Granić, A., Ćukušić, M.: Usability testing and expert inspections complemented by educational evaluation: a case study of an e-learning platform. J. Educ. Technol. Soc. **14**(2), 107–123 (2011)
7. Insfran, E., Fernandez, A.: A systematic review of usability evaluation in web development. In: Hartmann, S., Zhou, X., Kirchberg, M. (eds.) WISE 2008. LNCS, vol. 5176, pp. 81–91. Springer, Heidelberg (2008). https://doi.org/10.1007/978-3-540-85200-1_10
8. Islam, K.T., Islam, A.J., Pidim, S.R., Haque, A., Khan, M.T.H., Morsalin, S.: A smart metering system for wireless power measurement with mobile application. In: 2016 9th International Conference on Electrical and Computer Engineering (ICECE), pp. 131–134. IEEE (2016)
9. Itkonen, J., Mäntylä, M.V., Lassenius, C.: The role of the tester's knowledge in exploratory software testing. IEEE Trans. Softw. Eng. **39**(5), 707–724 (2012)
10. Jewell, C., Salvetti, F.: Towards a combined method of web usability testing: an assessment of the complementary advantages of lab testing, pre-session assignments, and online usability services. In: CHI 2012 Extended Abstracts on Human Factors in Computing Systems, pp. 1865–1870 (2012)

11. Lei, H., Ganjeizadeh, F., Jayachandran, P.K., Ozcan, P.: A statistical analysis of the effects of Scrum and Kanban on software development projects. Robot. Comput. Integr. Manuf. **43**, 59–67 (2017)
12. Melo, C.O., et al.: The evolution of agile software development in Brazil. J. Braz. Comput. Soc. **19**(4), 523–552 (2013)
13. Morawska, L., Cao, J.: Airborne transmission of SARS-CoV-2: the world should face the reality. Environ. Int. **139**, 105730 (2020)
14. Nielsen, J., Molich, R.: Heuristic evaluation of user interfaces. In: Proceedings of the SIGCHI Conference on Human Factors in Computing Systems, pp. 249–256 (1990)
15. dos Santos, E.B., da Costa, L.S., Aragão, B.S., de Sousa Santos, I., de Castro Andrade, R.M.: Extraction of test cases procedures from textual use cases to reduce test effort: test factory experience report. In: Proceedings of the XVIII Brazilian Symposium on Software Quality, pp. 266–275 (2019)
16. Santos, I., Melo, S.M., de Souza, P.S.L., Souza, S.R.: Towards a unified catalog of attributes to guide industry in software testing technique selection. In: 2020 IEEE International Conference on Software Testing, Verification and Validation Workshops (ICSTW), pp. 398–407. IEEE (2020)
17. Taylor, K., Silver, L.: Smartphone ownership is growing rapidly around the world, but not always equally. Pew Research Center 5 (2019)
18. Vidinich, R., Nery, G.A.L.: Pesquisa e desenvolvimento contra o furto de energia. Revista Pesquisa e Desenvolvimento da ANEEL-P&D 15 (2009). (in Portuguese)
19. Zaharias, P., Poylymenakou, A.: Developing a usability evaluation method for e-learning applications: beyond functional usability. Intl. J. Hum. Comput. Interact. **25**(1), 75–98 (2009)

Euros from the Heart: Exploring Digital Money Gifts in Intimate Relationships

Freya Probst(✉), Hyosun Kwon, and Cees de Bont

Loughborough University, Loughborough, UK
{f.x.probst,h.kwon,c.j.de-bont}@lboro.ac.uk

Abstract. Advances in financial technology open up new ways to exchange gifts in personal relationships across distance. Digitally transmitted monetary gifts are particularly relevant in our mobile centric lifestyle. As is widely known money has cultural restrictions in intimate exchange and is limited in its emotional appeal. Previous research in HCI studied currently practiced digital money gifts, their use, and technological developments. In this paper, we contribute by exploring plausible futures of such trends in intimate relationships with two speculative design scenarios, PregMoments and FriendBit. We conducted a series of focus group interviews to understand people's underlying values and concerns. The results underline a concern about the change in intimate relationships by the more efficient forms of exchange and its increased adoption. We found five key themes that need to be addressed to deepen intimacy: 'spent time and thought', 'get-together', 'intensified emotions', 'memory', and a 'distinction from the commercial'.

Keywords: Digital gifts · Digital money · Intimacy · Speculative design

1 Introduction

The exchange of digital money as a personal gift is a recent cultural phenomenon that developed a variety of forms with advances in financial technologies. Its convenience led to increased appliance also in closer personal relationships. Widely used examples include digital vouchers or monetary transfers in messaging and mobile payment applications [18, 19]. This shows growing interest and influence in this market and the change in the social cultural behavior of people. On the other hand, monetary gifts were found to be limited in their expression of care and connection. That elicits the question about the development of our personal relationships in the upcoming future.

In particular, we refer to money gifts that do not involve any physical transmission and that were adapted to be more personal by the means of new financial technologies. Such digital financial innovations were accelerated by the financial crisis in 2008, followed by more recent work in HCI emphasizing the social and cultural context of monetary exchange [e.g., 4, 12, 14, 23]. Research into existing digital gifting services and platforms led to a deeper understanding of emerging cultural practices and technologies [18, 21, 32, 33]. How money could be adapted for different social occasions was, for instance, examined by [18]. With the background of money as a gift, potential

C. Stephanidis et al. (Eds.): HCII 2021, LNCS 13094, pp. 342–356, 2021.
https://doi.org/10.1007/978-3-030-90238-4_24

cultural restrictions were explored. Money or also gift vouchers may feel impersonal in private relationships [19]. There are limits to communicate effort, surprise, indulgence, and similar emotional states [7, 9, 30]. There have been recommendations for platforms such as online wedding registries for a more creative arrangement to overcome the "taboo of asking for money" [21]. While these studies demonstrated the current practices and technological developments, we want to explore trends of digital money gifts and its potential influence on future gifting in our intimate relationships [26].

To understand how people judge current digital developments, we generated two speculative prototypes. They consist of two video scenarios: PregMoments and FriendBit. PregMoments is a smart pregnancy test device that can receive gifts at the time the recipient finds out about her pregnancy. It informs close contacts upfront through a linked application where family and friends can set up a gift in advance. FriendBit is a friendship bracelet that recommends a monetary gift amount based on the evaluation of previously recorded interactions. The prototypes were discussed with German participants. In this cultural context, money as a gift is unusual and more often exchanged in form of vouchers and restricted to specific occasions and relationships. We address the research question *'What are our underlying values and concerns in a future where digital money gifts could become more prevalent in intimate relationships?'* The findings show higher expectations in closer relationships. The participants preferred gifts that exhibited the invested time and thought, and that would be emotionally engaging. Social qualities in spending time together and the memory of the giver were appreciated. Participants also sought a separation from daily commercial transactions. We reflect on the relational consequences of the presented gift interactions. The insights have implications for informing a future design of digital gifts in financial and social applications to preserve the personal qualities of our relationships: e.g., emotional involvement and care.

2 Background

In the following we review anthropological and sociological literature on limitations of money as a gift and intimacy. We further provide an overview of research about developments of digital money and non-monetary files as a gift.

2.1 Intimacy and Money as Gift

To define intimacy, Prager differentiates between intimate relationships and intimate interactions. Intimate relationships can be broadly characterized by an "enduring affection", "mutual trust" and "partner cohesiveness" that derives from continuous intimate interactions and spending time together. Intimate interactions can involve that "partners share personal, private material", "feel positively about each other" and "perceive a mutual understanding" [26]. There are certain values and expectations connected to the idea of intimacy and of gifts that seem to be independent of material or monetary advantages. Bloch and Parry explained the "ideology of the gift has been constructed in antithesis to market exchange" [5]. From a societal perspective Simmel emphasizes money as an objective, rational evaluation that is not influenced by individual, personal feelings and that can lead to a greater independence of social bonds [28]. While money and market exchange have often been associated with 'instrumental rationality' and 'self-interest', altruism and care were expected to influence people's judgement on monetary questions in intimate relationships and interactions [35]. A connection of both worlds could result either in a 'corruption' of market relationships or "threaten social solidarity" [35]. Further work emphasized similarities and a connection of both realms, such as an underlying obligation that comes with the receipt of a gift or a good, that no gift is really free [24]. Zelizer described how in practice, people adapt money to manage intimate circumstances and to mark significant rites of passage, for instance through unique currencies or meaningful amounts [34]. Empirical research showed how the perception of money as a gift depends on the type and intimacy of the relationship. Webley et al. tested the appropriateness of money as a gift to mothers by students and found people responded with aversion to the idea [30]. On the other hand, in certain intimate relationships, it seems to be accepted to receive money as a gift such as from older to younger relatives [7].

Money has communicative difficulties to convey the significance and care required in social circumstances. Cheal observes that a gift or its wrapping are nowadays typically bought, which weakens their meaning and reveals a difficulty to convey effort [9]. Money can also likely be spent on daily purposes that will not allow the recipient any indulgence and there is a perceived lack of thought [30]. The previous work shows cultural and communicative limitations connected to physical monetary exchange. Yet, there have been some new practices that evolved with digital technologies. Digital money differs from physical money in its efficiency and the way it connects to other technologies. This leads to technology-related questions that we aim to address.

2.2 Digital Money Gifts

Recent work in HCI provided various insights on the phenomenon of digital money gifts. Research on more prevalent mobile payment and messaging services indicate the design encourages a use of digital money gifts transgressing traditional gifting. Wu and Ma analysed the exchange of digital red packets in group chats on WeChat (messaging application) [32]. The red envelopes often display characters adapted to emphasize the occasion. They offer a playful feature, where random amounts can be drawn to involve other members in activities: e.g., to raise the mood, to express an apology. It seems to cause less social pressure connected to the gifted amount. The mobile payment application Venmo combines transactions with emoji and text [8]. Its transferred amounts and purposes can be made public. This fun and spontaneous handling of money was found to reduce previous restrictions and ease the exchange in personal relationships. In some cases, the application was used for special occasions, such as birthdays. Light and Briggs described 'Patchwork', which is a wedding registry that helps the recipient to comfortably ask for money gifts. This platform can present the recipient's wish list in a more creative way with pictures and chosen funding amounts. It enables givers to fund wishes collaboratively as well as of larger scale. The amount is then transferred to the recipient's PayPal account [21]. The gift applications fuse money and social applications for a less formal and easy exchange.

Evolving practices link monetary donation and money to data-driven processes. For instance, a transaction to an environmental organization could become bound to the present temperature, or whenever a donor would engage in environmentally harmful behavior such as the purchase of a plastic bottle [13]. Curmi et al. suggest future research on whether the live streaming of a runner's heartrate data at a charity event could impact donations [10]. Online banking activities, such as savings or transfers have been personalized and automated by users. That is done by linking the application Monzo to other applications [12]. Those examples depict how digital monetary exchange increasingly faces the impact of data and artificial intelligence (AI).

Recent developments are digital money gifts in live streaming and the viewer-streamer relationship. In such a context gifts are often publicly displayed over the streaming video represented by text, special animations, and effects. They further engagement of viewers and enable an interplay with streamers or other viewers [29, 31]. From the viewer's perspective there were also some altruistic motivations of supporting the streamer out of gratitude or a personal liking [31]. While these relationships involve some social elements, they are maintained for commercial reasons.

The personalization of a digital gift experience has been of increased interest in HCI. The lack of invested effort was found as a central obstacle to the recognition of digital files as gift [19]. To address its limitations, Koleva et al. designed gifting applications that combined sensory, engaging qualities of a physical object with digital files chosen and modified by the giver [17]. Their applications were an enhanced experience of museum artworks, an augmented advent calendar, and a musically enhanced dinner or walk, suggesting the effort would become visible and delightful. For the improvement of the personal experience of the gift, the literature explored a connection of the physical and the digital.

Existing literature addressed current practices which have shown an increased connection to social applications, algorithmic and data-guided technology in a variety of contexts [e.g., 8, 12, 13, 18]. Up to now little is known about the potential influence of such trends on our intimate relationships, where the option to gift digitally and remotely appears to be of importance. Some anthropological, sociological perspectives suggest a limitation of money in intimacy [9, 28]. We present two plausible future designs to provoke people's opinions on digital money gifts in intimate relationships.

3 Study Design

Two speculative prototypes were developed in form of video scenarios. They were presented and discussed in multiple focus groups.

3.1 Critical Gifting Scenarios

Critical and speculative design involves a reflection on existing social and cultural norms and values [2, 11]. While the designer has a certain liberty in their intentions, the method requires an expression of motivations and their criticality [2, 25]. Speculative design aims to stir imagination and to reflect about ones' stance towards future technological developments on a personal and societal level. This works better with a subtle provocation than with ideas that adapt too well to current habits and expectations [1, 11]. The presented speculative prototypes challenge developments in technologically facilitated monetary transactions and their potential future impact on intimate relationships. Those prototypes are designed with respect to the described trends of linking money to social applications, data and algorithms [8, 13, 18]. The scenarios [27] were constructed to elicit empathy of participants. Both prototypes take place in intimate relationships and provoke a reflection about the conflict of digital money and intimacy. They are not intended to be implemented in normal social networks but to help show up problems and opportunities.

PregMoments –Smart Pregnancy Test

PregMoments (Fig. 1) is an application with a private blog page where a person shares significant moments in the pregnancy with invited personal contacts. A smart pregnancy test could be connected to share its results through automated blog posts. Money gifts in this scenario can be arranged in advance by friends and family, together with a recorded voice message to arrive at the point of the discovered pregnancy. This would surprise the gift recipient in the moment of a positive pregnancy result. With the derived credit, the recipient can choose to select a gift from an online shop in the application or keep the monetary amount. The pregnancy test serves later as a memento of the special moment that can play back the voice recordings. To share the incident of pregnancy as applied in this prototype is a sensitive topic. Not everyone may be willing to share this information right away with others. So, there should be additional options for people to have a delayed or secret receipt.

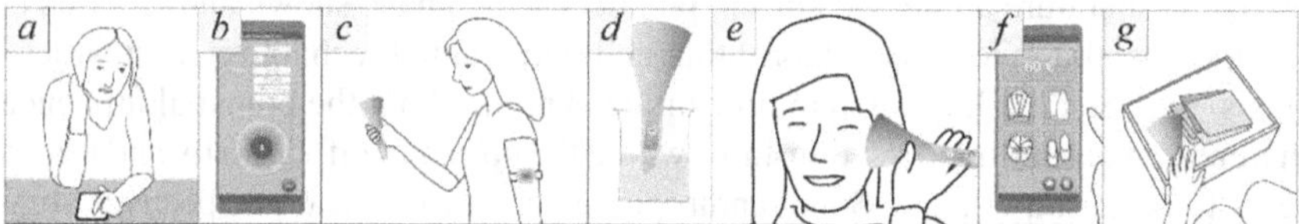

Fig. 1. a. Giver finds invitation to pregnancy blog, b. giver sets up gift and message, c. recipient uses smart devices that support pregnancy, d. positive pregnancy test posts update on blog, e. voice message and gifts are played back, f. gift amount and online store in pregnancy application, g. recordings stay as memory.

FriendBit – Smart Friendship Bracelet

FriendBit (Fig. 2) is a friendship bracelet that tracks friendship behavior. AI makes evaluations and recommendations to maintain the relationship. This aligns with the increased use of tracking technology in personal finances and their connection to algorithms [13]. Based on the records of exchanged gifts, messages, or the spent hours the algorithm determines the intensity and balance of the relationship. AI uses such data for behavioral hints, by reminding friends to engage at times. In the case a friend wants to send a digital monetary gift using the bracelet, it will recommend an amount based on their relationship and outstanding debts (e.g., forgotten congratulations, owed responses to messages). The fact that AI can evaluate friendship can be assumed in the context of today's AI developments. The question is whether society and individuals find it desirable.

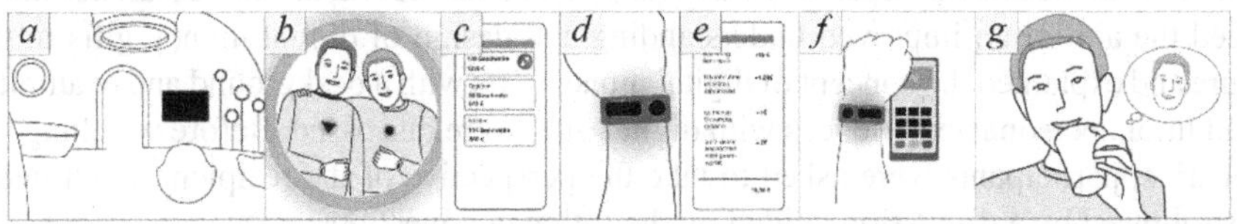

Fig. 2. a. Home setting in year 2030, b. friends use smart friendship bracelet, c. tracking of exchanged gifts, d. red glow reminds giver of negative balance in relationship, e. list of how the gift amount is calculated from parameters and past debts, f. recipient pays with gift amount, g. recipient thinks of giver over the bought gift.

3.2 Participants

We found 18 participants by convenience sampling, divided into 5 focus groups of 3–5 participants (see Table 1). They were selected with a common cultural background as Germans and their age (digital affinity). In this cultural background money as a gift is less common. Occasions such as weddings or communions or relationships such as from grandparents to grandchildren may be appropriate. Money needs a special adaptation in other circumstances, such as the more commonly gifted vouchers. In these cultural circumstances restrictions of money as a gift can be better understood.

Table 1. Focus group: gender and age.

Focus Groups [FG]	Female (n = 12)	Male (n = 6)	Mean age
FG 1	3	0	30.3
FG 2	3	1	25.5
FG 3	4	1	25
FG 4	0	3	26.3
FG 5	2	1	25

Our focus groups also consisted of participants aged 24–36 years, which are more familiar of digital communication and exchange technologies. Six of them reported to have experienced digital gifts including greeting cards, PC games over a game launcher, cinema or Amazon gift vouchers, a wine order and Apple music account. The compensation was a 10 Euro voucher.

3.3 Procedure

Participants were asked for their demographic information and consent before the focus groups took place. The focus group was conducted through a video conferencing platform. It took around one hour and was video and audio recorded. The researcher introduced the aim of an improved understanding and design of digital money gifts in the future and explained the concept of digital money gifts with a background and examples. In addition, personal experiences with digital gifts were discussed. Before watching the scenarios, participants were asked to take the perspective of the recipient. Each video was followed by a discussion guided by the following questions:

- What is the greatest advantage and disadvantage of the presented digital money gift (adapted from [20])?
- How could such a digital money gift affect or improve your relationship?
- In which alternative occasions or situations would such a digital money gift be desirable and why?
- What would you be missing or enjoying in comparison to a physical, handmade gift?
- How do you perceive money in the presented situation? What makes it special?

3.4 Analysis

A reflexive thematic analysis was applied, which is appropriate to develop patterns in qualitative data and to understand people's perceptions or values [6]. The analysis was focused on participant evaluation of the presented future. An inductive approach close to the collected data was chosen. Initially, the interviews were transcribed manually, which supported a familiarization with the data. The first author generated codes throughout the transcripts, for example "efficiency" or "authenticity: pre-recorded gift" were applied. These were grouped and mapped to develop themes. They were examined again to see whether they could reflect the data material from the interviews accurately, with the intention to show a holistic, rich account of what participants stated. The further discussion in the group of the authors led to a better compilation of the five presented themes.

4 Results

The focus group study led to the following results. Digital money gifts were appreciated for their usefulness, avoiding unneeded gifts and ease of receipt for the recipient. For the giver, it allows time-saving transmission especially across distance. The prototypes also encouraged participants to reconnect with old friends, on special events, by reminding them. However, monetary gifts were often deemed as impersonal. Particularly for closer relatives and friends, there were greater expectations that the gift would be more personal. We present values and concerns participants addressed in relation to the presented digital gift prototypes.

4.1 Spent Time and Thought

A key value of the gift is the devotion of the giver towards the recipient that has been addressed multifold by participants as the underlying *"intention"* (P7) or *"motivation"* (P10), expressed in the thought of how to please the recipient. This aspect however seems not well communicated by the properties of digital money: it is characterized by its ease, the lack of required thought, choice, and knowledge. It reminded more of work, like a *"to-do-list"* (P10) or *"duty"* (P5) to be executed rather than an act of care. The ease of the gift was associated with having fulfilled a task or obligation, because it felt like the giver was not very motivated to invest more time and thought (P1, P5, P15). This aligns with [3] and their description of 'sacred' money won from a labor that brings personal joy and derives from an inner motivation.

Another problem was whether it was the giver who initiates the gift. In the first scenario, the recipient invited friends and family to the blogging platform, where they are shown an opportunity to set up a gift, and in the second scenario the algorithm was making recommendations. This participant describes how this initiation by technology would clearly diminish the value of the gift from the perspective of the recipient: *"that is already worth a lot more from a friendship point of view, maybe also or even worth more for yourself, that you think by yourself and say 'I want to gift this now' and not that I'm already in a situation of pressure, because a device tells me 'oh you could order a bottle of wine now'"* (P10). The simplicity also elicited fears that it could lead to an increased frequency of gifts that are less caring (P1, P4): *"I would have the concerns that if it is then super extremely established... that you are just not so personally involved anymore. And that you just distance yourself from each other, because... suddenly everyone does it"* (P1).

To express this interest in the recipient, participants suggested to connect the gift to a purpose through supporting recipient wishes, or making propositions (P3, P5, P6, P14, P15, P18). P5 states: *"Maybe it would be better if I knew what the person is doing with it. So, if I knew 'ok the person needs a stroller' and I could say 'ok, I'm transferring so and so much money now', and with that I would know that I contribute a part to the stroller"*. If the giver has an affinity to what the gift intends to support, the emotion and underlying motivation will be stronger expressed. This becomes apparent in the following idea by P6: *"If you were to track... a specific aspect that is only dedicated to the friendship... like 'oh you have reached* 20 km, *today I also give you 20 Euros so you can buy a new T-shirt for jogging... like we know that we both like it and that we are interested in what the other is doing... If you then combine this with your gift, this emotional aspect that you share, then it is also such a sign of support"*. Beyond showing knowledge and reflection about individual characteristics of recipient and relationship, the giver can also incorporate creativity by a personal design or crafting, investing their special abilities (P17) or by wrapping (P9) [19]. This was seen as one of the advantages of physical gifts (P9, P10, P14): *"You can also wrap physical gifts of money creatively, so that somehow more is remembered by the gifted person than 'I just got money', whether that's a card or some other creative packaging, I think that's very difficult to implement digitally"* (P14).

4.2 Get-Together

This section derives from the missing face-to-face interaction in the digitally transmitted gifts, but also the personal conversations and interactions that usually come with events where gifts are exchanged. Out of its efficiency, it was feared to become an ordinary form of gift exchange left to less social contact and intimate exchange: *"if you don't get any presents or almost no presents anymore, where you're kind of out and about with someone going shopping and having an exchange, so to speak, then the social contact is also a bit missing... and then somehow you're more likely to have an exchange and you are more likely I'd say to talk about it more personally, and maybe also share more with each other"* (P3). Some participants could therefore imagine a money gift with a live component or in a live video call (P2, P7, P11): e.g., *"if you would wrap it with calling, and you see each other, then I think it's a good solution"* (P11).

One participant in our study cited her mother-in-law: she called to show her clothes that she could choose from. This was a fun joyful experience for both: *"[she] then called me from there with a video call with WhatsApp and showed me things together with the saleswoman so she had already made an effort and had fun with it... that we had so to speak a virtual shopping experience a bit together"* (P2). She missed this element where they would talk to one another in the shown scenarios.

Besides meeting live, three participants also suggested that a meeting proposition added to a monetary gift or an invitation to a joint activity for the future could be much more personal (P2, P7, P16). For example, P2 proposed an invitation to a conversation or a future shopping tour: *"here are 5 lb for now, something small, but then we meet again baby shopping"* or "with an invitation to talk". Some could not imagine to simply buy a voucher as a gift, but instead it should be personally connected, for instance through a meeting: *"that you kind of say that 'you like to go to the beer garden' or something, 'we haven't seen each other for a long time, and that's why I gift you that we spend an evening together'"* (P16). This participant disliked the idea to simply give a purchased voucher lacking the prospect of a joint meeting.

4.3 Intensified Emotion

An important part of the gift is its effect on the recipient, but also the giver's emotions play a role. Especially a significant occasion required a differentiation from more common and simple forms of congratulation: *"at the message 'pregnant' it is on a different level, and at least I think the reaction is on another level and not according to the motto on Facebook I am told it is the person's birthday, then I will write her briefly 'all the best'"* (P7). An automated transaction and congratulation message was perceived as too little to congratulate someone to a pregnancy. It reminded of a birthday on Facebook. Some participants also expected honest concern by the givers in case of their pregnancy, which was not possible with a pre-recorded message: *"the others are already thinking about what they are doing and are already recording messages when I am not yet pregnant, and so I find this whole feeling or this emotion, they are actually just pretending it to me"* (P5). Another participant describes how the strong feelings of excitement of awaiting presents under the Christmas tree, or receiving a long-craved thing could be lost: *"you... just look forward to the gift all day, so this feeling is then simply no longer there... with the app... I think it is also difficult for the recipient that he is so exuberantly happy about it. So, you are happy... but not exuberantly"* (P11). Voice messages were appreciated for their emotional appeal (P2, P16): *"I think that has a lot of personality... such a voice makes a big difference I think, so it can also convey emotions"* (P16). Another way to elicit stronger emotions was through creating anticipation, for instance with a gift voucher and a letter that would indicate to the recipient what it was (P16).

4.4 Memory

The memory played an important role in reinforcing the relationship over a longer period of time (P13, P18). For instance, the scenario of a smart pregnancy test showed how the recipient could later replay the received voice recordings: *"It just brings in a longer time dimension... so you can just remember who thought of you at that time... that would strengthen the relationship"* (P18). During the payment with the friendship bracelet, the recipient would be reminded of the friend who transmitted the gift to it: *"I found the point quite good that you can buy something with the bracelet and thereby be reminded of the person"* (P13). Participants appreciated the suggestion of products that could be purchased and later remind of the money gift: *"I could directly redeem the small amounts of money that my friends would have sent for something, and then have something I really need and where I would know at the same time 'ok that comes somehow from my friends'"* (P5). For others, the option for people to fund a gift collectively diminished the ability to later connect the monetary amount to an object and hence the memory of the individual: *"the fact that this is a gift from a lot of people at once. This may be the case with physical gifts, but it is probably not as strong then at least"* (P14).

There were different preferences how a gift should store and replay memories. While some preferred the more subconscious remembrance of physical gifts, others appreciated the fact that the memories in form of voice messages were digitalized, as they faced an overload in part of greeting cards. They were more likely to listen back if the memories were stored compactly (P16, P18): *"the possibility to remember... and to be able to just look at these voice messages again is certainly nice, rather than if somehow birthday cards lie around somewhere at first and then are still thrown away. Because just from the storage capacity this means something different than having to store it at home, and I think it's maybe quite nice when a lot is in one place and you kind of have a bit of time and want to reminisce"* (P18).

The ease and speed of exchanging money for a gift was seen to encourage a more purposeful spending. In case of the smart pregnancy device a user could retrieve the received amount in context of baby-related products. To exchange the received amount directly for a goods could still remind of the giver: *"if I only see it in my bank account afterwards... then I go to the supermarket afterwards and buy yogurt. That I will have some object afterwards that I can associate with it"* (P5). It led to a separation of monies with different purposes avoiding being missed or being dedicated to mundane daily spending [30].

4.5 Distinction from the Commercial

There is a need to keep a distance to commercial intentions from social values to avoid the feeling of intrusiveness. That became apparent in the PregMoments prototype and the temporal proximity of the intimate interaction with the possibility to select a gift in an online shop. P6 had the impression a commercial actor was interfering to take profit: *"with the data just this whole thing, I would just really be concerned that someone else wants to make money with it... that something bigger is interfering in an interaction... [which] is between the two of us and you gift each other something because you like each other and because it's important to you and because of the feeling 'ah she could really like that' and all of a sudden it's kind of someone wants to profit from it, and actually it has so little to do with it for me"* (P6). The answer illustrates the concern for the extensive collection of data in connection with commercial interests. This irritation could cause the avoidance of such social exchange platforms: *"Somehow you reveal so much data... that maybe I would shy away from it a bit"* (P11).

Another problem mentioned in relation to previously experienced digital money gifts was its mixing with more daily transactions. P18 recalled finding it difficult to distinguish receiving gift cards in her email account: *"So I once got an Amazon voucher just by e-mail, and somehow didn't find it cool at all, because I think I also get somehow quite a lot of automated things by e-mail"* (P18). This is supported by the statement of another participant that she preferred the *"idea that I do not find the gifted money in my account between my salary"* (P5). This seems to contradict with current applications, where money is often exchanged in context of chatrooms, daily payments, and emails [8, 18]. There was also a need to distinguish monetary from psychological or personal relationships. The FriendBit application showcased the gift exchange as a monetary-oriented exchange: *"I thought it was very bad that the relationship is balanced with money. That's so materialistic and a friendship relationship is more like 'I get along well with the person'"* (P16).

Our findings show how participants wished for care, positive intent, intrinsic motivation, and emotional involvement by the giver. Moreover, they sought a more personal and emotional experience of the gift. There was also a need for more social exchange in form of a meeting. There were still limitations in the perception of invested time and thought due to the simplicity of the gift as automation having undermined the initiation and creativity by the giver. Such gifts were also at risk to have a commercial impression, due to collected data or the option to acquire a gift.

5 Discussion

In this study we presented two speculative prototypes of plausible future digital monetary gifts. The prototypes were not meant as solutions but to provoke reflection about people's values concerning a trend of data-driven monetary exchange processes and digital gifting in intimate relationships. A common fear concerning future personal relationships was a growing influence and prevalence of such gifts. This impression was driven by the digital nature of applications due to their convenience and efficiency in usage. Digital money was seen to aid a more careless and disinterested form of gifting, and thus also a more careless interaction and engagement with one another. It was assumed this could lead to an increased frequency of gifts, which would in consequence undermine its meaning and impression as a gesture. This particular finding contrasts with current trends, where money merges with daily applications like any other form of data [12]. Money is displayed in a playful informal way in platforms that facilitate an easier exchange, partially working with game mechanisms such as random drawings of money [8, 18]. Yet, multiple of our participants expressed discomfort at the idea of such a frequent exchange of money and of gifts, seeming culturally inappropriate, as money has a certain meaning dedicated to occasions, making it less significant as a gesture. It would also not allow for a more intense emotional experience. Design would need to support a greater involvement of the giver and provide a more memorable experience that can be distinguished from everyday exchanges. Our research aligns herein with previous research, where effort was emphasized as a quality to enhance. This may be expressed in the crafting and creativity of the giver [16, 17], by spending time together, or by addressing their individual characteristics and expressing the purpose of the gift to delight the recipient.

Also, the participants feared the loss of need to meet a person. The gift seemed to encourage a delayed exchange and receipt where the giver might not be presently aware of the recipient in the moment. Thus, we pronounce the need for a more social experience. Little literature in digital gifting has previously addressed live interaction across distance. Literature on money gift exchange in live streaming described possible ways to connect a money gift with live video. Viewers could send money that was animated with special effects on the screen [22]. We thus suggest the incorporation of a live meeting at some point of the gifting process. This suggestion is more related to intimate relationships [15]. It helps to personally thank a giver, to have a conversation or to raise the significance of the special event. Such a live meeting may be postponed to a later date or held before and is not bound to the time of transfer.

To conclude this section, there is a need for design to support relational values in maintaining a caring way of interaction in our personal relationships. Therefore, we emphasize to include some form of invested time and thought by the giver. Additionally, we found the need for a more personal interaction, to spend time together and recommend a direct meeting as part of the gift.

6 Limitations

Aim of this research was to understand how people would evaluate a future of digital money gifts based on current technological trends. Our focus was only on a small sample and is not generalizable to the German or Western cultural perspective. In addition, we focused on a younger group of participants that could relate better to current technological developments, and responses by an older generation might differ severely in their perception and needs. The designs were not explored in real relationships that might have elicited values in more detail, but in a laboratory context. Future work may explore the application of such values in gifting prototypes to see how they function in a real user setting.

7 Conclusion

An increasingly convenient way to exchange gifts digitally is taking place in personal relationships. Yet, money has been described as limited embodying and expressing care and intrinsic motivation. Recent developments in the realm of digital money gifts have shown up digital trends of connection to data and algorithms, that raise new questions how digital money could affect our personal bonds. To address this, we explored with speculative design how such gifts were connected to people's important concerns and values. With the presented values we hope to serve as orientation for future design.

Acknowledgement. The research was funded by Loughborough University. We thank the participants for sharing their perspectives with us and our colleague and supervisor Martin Maguire for his helpful support and valuable comments. We would also like to express our thanks to the anonymous reviewers, who helped us in the improvement of the paper.

References

1. Auger, J.: Speculative design: crafting the speculation. Digit. Creat. **24**(1), 11–35 (2013). https://doi.org/10.1080/14626268.2013.767276
2. Bardzell, J., Bardzell, S.: What is "critical" about critical design? In: Proceedings of the SIGCHI Conference on Human Factors in Computing Systems, pp. 3297–3306. ACM, New York (2013). https://doi.org/10.1145/2470654.2466451
3. Belk, R.W., Wallendorf, M.: The sacred meanings of money. J. Econ. Psychol. **11**(1), 35–67 (1990). https://doi.org/10.1016/0167-4870(90)90046-C
4. Bellotti, V.M.E., et al.: Towards community-centered support for peer-to-peer service exchange: rethinking the timebanking metaphor. In: Proceedings of the SIGCHI Conference on Human Factors in Computing Systems, pp. 2975–2984. ACM, New York (2014). https://doi.org/10.1145/2556288.2557061
5. Bloch, M., Parry, J.: Introduction: money and the morality of exchange. In: Parry, J., Bloch, M. (eds.) Money and the Morality of Exchange, pp. 1–32. Cambridge University Press (1989)
6. Braun, V., Clarke, V.: Reflecting on reflexive thematic analysis. Qual. Res. Sport Exerc. Health **11**(4), 589–597 (2019). https://doi.org/10.1080/2159676X.2019.1628806
7. Burgoyne, C.B., Routh, D.A.: Constraints on the use of money as a gift at Christmas: the role of status and intimacy∗. J. Econ. Psychol. **12**(1), 47–69 (1991). https://doi.org/10.1016/0167-4870(91)90043-S
8. Caraway, M., Epstein, D.A., Munson, S.A.: Friends don't need receipts: the curious case of social awareness streams in the mobile payment app Venmo. Proc. ACM Hum.-Comput. Interact. **1**, CSCW, Article 28, 17 (2017). https://doi.org/10.1145/3134663
9. Cheal, D.: 'Showing them you love them': gift giving and the dialectic of intimacy. Sociol. Rev. **35**(1), 150–169 (1987). https://doi.org/10.1111/j.1467-954X.1987.tb00007.x
10. Curmi, F., Ferrario, M.A., Southern, J., Whittle, J.: HeartLink: open broadcast of live biometric data to social networks. In: CHI 2013 Extended Abstracts on Human Factors in Computing Systems (CHI EA 2013), pp. 2793–2794. ACM, New York (2013). https://doi.org/10.1145/2468356.2479515
11. Dunne, A., Raby, F.: Speculative Everything: Design, Fiction and Social Dreaming. Massachusetts Institute of Technology (2013)

12. Elsden, C., Feltwell, T., Lawson, S., Vines, J.: Recipes for programmable money. In: Proceedings of the 2019 CHI Conference on Human Factors in Computing Systems, Paper 251, 13 p. ACM, New York (2019). https://doi.org/10.1145/3290605.3300481
13. Elsden, C., Trotter, L., Harding, M., Davies, N., Speed, C., Vines, J.: Programmable donations: exploring escrow-based conditional giving. In: Proceedings of the 2019 CHI Conference on Human Factors in Computing Systems, Paper 379, 13 p. ACM, New York (2019). https://doi.org/10.1145/3290605.3300609
14. Ferreira, J., Perry, M., Subramanian, S.: Spending time with money: from shared values to social connectivity. In: Proceedings of the 18th ACM Conference on Computer Supported Cooperative Work & Social Computing (CSCW 2015), pp. 1222–1234. ACM, New York (2015). https://doi.org/10.1145/2675133.2675230
15. Hassenzahl, M., Heidecker, S., Eckoldt, K., Diefenbach, S., Hillmann, U.: All you need is love: current strategies of mediating intimate relationships through technology. ACM Trans. Comput.-Hum. Interact. **19**(4), Article 30, 19 (2012). https://doi.org/10.1145/2395131.2395137
16. Kelly, R., Gooch, D., Patil, B., Watts, L.: Demanding by design: supporting effortful communication practices in close personal relationships. In: Proceedings of the 2017 ACM Conference on Computer Supported Cooperative Work and Social Computing, pp. 70–83. ACM, New York (2017). https://doi.org/10.1145/2998181.2998184
17. Koleva, B., et al.: Designing hybrid gifts. ACM Trans. Comput.-Hum. Interact. **27**(5), Article 37, 33 (2020). https://doi.org/10.1145/3398193
18. Kow, Y.M., Gui, X., Cheng, W.: Special digital monies: the design of Alipay and WeChat wallet for mobile payment practices in China. In: Bernhaupt, R., Dalvi, G., Joshi, A., K. Balkrishan, D., O'Neill, J., Winckler, M. (eds.) INTERACT 2017. LNCS, vol. 10516, pp. 136–155. Springer, Cham (2017). https://doi.org/10.1007/978-3-319-68059-0_9
19. Kwon, H., Koleva, B., Schnädelbach, H., Benford, S.: It's not yet a gift. In: Proceedings of the 2017 ACM Conference on Computer Supported Cooperative Work and Social Computing (CSCW 2017), pp. 2372–2384. ACM, New York (2017). https://doi.org/10.1145/2998181.2998225
20. Lawson, S., Kirman, B., Linehan, C., Feltwell, T., Hopkins, L.: Problematising upstream technology through speculative design: the case of quantified cats and dogs. In: Proceedings of the 33rd Annual ACM Conference on Human Factors in Computing Systems, pp. 2663–2672. ACM, New York (2015). https://doi.org/10.1145/2702123.2702260
21. Light, A., Briggs, J.: Crowdfunding platforms and the design of paying publics. In: Proceedings of the 2017 CHI Conference on Human Factors in Computing Systems, pp. 797–809. ACM, New York (2017). https://doi.org/10.1145/3025453.3025979
22. Lu, Z., Xia, H., Heo, S., Wigdor, D.: You watch, you give, and you engage: a study of live streaming practices in China. In: Proceedings of the 2018 CHI Conference on Human Factors in Computing Systems, Paper 466, 13 p. ACM, New York (2018). https://doi.org/10.1145/3173574.3174040
23. Mainwaring, S., March, W., Maurer, B.: From Meiwaku to Tokushita! lessons for digital money design from Japan. In: Proceedings of the SIGCHI Conference on Human Factors in Computing Systems (CHI 2008), pp. 21–24. ACM, New York (2008). https://doi.org/10.1145/1357054.1357058
24. Mauss, M.: The Gift: The Form and Reason for Exchange in Archaic Societies (Halls, W.D., Trans. 1st edn.). Routledge (1990). https://doi.org/10.4324/9780203407448
25. Pierce, J., Sengers, P., Hirsch, T., Jenkins, T., Gaver, W., DiSalvo, C.: Expanding and refining design and criticality in HCI. In: Proceedings of the 33rd Annual ACM Conference on Human Factors in Computing Systems, pp. 2083–2092. ACM, New York (2015). https://doi.org/10.1145/2702123.2702438

26. Prager, K.J.: The Psychology of Intimacy. The Guilford Press, New York (1995)
27. Rosson, M.B., Carroll, J.M.: Scenario-based design. In: Sears, A., Jacko, J.A. (eds.) Human-Computer Interaction: Development Process, 1st edn., pp. 145–164. CRC Press (2009)
28. Simmel, G.: The Philosophy of Money. Routledge, London (2004)
29. Wang, D., Lee, Y., Fu, W.: "I love the feeling of being on stage, but I become greedy": exploring the impact of monetary incentives on live streamers' social interactions and streaming content. Proc. ACM Hum.-Comput. Interact. **3**, CSCW, Article 92, 24 (2019). https://doi.org/10.1145/3359194
30. Webley, P., Lea, S.E.G., Portalska, R.Z.: The unacceptability of money as a gift. J. Econ. Psychol. **4**(3), 223–238 (1983). https://doi.org/10.1016/0167-4870(83)90028-4
31. Wohn, D.Y., Freeman, G., McLaughlin, C.: Explaining viewers' emotional, instrumental, and financial support provision for live streamers. In: Proceedings of the 2018 CHI Conference on Human Factors in Computing Systems, Paper 474, 13 p. ACM, New York (2018). https://doi.org/10.1145/3173574.3174048
32. Wu, Z., Ma, X.: Money as a social currency to manage group dynamics: red packet gifting in Chinese online communities. In: Proceedings of the 2017 CHI Conference Extended Abstracts on Human Factors in Computing Systems (CHI EA 2017), pp. 2240–47. ACM, New York (2017). https://doi.org/10.1145/3027063.3053153
33. Yang, J., Ackerman, M.S., Adamic, L.A.: Virtual gifts and Guanxi: supporting social exchange in a Chinese online community. In: Proceedings of the ACM 2011 Conference on Computer Supported Cooperative Work (CSCW 2011), pp. 45–54. ACM, New York (2011). https://doi.org/10.1145/1958824.1958832
34. Zelizer, V.A.: The Social Meaning of Money: Pin Money, Paychecks, Poor Relief, and Other Currencies. Princeton University Press (1997)
35. Zelizer, V.A.: The Purchase of Intimacy. Princeton University Press (2009)

Impact of Social Media Marketing on University Students - *Peru*

Julissa Elizabeth Reyna González[1](✉), Víctor Ricardo Flores-Rivas[2], and Irene Merino Flores[3]

[1] Hermilio Valdizan National University, Huanuco, Peru
[2] Continental University, Lima, Peru
[3] Cesar Vallejo University, Piura, Peru

Abstract. The world of social networks is as complex, as the world is today, Social Media Marketing (MMS), represents an opportunity for companies that want to strengthen the emotional bond with their customers and boost their sales. The general objective of the research was to Analyze the Impact of Social Media Marketing on university students, among the methods that support the research we have the theoretical method, related to the dependent variable MMS, whose authors are Jadhav, Kamble & Patil, the research addresses a mixed approach (quantitative-qualitative), the study subject population consisted of 253 students and the sample of 110; In addition, empirical methods such as the proposed instruments, the questionnaire for the quantitative research process, the focus group guide for the qualitative research process and the statistical method of the data obtained from the questionnaire processed with the SPSS 27 software, allowed to know the impact of social media marketing on university students, reporting a deficient level of MMS for which the MMS should be reinforced through messages, communication in real time where students stay informed and dispel their doubts in time.

Keywords: Internet · Marketing · Social media

1 Introduction

Social media (MS) have become the daily work in people's lives, as well as very useful tools for university students, companies in the education sector through which knowledge, values are imparted, based on various strategies teaching. These tools are of great importance to facilitate and improve communication between the university community, educators, students and parents.

Sharma et al. (2021) in their study argues that the "customer" - brand relationship has a positive impact on the purchase intention of "customers" through MS. It implies that when "customers" develop trust in fashion brands, greater satisfaction and engagement with specific brands is observed, indicating a strong purchase intention among customers. In keeping with the practical implications, marketing managers should focus on building trust, satisfaction, and engagement through various brand-related activities so that customers' purchase intent can be induced (p. 614).

C. Stephanidis et al. (Eds.): HCII 2021, LNCS 13094, pp. 357–366, 2021.
https://doi.org/10.1007/978-3-030-90238-4_25

MMS creates value beyond financial value. Non-measurable data, such as customer emotions and opinions across channels, is considered important, especially for understanding customer engagement (Kumpu et al. 2021).

In the study by Kübler et al. (2020) argues to collect data from social networks, they developed a tracker to extract all public information on the official Facebook page of each brand. Although future research may compare SET on other platforms, Facebook is a good option to keep the focus of the analysis on the same platform. As the largest social media platform, Facebook provides a dynamic environment for brand-consumer interactions. Social Media has become an important data source for organizations to monitor how they are perceived. Obtaining useful and consistent information for this data requires careful selection of the right opinion extraction tool.

In the educational sector, both public and private, the acceptance and application of tools in social networks is minimal, it must be taken into account that the knowledge, use and propagation of said tools must involve each of the members of the organization.

"Consumers will always have more confidence in their friends than in a television advertisement. For this reason, the messages sent on social media arrive with more certainty and greater forcefulness" (Betech 2008, p. 115).

According to Ramírez (2015), "Social Media has caused a revolution in education, since it is a reality that educational institutions of both regular basic education and higher education are adapting", generating more advantages than annoyances, so it is you should take advantage of it positively.

In the research carried out by Curiel et al. (2010), the need for a re-education among students was published in the V International Research Forum, that is, that they understand and understand that the use of the university platform "is for informative purposes specific to their iteres as students", therefore certain rules must be respected and teachers, managers and students involved on Facebook to integrate the content of the website to social networks, having a greater participation in society and more involvement in the website. The company must know how to take advantage of its presence on the web and the comments of its followers. Emphasize the positives to create a strong positioning; as well as solving complaints and responding to suggestions, so that the image of the company does not decline. Meanwhile, now the MMS has revolutionized this massive space to become personalized and it is about giving value and useful answers to the user in order to establish a long-term connection with it. Now you have to exchange value instead of just sending messages. Kübler (2020) argues in his study that marketers can use various tools to extract sentiment from textual data. Because some of these tools are not known to a wide audience of marketing researchers, the range of tools available is discussed first before focusing on the specific tools used in different research schools. The University Reality and the development of information through social media linked to education, allow us to recognize its potential and value. It is not enough to create an official account on Twitter or be present on Facebook. The institutional profiles in each social network must be managed by people who know how to do it, it is as important to have a good website as it is to offer meeting, exchange and personalized contact points, being the aforementioned one of the main objectives of university education, and it is pertinent to know.

To achieve the objectives of the study, two aspects were chosen: a quantitative approach and a qualitative approach, that is, a mixed model, where throughout the entire investigation the two approaches were combined, a quantitative technique was used that consisted of a survey for the which was used a representative sample of 110 students, being the instrument used for this purpose, the questionnaire composed of closed questions with a Likert-type ordinal scale, which allowed determining the level of MMS in university students, in addition to a parallel qualitative technique to analyze data, that is, the focus group with 18 volunteer students, both men and women, who were enrolled in the University and belonged to the professional school under study, also participated in a driver of the questions posed in the guide focus group and an observer standing behind the Gesell camera behind of a polarized glass, all the responses of the focus group were recorded, which were noted by the observer and recorded through the video camera located in the Gesell camera, the questions asked to the focus group mainly explored issues related to how you feel at the university, social media you use, time of connection to a social network, behaviors, conduct among others, as well as language experiences in the study subjects. Afterwards, the statistical analysis was carried out with the survey data where an interpretive analysis of the results was applied. Quantitative research offers the possibility of generalizing the results more widely, gives control over the phenomena, as well as a point of view based on counts and magnitudes. (Hernández et al. 2014, p. 15), for its part, qualitative research provides depth to the data, dispersion, interpretative richness, contextualization of the environment, details and unique experiences. It also provides a "fresh, natural and holistic" point of view of phenomena. (Hernández et al. 2014, p. 16), Therefore, the main research question is to know what is the contextualized impact of Social Media Marketing on university students.

2 Theoretical Framework

The epistemological foundation of the research is based on the Gestalt Theory and neurosciences; they are theoretical foundations that contributed to the support of the present research.

As mentioned by Jadhav et al. (2012) Social media marketing (MMS), "are the marketing strategies that smart businesses are using in order to be part of a network of people online." Just as friends gather in public pubs, coffee shops, or barbershops, groups of people meet and connect through various online tools and websites. These people rely on their online network of friends for advice, sharing, and socializing.

Social Media

Jadhav et al. (2012) Social media technologies take many different forms including magazines, Internet forums, weblogs, social blog, microblogging, wikis, social networks, podcasts, photos or images, video, ranking, and social bookmarking.

Social Media is: "a set of supports that allow linking information through the Internet and that make up social networks and virtual communities." Social Media allows us to be active or passive actors in the generation of information on the Internet (Jadhav et al. 2012).

Types of Social Media

Six different types of social media are defined: collaborative projects (Wikipedia), blogs and microblogs (e.g., Twitter), content communities (e.g., YouTube), social networking sites (e.g., Facebook), virtual game worlds (for example, World of Warcraft) and virtual social worlds (for example, Second Life). Technologies include: blogging, picture sharing, wall posting, email, instant messaging, music sharing, crowdsourcing, and voice over IP, to name a few. Many of these social media services can be integrated through social media aggregation platforms. Social media networking websites include sites like Facebook, Twitter, Bebo, and MySpace.

Dimensions of Social Media Marketing Visibility

Barker et al. (2015) point out that visibility is necessary to be able to maintain a good relationship with users of the digital community, since this influences brand recognition, and for this it is important to increase the number of publications and articles that are disseminated on social platforms.

Interaction

The interaction is the communication that exists between the brand and the user, however, this communication must always be transparent and true, since users highly value dialogue and truthful information, and their instead they punish brands that do not base their strategy on these two conditions.

Loyalty

Loyalty consists of ensuring that the consumer who has already purchased our product or service, becomes a regular or frequent customer, loyal to the brand, generating repetitions in the purchase. On the other hand, it is shown that it is more effective to retain our clients, compared to attracting new ones, since reaching other consumers is much more expensive and is a slower process.

2.1 General Objecives

Analyze the impact of Social Media Marketing on university students.

Specific objectives

- Determine the level of Social Media Marketing in university students.
- Determine the perception of Social Media Marketing in university students.
- Identify the main advantages and disadvantages of Social Media in university students.

3 Methodology

To achieve the objectives of the study, a mixed model approach was used, mixed methods represent a set of systematic, empirical and critical research processes and involve the collection and analysis of quantitative and qualitative data (Hernández et al. 2014), that is, in the study Theoretical level methods were used where the theories that support the present study were addressed, an empirical method supported by the use of quantitative techniques (an instrument that measured MMS and the survey technique), applied to a significant sample of 110 students of the faculty In the study, a simple random probability sampling was used. The study subjects surveyed were both male and female students. The qualitative technique used was the focus group where 18 student volunteers participated, who were asked a set of questions noted in the focus group guide, for which they had the support of a presenter and an observer, located in the Gesell camera behind a tinted glass, a video camera was also placed in the place to record the responses and reactions of the students with the main objective of not losing essential information. The epistemological support of the mixed method is based on pragmatism, where the position is decided to use the most appropriate method (Hernández et al. 2014, p. 539).

In turn, there is not a single mixed process, but rather in a hybrid study where various processes concur (Hernández-Sampieri and Mendoza 2008).

Based on the literature, 3 main dimensions were identified for the questionnaire, which were visibility, interaction and loyalty. Jadhav et al. (2012), the questions of the questionnaire were all made in Spanish, in order to focus on the question of the survey by addressing social media issues, especially Facebook.

In a methodological framework, several forms can be used in order to maintain the sense of what is being investigated, the concurrent triangulation design (DITRIAC) was used, simultaneously quantitative and qualitative data about the research problem are collected and analyzed approximately in the same time (Hernández et al. 2014, p. 557), this design can address the entire investigative process or only the collection, analysis and interpretation part (Hernández et al. 2014, p. 557).

The data were processed with the help of the SPSS 24 software and the Ms Excel spreadsheet. The data analysis was carried out based on the respective tables and graphs that contain quantitative information on the results obtained in the questionnaire.

4 Results

Table 1. Nivel de social media marketing

Nivel	Frecuencia	Porcentaje
Deficiente	51	51,77
Regular	35	36,04
Excelente	14	12,18
Total	110	100,0

Table 1 shows the levels of the MMS, organized through 3 levels excellent, fair and poor, these levels were collected through the questionnaire, first the collection of information based on the likert scale was carried out, then the database in IBM SPSS

Statistics 27 from the data obtained, from the 110 respondents and the MMS variable with its dimensions distributed in the questionnaire with a set of 15 questions, the responses according to the Likert scale were established a range of categories from 1 to 5, where the value 1 means totally disagree, 2 disagree, 3 neither agree nor disagree, 4 means agree and a response with a value of 5 implies totally agree. From these data, the crombach alpha was found, where the instrument's reliability value of 0.974 was obtained. 3 measurement categories are established (poor, regular and excellent) for the MMS variable, from the scores obtained the maximum and minimum values are grouped, from which a deficient category of MMS was obtained, which means that it should be rethink the content of the MMS through messages and communication in real time where students stay informed and dispel their doubts in real time.

For the perception of Social Media Marketing in university students, the qualitative technique was applied through the focus group where 18 student volunteers participated, who were asked a set of questions with the support of a presenter and an observer who remained located In the Gesell chamber behind a polarized glass, the information obtained through the transcription and annotations of important findings was collected, such as, that if they feel comfortable in college, that the social media of greatest interest are preferably Facebook and Instagram, being the most used social media Frequently used WhatsApp, Facebook and Instagram, all the focus group participants indicated that they are permanently connected through their mobile, laptop or PC, at night they connect to carry out and share information about a job assigned in group. They use blogs and other sites to share photos, videos, among others such as YouTube, Pinterest and Google, in turn they mentioned that they have copied some jargon or memes from countries such as Mexico which are disseminated on social media such as WhatsApp and Facebook. It is important to mention that university students consider that social media has allowed them to set a dressing style, adopting new fashion trends, one of the reported inconveniences and the most important is the scarce communication that there is acquired in their homes, they have lost the face-to-face communication and now use gestures or virtual icons to convey their emotions.

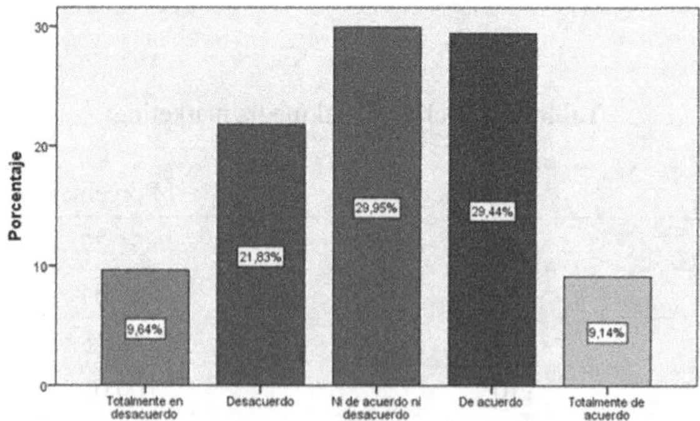

Fig. 1. You visit the University's social media frequently to keep me informed of your comments and events.

In Fig. 1, it is observed that of the total of respondents it is observed that there is a high percentage of students who disagree with the University's social media to keep informed of their comments and events, and very few of the students totally agree with the University's social media (Figs. 2, 3 and 4).

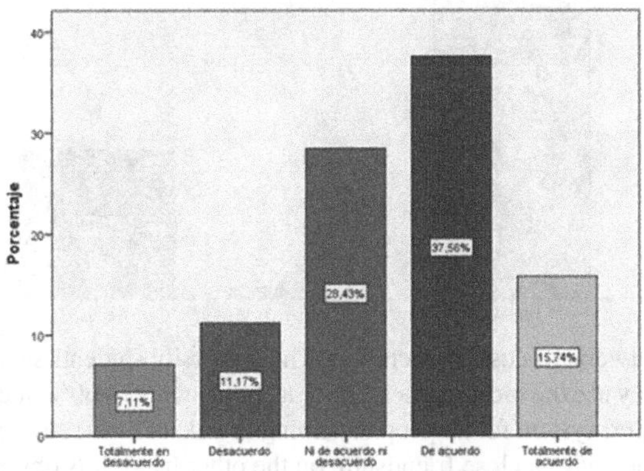

Fig. 2. The fan page of the University is always updated. In figure, it is observed that of the total of respondents it is observed that there is a high percentage of students who consider that the university's fan page is always up-to-date, on the other hand, there is a low percentage of students who indicate they do not agree with the University's fan page.

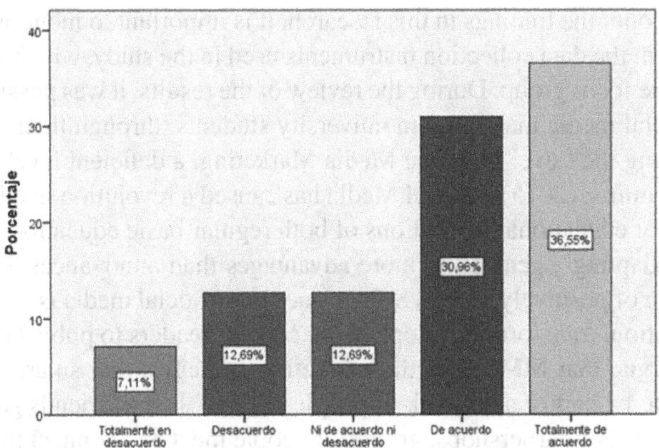

Fig. 3. You frequently use social media to communicate with friends. The figure shows that a high number of university students frequently use social media to communicate with friends and to carry out their university work and use it effectively.

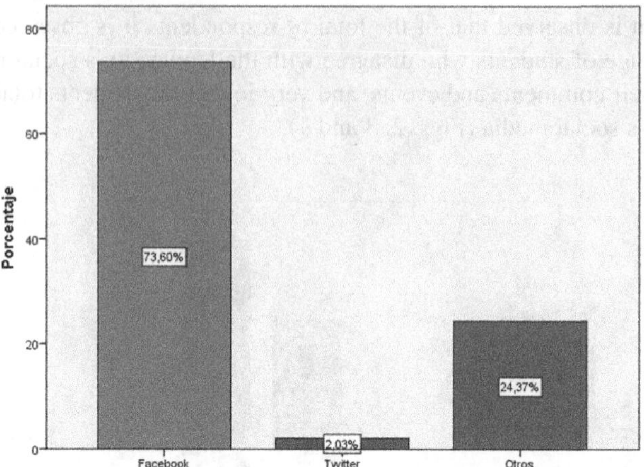

Fig. 4. The social network you use frequently is. The university students surveyed answered that the social network they use the most frequently is Facebook, with 73.60%, it can be indicated that this network is a support system for all people, helping to link and have stronger interactions with people with whom they have a close friendship, on the other hand, it is observed that 2.03% use Twitter and others 24.37%. In conclusion, it can be indicated that Facebook continues to be the most active network in terms of the participation of its users.

5 Discussion

Taking into account the findings in this research, it is important to mention the inquiries obtained through the data collection instruments used in the study, which were the questionnaire and the focus group. During the review of the results, it was possible to identify the level of social media marketing in university students, through their analyzed categories. Regarding the Level of Social Media Marketing, a deficient level was obtained, according to Ramírez (2015), "Social Media has caused a revolution in education, since it is a reality that educational institutions of both regular basic education and education superiors are adapting", generating more advantages than annoyances, so it should be taken advantage of positively, in turn Solis argues that "Social media is the democratization of information, transforming people from content readers to publishers…". Jadhav et al. (2012) argue that MMS "are the marketing strategies that smart businesses are using in order to be part of a network of people online." Just as friends gather in public pubs, coffee shops, or barbershops, groups of people meet and connect through various online tools and websites. In turn, Kübler et al. (2020) argue in their study that marketers can use various tools to extract sentiments from textual data. On the other hand, it is said that users can generate a great influence with comments on social networks due to the great power of choice and vote they have and following the research sustained by Freire (2016) shows that Social Media Marketing is a valuable tool for the use of advertising campaigns, which allows you to have a better reach or reach your consumers in a positive way, thus increasing the number of sales. According to the research carried out by Kumpu et al. (2021), where he argues that non-measurable data, such as customer emotions and opinions in various channels, is considered important, very important for understand customers.

From the results obtained in the focus group, it is considered that social media have limited communication between people, we are no longer very expressive, and their emotions are transmitted virtually based on icons or virtual gestures. Adequate person-to-person and family communication has been lost. However, they are an important means of keeping in touch. Therefore, the epistemological foundation theory of the research is supported by the Gestalt Theory and the neurosciences, they are theoretical foundations that contributed to the support of the present research, as well as the theory of Jadhav et al. (2012) and results obtained in the research arriving at the importance that social media have today for communication, education and information between students, parents and the community. Finally, social networks act as a tool to know people's preferences and feelings.

6 Conclusions

It was possible to analyze the impact of Social Media Marketing on university students through the mixed method, it was possible to demonstrate the analysis of quantitative and qualitative data.

The level of Social Media Marketing in university students was determined through the quantitative method using the survey technique, having obtained a deficient level in the analyzed variable.

The perception of Social Media Marketing in university students was determined, through the qualitative method using the focus group technique where a transcription and interpretation of the responses and language described during the development of the technique was made, being one of the findings more important than the social medium of interest is Facebook and that the continuous use of MMS has decreased direct person-to-person communication and feelings and emotions are transmitted through gestures or virtual icons.

The main advantages and disadvantages of Social Networks in university students were identified, one of the most important advantages being easy access to virtual communication and the disadvantage that communication styles such as jargon and dress styles have been copied.

Recommendations

For future research, the results of this study open up new possibilities and directions to delve into the analyzed study variable MMS.

References

Barker, M., Barker, D., y Borman, N.: Social Media Marketing. A Strategic approach. South-western cengage learning (2015). ISBN-13: 978-0538480871. ISBN-10: 0538480874

Betech, R.: Let others advertise your brand. Entrepreneur México **16**(9), 12–115 (2008)

Curiel, R., Gándara, M., García, F.: Importance of the use of social networks in higher education. V International Research Forum 102 Education. University of Guadalajara University Center for Economic and Administrative Sciences, Guadalajara, Mexico (2010). https://www.academia.edu/7051557/IMPORTANCIA_DEL_USO_DE_LAS_REDES_SOCIALES_EN_LA_EDUCACION_SUPERIOR

Freire, F.: Social media marketing strategy to increase sales of party items at DISPROEL company. [Postgraduate thesis, Vicente Rocafuerte Secular University of Guayaquil]. Institutional Repository UNLVR (2016). http://repositorio.ulvr.edu.ec/handle/44000/1346

Hernández, R., Fernández, C., Baptista, P.: Metodología de la Investigación. México: Editorial McGraw – Hill Interamericana (2014)

Hernández-Sampieri, R., Mendoza, C.P.: The qualitative quantitative marriage: the mixed paradigm. In: Álvarez Gayou, J.L. (ed.) 6th Congress of Research in Sexology. Congress held by the Mexican Institute of Sexology, A. C. and the Universidad Juárez Autónoma de Tabasco, Villahermosa, Tabasco, Mexico (2008)

Jadhav, N., Kamble, R., Patil, M.: Social media marketing: the next generation of business trends. IOSR Journal of computer Engineering (IOSR-JCE), 45–49 (2012). http://www.iosrjournals.org/iosr-jce/papers/sicete-volume2/21.pdf

Kübler, R., Colicev, A., Pauwels, K.: Social media's impact on the consumer mindset: when to use which sentiment extraction tool? J. Interact. Mark. **50**, 136–155 (2020). https://doi.org/10.1016/j.intmar.2019.08.001. https://www.sciencedirect.com/science/article/pii/S1094996819300933. ISSN 1094-9968

Kumpu, J., Pesonen, J., Heinonen, J.: Measuring the value of social media marketing from a destination marketing organization perspective. In: Wörndl, W., Koo, C., Stienmetz, J.L. (eds.) Information and Communication Technologies in Tourism 2021, pp. 365–377. Springer, Cham (2021). https://doi.org/10.1007/978-3-030-65785-7_35

Ramírez, P.: Social media marketing as a sales strategy for the company My Shoes in the city of Ambato. [Graduate thesis, Technical University of Ambato]. Institutional Repository (2015). https://repositorio.uta.edu.ec/handle/123456789/12221

Sharma, S., Singh, S., Kujur, F., Das, G.: Social media activities and its influence on customer-brand relationship: an empirical study of apparel retailers activity in India. J. Theor. Appl. Electron. Commer. Res. 602–617 (2021). https://doi.org/10.3390/jtaer16040036

Dynamic Difficulty Adjustment Using Performance and Affective Data in a Platform Game

Marcos P. C. Rosa[1(✉)], Eduardo A. dos Santos[1], Iago L. R. de Moraes[1], Tiago B. P. e Silva[1], Mauricio M. Sarmet[2], Carla D. Castanho[1], and Ricardo P. Jacobi[1]

[1] University of Brasília, Brasília, Brazil
[2] Federal Institute of Education, Science and Technology of Paraíba, Paraíba, Brazil

Abstract. The Dynamic Difficulty Adjustment (DDA) of games can play an important role in increasing the player engagement and fun. Gameplay difficulty can be adapted according to the player's performance, its affective state or by using a hybrid model that combines both approaches. This work investigates a hybrid DDA mechanism for a platform game to appropriately adapt its difficulty level and keep the player in a state of flow. The three approaches are compared to verify the efficiency of each model. An open source platform game was adapted to support the hybrid DDA algorithms. Game telemetry was introduced to acquire performance data and the affective state of the player is estimated through physiological data obtained from the Electrodermal Activity (EDA) of the skin. A method that estimates game difficulty when varying platform size and jump height was developed to support the DDA process. Besides playing with the different DDA models, each participant answered questionnaires and had their data collected for inquiry purposes. The results indicate that the DDA models were able to adjust the gameplay difficulty to the players, increasing the number of completed levels and reducing the variation in the playing time.

Keywords: Platform game · Dynamic difficult adjustment · Affective game · Game telemetry · Flow

1 Introduction

The video game industry has seen huge growth over the last few decades, with game developers appealing to a broad audience to sell their products. Thereby, the rules and requirements to make good games need to fit new scenarios, creating fun games for each consumer profile.

Adaptive systems strive to achieve the balance of the game for different players in a consistent, fair and fun way [17]. Such systems provide a personalized experience, adjusting the game according to identifiable, measurable and

C. Stephanidis et al. (Eds.): HCII 2021, LNCS 13094, pp. 367–386, 2021.
https://doi.org/10.1007/978-3-030-90238-4_26

influential goals [16]. They are usually defined by a dynamic factor and make adjustments based on those goals [3].

Non-adaptive games can cause boredom or frustration by not evaluating the match between the degree of challenge and the player profile [7]. For this, the activity can keep the player in a state of flow, offering a continuous challenge according to the evolution of the player's skills, so that he/she feels motivated to continue playing with a challenge proportional to his/her performance [18].

The concept of flow offers a theoretical explanation for pleasure [10]. During the state of flow, attention is free to focus on achieving the player goals as there is no clutter to strengthen or threat to defend. Thus, the flow occurs when there is a balance between the abilities of the individual and the undergoing task. Each player has an experience curve obtained by playing a segment of the game while remaining in a safe zone. If the experience moves away from the flow zone, negative psychic entropy (anxiety, boredom) will interrupt the flow state.

In contrast to the pre-established difficulty levels (commonly set as easy, normal, and hard), Dynamic Difficulty Adjustment (DDA) is an emerging technique in electronic games that seeks to adapt the level of difficulty of a game while to suit the skills of the player during the gameplay [1]. There are three requirements for dynamic game adaptivity [1]:

1. Identify and adapt to the player level as fast as possible;
2. Perceive and record changes in player performance; and
3. Maintain the behavior of the game, during the adaptation, credible and discreet, so that the player does not perceive the adaptive system.

In short, DDA modifies, in runtime, game factors and variables to match the challenge to the player. It should act as a control mesh that monitors player and game variables and acts accordingly, adapting avatar attributes, NPCs or level variables.

The DDA can be based on the player's performance data, on the player's physiological data or on the combination of these, thus characterizing a hybrid DDA [5]. Most studies are related to the player's perceived difficulty, where the proposed ADD models are based on performance data [11,12] or on the player's physiological responses [6,13,23]. Besides, there is scant in-depth research into the efficiency of using sensors and hybrid adjustments to make the player feel in a state of flow, especially when considering the platform game genre.

In this work, a hybrid DDA model is implemented and evaluated for a platform game to verify if the combined approach can provide a better difficulty adjustment for the player. Also, this work aims to compare the state of flow with the adaptation of distinct variables of the game, such as platforms or the jump height. Thus, the study is divided into three stages of analysis, comparing (1) the difficulty estimated by algorithms and the difficulty experienced, (2) the DDA models, and (3) the performance model applied to distinct game variables.

The adaptations are applied to a 2D platform game, which has a player-controlled character, called an avatar, that runs and jumps to avoid obstacles and/or defeat enemies. Based on the most common mechanics, there are 3 main aspects to analyze platform games [19]:

1. Movement, which expresses the range of movements included in the avatar and the respective control over those movements;
2. Confrontation, which expresses the importance given to an environment with opponents; and
3. Interaction with the environment, representing additional gameplay features that are not directly related to the original idea of the game.

According to the proposed objectives, the game adaptation prioritises movement rather than confrontation or interaction with the environment. Thereby, the game will adapt platforms and player's jump height configuration through a hybrid DDA system that considers the player's performance correlated to an assessment of their affective state. The difficulty measurement of a component will be based on the mathematical equations proposed by Mourato [21], related to the success in executing jumps. Thus, the probability of success or failure of a jump will determine the relative difficulty.

Furthermore, a data acquisition device (E4 Wristband) is used to collect physiological data from the player. This sensor checks the skin conductance that provides a reference for interpreting the level of frustration or boredom of the player. This type of measure (Electrodermal Activity (EDA)) serves as input to the DDA system and is usually related to some sympathetic nervous system reaction, indicating a change in the level of enthusiasm [22].

In each of the stages of analysis defined, tests were made with volunteers and the flow state and adequacy of the difficulty were evaluated, according to the answers to the questionnaire. In addition, these data were compared to the performance of the player according to game variables captured during the test sessions.

2 Related Work

Araújo conducted a study on player modeling and a survey on commercial and academic games that use DDA concepts [3], both commercial and academic. With a 2D shooter, it was analyzed the influence of adaptability on the gaming experience from the perspective of the flow theory [10]. In this case, the enemy's variables (speed, shooting time and range) are modified according to the difficulty, with a finite number of lives for the player and pre-defined difficulties.

The work found that changes that are easier to perceive impact the player's experience negatively. It has been observed that a poor performer may feel unmotivated when changes in the game are evident or when he understands how the adjustment system works. In addition, it becomes possible to break immersion and abuse the system by the player.

On the other hand, the difficulty in automatically generated 2D platform games was the research theme of Mourato and Santos [21]. For this, a method of measuring difficulty was proposed from the probability of success to complete a level. What they considered an obstacle was simplified and it was not considered the presence of checkpoints or lives. Moreover, they used the concept of levels as a path between a start and an end point that may contain components that hurt the player [9].

The measurement of a level difficulty was examined by the probability of the player giving up, repeating the challenge and succeeding in a jump. Besides, failure can occur when taking damage and dying or becoming frustrated and giving up. In turn, the probability of success of the jump is estimated by measuring the difficulty introduced by each individual component, based on formulas from kinematic physics.

Therefore, the combination of the probabilities of success of each jump situation produces an estimation of the level of difficulty. The height and position of a platform, the distance between platforms and the position of the enemies can be parameters that determine the success of the jump and facilitate or hinder the level.

Following the studies carried out in [21] and [20], Mourato discussed the improvement of Procedural Level Generation (PLG) in platform games [19]. Content adaptation is guided by the user, analyzing patterns of success and failure. Thus, the proposed set of difficulty estimation metrics is used to increase the capacity of the PLG algorithms and to customize the experiment. From this, it has implemented an algorithm for generating global structures and another for adapting the content, transforming simple paths into challenges with complex structures when changing components in strategic locations.

Difficulty can be adapted according to the player's performance, its affective state or the combination of both models. Considering this scenario, a facial expression analysis was applied as a means to predict difficulty through a classification task [4]. In this sense, it was needed to define metrics to measure perceived difficulty through affective data analysis, as well as model player behaviour and predict their affective state from this data.

The study was able to predict difficulty by correlating facial expressions to distinct player emotions. Despite that, these methods need a wider range of testing and improvements in accuracy. Also, they may evolve by integrating performance and physiological data, such as the Electroencephalography (EEG) and the Electrodermal Activity (EDA) of the skin.

3 Approach for Dynamic Difficulty Adjustment

The present work deals with the development of a hybrid Dynamic Difficulty Adjustment system for platform games. Since the purpose is not to develop a new electronic game, efforts were focused on adapting an existing game to meet the needs of the research. Thus, the platform game *The Explorer: 2D*, developed by Unity Technologies, was chosen, where the avatar can move horizontally, jump, crouch, attack and shoot.

The ready-made game levels have enemies of different ranges and non-linear paths, making it necessary to solve puzzles such as locating collectibles or standing on pressure panels to open doors. Regarding this study, among the three main components of the platform genre [19], the focus is on movement rather than confrontation and interaction with the environment. In this way, the game was adapted for level generation by bitmaps. Also, when creating levels, platforms and player variables are prioritized, in view of their more deterministic nature [14], such as in a classic platform game.

All of the game components were categorized, as the avatar, end point, checkpoints, collectibles, platforms, destructible or pushable objects, death areas, and enemies. Each level is organized into a map of components through the bitmap input, with each component changeable by applying the DDA mechanism.

3.1 Data Extraction

Gameplay and player data are needed for Dynamic Difficulty Adjustment and analysis. Performance data was captured directly from the game and real-time physiological data was acquired with the Empatica E4 wristband, a wearable sensor device. The Electrodermal Activity (EDA) signal was chosen due to its correlation with the levels of excitation, tension and frustration of the player. This type of sensor enables DDA to achieve immersion [15], and the reliability of this type of measure has been tested in previous studies [22].

Empatica Inc. provides Android and iOS applications designed to enable real-time data flow from the E4 devices, and a Bluetooth connection with cell phones. As a result, the default developer application has been modified to send the raw data to a database server over an internal network. Also, the game was adapted to request the recent data orderly, with the raw data of the EDA processed to generate two derived data: tonic (underlying slow changes) and phasic (rapidly changing peaks).

On the other hand, the gameplay data was organized into time-related, quantitative and death-based data. These are, respectively, the time to end each level, the quantity of successful jumps, and the distance travelled before dying. Furthermore, the success of a jump is related to two main goals established for the player: respectively, to reach the end of the level, and to obtain collectibles.

Considering these goals, the type of platform is divided according to the expected origin and destination of the avatar in relation to the position of the platform. In this sense, each platform carries out informations such as if the player must traverse from left or right, and from below, above or on the same height, always aiming to reach the end of the level. From this, the jumps are categorized into successful, flawed or randomic when considering the factors:

1. Fall into an chasm or die through an enemy or a trap;
2. Change of platform, considering the origin, the destination, the distance traveled and the collisions of the avatar;
3. Surpass a gap or an chasm;
4. Get collectible.

With this type of data, it is possible to measure the player's performance while traveling a level, regarding the jumps that are made and the consequent interactions with platforms, gaps, collectibles, and death areas.

3.2 Difficulty Measuring

To evaluate the difficulty of a level in a platform genre game, this work implements a version of the Mourato method [19,21]. His method evaluates the generation of levels and the adaptation to generate more complex and non-linear structures, such as gates and levers. Thus, it performs a procedural content generation associated with the variation of the environment complexity, rather than an adjustment of the difficulty of the components of the game.

In relation to the representation of difficulty, it follows the idea of Aponte [2] which is related to the probabilities of success and failure during the execution of tasks, whether analyzed independently or at an entire level. Thus, the approach is based on two steps of level analysis:

1. First, the level is decomposed into segments that represent independent parts of a level, with possible transitions between them.
2. Then an analysis is made of each individual component, based on its probability of success.

To determine the difficulty of a level, the sum of the difficulty of specific components is made based on its probability of success. The analyzed components are those that have jump as an integral part of their challenge, being these: platforms, gaps, and chasms. Moreover, the jump corresponds to a parabola of the launch of a projectile from an origin to a destination platform, with the probability of success coming from the horizontal and vertical margin error of the parabola in relation to the destination point.

For moving platforms, the failure is defined when the user collides with the entity in the period of loss or if its movement is late and does not fit into the entity's overcoming interval. Combining the two, the spatiotemporal difficulty is measured by the multiplication of the success probabilities of the previously mentioned situations.

To automate this process at runtime, for each platform component in the segment, a search is made to find the shortest path from the origin platform (P_0) to a destination platform (P_1). There must always be a reachable destination when adapting, according to the desired movement stored in the component.

3.3 Dynamic Difficulty Adjustment

The DDA of the game is adjusted: (i) according to factors of player performance, (ii) another by the affective state evaluation of the EDA and (iii) a hybrid of both. In the three models described, the difficulty is changed in the same way. In short, there are global values that measure the player's performance and the affective state, varying from 0 to 1. The difficulty for the hybrid model corresponds to the average of the others.

The implementation of the three DDA models followed a cyclic pattern with 4 interconnected systems, based on Chen [8] and Bontchev [5] studies. The systems refer to the Player's raw data, the Monitoring of these metrics, the corresponding

Analysis and the Control of the components that will be adjusted. The Fig. 1 shows the diagram of the systems used to apply hybrid DDA and a summary of the associated steps.

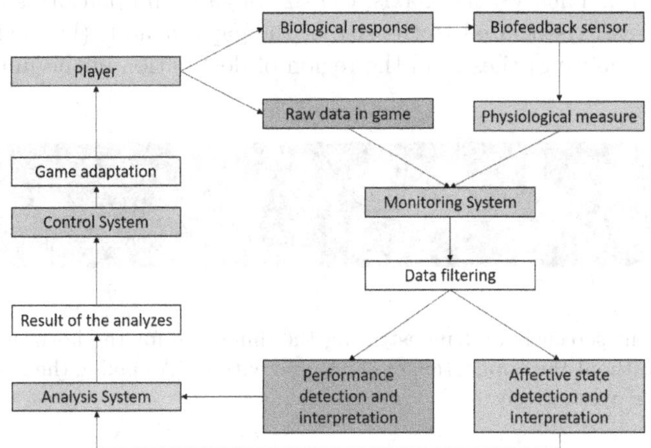

Fig. 1. Cyclic diagram of the system-driven hybrid DDA. The systems are in green, the steps related to the player's performance in red and the affectivity in yellow.

The Analysis System applies heuristics for both performance and affective data, generating a global difficulty value used to define the desirable mean probability of jump success. For the affective case, the number of arousal levels is extracted based on the normalization of individual peaks by the asset from Project Rage[1]. At each time period, the ratio is recalculated and its variation is added to the overall difficulty.

As for the performance case, heuristics were defined. When death is related to a fall, a counter is incremented and the position of death is saved. If death is in a previous or equal position to that of the current death, the overall difficulty decreases (facilitating). In the opposite case, the global difficulty increases (hampering), as the player was able to further advance in the level.

When a segment is completed and at least one death occurs, the difficulty decreases. Meanwhile, the difficulty grows when segments are completed without the occurrence of death. Finally, after each death or segment completion, a ratio is made between the number of successful and fault jumps. Thus, it makes harder when there is a greater incidence of success, and facilitates when the number of failures increases.

The DDA adaptations were programmed to be made when creating segments and levels, and when dying. It is observed that they are not done at every moment to avoid that the player perceives the visual modifications. Moreover, to maintain an increasing difficulty throughout the game, the inferior and superior global

[1] https://github.com/ddessy/RealTimeArousalDetectionUsingGSR.

difficulty limits are defined according to the number of levels in the game. For example, it must be between 0 and 0.75 for the first level, and 0.25 and 1 for the last level.

The Control System is responsible for applying the DDA to the game and player variables. Thus, on the x-axis, the size of gaps and platforms is modified, as seen in Fig. 2. Associated to the size of the gaps, it adds (by facilitating) or removes (in hindering) tiles from the region of destination of the jump.

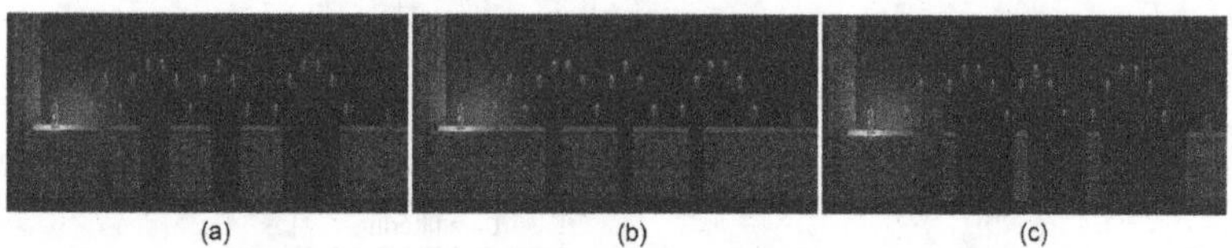

Fig. 2. Segment screenshots demonstrating the difference for the horizontal platform distance (a) without the application of DDA, and with DDA making the level (b) easier and (c) harder, respectively.

On the y-axis, the position or size of the platforms is changed, shown in Fig. 3. In the case of column-shaped platforms, it is changed by removing or adding tiles, otherwise the position of the platform that is modified. Moreover, the player's jump amplitude was adapted in parallel or together with game variables in a composite Control System. As seen in Fig. 4, the amplitude can be increased to make it easier to access more distant platforms.

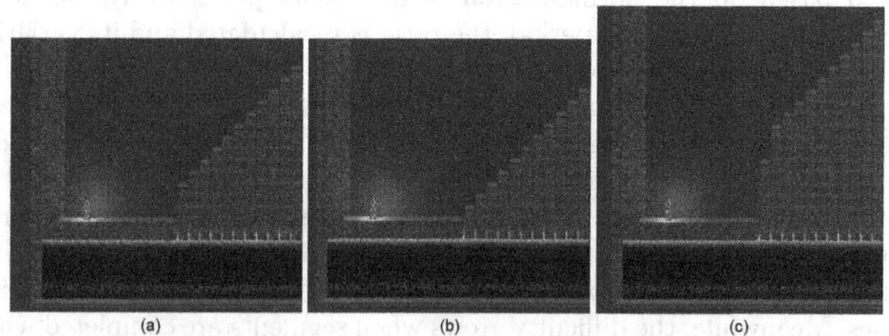

Fig. 3. Segment screenshots demonstrating the difference for the vertical platform distance (a) without the application of DDA, and with DDA making the level (b) easier and (c) harder, respectively.

At the end of the process, the difficulty of each component is updated, as well as the level difficulty. Note that the Control System of the DDA adjust the game variables and the player configuration guaranteeing that all game content remain reachable during adaptation.

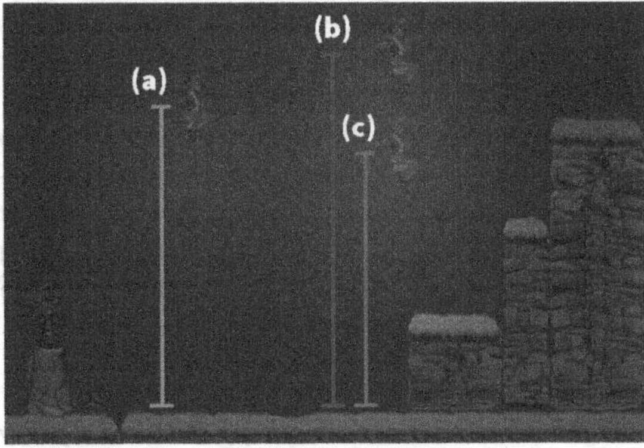

Fig. 4. Segment screenshots demonstrating the difference for the jump amplitude (a) without the application of DDA, and with DDA making the level (b) easier and (c) harder, respectively.

4 Experiments

Three test batteries were made with sample groups. In each battery, the player could pause, review the controls and give up on a level. Moreover, lives were infinite and levels were divided into segments of same length delimited by checkpoints, in which the progress was saved. The game session was finalized at the end of all levels and when the questionnaire was fully answered.

In the first battery, the game presented 12 levels, each containing one or two segments. The difficulty during the game was increasing and analyzed, in each level, different components of the game (or the conjunction of these). It was expected to determine a balance of the difficulty of the game and to analyze the method of estimation of difficulty proposed and implemented in the platform game *The Explorer: 2D*. Therefore, the objective was to verify if the estimated difficulty is able to relate to the difficulty experienced by the players.

In the second battery, the game featured 12 levels, each containing four segments with similar lengths. It followed the pattern: (1) level without the application of DDA, (2) with DDA based on player performance, (3) with affective DDA, and (4) with hybrid DDA. This pattern was repeated 3 times, whereas the components and base difficulties were similar in each set of levels, the adjustment could change the general difficulty according to the DDA model applied in the level.

The four cases were compared and similar components and obstacles in each set were selected in order to avoid particular factors biasing the conclusions, such as the usage of different game components or exploratory paths. From this, the following hypothesis was defined:

Theorem 1. *A hybrid DDA system with performance and affective data can provide suitable difficulty for the player.*

In the third battery, the game also featured 12 levels, simplifying the general presentation of those used in the second battery to allow remote tests. In this case, a Control System adaptation was randomized for each volunteer, and it was guaranteed that each played only once. The available adaptations were: (1) game without the application of DDA, (2) with performance-based platform adaptation, (3) with performance-based jump height adaptation, and (4) with performance-based combined adaptation of both platform and jump height. In this sense, all the levels were played entirely by each volunteer with a single version of the Control System adaptation. Also, the following hypothesis were defined:

Theorem 2. *A combined DDA system with platform and jump height adaptation maintains the player in a state of flow.*

Theorem 3. *A combined DDA system with platform and jump height adaptation has a suitable difficulty for the player.*

The tests were validated by the application of questionnaires during the game session. The questions applied before the start of the game concerned sociodemographic data. During the game, they checked the difficulty perceived at each level. In the end, they analyzed the perceived difficulty, and whether the player remained in a state of flow. For the first battery it was also observed what influences the player's performance and the challenge provided by different components. Meanwhile, for the third battery, there was no in-game questionnaires, so that the flow state was not harmed.

In all test cases, the performance data were collected and saved. The test conditions ensured that all participants had the same game information, and, for the non-remote first and second batteries, the same environment with no external contact.

5 Discussion

In the first test battery, 20 university students participated as volunteers, ranging from 18 to 24 years old, with a higher male participation (70%). The players had a predilection for games with easy difficulty (45%) rather than normal (30%) or hard (25%).

In the second test battery, the same university context was maintained and 36 volunteers participated, ranging in age from 18 to 25 years old, and with a greater participation of men (61.1%). There was a predilection for easy difficulty (50%) than normal (33.3%) or hard (16.7%). Furthermore, 16 of the 20 participants of the first battery participated again, noting that new levels were created and several DDA models were applied that should adjust the game regardless of whether the player knows the game dynamics.

In the third test battery, 155 volunteers participated remotely, ranging in age from 15 to 65 years old, and with a higher participation of men (81.69%). 12 of the 155 volunteers were discarded for not completing the game session, and 11 for being categorized as an outlier, finishing levels with less than 10 s and with one death at most, stating that the player was not committed to the attempt. The players only had one DDA adaptation applied (37 without DDA, 29 with platform DDA, 35 with jump DDA, and 31 with combined DDA), having a total of 132 volunteers considered. Also, there was a predilection for normal difficulty (51.41%) than hard (32.39%) or easy (16.20%).

5.1 Balancing and Checking Difficulty Measurement

The difficulty of each level was numerically described by the participants in a questionnaire after playing it in the first battery, according to Fig. 5. The levels tested distinct components and their interactions, and it was possible to identify the proportionality between the normalized value calculated by the algorithm and the one obtained by the players' average, between 0 and 10. Therefore, this estimation method can be used in the application of DDA. However, level 6 showed that the game design made the difficulty to not vary in agreement. In level 6, the challenge is created through the proposed design, whereas the algorithm only counts the positions and specifications of the components in the construction of the difficulty of the levels.

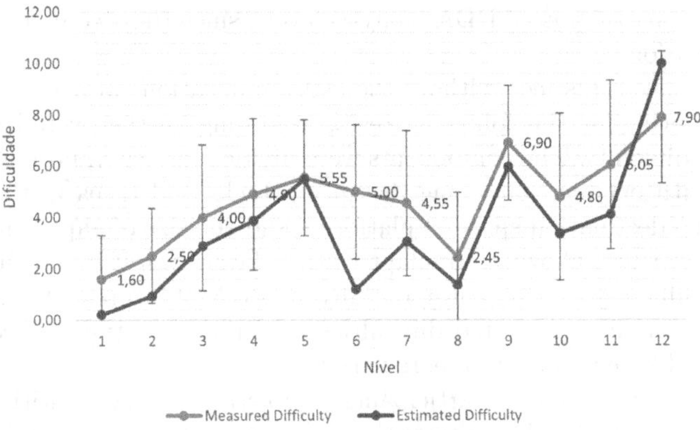

Fig. 5. Comparison, in each level of the first battery, between the average of the difficulties measured by the questionnaire and the value estimated by the developed algorithm.

There was a peak in levels 5, 9 and 12 for having checkpoints and double the standard size, thus bringing greater difficulty for the accumulation of challenges and components. In this way, it was possible to corroborate the relationship between the amount of components and the difficulty perceived by the players. Besides, greater difficulty has been observed in separate components working together, and that the modification of specific components alters the user's perception of difficulty, such as the complementary levels 2 and 7 that alter the difficulty only by increasing the size of gaps. Therefore, it was decided that the levels should have the same size and do not present challenge based on the game design, as the blocking of the path by a riddle.

At the end of the game session, it was questioned about the actions that most influenced the performance of the players, with the highest goal being to complete the level (with 80% of frequency among the participants) as opposed to completing the levels quickly (25%). In addition, there was a low concern in exploring (30%), and defeating the maximum of enemies (30%). Thus, the DDA was based on more general components (such as platforms and gaps) and was tested with linear paths.

Additionally, it was found that different characteristics of a platform influence the perception of challenge by the participants in a similar way. The volunteers were questioned about the difficulty that different characteristics of a platform can provide, such as size and distance, with all getting an average between 3.8 and 3.9 on a scale of 1 to 5. From this, it was decided that the DDA would adapt both the size and the vertical and horizontal distances of platforms, gaps and chasms.

Likewise, the moving platforms had similar difficulty considering the direction of movement (horizontal, vertical or diagonal), and components with the same purpose and different visual provided similar difficulty, such as the gap and the acid (average of 3.9). On the other hand, enemies with distance attacks and traps increased the difficulty (mean of 4.05 and 4.4, respectively), avoiding the use of these in the construction of DDA analysis levels, since they were not considered in the adaptation.

The basic levels (as seen without the DDA application) were divided in three stages of 4 levels for the following tests, according to the overall component experienced difficulty. The components were divided into easy, normal, and hard level sets, with enemies with long distance attacks and traps in the hard set. Note that the destination space of platforms and the size of gaps is modified by the DDA algorithm, allowing to adapt to the player even if its not in a specific component difficulty. In this sense, initially all levels of future tests need to have each set with similar estimated difficulties, and the same types of components organized in different ways to create variation.

There was a high rate of participants completing the levels, with a decrease according to the greater difficulty of the levels, and a high variation in the dwell time at each level, especially when analyzing levels with hard components. It was observed that the possibility of giving up a level was chosen after many deaths or with a high level of frustration of the participant at the level. These factors are intended to be normalized with the presence of the DDA and by avoiding nonlinear paths.

5.2 Comparison Between the DDA Models

The performance data of the second battery was tested with the non-parametric Friedman test, which does not assume the independence of observations, as every volunteer played for all DDA models applied during gameplay and the data is not normally distributed. The analysis were made for each case, comparing the four models for easy (1–4), normal (5–8) and hard (9–12) level sets. The data was verified with the Shapiro-Wilk test, having significant p-values, failing to reject the null hypothesis and concluding that data is not drawn from normal distribution for all the analyzed data. Moreover, it was observed with the Bartlett's test that samples from populations have equal variances for most of the questionnaire data and not for performance data.

As seen in Fig. 6, players numerically described the difficulty of levels for three sets of four levels. From this, it was possible to observe that the mean value grows according to the progression of the easy and medium levels, which is expected in view of the difficulty to be increasing in platform games and the algorithm for adjusting the levels to follow this pattern. However, there is a peak at levels 5 and 9 compared to the previous growth pattern, as there is no application of the DDA in them.

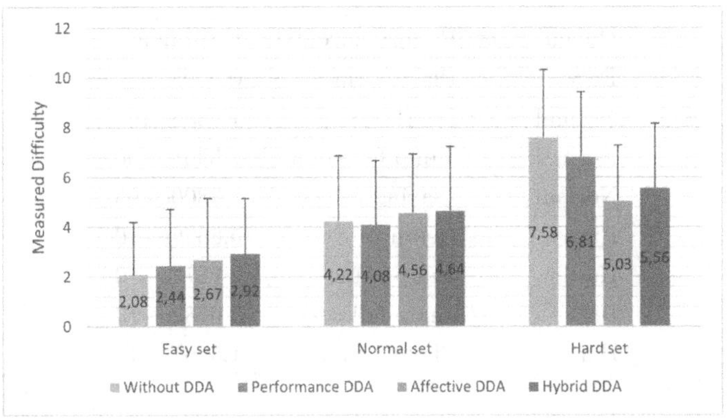

Fig. 6. Mean value and standard deviation of the difficulties measured by the questionnaire at each level of the second battery, divided into the three difficulty sets tested and grouped by the four models.

According to the progression of the game, more data is captured and a better estimate of the player's performance and affective state occurs. Thus, in the normal and hard level sets, a greater standard deviation is observed when the application of the DDA does not occur. In addition, the difficulty of these levels stands out in relation to the immediately following (6 and 10) because they fit the player's performance during the game until then, including the level without DDA, and thus decrease the difficulty for the players average.

Additionally to the questionnaires, 7 of the 12 levels were completed by all volunteers. Besides, the levels with lowest percentage of completion of each set were those without the application of DDA (levels 5 and 9 with 94.4% and 63.9%, respectively). Also, there were peaks in the affective and hybrid levels, emphasizing that all participants completed the levels with affective adjustment. The lowest percentage of completion with DDA was in the performance-based level 10 (80.5%), which is still much higher than the lowest percentage without DDA in level 9 (63.9%).

In addition to the increasing rate of players completing the levels, considering performance data, a wider range of models medians difference were statistically significant, only not for death rate and jump count on easy levels, considering a significance level of 0.05. Table 1 presents a clipping of the cases for measured performance data p-value with the Friedman test. Furthermore, the most consistent significantly different performance data for different levels was playing time by the participant on each level, shown in Fig. 7. In this sense, statistics of the players show a better performance with DDA models, specially considering a lower time to complete the game and with less dispersion.

Table 1. Friedman measured p-value, according to performance data to each set of levels by difficulty.

Difficulty set	Performance data	Friedman
Easy	Playing time	0.0005
Normal	Death rate	1.438e−10
Normal	Jump fail count	2.5361e−06
Normal	Playing time	1.3800e−05
Hard	Death rate	5.8626e−10
Hard	Jump count	2.0277e−10
Hard	Jump fail count	2.0782e−13
Hard	Playing time	1.5065e−13

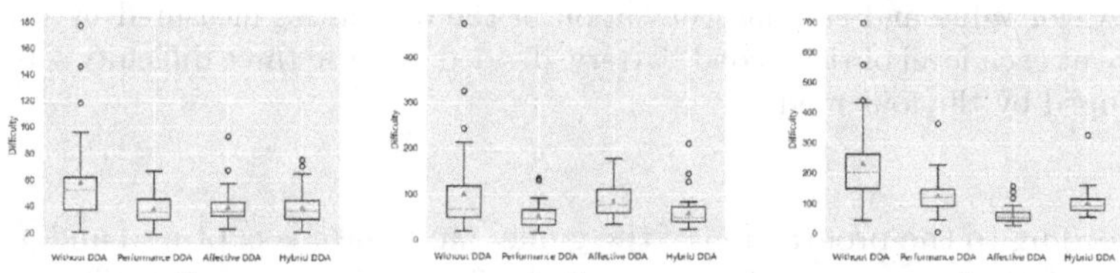

Fig. 7. Box plot for playing time, respectively, for (a) easy, (b) normal, and (c) hard levels, comparing for the four different models.

The standard deviation in games without DDA was considerably higher, because they do not fit to each player and consequently bring a greater variation in the performance of distinct players in such levels. In most cases, the mean times in levels with different models of DDA are approximated in each analyzed set. Besides, there is a negative peak in the standard deviation in performance DDA in the easy and normal difficulty sets and in the hybrid DDA in the hard set.

Moreover, at the end of each level, the player was questioned about which one had a more adequate difficulty, shown in Fig. 8. As each level had a different model, the models could be compared and had the following order of better perception for the players: hybrid DDA, affective DDA, performance DDA, and without DDA. It is observed that, for hard levels, it is easier to note a difference between models because of greater disparities in player's abilities than in easy and normal levels. However, the perception of the player to differentiate the difficulty adequacy decreases in higher level sets, because of the increasing difficulty.

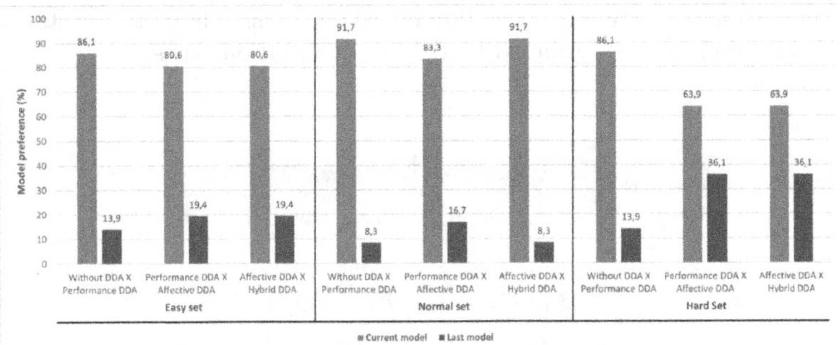

Fig. 8. Frequency of answers about the difficulty being more appropriate in the current or previous level with frequencies grouped by difficulty sets, asked at the end of the levels in the second battery.

Considering the data from questionnaires and from the performance data presented above, the corresponding analysis can validate the presented hypothesis. This is specially observed in the reduction of dispersion, higher performance results and the shown perception of the player about the difficulty. Therefore, a hybrid DDA system have a suitable difficulty adaptation for the player. Also, it was perceived that future tests of comparison should be made separately to be able to measure the difference between models through questionnaires and maintain the player's flow state.

5.3 Comparison Between DDA Distinct Adaptations

Questions were made at the end of the third battery to verify if the player was in a state of flow, and if the difficulty was adequate in their perception, as seen respectively in Figs. 9 and 10. In this sense, volunteers reported the combined and platform adaptations as the best in average to maintain the player's state of flow, considering the focus on the game, losing track of time, and automatic actions. The cases where no ADD was applied come next, with the adaptation only on the jump having a lower rating.

The difficulty was found to be most appropriate for the platform adaptation, followed by the combined adaptation and with lowest result for the game without

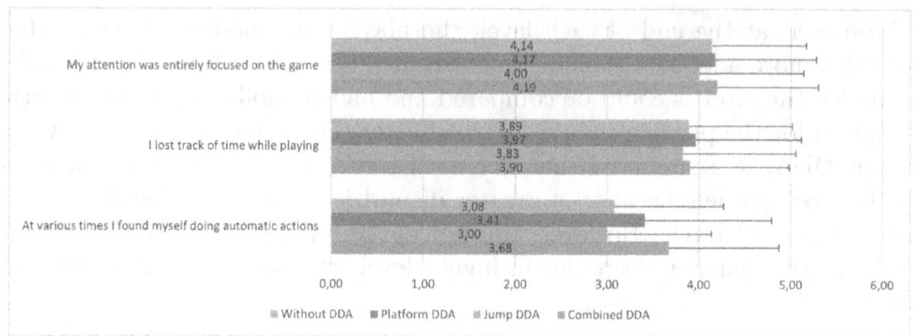

Fig. 9. Mean value and standard deviation of the questions related to the state of flow of the third battery of tests, grouped by the four adaptation types.

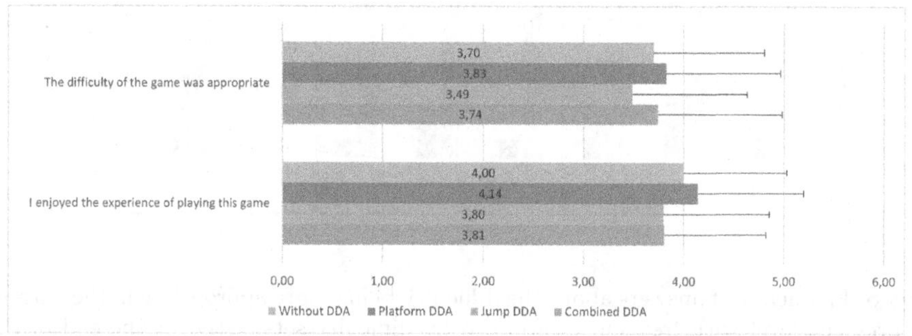

Fig. 10. Mean value and standard deviation of the questions related to the adequacy of difficulty of the third battery of tests, grouped by the four adaptation types.

DDA. Also, for the quality of the experience perceived by users when playing, the DDA applied only to platforms had a better performance compared to other models. It is also noticed that playing without any DDA application still yields a better experience than the DDA applied to the jump or combined.

However, the data was also tested with the non-parametric Kruskal-Wallis test, which assumes the independence of observations, as the data is not normally distributed and it was guaranteed that each volunteer played only with one DDA adaptation. For a significance level of 0.05, the differences between the medians are not statistically significant for the majority of the questionnaire data.

As game levels are designed to grow with player performance, levels with higher extremes are needed to actually analyze a difference when models are applied in relation to player perceptions. In this sense, there must have been a co-variance between the levels and the ADD model, with the levels varying in a fixed way and ADD varying on top of that. To avoid this, future tests will have levels revised to provide greater variability of adaptations.

In parallel, the players' perception regarding the jumping mechanics in the game was analyzed, as seen in Fig. 11. The combined DDA turned out to be the one whose jumping would be the most difficult to control and the easiest would be with the jump DDA. In terms of the responsiveness and satisfaction generated by the jump, platform DDA outperforms and without ADD was the worst case, indicating that adaptation had an effect on the perception of controls.

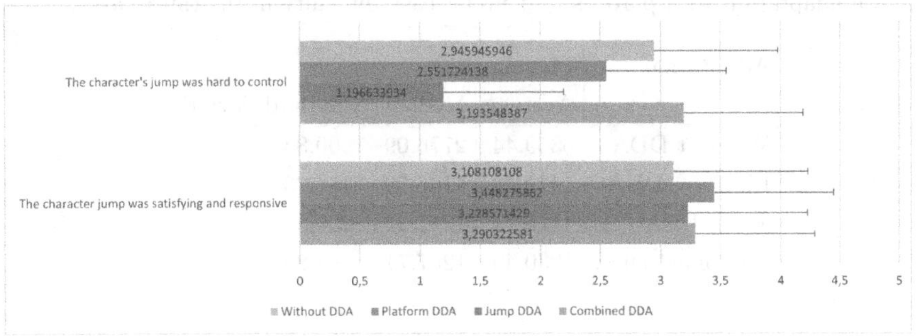

Fig. 11. Mean value and standard deviation of the questions related to the jump control of the third battery of tests, grouped by the four adaptation types.

In this sense, as in the second battery, the most consistent divergence of data for different levels was the playing time to complete each level, shown in Table 2. The playing time in all levels had a lower average and standard deviation for the adaptations, with lowest average for the jump adaptation and lowest standard deviation for the platform adaptation. Therefore, the adaptations were able to decrease the variation in the performance of distinct players.

Table 2. Average and standard deviation of the playing time to complete the game for each adaptation in the third battery.

Adaptation	Playing time		
	Average	Median	Standard deviation
Without DDA	1164.00	808.39	1110.87
Platform DDA	1038.46	884.63	517.84
Jump DDA	1012.95	830.73	689.65
Combined DDA	1080.32	924.24	660.62

The DDA adaptations aim to lower the dispersion and the average playing time or death rate by fall, which is related to chasms or death areas. The results have shown better adequacy related to the player preference for easy games. Thus, Tables 3 and 4 present the average and standard deviation values for playing time and death rate by fall. When considering performance data for player's with preference for easy games, the differences for playing time and death rate were statistically significant when using Kruskal-Wallis test with 0.5 of significance level, and comparing the game without DDA with each distinct adaptation (p-value between 0.018 and 0.023).

Table 3. Average and standard deviation of the playing time to complete the game for each adaptation with player's that prefer easy difficulty in the third battery.

Adaptation	Playing time		
	Average	Median	Standard deviation
Without DDA	3359.44	2776.09	1900.83
Platform DDA	1471.35	1166.59	842.73
Jump DDA	1119.11	1212.23	236.83
Combined DDA	1230.49	1257.71	267.89

Table 4. Average and standard deviation of the total deaths by falls for each adaptation with player's that prefer easy difficulty in the third battery.

Adaptation	Total deaths		
	Average	Median	Standard deviation
Without DDA	50	85	25.02
Platform DDA	30	68	21.98
Jump DDA	31	75	14.83
Combined DDA	33	73.5	10.15

The platform and composite adaptations generated a better experience related to the state of flow, but without significant difference. Therefore, it was inconclusive if the combined DDA was able to improve the maintenance of the player in a state of flow when using the game without DDA as basis. On the other hand, the combined DDA system was able to adequately tailor the difficulty level, decreasing the dispersion of performances. The result was positive specially for players that prefer the easy difficulty, showing that the system may be improved by adapting the game according to the player profile.

6 Conclusion

The DDA studies are growing in scope of themes and studies carried out, considering several variables such as the player's performance or the affective state. However, few studies in the literature combine both factors, categorizing a hybrid model. Specifically on the platform genre, no research with this model was found. Furthermore, in-depth investigations of the efficiency of sensor use are scarce, especially when considering real-time adaptation, and of hybrid adaptation to keep the player in a state of flow.

Therefore, this work aimed to investigate a hybrid DDA model in platform games. More precisely, the focus was on adapting the difficulty to each player and keeping it in a state of flow, while the game was modified based on the performance data and the player's affective state, specifically through the analysis of the Electrodermal Activity (EDA) of the skin. In addition, the performance-based model was divided into three possible adaptations: the size and position of platforms and gaps, jump height and the combination of those.

From the analysis of the obtained results, the method of computation of difficulty associated with the developed DDA was corroborated. Thus, it was observed a relationship between the difficulty estimated by the algorithm and experienced by the players participating in the test. Also, it was possible to balance the game in relation to the components used. Thus, the DDA test levels remained linear and without the application of conditions that could bias the player's measure of difficulty and affective state, including level planning that did not bring challenge by puzzles or traps.

Based on the results, limitations were observed when comparing few participants between the analyzed models, and the models being only tested together. Still, it was observed that a DDA system was able to successfully tailor the difficulty of a platform game to heterogeneous players, which was demonstrated by the larger number of completed levels, and the lower average and smaller dispersion of the time to complete them. In particular, the difficulty level was more adequate when using the hybrid model, making the experience more suitable for each player.

Specifically about the performance-based DDA system, all of the adaptations were tested independently and with a wider range of participants. Thereby, the limitations of the previous test have been overcome, and it was possible to verify the adequacy of difficulty in the adaptations, according to the players performance. However, the improvement of the player's state of flow was not verified, when considering that the questionnaire data was not significantly different from the game without DDA for the gameplay characteristics of the state of flow.

Further studies will apply levels with extremes to augment the possibilities of distinct adaptations and players perceptions. Moreover, fields to increase the efficiency of the DDA model are the development of alternative EDA computation through both normalization and peak recognition, and the recognition of the style of play by means of machine learning. Finally, one can analyze the difficulty as influenced by the game design and not only by the presence and characteristics of the components. In this way, the use of traps and puzzles could be investigated, as well as the presence of specific missions and contents to challenge the player.

References

1. Andrade, G., Ramalho, G., Gomes, A.S., Corruble, V.: Dynamic game balancing: an evaluation of user satisfaction. American Association for Artificial Intelligence (2006)

2. Aponte, M.V., Levieux, G.: Difficulty in video games: an experimental validation of a formal definition. In: Proceedings of the 8th International Conference on Advances in Computer Entertainment Technology (2011)
3. de Araujo, B.B.P.L., Feijó, B.: Evaluating dynamic difficulty adaptivity in shoot'em up games. In: XII Brazilian Symposium on Computer Games and Digital Entertainment, October 2013
4. Blom, P., Bakkes, S., Spronck, P.: Modeling and adjusting in-game difficulty based on facial expression analysis. Entertainment Comput. **31**, 100307 (2019)
5. Bontchev, B.: Adaptation in affective video games: a literature review. Cybern. Inf. Technol. **16**(3), 3–34 (2016)
6. Chanel, G., Lopes, P.: User evaluation of affective dynamic difficulty adjustment based on physiological deep learning. In: Schmorrow, D.D., Fidopiastis, C.M. (eds.) HCII 2020. LNCS (LNAI), vol. 12196, pp. 3–23. Springer, Cham (2020). https://doi.org/10.1007/978-3-030-50353-6_1
7. Chang, D.M.J.: Dynamic difficulty adjustment in computer games. In: Proceedings of the 11th Annual Interactive Multimedia Systems Conference (2013)
8. Chen, J.: Flow in games (and everything else). ACM Mag. **50**(4) (2007)
9. Compton, K., Mateas, M.: Procedural level design for platform games. American Association for Artificial Intelligence (2006)
10. Csikszentmihalyi, M.: Flow: The Psychology of Optimal Experience. Harper & Row, New York (1992)
11. Denisova, A., Cairns, P.: Adaptation in digital games: the effect of challenge adjustment on player performance and experience. In: Proceedings of the 2015 Annual Symposium on Computer-Human Interaction in Play, pp. 97–101 (2015)
12. Hintze, A., Olson, R.S., Lehman, J.: Orthogonally evolved AI to improve difficulty adjustment in video games. In: Squillero, G., Burelli, P. (eds.) EvoApplications 2016. LNCS, vol. 9597, pp. 525–540. Springer, Cham (2016). https://doi.org/10.1007/978-3-319-31204-0_34
13. Imre, D.: Real-time analysis of skin conductance for affective dynamic difficulty adjustment in video games. Doctoral thesis, Algoma University (2016)
14. Koens, E.: Generating non-monotone 2D platform levels and predicting difficulty. Master's dissertation, Utrecht University (2015)
15. Liu, C., Agrawal, P., Sarkar, N., Chen, S.: Dynamic difficulty adjustment in computer games through real-time anxiety-based affective feedback. Int. J. Hum.-Comput. Interact. **25**, 506–529 (2009)
16. Lopes, R., Bidarra, R.: Adaptivity challenges in games and simulations: a survey. IEEE Trans. Comput. Intell. AI Games **3**, 85–99 (2011)
17. Novak, J.: Desenvolvimento de Games. Cengage Learning (2011)
18. Schell, J.: The Art of Game Design: A Book of Lenses, 2nd edn. AK Peters (2014)
19. Mourato, F.J.S.V.: Enhancing automatic level generation for platform videogames. Doctoral thesis, Universidade Nova de Lisboa (2015)
20. Mourato, F.J.S.V., Birra, F., dos Santos, M.P.: Difficulty in action based challenges: success prediction, players' strategies and profiling. In: Proceedings of the 11th Conference on Advances in Computer Entertainment Technology, pp. 1–10 (2014)
21. Mourato, F.J.S.V., dos Santos, M.P.: Measuring difficulty in platform videogames. 4ª Conferência Nacional em Interação Pessoa-Máquina (2010)
22. Soares, R.T.: Biofeedback sensors in game telemetry research. In: SBC - Proceedings of SBGames 2016 (2016)
23. Vollmers, C.G.K.: Dynamic difficulty adjustment in games using physiology. Master's dissertation, Aalborg University Copenhagen (2018)

Exploring the Effect of Resolution on the Usability of Locimetric Authentication

Antonios Saravanos[(✉)], Dongnanzi Zheng, Stavros Zervoudakis, and Donatella Delfino

New York University, New York, NY 10003, USA
{saravanos,dz40,zervoudakis,dd61}@nyu.edu

Abstract. Locimetric authentication is a form of graphical authentication in which users validate their identity by selecting predetermined points on a predetermined image. Its primary advantage over the ubiquitous text-based approach stems from users' superior ability to remember visual information over textual information, coupled with the authentication process being transformed to one requiring recognition (instead of recall). Ideally, these differentiations enable users to create more complex passwords, which theoretically are more secure. Yet locimetric authentication has one significant weakness: hot-spots. This term refers to areas of an image that users gravitate towards, and which consequently have a higher probability of being selected. Although many strategies have been proposed to counter the hot-spot problem, one area that has received little attention is that of resolution. The hypothesis here is that high-resolution images would afford the user a larger password space, and consequently any hot-spots would dissipate. We employ an experimental approach, where users generate a series of locimetric passwords on either low- or high-resolution images. Our research reveals the presence of hot-spots even in high-resolution images, albeit at a lower level than that exhibited with low-resolution images. We conclude by reinforcing that other techniques – such as existing or new software controls or training – need to be utilized to mitigate the emergence of hot-spots with the locimetric scheme.

Keywords: Locimetric authentication · Hot-spots · High-resolution graphical passwords

1 Introduction

Locimetric authentication (also known as click-based authentication) is a graphical mechanism that verifies users' identity through their selection of a series of predetermined points on an image in a particular order. Initially described by Blonder [4] in his patent filing (US5559961A), it serves as the first form of graphical authentication. Over the years, several other implementations of the scheme have been developed, such as PassPoints [31], Cued Click Points [6], and Persuasive Cued Click-Points [7]. However, none of these implementations enjoy the level of diffusion as Microsoft's Picture Password, which is installed by default on any machine running the Windows

C. Stephanidis et al. (Eds.): HCII 2021, LNCS 13094, pp. 387–396, 2021.
https://doi.org/10.1007/978-3-030-90238-4_27

8 operating system or higher. In actuality, Picture Password is a combination of two schemes, locimetric and drawmetric, with the user empowered to select how much of each method they prefer to use. Thus, the created password could be fully locimetric, fully drawmetric, or a combination of both schemes. Drawmetric authentication is a form of graphical authentication that validates users by requiring them "to draw a preset outline figure, either on top of an image or on a grid" [10]. Given the Windows operating system's prominence, especially in the desktop market, insight into the potential weaknesses inherent with locimetric authentication is valuable.

In this paper, we focus on one of the accepted weaknesses: the users' propensity to select the same point on images to form their passwords, known colloquially as hot-spots [6] (also sometimes known as click-point clustering [25]). Strategies have been proposed to counter the hot-spot problem by modifying the scheme (see Cued Click-Points [6] and Persuasive Cued Click-Points [8]). However, one area that has not been extensively studied pertains to the resolution of the image that is used [13], the postulation being that high-resolution images would present the user with a larger password space, and therefore any hot-spots would dissipate. Although higher-resolution images have been explored in the past within the context of graphical authentication (e.g., gaze-based authentication [5]), locimetric authentication per-se has received little attention [13]. Taking an experimental approach, we seek to establish whether clustering persists with high-resolution images for locimetric authentication.

2 Background

The existence of hot-spots was initially speculated by Wiedenbeck et al. [31], who wrote, "logically, it seems that many users may be attracted to incongruous or unexpected elements in an image". Indeed, while theoretically locimetric authentication has the potential to be superior to text-based authentication [31], if users only select from specific regions, the effectiveness of the scheme drops. Several authors have reported the presence of hot-spots when studying the usability of locimetric authentication [29]. Others have attempted to evaluate whether the weakness can be exploited [11]. The first study we could find examining the usability of locimetric authentication was that by Wiedenbeck et al. [29], who investigated using the ClickPoints implementation while relying on images with a resolution of 451 by 331 pixels. When their study [29] was conducted in 2005, this resolution could be described as adequate. At the time of writing, it is considered a particularly low resolution. To account for backward compatibility, later studies retained the low-resolution specification. This includes other evaluations using PassPoints [30], web-based simulations inspired by PassPoints [26], Java-based simulations inspired by PassPoints [11], and Persuasive Cued Click-Points [8].

Dirik et al. [11] acknowledge the importance that resolution plays in the suitability of an image for use with locimetric authentication, writing that there are "a vast number of possibilities, if the image is large and complex, and if it has good resolution". Accordingly, increasing the resolution should then resolve the hot-spot problem. Indeed, as the resolution increases, there would hypothetically be more potential points for users to click on for their password. We were able to find one paper that examines high-resolution images within the context of the Picture Password mechanism: Gao et al.

[13], who undertook a holistic evaluation of the usability of Microsoft's Picture Password. Simulating the Windows 8 operating system, the authors do not explicitly state the size of the images used, although they do disclose that their experiment was conducted on "a PC with a 19-inch screen and 1024 x 1280 screen resolution". The authors do report the presence of hot-spots in all three of the images that they studied.

3 Methodology

We employed an experimental approach to evaluate the effect that usability plays on the security of locimetric authentication, looking at two image resolutions: low-resolution (451 by 331 pixels) and high-resolution (1280 by 720 pixels). A series of web-based experiments were held in which participants were each asked to generate seven locimetric passwords based upon preselected images, using software designed to simulate the password setup phase (see Fig. 1). The images selected were of a complex composition in order to ensure that participants had adequate points throughout each image to choose, thereby addressing the concern raised by Dirik et al. [11] vis-à-vis image suitability. Each password was comprised of five points; following its creation, participants were asked to reinput their password for verification. The number five was selected as the appropriate password length based on the majority of existing research examining the usability of locimetric authentication, which uses as a base Wiedenbeck et al.'s [31] PassPoints.

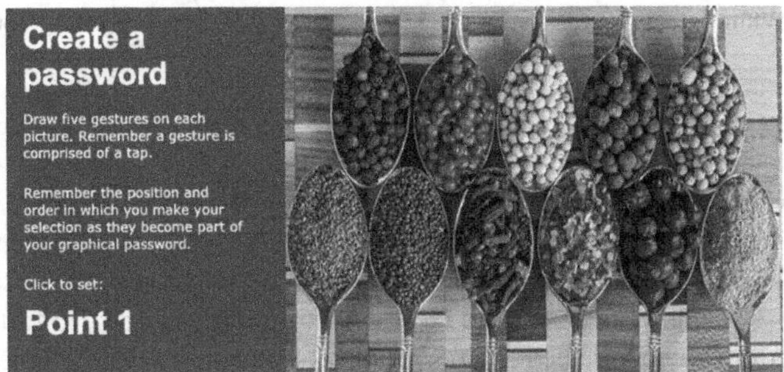

Fig. 1. Screen capture of software designed to simulate the creation of locimetric passwords.

3.1 Participant Recruitment and Profile

Participants were recruited using Amazon Mechanical Turk, which has become quite popular for usability studies. Our sample was comprised of a total of 204 participants from the United States. The first 102 participants were presented with the high-resolution condition and the subsequent 102 with the low-resolution condition. A range of requirements and level of compensation were used, as attracting participants varied in difficulty during the recruitment process. The breakdown by characteristic and condition (i.e., low- vs. high-resolution) is outlined in Table 1. A series of chi-squared tests indicates that there is no statistically significant difference in the structure of the groups with respect to factor.

Table 1. Participant profile

Factor	Category	Low resolution	High resolution	Combined
Gender	Female	39 (38.24%)	41 (40.20%)	80 (39.22%)
	Male	63 (61.76%)	61 (59.80%)	124 (60.78%)
Age	18–30	27 (26.47%)	28 (27.45%)	55 (26.96%)
	31–45	52 (50.98%)	51 (50.00%)	103 (50.49%)
	46 or older	23 (22.55%)	23 (22.55%)	46 (22.55%)
Education	<Undergraduate degree	24 (23.53%)	31 (30.69%)	55 (27.09%)
	Associate's degree	8 (7.84%)	6 (5.94%)	14 (6.90%)
	Bachelor's degree	47 (46.08%)	50 (49.50%)	97 (47.78%)
	Postgraduate	23 (22.55%)	14 (13.86%)	37 (18.23%)
Income	$10,000–$39,999	29 (29.00%)	34 (33.66%)	63 (31.34%)
	$40,000–$79,000	43 (43.00%)	45 (44.55%)	88 (43.78%)
	>$80,000	28 (28.00%)	22 (21.78%)	50 (24.88%)
Race	Asian	11 (10.78%)	5 (4.90%)	16 (7.84%)
	Black or African American	6 (5.88%)	14 (13.73%)	20 (9.80%)
	Other	0 (0.00%)	2 (1.96%)	2 (0.98%)
	White	85 (83.33%)	81 (79.41%)	166 (81.37%)

4 Analysis and Results

To identify whether clustering was present, we first generated scatterplots (see Fig. 2) to visualize where each of the password points was located on each of the images. This was done using the seaborn visualization package (version 0.11.1) [28]. We then inspected those scatterplots and found clear evidence of clustering. To further support this initial finding, we conducted a series of Clark-Evans tests [9], designed explicitly to identify spatial randomness, using R (version 4.0.3) [1, 2, 24] and, in particular, the spatstat package [3]. One can interpret the results of the Clark-Evans test by examining the R index, as "when $R = 0$, there is a limit situation of complete aggregation" and then "when $R = 1$ the pattern of distribution of individuals is random" [15]. To address the edge bias [23] inherent in the original Clark-Evans test, we relied on Donnelly's [12] correction. The results support the conclusion that clustering was present in all of the high-resolution images tested, as the R values were all between 0 and 1 and were statistically significant (see Table 2). Furthermore, the locimetric passwords based on high-resolution images exhibit a slightly lower level of clustering ($\mu = 0.40816$) than the corresponding low-resolution images ($\mu = 0.57608$).

4.1 Password Length

We also considered the possibility that the use of longer locimetric passwords (i.e., those with more points) might be influencing the presence of hot-spots and resulting in the

Fig. 2. Scatterplot of points that users selected on high-resolution images to form their locimetric password.

Table 2. List of images and corresponding spatial randomness

Image	Description	N	R: Low Resolution	R: High Resolution	
			5-points	3-points	5-points
1	Home Interior [18]	102	0.60042	0.41253	0.40277
2	Vegetables [22]	102	0.53065	0.38763	0.37804
3	Landscape [21]	102	0.63515	0.40670	0.39522
4	Vehicle [17]	102	0.52444	0.37172	0.33962
5	Spices [20]	102	0.54077	0.51709	0.51457
6	Hot Air Balloons [19]	102	0.52618	0.35804	0.33930
7	Drawing Tools [16]	102	0.67496	0.47499	0.48757
Average			0.57608	0.41839	0.40816

Note: For all R values, $p < 0.01$.

reuse of the same point(s) multiple times. To make that determination, we compared the presence of clustering within high-resolution images, looking at both the first three points selected by users as part of their passwords and at all five points (see Table 2). We conclude that increasing the points for passwords from three points to five points does not result in a level of clustering that is radically different (i.e., $\mu = 0.41839$ for three-point passwords vs. $\mu = 0.40816$ for five-point passwords). Thus, there is no evidence that longer passwords would lead to the reuse of password points. Conversely, longer locimetric passwords led to slightly less clustering, indicating that the passwords were marginally stronger.

Table 3. Number of users who apply patterns or reuse the same points

Image	Low Resolution			High Resolution		
	x-dim	y-dim	Both dimensions	x-dim	y-dim	Both dimensions
1	72	85	8	56	67	4
2	28	83	5	15	57	3
3	47	64	8	38	34	5
4	31	78	8	20	43	8
5	12	99	4	7	85	5
6	32	68	5	18	44	4
7	48	40	4	26	34	2
Average	38.57143	73.85710	6.00000	25.71430	52.00000	4.42857

A further investigation, this time to identify the number of users who reuse password points within a 10-pixel threshold, revealed that this figure varies by image. For high-resolution images it ranges between 2 and 8 ($\mu = 4.42857$) users. Similarly, for low-resolution images, the number of users was between 4 and 8 ($\mu = 6.00000$) with respect to image. Additionally, we sought to understand whether there were any patterns in the way in which users were selecting their password points. Specifically, we wanted to ascertain whether users would pick points following an imaginary horizontal or vertical line (i.e., within the same column or row of the image), again allowing for a 10-pixel threshold. For high-resolution images, we found that, depending on the image, anywhere from 7 to 85 users would generate passwords in the form of a horizontal line ($\mu = 25.71430$ users) or vertical line ($\mu = 52.00000$ users). For low-resolution images, the use of patterns by users was even more pronounced (this time between 12 and 99 users), with respect to both the horizontal ($\mu = 38.57143$ users) and vertical ($\mu = 73.85710$ users) dimensions. These results are outlined in Table 3.

4.2 Considering Demographics

The impact of demographics (i.e., age and gender) on the usability of locimetric passwords was explored for high-resolution five-point passwords. To accomplish this, we once again relied on the Clark-Evans test using Donnelly's edge correction (see Table 4). In the first instance, we compared participants who were over 35 years of age ($\mu = 0.43369$) to those who were 35 or younger ($\mu = 0.42161$). Age was found not to affect randomness, though younger participants generate slightly stronger passwords. In the second instance, we compared the level of clustering between males ($\mu = 0.41029$) and females ($\mu = 0.46251$) but found no consistent pattern amongst all the images evaluated; however, the mean scores indicate that men generate slightly stronger passwords.

Table 4. Clustering with respect to age and gender

Image	R: Age ≤ 35	R: Age >35	R: Male	R: Female
1	0.41439	0.48897	0.45468	0.45029
2	0.38615	0.41509	0.33840	0.44921
3	0.40558	0.37023	0.37482	0.49303
4	0.34447	0.35000	0.33546	0.44325
5	0.49613	0.51432	0.52483	0.47539
6	0.38198	0.38391	0.30218	0.40336
7	0.52260	0.51333	0.54163	0.52302
Average	0.42161	0.43369	0.41029	0.46251

Note: For all R values, $p < 0.01$

5 Discussion, Limitations, and Future Work

Through this work we established that hot-spots persist even with high-resolution images, exploring an extensive collection of images to come to this conclusion. Although the level of clustering was considerable, the findings do indicate that as resolution increases, the level of clustering decreases. The implications of these findings are two-fold. In the short term, they indicate that the hot-spot problem should be taken into consideration and addressed by both users and systems administrators so that approaches can be put in place to mitigate the presence of hot-spots. These could be either software-based (as demonstrated by the Cued Click-Points [6] and Persuasive Cued Click-Points [8] solutions) or training-based solutions (which have been demonstrated to be effective in the field of computer security [14, 27, 32]). In the longer term, these findings suggest that, at some point, as screen resolutions increase and locimetric authentication mechanisms adjust to support those higher resolutions, the hot-spot problem may dissipate.

Our research also endeavored to identify whether there were any patterns of user behavior that we could identify as being responsible for the manifestation of hot-spots in the high-resolution images that we were using for the experiment. First, we sought to determine whether user characteristics (i.e., age and gender) influence the formation of the hot-spots that were observed. The benefit of finding such a pattern would inform the allocation of resources (such as training) exclusively to those users. However, no such relationship was found; users generally appear to gravitate towards hot-spots equally. Second, we investigated whether users were reusing points or employing simple (i.e., horizontal or vertical) line patterns as part of their passwords, thereby introducing an element of predictability and further weakening the locimetric scheme. These are practices that we found to vary considerably by image, and slightly by resolution. In particular, point reuse in five-point passwords in the high-resolution condition varied between 1.96% and 7.84%, and pattern use ranged between 6.86% and 83.33%, depending on the image. These practices were more prominent in the low-resolution condition, where point reuse varied between 3.92% and 7.84% and pattern use varied between 11.76% and 97.06%, depending on the image.

However, we would be remiss if we did not point out that further research is necessary to confirm the aforementioned conclusions, specifically the finding that as resolution increases: 1) users rely less on the same points, and consequently hot-spots slowly fade away; 2) point reuse by users falls; 3) password points are less likely to appear in the form of a horizontal or vertical line. Accordingly, we identify two ways in which this line of inquiry could be advanced in the future. The first is to test locimetric authentication with even higher resolutions in order to confirm that the benefits realized from the increase of resolution continue and do not plateau. The second involves carrying out a field study that examines usage 'in the wild'. It is also evident that the composition of the image influences the level of clustering; this has been previously highlighted as a consideration regarding the suitability of images for use with locimetric authentication [11]. In our work, we controlled for complexity by taking considerable efforts to ensure that all of our images had a plethora of points for users to select. Nevertheless, differences were observed in the level of clustering between images. Consequently, a superior understanding of the interplay between composition and resolution is needed to understand better the benefits that can be realized by the increasing of image resolution.

Should the hot-spot problem be successfully addressed, the locimetric scheme has considerable potential to serve as a competitive alternative to traditional forms of authentication.

References

1. Baddeley, A., Turner, R., Mateu, J., Bevan, A.: Hybrids of Gibbs point process models and their implementation. J. Stat. Softw. **55**(11), 1–43 (2013). https://doi.org/10.18637/jss.v055.i11
2. Baddeley, A., et al.: Spatial Point Patterns: Methodology and Applications with R. Chapman and Hall/CRC Press, London (2015)
3. Baddeley, A., Turner, R.: Spatstat: an R package for analyzing spatial point patterns. J. Stat. Softw. **12**(6), 1–42 (2005)
4. Blonder, G.E.: Graphical password. Patent number: 5559961. United States Patent and Trademark Office (1996)
5. Bulling, A., et al.: Increasing the security of gaze-based cued-recall graphical passwords using saliency masks. In: Proceedings of the SIGCHI Conference on Human Factors in Computing Systems, pp. 3011–3020 ACM Inc., New York (2012). https://doi.org/10.1145/2207676.2208712
6. Chiasson, S., van Oorschot, P.C., Biddle, R.: Graphical password authentication using cued click points. In: Biskup, J., López, J. (eds.) ESORICS 2007. LNCS, vol. 4734, pp. 359–374. Springer, Heidelberg (2007). https://doi.org/10.1007/978-3-540-74835-9_24
7. Chiasson, S., et al.: Influencing users towards better passwords: persuasive cued click-points. In: Proceedings of the 22nd British HCI Group Annual Conference on People and Computers: Culture, Creativity, Interaction, vol. 1, pp. 121–130. BCS Learning & Development Ltd., Swindon (2008)
8. Chiasson, S., et al.: Persuasive cued click-points: Design, implementation, and evaluation of a knowledge-based authentication mechanism. IEEE Trans. Dependable Secure Comput. **9**(2), 222–235 (2012). https://doi.org/10.1109/TDSC.2011.55
9. Clark, P.J., Evans, F.C.: Distance to nearest neighbor as a measure of spatial relationships in populations. Ecology **35**(4), 445–453 (1954). https://doi.org/10.2307/1931034
10. De Angeli, A., et al.: Is a picture really worth a thousand words? Exploring the feasibility of graphical authentication systems. Int. J. Hum. Comput. Stud. **63**(1–2), 128–152 (2005). https://doi.org/10.1016/j.ijhcs.2005.04.020
11. Dirik, A.E., et al.: Modeling user choice in the PassPoints graphical password scheme. In: Proceedings of the 3rd Symposium on Usable Privacy and Security, pp. 20–28. ACM Inc., Pittsburg (2007). https://doi.org/10.1145/1280680.1280684
12. Donnelly, K.: Simulation to determine the variance and edge-effect of total nearest neighbour distance. In: Hodder, I. (ed.) Simulation Methods in Archeology, pp. 91–95. Cambridge University Press, Cambridge (1978)
13. Gao, H., Jia, W., Liu, N., Li, K.: The hot-spots problem in Windows 8 graphical password scheme. In: Wang, G., Ray, I., Feng, D., Rajarajan, M. (eds.) CSS 2013. LNCS, vol. 8300, pp. 349–362. Springer, Cham (2013). https://doi.org/10.1007/978-3-319-03584-0_26
14. Huang, D.-L., Rau, P.-L.P., Salvendy, G.: A survey of factors influencing people's perception of information security. In: Jacko, J.A. (ed.) HCI 2007. LNCS, vol. 4553, pp. 906–915. Springer, Heidelberg (2007). https://doi.org/10.1007/978-3-540-73111-5_100
15. Petrere, M.: The variance of the index (R) of aggregation of Clark and Evans. Oecologia **68**(1), 158–159 (1985). https://doi.org/10.1007/BF00379489

16. Pixabay: Brush Chalk Color Atelier Paint. https://pixabay.com/photos/brush-chalk-color-atelier-paint-2927793/. Accessed 2 Feb 2021
17. Pixabay: Car Vehicle Motor Transport. https://pixabay.com/photos/car-vehicle-motor-transport-3046424/. Accessed 2 Feb 2021
18. Pixabay: Home Interior Room House Furniture. https://pixabay.com/photos/home-interior-room-house-furniture-1438305/. Accessed 2 Feb 2021
19. Pixabay: Hot Air Balloons Adventure Balloons. https://pixabay.com/photos/hot-air-balloons-adventure-balloons-1867279/. Accessed 2 Feb 2021
20. Pixabay: Mat Spices. https://pixabay.com/photos/mat-spices-3251064/. Accessed 2 Feb 2021
21. Pixabay: Santorini City Greece Tourism. https://pixabay.com/photos/santorini-city-greece-tourism-4044972/. Accessed 2 Feb 2021
22. Pixabay: Vegetables Carrots Garlic Celery. https://pixabay.com/photos/vegetables-carrots-garlic-celery-1212845/. Accessed 2 Feb 2021
23. Pommerening, A., Stoyan, D.: Edge-correction needs in estimating indices of spatial forest structure. Can. J. For. Res. **36**(7), 1723–1739 (2006). https://doi.org/10.1139/x06-060
24. R Core Team: A language and environment for statistical computing. R Foundation for Statistical Computing, Vienna, Austria (2013). http://www.R-project.org/
25. Stobert, E., et al.: Exploring usability effects of increasing security in click-based graphical passwords. In: Proceedings of the 26th Annual Computer Security Applications Conference, pp. 79–88. ACM Inc., New York (2010). https://doi.org/10.1145/1920261.1920273
26. Thorpe, J., Van Oorschot, P.C.: Human-seeded attacks and exploiting hot-spots in graphical passwords. In: Proceedings of the 16th USENIX Security Symposium, pp. 103–118 (2007). https://www.usenix.org/legacy/events/sec07/tech/full_papers/thorpe/thorpe.pdf. Accessed 11 June 2021
27. Ur, B., et al.: How does your password measure up? The effect of strength meters on password creation. In: Proceedings of the 21st USENIX Security Symposium, pp. 65–80 (2012). https://www.usenix.org/system/files/conference/usenixsecurity12/sec12-final209.pdf. Accessed 11 June 2021
28. Waskom, M., et al.: mwaskom/seaborn: v0.11.1. Zenodo (2020). https://zenodo.org/record/4379347#.YMQHVjZKh6M. Accessed 11 June 2021
29. Wiedenbeck, S., et al.: Authentication using graphical passwords: Basic results. In: Proceedings of the 11th International Conference on Human-Computer Interaction International (HCII 2005), Las Vegas, NV (2005). http://www.jimwaters.info/pubs/Graphical-Password-Basic-Results-2005.pdf. Accessed 11 June 2021
30. Wiedenbeck, S., et al.: Authentication using graphical passwords: Effects of tolerance and image choice. In: Proceedings of the 2005 Symposium on Usable Privacy and Security (2005), pp. 1–12 (2005). https://doi.org/10.1145/1073001.1073002
31. Wiedenbeck, S., et al.: PassPoints: design and longitudinal evaluation of a graphical password system. Int. J. Hum. Comput. Stud. **63**(1), 102–127 (2005). https://doi.org/10.1016/j.ijhcs.2005.04.010
32. Yıldırım, M., Mackie, I.: Encouraging users to improve password security and memorability. Int. J. Inf. Secur. **18**(6), 741–759 (2019). https://doi.org/10.1007/s10207-019-00429-y

Usability Assessment of the GoPro Hero 7 Black for Chinese Users

Guo Sheng-nan(✉), Chen Jia(✉), Chang Le(✉), Jiayu Zeng(✉), and Marcelo M. Soares(✉)

School of Design, Hunan University, Hunan 410000, People's Republic of China

Abstract. In today's Internet age, video recordings of life called vlogs are popular. Most of the vlogs were shot with a motion camera, and it is necessary to improve some specific requirements of sports cameras. This study chooses GoPro Hero 7 Black and explores the usability assessment of this product for Chinese users. The research also aims to measure usability through a series of questionnaires. Analytical methods are used to determine its advantages and any existing functional problems. Based on the results of the usability assessment, some suggestions were given.

Keywords: Usability · Sports camera · Chinese users · Measurement · Design

1 Introduction

With the development of technology and the improvement of people's standard of living, the requirements for sports cameras are increasing. A wide variety of different sports cameras have emerged in the market, but most sports camera products still fall short in terms of usability. By using usability research, it's easy to discover some of the problems in the interaction design of existing brands of sports cameras. Then designers can solve these problems step by step and suggest improvements to bring a better user experience.

The purpose of this study is to explore whether the features of the GoPro Hero 7 Black model are suitable for Chinese users and whether there are usability issues with this sports camera. The GoPro Hero 7 Black is small and portable and very high in pixels and takes very clear shots. It's waterproof and anti-shake, making it perfect for shooting in sports. Although this brand's products are currently well received in the market, it still has some issues that can be addressed. Therefore, this study plans to conduct experimental usability studies, related theoretical studies, and technical analyses and try to identify some specific problems that will lead to recommendations for improvement [1].

2 Experiment Method

The study is based on user experience measurement and usability metrics. According to the international organization for standardization, the ISO International Organization

C. Stephanidis et al. (Eds.): HCII 2021, LNCS 13094, pp. 397–411, 2021.
https://doi.org/10.1007/978-3-030-90238-4_28

for Standardization mentioned that the usability of something should be measured in three dimensions: effectiveness, efficiency, and satisfaction [2].

The definition of effectiveness refers to whether the user task is completed or not and whether the task is completed valid. Efficiency is defined as the amount of effort required to accomplish a goal. Satisfaction refers to the level of comfort that the users feel when using a product and how acceptable the product is to users to achieve their goals [3]. To more accurately analyze the performance of the camera, we adopted the way of comparative experiment in the study. We selected another camera called YI 4K, a competing product for analysis to compare the performance differences between the two [4].

2.1 Materials

Electronic Equipment. For this usability test experiment, the study chose the model GoPro Hero 7 Black. Since this sports camera is equipped with a mobile phone app that allows for the transfer and saving of images, the other device used in the experiment was an iPhone 7 (Fig. 1).

Fig. 1. GoPro Hero 7 Black

Experimental Informed Consent. In many settings, obtaining explicit consent from participants is legally necessary. A consent form usually specifically notes that the study is anonymous. Therefore, before the experiment begins, a signed consent form needs to be prepared for the testers to live (Fig. 2).

Instruction Sheet. The content of the instruction sheet is a brief introduction to the product's functions and basic function operation steps. The purpose is to allow testers to understand the product GoPro Hero 7 Black in the shortest time and know the basic operating steps, which is conducive to the smooth completion of the later task experiment.

Experimenter Notes Page. In order for staff to record user data and responses more effectively during the course of the experiment, an experimental record page needs to be developed in advance to save time and efficiently record the data needed to facilitate later experimental analysis and problem improvement.

Experimental Informed Consent

Dear Participant:

You will be participating in a study of a course experiment and this note will give you some information to help you decide whether to participate in this study. Please read the following carefully and if you have any questions, please ask the researcher in charge of this experiment: participation in this experiment is a voluntary process and you can opt out at any time. The data results obtained from this experiment are for course research purposes only, will be anonymized prior to data use, will not have any effect on your personal data, and all records will be kept confidential to the highest degree possible by law.

Objective of the experiment: to compare the usability of the study products

Experimental process: after reading the instructions carefully, perform the task, complete the task, fill in the questionnaire and submit it to the experimental director.

Signature: ____________

Fig. 2. Experimental Informed Consent

2.2 Participants

Eight participants were aged from 18 to 20, including four men and four female. They were divided into two groups after they signed the Informed Consent Form: novice and expert [4]. For the novice group, participants were asked to read the instruction manual carefully and then handed out task manuals to guide them through the task. The Group of Experts, in turn, received the assignment directly and began the experiment. During the course of the experiment, the research kept a constant record of the testers' reactions, the number of requests for help, etc. Immediately after the experiment, participants were all asked to complete a questionnaire. These questionnaires are used to collect subjective and qualified data for the data analysis of the usability of the products.

2.3 Questionnaire Design

At the end of the experiment, users were asked to fill out two types of questionnaires: the user information table questionnaire and the post-mission satisfaction questionnaire (Fig. 3).

User Information Table. The information user should fill user information questionnaire (Fig. 3)**. The questionnaire** included participants' basic information and their experience with GoPro. The researcher could become familiar with the participants and easily divided them into novice or expert. It helps the follow-up study of different types of tests used to analyze the change.

Post-Mission Satisfaction Questionnaire. The post-mission questionnaire (Fig. 3) was based primarily on the Likert scale [5], which sets up a series of feedback about how participants interact with the GoPro's interface, buttons, operations, and so on as they complete the task [6]. Participants give their rating according to how much they agree with each of the following statements. Participants' scores for each question were quantified and analyzed during subsequent analysis.

After doing the experiment, give your rating based on how much you agree with each of the following statements

1. I think it is easy to accomplish these tasks.

Strongly disagree -3 -2 -1 0 1 2 3 Strongly agree

2. I did not make an operational error.

Strongly disagree -3 -2 -1 0 1 2 3 Strongly agree

3、I was able to quickly identify and revert to the correct operation when a mistake was made.

Strongly disagree -3 -2 -1 0 1 2 3 Strongly agree

4、I really like the styling of this product.

Strongly disagree -3 -2 -1 0 1 2 3 Strongly agree

5、I think the design of this product is very user-friendly.

Strongly disagree -3 -2 -1 0 1 2 3 Strongly agree

6、I think this product buttons (such as switch, photo button) set very reasonable.

Strongly disagree -3 -2 -1 0 1 2 3 Strongly agree

7、I have a comfortable operating experience with no physical discomfort.

Strongly disagree -3 -2 -1 0 1 2 3 Strongly agree

8、I think this product is very small and light, easy to carry.

Strongly disagree -3 -2 -1 0 1 2 3 Strongly agree

9、I think this product is easy to understand how to operate and use.

Strongly disagree -3 -2 -1 0 1 2 3 Strongly agree

10、I think the images and text on the product are very clear and easy to understand.

Strongly disagree -3 -2 -1 0 1 2 3 Strongly agree

11、I think the product can give me clear information to help me when using it.

Strongly disagree -3 -2 -1 0 1 2 3 Strongly agree

12、I was able to do what I wanted to do very quickly.

Strongly disagree -3 -2 -1 0 1 2 3 Strongly agree

13、I think the product is responsive and fast in use.

Strongly disagree -3 -2 -1 0 1 2 3 Strongly agree

14、I think the design of the font size and color on the product screen is very reasonable.

Strongly disagree -3 -2 -1 0 1 2 3 Strongly agree

15、I really like the overall color scheme of the product (including the shape and screen).

Strongly disagree -3 -2 -1 0 1 2 3 Strongly agree

16、I think this product takes very good photos and videos.

Strongly disagree -3 -2 -1 0 1 2 3 Strongly agree

17、It was easy for me to shoot the effect I wanted.

Strongly disagree -3 -2 -1 0 1 2 3 Strongly agree

18、Would you consider buying this product if you have a need for it afterwards (taking daily videos, recording sports moments, etc.)? Why?

19、For this product, do you think it has any shortcomings?

20、What do you think about the whole process of using it?

Fig. 3. Post-mission satisfaction questionnaire

2.4 Task Analysis

The study defined camera operation into five steps, including 1) opening and connecting, 2) fixing, 3) taking a photo, 4) running and taking a video, 5) reviewing and saving to the phone. Firstly, the study analyzes the subtask in detail and optimizes the process [7].

2.5 Design Method

Field Observation. Field research is defined as a qualitative method of data collection that aims to observe, interact and understand people while they are in a natural environment (cite the reference). The purpose of the field observation method is to obtain participants' data, such as the time required for each participant to complete the task and the number of times the participant asked for help in completing the task without interfering with the participant's test.

During the experiment, the observers need to encourage users to communicate their feelings about their use, as well as the participants to communicate with each other and raise questions. On the one hand, the researchers should minimize the output of subjective opinions, on the other hand, they should guide the participants to complete the task, and at the same time, the researchers should record the objective data of the experiment to facilitate the subsequent analysis.

The key of field observation should answer the following questions: 1) Did the user ask for help when completing the task? If they ask for help, the product task can't be done on its own and may contain usability issues. 2) Did the novice and expert groups spend too much time on tasks? If so, there is a problem with the efficiency of the product. 3) Does the user feel dissatisfied or uncomfortable when completing the task? If so, the product may determine low user satisfaction on a particular operation.

Next, the researcher will analyze three task steps through field observation: effectiveness, efficiency, and learnability. The effectiveness and efficiency of the experiment can be analyzed by studying the time and number of calls for help of different participants. Meanwhile, the user's evaluation of the product satisfaction obtained by the experimenter during site observation will also help this research analyzing the product satisfaction. And it is worth noting that some participants have less communication with the experimenter due to personality reasons, which may result in a lack of data. At this point, the experimenter needs to further encourage and communicate with the participant in the experiment.

Meanwhile, the completion time of the experiment may be affected by different external reasons, so the experimenter needs to ensure that each user can be as focused and free from interference as possible when completing the experiment.

Questionnaire and Interview. The questionnaire was filled out immediately after the experiment. According to the method of the Likert six-point scale, the questionnaire of the experiment set different options for effectiveness, learnability, and fault tolerance. Participants were asked to score each option to obtain qualitative data.

The interview was conducted after the user completes the questionnaire, and the participants will be asked questions about the effectiveness and satisfaction of the product, respectively. Users are encouraged to express their opinions in the interview freely. The answers were registered for each participant. The participants' answers in the interview help us to evaluate the product qualitatively.

3 Analysis

This study collects data on user's behavior and attitude during the experiment and the time spent to analyze the user's task and analyze data to obtain indicators that include learnability, effectiveness, fault tolerance, and user satisfaction. Through descriptive analysis, the overall data was used to describe the basic characteristics of the variables, and quantitative data such as mean, standard deviation, and deviation were obtained. Through the mean test, the differences between the questions affecting each evaluation criterion (learnability, validity, fault tolerance, and user satisfaction) and specific factors affecting the evaluation criteria were analyzed [8].

3.1 Materials

Qualitative data [9] include user-user attitudes towards the various tasks when conducting the experiment, errors made, and time spent carrying out the experiment. These data are objective evaluations that show the problems that exist when the product is used.

In this part of the study, user data was collected through observations and interviews. It was recorded and observed the questions and mistakes made by the users in the course of the experiment Table 1, interview and record the users' scores on different aspects of the experiment task and calculate the average Table 2.

Table 1. Problems arising from the experiment.

	Open GoPro	Connect app	Fix	Take a photo	Take a video	Review	Save to the phone
Task success time		The connection to the phone is slow		Voice control slow	Voice control slow		The connection to the phone is slow
Error	The shoot button is confused with the on/off button		Can't find a way to use the fastener.	The delay of the shooting mode cannot be selected			
Efficiency Learnability			The fixing frame is difficult to use				

Table 2. Problems arising from the experiment.

	Turn on	Connect app	Fix	Take a photo	Take a video	Review	Save to the phone
GoPro	2.85	2.625	2.5	2.625	2.65	2.95	2.85
YI 4K	2.675	2.825	2.8	2.575	2.625	2.8	2.9

Based on the analysis, it is known that the fixed performance of GoPro is significantly weaker, users can obviously feel inconvenience when using it, and the connection efficiency with mobile phones and apps is low. Users who use GoPro's buttons and features feel significantly better than competitors, but it's important to note that GoPro still scores poorly when taking photos and videos.

3.2 Quantitative Data

Quantitative data included information obtained from the reconciliation questionnaire and the time taken for subjects to complete the experiment.

This study divided the questions in the sub-volumes into four categories, reflecting the learnability, effectiveness, fault tolerance, and user satisfaction of the product [10]. Product learnability includes Q1, Q2, Q9 Table 3, effectiveness includes Q6, Q10, Q11, Q12, Q14, Q17 Table 4, fault tolerance includes Q3 Table 5, user satisfaction includes Q4, Q5, Q7, Q8, Q13 Table 6, Q15, Q16 Table 7. An independent sample t-test analyzed comparative products.

Fundamental analysis. The amount of time subjects took to complete the task and the number of times they sought help reflected the learnability of the product. The data in the Table 3 shows that the minimum time to complete the task was 1, the maximum was 7, and the mean was 3.37, which is in line with the normal completion time, although the difference in the completion time was large, all were completed within the specified time. The bias for both variables was positive, indicating that the majority of subjects completed the task in a lower-than-average time and number of requests for help, suggesting that the product is more learnable.

Table 3. Descriptive analysis of product learnability

	N	Range	Minimum	Maximum	Mean	Std. deviation	Skewness	
	Statistic	Statistic	Statistic	Statistic	Statistic	Statistic	Statistic	Std. Error
Q1	8	1	2	3	2.87	.354	−2.828	.752
Q2	8	4	−1	3	2.13	1.356	−2.126	.752
Q9	8	1	2	3	2.88	.354	−2.828	.752
Valid N (listwise)	8							

Based on the analysis of the questionnaire, the results obtained from questions 6, 10, 11, 12, 14, and 17 reflect the effectiveness of the product. The data in the Table 4 shows that the minimum value of the six variables is 2, 1, 1, 0, 1, 1, maximum value is 3, the average value is 2.63, 2.63, 2.38, 2.00, 2.00, 2.25, the overall average value is 2.32, it can be seen that users are more satisfied with the effectiveness of this product. There were five variables with a negative bias, indicating that relative to the normal distribution, the distribution graph in this case of the data was trailing to the left and the peak to the right, indicating that the subjects mostly scored higher on the effectiveness of the product.

Based on the analysis of the questionnaire, the results obtained from question 3 reflect the fault tolerance of the product. The data in the Table 5 shows that the minimum value of this variable is 1. The maximum value is 3, and the mean value is 2.63, respectively, which shows that the subjects are satisfied with the fault tolerance of the product. The bias of this variable is negative, indicating that relative to the normal distribution, the distribution graph of the data, in this case, is trailing on the left, and the peaks are to the right, indicating that the subjects mostly scored high on product fault tolerance.

Table 4. Descriptive analysis of product effectiveness

	N	Range	Minimum	Maximum	Mean	Std. deviation	Skewness	
	Statistic	Statistic	Statistic	Statistic	Statistic	Statistic	Statistic	Std. Error
Q6	8	1	2	3	2.63	.518	−.644	.752
Q10	8	2	1	3	2.63	.744	−1.951	.752
Q11	8	2	1	3	2.38	.744	−.824	.752
Q12	8	3	0	3	2.00	.926	−1.440	.752
Q14	8	2	1	3	2.00	.756	.000	.752
Q17	8	2	1	3	2.25	.707	−.404	.752
Valid N (listwise)	8							

Table 5. Descriptive analysis of product fault tolerance

	N	Range	Minimum	Maximum	Mean	Std. Deviation	Skewness	
	Statistic	Statistic	Statistic	Statistic	Statistic	Statistic	Statistic	Std. Error
Q3	8	2	1	3	2.63	.744	−1.951	.752
Valid N (listwise)	8							

According to the analysis of the questionnaire, the results obtained from questions 4, 5, 7, 8, 13, 15, and 16 reflect user satisfaction with the product. The data in the Table 6 show that the minimum values of the seven variables are −1, −1, 2, −1, 1, 0, 2, and the maximum value is 3. The mean values are 2.00, 1.88, 2.87, 2.50, 1.88, 2.38, 2.63, and the overall mean value is 2.31, which shows that the subjects are more satisfied with the product. Five variables have a negative bias, indicating that the distribution graphs, in this case, are trailing on the left. The peaks are to the right, meaning that subjects mostly scored higher on their satisfaction with the product, relative to the normal distribution, and Q13 has a positive bias, indicating that subjects mostly scored lower on their reaction rate to the product.

Comparative User Analysis. The data in the Table 7 shows that for novices, the mean values of these three variables are 2.75, 1.25, 3.00, and the overall mean value is 2.33; for experts, the mean values of these three variables are 3.00, 3.00, 2.75 and the overall mean value is 2.92. It can be seen that experts are more likely to use the product and understand how to use it. In Q2, the lowest score given by the novice is −1, which shows that the novice has made a mistake in using the product.

Table 6. Descriptive analysis of user satisfaction

	N	Range	Minimum	Maximum	Mean	Std. deviation	Skewness	
	Statistic	Statistic	Statistic	Statistic	Statistic	Statistic	Statistic	Std. Error
Q4	8	4	−1	3	2.00	1.414	−1.616	.752
Q5	8	4	−1	3	1.88	1.246	−2.056	.752
Q7	8	1	2	3	2.87	.354	−2.828	.752
Q8	8	4	−1	3	2.50	1.414	−2.828	.752
Q13	8	2	1	3	1.88	.835	.277	.752
Q15	8	3	0	3	2.38	1.061	−1.960	.752
Q16	8	1	2	3	2.63	.518	−.644	.752
Valid N (listwise)	8							

Table 7. Data on product learnability for different users

Novice	Mean	2.75	1.25	3.00
	N	4	4	4
	Minimum	2	−1	3
	Maximum	3	2	3
Expert	Mean	3.00	3.00	2.75
	N	4	4	4
	Minimum	3	3	2
	Maximum	3	3	3
Total	Mean	2.87	2.13	2.88
	N	8	8	8
	Minimum	2	−1	2
	Maximum	3	3	3

The data in the Table 8 shows that for novices, the mean value of these six variables is 2.75, 2.50, 2.50, 1.50, 2.25, 2.50, and the overall mean value is 2.33; for experts, the mean value of these six variables is 2.50, 2.75, 2.25, 2.50, 1.75, 2.00 and the overall mean value is 2.29. It can be seen that experts have higher requirements for product effectiveness. With multiple data subjects giving lower breakups, it can be seen that there is room for improvement in the effectiveness of this product.

The data in the Table 9 shows that the mean value of this variable is 2.25 for novices and 3.00 for experts, which shows that the product is more fault-tolerant.

Table 8. Data on product validity for different users

		Q6	Q10	Q11	Q12	Q14	Q17
Novice	Mean	2.75	2.50	2.50	1.50	2.25	2.50
	N	4	4	4	4	4	4
	Minimum	2	1	1	0	1	2
	Maximum	3	3	3	2	3	3
Expert	Mean	2.50	2.75	2.25	2.50	1.75	2.00
	N	4	4	4	4	4	4
	Minimum	2	2	2	2	1	1
	Maximum	3	3	3	3	2	3
Total	Mean	2.63	2.63	2.38	2.00	2.00	2.25
	N	8	8	8	8	8	8
	Minimum	2	1	1	0	1	1
	Maximum	3	3	3	3	3	3

Table 9. Data on product fault tolerance for different users

	Mean	N	Minimum	Maximum
Novice	2.25	4	1	3
Expert	3.00	4	3	3
Total	2.63	8	1	3

The data in the Table 10 shows that for novices, the mean value of these six variables is 2.75, 2.50, 2.50, 1.50, 2.25, 2.50, and the overall mean value is 2.33; for experts, the mean value of these six variables is 2.50, 2.75, 2.25, 2.50, 1.75, 2.00 and the overall mean value is 2.29. It can be seen that experts have higher requirements for product effectiveness. With multiple data subjects giving lower breakups, it can be seen that there is room for improvement in the effectiveness of this product.

Comparative Product Analysis. Based on the comparison of the data in the Table 11, the average of the scores obtained by GoPro 7 black and YI4K Action on the three questions representing the learnability of the products is not significantly different, indicating that the learnability of the two products is not significantly different.

According to the data comparison in the Table 12, the overall average of the six issues of YI4K Action's product effectiveness is smaller than GoPro 7 black, with relatively large differences in the average of Q10, Q11, Q12, and Q13. Q10's score GoPro 7 black is larger than YI4K Action. The remaining three YI4K Actions are larger than GoPro 7 black, indicating that the clarity of images and text on GoPro 7 black is relatively good. The clarity of YI4K Action's prompt information and the design of the font size and

Table 10. Data on User satisfaction for different users

		Q4	Q5	Q7	Q8	Q13	Q15	Q16
Novice	Mean	2.00	1.75	3.00	3.00	1.75	2.25	2.75
	N	4	4	4	4	4	4	4
	Minimum	−1	−1	3	3	1	0	2
	Maximum	3	3	3	3	3	3	3
Expert	Mean	2.00	2.00	2.75	2.00	2.00	2.50	2.50
	N	4	4	4	4	4	4	4
	Minimum	1	2	2	−1	1	2	2
	Maximum	3	2	3	3	3	3	3
Total	Mean	2.00	1.88	2.87	2.50	1.88	2.38	2.63
	N	8	8	8	8	8	8	8
	Minimum	−1	−1	2	−1	1	0	2
	Maximum	3	3	3	3	3	3	3

Table 11. Compare and contrast product learnability

		N	Mean	Std. Deviation	Std. Error Mean
Q1	GoPro 7 black	8	2.88	.354	.125
	YI 4K Action	8	2.75	.463	.164
Q2	GoPro 7 black	8	2.13	1.356	.479
	YI 4K Action	8	2.00	1.414	.500
Q9	GoPro 7 black	8	2.88	.354	.125
	YI 4K Action	8	2.75	.463	.164

color on the product screen is relatively good, which can be focused on improving these two aspects of GoPro 7 black.

According to the data comparison in the Table 13, there is no significant difference between the YI4K Action and GoPro 7 black in terms of product fault tolerance.

According to the data comparison in the Table 14, the difference between YI4K Action and GoPro 7 black in the scores of the six questions representing user satisfaction is significant. Specifically, the average of Q13 GoPro 7 black is smaller than YI4K. The average of the remaining question variables GoPro 7 black is larger than YI4K Action, The YI4K Action's responsiveness is good, and the GoPro 7 black's responsiveness can be improved.

Table 12. Compare and contrast product validity

		N	Mean	Std. Deviation	Std. Error Mean
Q6	GoPro 7 black	8	2.63	.518	.183
	YI 4K Action	8	2.63	.518	.183
Q10	GoPro 7 black	8	2.63	.744	.263
	YI 4K Action	8	2.38	.744	.263
Q11	GoPro 7 black	8	2.38	.744	.263
	YI 4K Action	8	2.63	.518	.183
Q12	GoPro 7 black	8	2.00	.926	.327
	YI 4K Action	8	2.38	.744	.263
Q14	GoPro 7 black	8	2.00	.756	.267
	YI 4K Action	8	2.75	.463	.164
Q17	GoPro 7 black	8	2.25	.707	.250
	YI 4K Action	8	2.13	.835	.295

Table 13. Compare and contrast product fault tolerance

		N	Mean	Std. Deviation	Std. Error Mean
Q3	GoPro 7 black	8	2.63	.744	.263
	YI 4K Action	8	2.63	.744	.263

Table 14. Compare and contrast User satisfaction

		N	Mean	Std. Deviation	Std. Error Mean
Q4	GoPro 7 black	8	2.00	1.414	.500
	YI 4K Action	8	1.00	1.069	.378
Q5	GoPro 7 black	8	1.88	1.246	.441
	YI 4K Action	8	1.13	1.126	.398
Q7	GoPro 7 black	8	2.88	.354	.125
	YI 4K Action	8	2.00	.756	.267
Q13	GoPro 7 black	8	1.88	.835	.295
	YI 4K Action	8	2.75	.463	.164
Q15	GoPro 7 black	8	2.38	1.061	.375
	YI 4K Action	8	2.13	.835	.295
Q16	GoPro 7 black	8	2.63	.518	.183
	YI 4K Action	8	1.75	.707	.250

4 Findings

4.1 Preliminary Ideas

From the data this study took on user reactions during the task, this study shows that installing GoPro is not easy, and most users get stuck. Simultaneously, some users express confusion about the parts indicated in the manual, so they often make mistakes at this step.

This research also used interviews to go into more detail, asking users how they felt about the experience and asking them to bring up some of their perceived shortcomings. While novices tend to find the product novel and interesting, experts have made some ideas about the design of the product based on their own previous experience.

4.2 Problems and Improvements

Based on the data analysis results and user interviews, this research collated some of the problematic points to be solved and conducted group discussions to address them, suggesting some possible improvement measures.

Since outdoor sports electronics are small, this also results in a small font being displayed. This is very unfriendly for some people with thicker fingers or poor eyesight [11]. Therefore, this study proposes to add a font adjustment in the user preferences, which can be adjusted to the large font, medium font, and small font, so that users can choose the font that suits them.

Simultaneously, the meaning of specific icons is somewhat challenging to understand and remember, and users always forget the meaning of certain icons during use. One solution to this problem is that when a user presses an icon for a long period of time, the meaning of the icon will pop up, making the process smoother for the user.

Some experienced users mentioned in interviews that they had difficulty setting certain parameters of the camera, such as the data adjustment page being too full of content and not knowing which parts could be set for which values. Therefore, this study envisions adding a guide arrow next to the adjustable content of this data interface, and when the user clicks on the arrow, the display will pull up the adjustable value pairing, allowing the user to make a quick selection.

One female user found it difficult to view the image when posing for a selfie with the camera, and another female tester suggested after completing the task that the selfie did not work well because she could not see the image while taking a selfie. This study came up with a way, that is, the small screen next to the lens can also display the shooting screen. When the users take a selfie, they can slide the small screen from the main interface to the image to adjust the position and angle, shooting a more satisfactory.

Although the material chosen for this product is frosted and has a certain anti-slip effect, it still has the disadvantage of being easy to slip off. Friction can be increased by designing a layer of camera housing texture to prevent the camera from dropping.

5 Conclusion

This article explores the usability of the sports camera GoPro 7 black, working as follows:

- Firstly, this paper selects the YI 4K Action Camera, a comparison product with similar features for the target product, and conducts initial research on two different sports cameras.
- Secondly, this paper designed a list of usability testing tasks based on user operations, produced instruction manuals, and user task books, and designed questionnaires in a targeted manner.
- Thirdly, based on the data obtained from the experiment, various forms of data analysis were conducted, and based on the results of the study, the GoPro 7 black was found to have problems in the design of functional modules and user interface, and compared with the YI 4K Action Camera to find out the advantages and problems of GoPro 7 black.

Finally, based on the results of the analysis, the problems that need to be solved are analyzed, and the ideal way to improve them is arrived at through an in-group assessment.

The main conclusions drawn from this paper are the following:

- Firstly, it is known through analysis that expert users will have higher requirements for the effectiveness of the product than novice users. In order to meet the needs of users who are constantly improving their standards, the sports camera industry should actively develop new systems to adapt to the needs of different users.
- Secondly, the learnability of the product is very important for novice users. If the interactive features are complicated and difficult to understand, it will make novice users irritable, thus affecting the user experience. It can add a user tutorial function to the camera before using the camera.
- Thirdly, the responsiveness of sports cameras affects user satisfaction more than the imaging results. Although responsiveness is not a problem that this study as designers can solve, this study still hopes that the sports camera industry, especially GoPro, will focus on this issue and enhance the hardware input of cameras to improve the responsiveness of cameras.

Finally, because of its small screen, the interface of sports cameras is not much different, but users of sports cameras have their own different needs. Sports camera brands should enhance the diversity of their user interface and add more user personalization features to give users a better experience.

Many of the research methods in this paper are based on the results of previous research, drawing on the successful experiences of previous people and enhancing the depth of the research. This study hopes that the research in this paper will inform subsequent relevant designs.

Due to the limited research time and the limited capacity of the members of the subject group, there are still many shortcomings in this paper that need to be added and improved:

- Firstly, this study is not professional technicians, and the analysis of some tasks and results may not be accurate. This study welcomes all professionals or sports camera enthusiasts to correct.
- Secondly, the specific usability problems are investigated and evaluated in this research. However, due to the limited time, limited access to new techniques, and some other factors, there are still limitations in the following two aspects. One is that this research was not able to recruit enough participants, especially for the questionnaire. The other is that there is a shortcoming in the methodology of the field observation used in the experiment.

Sports cameras are loved by sports fans and video-sharing bloggers around the world because of their small and convenient size. GoPro is the strongest company in the sports camera market right now. In fact, its features are more complete in all aspects, its anti-shake and waterproof performance are unmatched in the sports camera market. With the expanding number of users and the expanding product market, sports camera brands should invest more in research and adapt to the latest market, so that research can be better and meet the needs of more users and discover their own irreplaceability.

References

1. Zhang, Y., Zeng, Y.: Research on product usability design methods. Eng. Constr. Design (01), 137–139 (2013). (in Chinese)
2. Madan, A., Kumar, S.: Usability evaluation methods: a literature review. Int. J. Eng. Sci. Technol. **4**(2) (2012)
3. Jordan, P.W.: An Introduction to Usability. CRC Press, Boco Raton (1998)
4. Huang, K.Y.: Challenges in human-computer interaction design for mobile devices. Lect. Notes Eng. Comput. Sci. **2178**(1), 235–236 (2009)
5. Likert, R.: A technique for the measurement of attitudes. Arch. Psychol. **22**(140), 55 (1932)
6. Wang, B., Sheng, J., Li, Y.: A review of human-machine interface usability testing and evaluation research. Mod. Comput. (Prof. Ed.) **000**(011), 26–28 (2012). (in Chinese)
7. Nielsen, J.: The usability engineering lifecycle. Computer **25**(3), 12–22 (1999)
8. Davies, I.K.: Task analysis. Educ. Tech. Res. Dev. **21**(1), 73–86 (1973)
9. Cowan, G.: Statistical Data Analysis. Oxford University Press, Oxford (1998)
10. Ye, D., Li, S.: Hierarchy of requirements and design strategies in interaction design. Packag. Eng. **34**(08), 75–78 (2013). (in Chinese)
11. Jiang, X.: Digital product interface design discussion. Packag. Eng. **06**, 188–190 (2008). (in Chinese)

A Study on Dual-Language Display Method Using the Law of Common Fate in Oscillatory Animation on Digital Signage

Takumi Uotani[1(✉)], Yuki Takashima[1], Kimi Ueda[1], Hirotake Ishii[1], Hiroshi Shimoda[1], Rika Mochizuki[2], and Masahiro Watanabe[2]

[1] Graduate School of Energy Science, Kyoto University, Kyoto, Japan
uotani@ei.energy.kyoto-u.ac.jp

[2] Service Evolution Laboratories, Nippon Telegraph and Telephone Corp., Chiyoda, Japan

Abstract. In recent years, the number of international tourists had been increasing until 2019 and the use of digital signage has become widespread as a means of displaying multiple languages. In this paper, we propose a dual-language display method using oscillating animation for digital signage (bilingual oscillatory display) to solve the problem of conventional multilingual display method on digital signage. The proposed method improves the amount of information per area by inserting sentences written in one language between lines of the text written in another language and maintains readability with the effect of the law of common fate by providing two languages with independent animations that oscillate vertically and horizontally. In order to evaluate the readability of the proposed method, we measured the reading speed of Japanese speakers using the proposed method and the monolingual version of the text without animation (monolingual static display) and compared the results. Experimental results showed that the reading time in the bilingual oscillatory display was 3% longer than that in the monolingual static display while almost doubling the amount of information per area.

Keywords: Digital signage · Multilingual · Law of common fate · Bilingual oscillatory display

1 Introduction

In recent years, the number of international tourists in the world had continued to increase due to the improvement of information and communication technology until 2019, the development of transportation systems, and the advance of globalization. According to the United Nations World Tourism Organization, the number of international tourist arrivals reached 1.46 billion in 2019 [1]. Since

C. Stephanidis et al. (Eds.): HCII 2021, LNCS 13094, pp. 412–423, 2021.
https://doi.org/10.1007/978-3-030-90238-4_29

international large-scale events in the future are expected to attract tourists from various countries, it is desirable to provide various contents such as tourist information, product descriptions, and traffic information in the native language of the tourists to improve their convenience. Therefore, digital signage is widely used as a media that supports multilingual display to smoothly transmit information. In Japan, digital signage is widely used as flat panel display devices are becoming larger in size and lower in price, and the market size of digital signage has been expanding in recent years [2].

When using digital signage to provide information to people with different native languages at the same time, the current multilingual display methods can be roughly divided into two categories: language switching display (LSD) and multilingual simultaneous display (MSD). LSD is a method of displaying content in only one language on a single screen and switching the screen at regular intervals to change the displayed language. MSD is a method of simultaneously displaying content in multiple languages on a single screen. However, the conventional methods, LSD and MSD, have the following problems. LSD takes longer time to display all the information than monolingual displays. In addition, LSD requires the viewer to wait when a screen that is not in the viewer's native language is being displayed. In MSD, the text size is smaller than that of monolingual display if the size of the screen on which the contents are displayed is constant, because sentences written in multiple languages are written together on one screen. In order to solve these problems, the authors propose a bilingual display method using oscillating animation (Bilingual Oscillatory Display: BOD).

BOD is a multi-language display method that simultaneously describes two languages. In BOD, sentences written in one language are displayed between the lines of text written in the other language, and each sentence is animated in a different motion. It is said that sentences are easier to read in a layout where the line spacing is about half the height of the text [3]. If a sentence in one language is inserted between the lines of a sentence in the other language, and the viewer can select one of these two sentences and read it without delay, the amount of information per area is approximately doubled compared to the mono-language sentence without changing the size of the text. However, simply inserting different sentences between the lines would reduce readability due to the extremely narrow line spacing and the overlapping of the top and bottom edges of adjacent lines. Therefore, in the proposed method, the readability of the text in the two languages is ensured by adding animations with different movements to the text in the two languages. In this way, the BOD achieves the simultaneous display of sentences in two languages while maintaining the size of the text.

It is expected that the amount of information per area can be approximately doubled by using BOD, however, its readability is unknown. Therefore, the purpose of this study is to evaluate the readability of sentences written in BOD. In this paper, high readability is defined as a fast reading speed, an objective measure.

2 Proposed Dual-Language Display Method

2.1 Previous Studies

Animated text on a monitor has been studied in a field called kinetic typography [4]. The advantage of kinetic typography, according to Uekita et al., is that it can convey additional information such as emotions and tone of voice by moving the characters [5]. Lee et al. introduced kinetic typography into a messenger service in order to express detailed emotions that cannot be conveyed by emojis, such as the tone of the speaker's voice and the intensity of the emotion. They showed that kinetic typography can convey specific emotions by investigating the emotions people feel when they see it [6]. Minakuhi et al. measured the intensity of the emotions felt by participants when they viewed emotion words that were given nine basic types of animations, such as moving, zooming in and out, vibrating, and fading. The results showed that the animations and emotion words could be classified into three groups, and that the animations could be divided into two groups: those that strengthened the emotion expressed by each emotion words group and those that weakened it [7].

Studies on multilingual digital signage have been conducted in various fields, such as multilingual digital signage that displays contents by switching to the user's native language, and multilingual display on the Web. Ogi et al. have developed a system that automatically responds to the user's native language by exchanging information between the user's smartphone and multilingual digital signage through iBeacon communication [8]. Fujiki et al. conducted an evaluation experiment on the menu, explanatory text, and character scrolling of a web page that supports Japanese and Korean in a multilingual simultaneous display. As a result, they found that the subjective evaluation results of the multilingual simultaneous display were higher than those of the individual display in terms of page viewability and interest, and that the scrolling that stops at each clause was higher than the normal scrolling in terms of character understanding [9].

Although there have been many studies on text display with animation and multilingual display on monitors, multilingual display with animation has not yet been proposed to increase the amount of information per unit area and to communicate information efficiently. If the recognition of sentences can be made easier by adding animation, the proposed multilingual display method will be able to convey more information in a shorter time than the conventional method.

2.2 Overview of Bilingual Oscillatory Display

The proposed BOD places sentences written in one language between the lines of sentences written in the other language and animates each sentence with a different oscillating motion. As shown in Fig. 1, the BOD aims to enable multiple viewers of different native languages to read sentences written in their native languages without delay when they view the display simultaneously. The reason for placing sentences of one language between the lines of the other is to prevent a decrease in readability by reducing the area of overlap between the characters of two languages. The reason for adding animation is to improve readability by using the effect of the law of common fate to make the viewers easily recognize the two sentences as different groups.

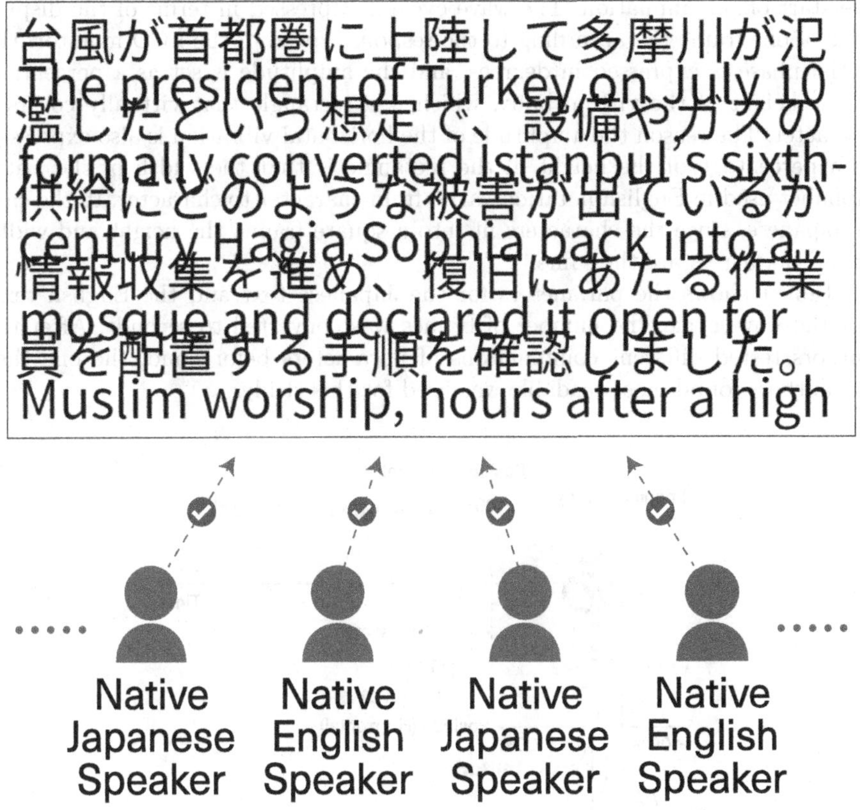

Fig. 1. Concept of bilingal oscillatory display.

In this study, Japanese and English were adopted as the languages used in the BOD. The reason for choosing Japanese is that the participants of the experiment were limited to native speakers of Japanese. The reason for choosing English is that English is the most preferred language for multilingual displays in Japan.

2.3 Details of the Animation

The oscillatory animation is represented by the time variation of the horizontal and vertical positions following sine waves, as shown in Fig. 2. The two sine waves are both controlled by three parameters: amplitude, initial phase, and wave cycle. Therefore, the oscillatory animation given to a language is controlled by a total of six parameters. The initial phase is the phase shift of the sine wave at 0 s after the start of the animation. The wave cycle is expressed in terms of the display time of one frame (F) according to the response speed of the LCD monitor, the initial phase is expressed in degrees, and the amplitude is set as a percentage of the height of the virtual body, which is a rectangle that virtually covers a character. The reason the amplitude of the horizontal vibration is also expressed as a percentage of the height of the rectangle is that the width of the Latin alphabet used in English notation varies from character to character. In the case of Japanese, since the characters fit into a square frame, the height and width of the characters are the same.

Table 1 shows the parameters for the Japanese text and the English text. For the wave cycle, the number of frames was converted to seconds. After the authors tested different combinations of parameters beforehand, the one that the authors found most readable was used for this study.

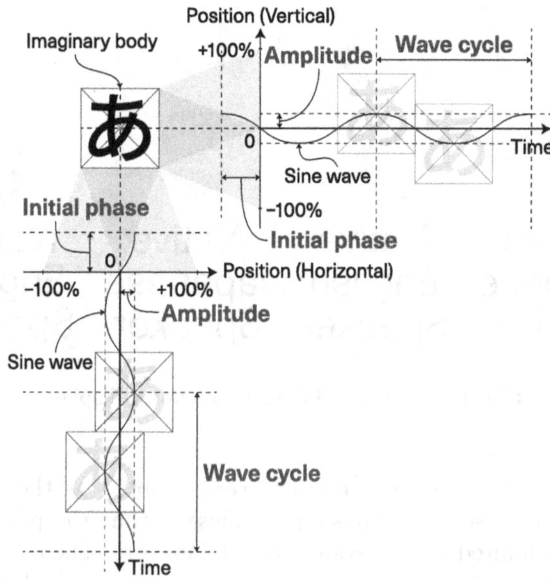

Fig. 2. Parameters of oscillation animation for Japanese text.

Table 1. Parameters of oscillation animation for Japanese and English text

Direction	Type of parameters	Value		Unit
		Japanese	English	
Horizontal	Amplitude	6	6	%
	Initial phase	−90	90	deg
	Wave cycle	1.33	1.33	s
Vertical	Amplitude	6	6	%
	Initial phase	0	180	deg
	Wave cycle	1.33	1.33	s

3 Experiment

3.1 Purpose and overview

The purpose of this experiment was to investigate the extent to which the reading speed of sentences displayed on BOD is decreased compared to that of a static display of Japanese only (Monolingual Static Display: MSD). Therefore, the task of measuring the reading time on the BOD and the MSD was performed. In the task, participants read sentences in BOD or MSD presented on a monitor, and when they finished reading the sentence, they clicked a button on the monitor with the mouse. Since the participants were Japanese, in the BOD, the participants selected and read only Japanese sentences from the Japanese and English sentences displayed simultaneously. In the MSD, the participants read the sentences written only in Japanese. The measurement time was from the time the sentence was presented on the screen to the time the button was pressed.

The experiment consisted of one practice task and one performance task per participant. In the practice task, participants were presented with two BOD sentences and two MSD sentences. In the performance task, participants were presented with six BOD sentences and six MSD sentences, thus the total reading time of 12 sentences per participant was measured.

3.2 Participants and Experimental Environment

Eight university students or graduate students whose native language is Japanese participated in this experiment. The experiment was conducted online, with the participants and the experimenter connected via Zoom web calls. Therefore, the size and resolution of the display, the brightness of the screen and room, and the viewing distance were not controlled.

3.3 Proposed Contents

The BOD sentences and the MSD sentences on the LCD display are shown in Fig. 3 and Fig. 4. The drawing area of the text was set to 1,000 × 500 px due to the limitation of the screen size. In the BOD and MSD, since the font size was set to 60 px, the number of Japanese characters per line was 16, and the text was drawn in five lines. Therefore, the number of Japanese characters displayed in the drawing area was 80. The length of the English text displayed in the BOD was set to be the same as the length of the 80-character Japanese text. Hyphenation was applied to the line breaks in order to reduce the amount of white space in the line breaks of the English text.

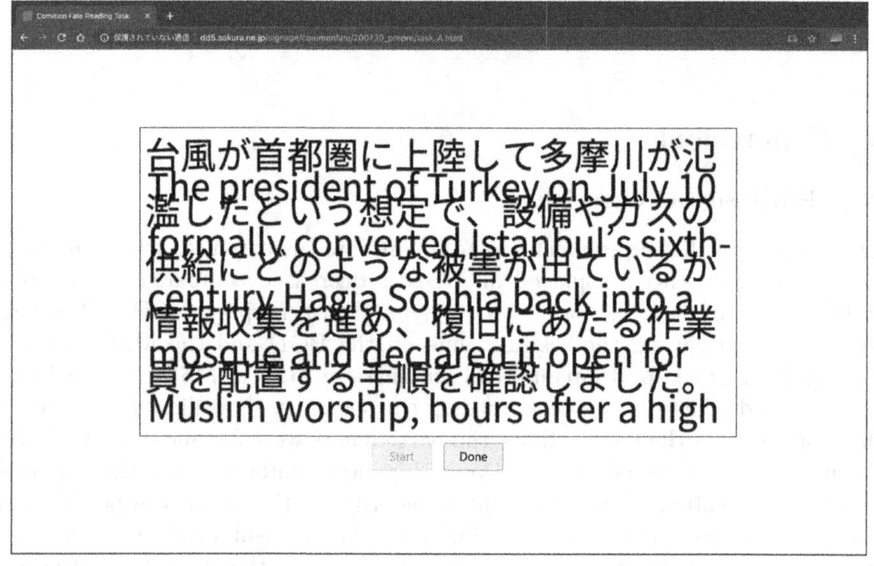

Fig. 3. Example of BOD text presented in the task.

The line spacing is set to 0.5 times the vertical width of the text, as shown in Fig. 5. The line spacing is the distance from the bottom of the text on one line to the top of the text on the next line. Noto Sans JP was used as the font for both Japanese and English.

For the Japanese text, 12 sentences were quoted from NHK News for the performance task and 4 sentences were quoted for the practice task. For the English text, 6 sentences were quoted from the Japan Times for the performance task and 2 sentences were quoted for the practice task. Only the Japanese sentences were partially modified so that the punctuation mark is placed at the 80th character. The 12 Japanese sentences in the performance task were assigned symbols from a to l.

新型コロナウイルスの感染者が増加
する中、飲食店の間で、混雑を避け
て営業を続けるため、利用客に来店
の時間や順番を割りふるシステムを
取り入れる動きが広がっています。
Start
Done

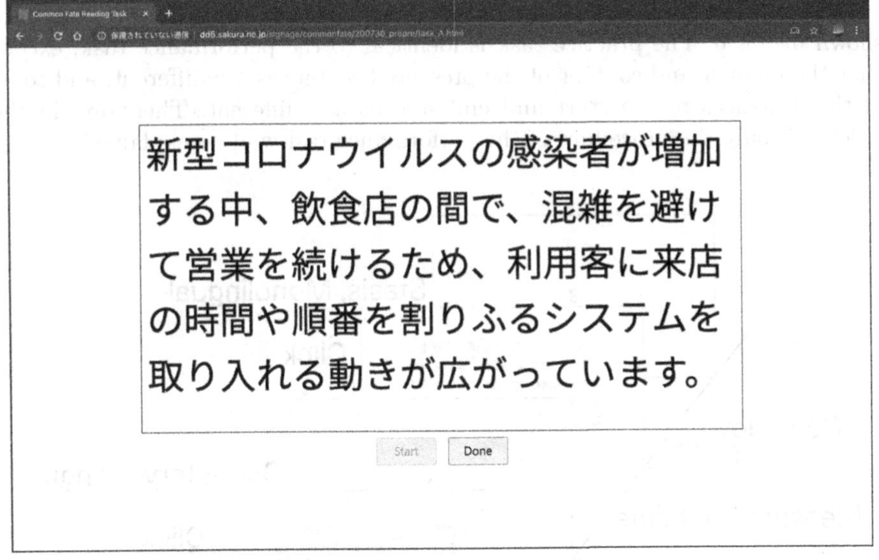

Fig. 4. Example of MSD text presented in the task.

60px
60px
30px
新型コロナウイルスの感染者が増加
する中、飲食店の間で、混雑を避け
て営業を続けるため、利用客に来店
の時間や順番を割り振るシステムを
取り入れる動きが広がっています。

Fig. 5. Line space of text stimulus presented in the task.

3.4 Procedure

The experiment was carried out in the following order: experimental explanation, practice task, and performance task. The procedure of the experiment is shown below, using the case where MSD sentences are displayed in odd-numbered order and BOD sentences in even-numbered order.

Before starting the task, the following three points were explained to the participants both in writing and orally to teach them how to proceed with the task and how to read the text.

- Read the displayed Japanese text silently, and when you have finished reading it, click the Done button with your mouse.
- Ignore the English text that may appear on the same screen, and read only the Japanese text silently.
- Be sure to read the text from the beginning, and read it aloud in your mind, and do not skip over the text.

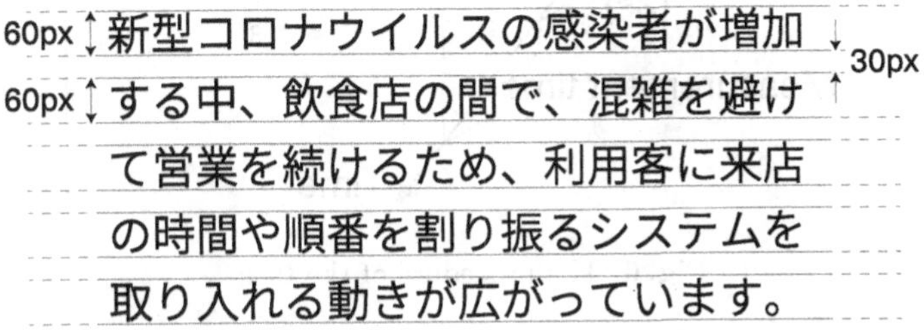

The practice and performance tasks proceeded according to the procedure shown in Fig. 6. The practice task is identical to the performance task, except that the number and content of the presented sentences are different, and some of the sentences in the start and end screens are different. Therefore, in the following, only the procedure of the performance task will be explained.

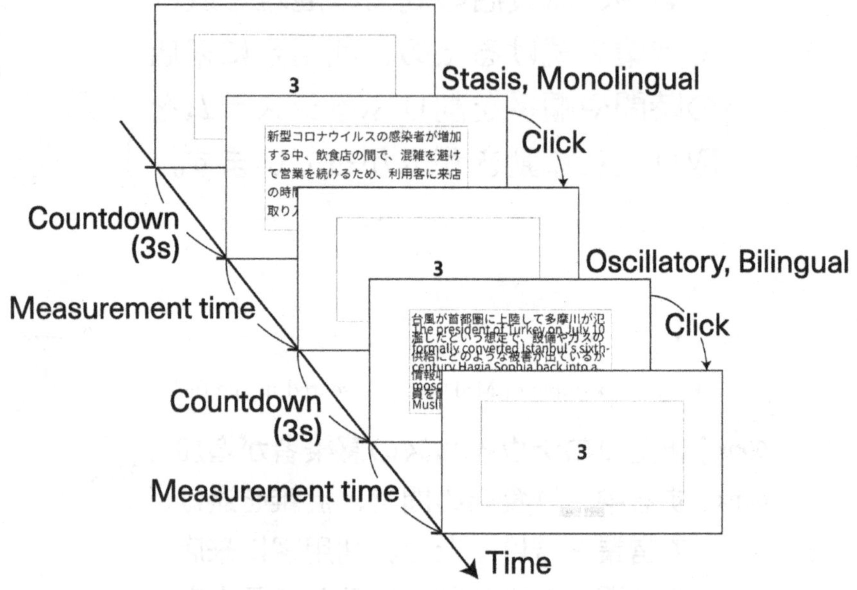

Fig. 6. The procedure of the task.

In the task, the participant started the task by clicking the Start button on the start screen. Next, a 3-second countdown was automatically displayed on the screen, and then an MSD sentence was displayed. The participant read the MSD sentence silently, and when they had finished reading the sentence, they clicked the Done button. After a 3 s countdown was displayed again, the BOD sentence was displayed. The participant read the Japanese sentence in the BOD silently, and when they finished reading the sentence, they clicked the Done button. The participant repeated the above operation six times. As a result, the time from the beginning to the end of reading was measured for each of the 12 sentences. When the participant finished reading the sentence displayed for the 12th time and clicked the Done button, the end screen was displayed, and the task was completed.

The first four sentences displayed are a dummy set to suppress the learning effect in the performance task. Therefore, only the latter eight sentences were analyzed. In addition, four types of counterbalanced tasks were prepared in terms of the order in which the BOD and MSD were presented and the combination of display method and sentences. One type of task was conducted on two participants. Table 2 shows the order in which the display methods and sentences were presented in the four types of tasks.

Table 2. The order of the display methods and sentences presented in the task

Order	Task A		Task B		Task C		Task D	
	Method	Sentences	Method	Sentences	Method	Sentences	Method	Sentences
1	BOD	a	MSD	b	MSD	a	BOD	b
2	MSD	b	BOD	a	BOD	b	MSD	a
3	BOD	c	MSD	d	MSD	c	BOD	d
4	MSD	d	BOD	c	BOD	d	MSD	c
5	BOD	e	MSD	f	MSD	e	BOD	f
6	MSD	f	BOD	e	BOD	f	MSD	e
7	BOD	g	MSD	h	MSD	g	BOD	h
8	MSD	h	BOD	g	BOD	h	MSD	g
9	BOD	i	MSD	j	MSD	i	BOD	j
10	MSD	j	BOD	i	BOD	j	MSD	i
11	BOD	k	MSD	l	MSD	k	BOD	l
12	MSD	l	BOD	k	BOD	l	MSD	k

Note: MSD—Monolingual Static Display, BOD—Bilingual Oscillatory Display

4 Results and Discussion

As a result of the experiment, 96 reading time data were obtained from eight participants. The total reading times on the BOD and MSD are shown in the left of Table 3. The reading times shown in the left of Table 3 are the sum of the reading times for the six sentences displayed on the BOD or MSD. The mean of the total reading time for all participants was 59.4 s on the BOD and 55.3 s on the MSD. The reading time on the BOD increased by 7% compared to the reading time on the MSD.

The reading time data obtained in the experiment included results with extremely long and short reading times, which clearly indicated that the participants were not concentrating, and thus data cleansing was performed. Data with minimum or maximum reading times in the MSD and BOD per participant were excluded from the analysis. Thus, 64 data were included in the analysis. The total reading times on the BOD and MSD after data cleansing are shown in the right of Table 3. The reading times shown in the right of Table 3 are the sum of the reading times for the four sentences displayed on the BOD or MSD. The mean of the total reading time for all participants was 38.0 s on the BOD and 36.7 s on the MSD. The reading time on the BOD increased by 3% compared to the reading time on the MSD.

These results suggest that the amount of information on digital signage per unit area can be doubled by changing from MSD to BOD if a decrease in reading speed of about 5% is allowed.

Table 3. The comparison of reading time of text in BOD and MSD

Participant No.	Before data cleansing			After data cleansing		
	Reading time (s)			Reading time (s)		
	MSD	BOD	BOD/MSD	MSD	BOD	BOD/MSD
1	74.1	75.5	1.02	49.1	50.6	1.03
2	35.5	39.4	1.11	24.3	26.3	1.08
3	70.2	82.4	1.17	47.7	54.7	1.15
4	55.7	53.5	0.96	37.0	35.2	0.95
5	75.3	74.0	0.98	49.8	48.4	0.97
6	69.6	84.9	1.22	44.9	44.9	1.00
7	41.0	41.5	1.01	27.3	27.5	1.01
8	21.2	24.3	1.15	13.6	16.4	1.20
Mean	55.3	59.4	1.07	36.7	38.0	1.03
S.D.	20.5	22.8	–	13.6	13.7	–

Note: MSD—Monolingual Static Display, BOD—Bilingual Oscillatory Display

5 Future Schedule

An experiment with participants in a real laboratory will be conducted in the future in order to control the size and resolution of the display, the brightness of the screen and room, and the viewing distance. In addition, since the BOD aims to be readable without delay by both types of viewers whose native language is different, it is necessary to verify the readability of the BOD by conducting experiments with two types of viewers whose native language is different. Therefore, an experiment similar to the present study will be conducted on the BOD with participants whose native language is Japanese and those whose native language is other than Japanese.

References

1. UNWTO, UNWTO Tourism Highlights, 2020 Edition (2021)
2. Inaba, S.: Domestic Market Trend and Future Outlook of Digital Signage (2010). https://digital-signage.jp/files/download1/share/66756a697072696e74.pdf
3. Ohashi, T., Miyazaki, M.: A fundamental study of character layout in newspapers, no. 4: a study of character size, tracking and line spacing (32nd annual conference). Des. Stud. **1985**(52), 26 (1985)
4. Ford, S., Forlizzi, J., Ishizaki, S.: Kinetic Typography: Issues in time-based presentation of text. In: Extended Abstracts on Human Factors in Computing Systems, pp. 269–270, CHI 1997. Association for Computing Machinery, New York, NY, USA (1997)
5. Uekita, Y., Aihara, K., Minakami, N., Tanaka, K.: Characters of Japanese kinetic typography no. 1 research of Japanese kinetic typograpy. In: Abstracts of the Annual Conference of Japanese Society for the Science of Design, pp. 94–95. Japanese Society for the Science of Design (1998)

6. Lee, J., Jun, S., Forlizzi, J., Hudson, S.E.: Using kinetic typography to convey emotion in text-based interpersonal communication. In: Proceedings of the 6th Conference on Designing Interactive Systems, pp. 41–49. Association for Computing Machinery, New York, NY, USA (2006)
7. Minakuchi, M., Ueda, A., Yamamoto, K., Kuramoto, I., Tsujino, Y.: Preliminary study on influence of motion patterns by kinetic typography on expressed emotion. J. Hum. Interface Soc. **14**(1), 9–20 (2012)
8. Ogi, T., Ito, K., Konita, S.: Multilingual digital signage using iBeacon communication. In: 2016 19th International Conference on Network-Based Information Systems (NBiS), pp. 387–392 (2016)
9. Fujiki, T., Ashizuka, S., Noguchi, A., Morita, Y.: Consideration about multi language synchronous display on web pages for distance exchange between Japan and Korea. J. Jpn Soc. Educ. Technol. **29**(Suppl), 217–220 (2006)

Research and Analysis of the Office Socket Design Based on User Experience

Xiangrong Xu, Yuanlong Gui(✉), Bo Fu, and Naizheng Liao

Guangdong Industry Polytechnic, Guangzhou, People's Republic of China
{1994105012,2005105077}@gdip.edu.cn

Abstract. The socket is necessary for office electricity, and the design of it directly affects the user experience of people's office electricity. According to the design theory of user research, a user survey questionnaire has been designed and issued, and the problems of current office sockets have been extracted through the questionnaire for the office environment problems that wire is easily wound, plug interference, can not be placed at will, occupy space, and not coordinated with the surrounding products. The demands for the use of the office socket have been put forward through the research on the problems found by offline interview, observation, real user simulation and so on. The opportunities and new directions of the design of office sockets are put forward from the product appearance, product use humanization, safety through in-depth analysis of the user needs.

Keywords: Socket · User · Experience · Product design · Research

1 Introduction

The electrical appliances need to have a power plug to cooperate with it in order to obtain the required electric energy. People had to screw the wires directly to the power terminals for electrical connection before the invention of plugs and sockets [1]. The use of direct electricity has caused a large number of safety problems, serious will cause electric shock accidents with the invention of a large number of electrical appliances, and that put forward the requirements of safety, convenience and quickness for electrical connection products, so the plugs and electrical products come into being. Socket, known as the plug board, put a plurality of sockets together with a wire. The socket is a simple and practical invention, and can be seen everywhere in the office, which mainly consists of socket head, power cord and plug body. It can connect multiple power plugs and can save space and the wires.

C. Stephanidis et al. (Eds.): HCII 2021, LNCS 13094, pp. 424–433, 2021.
https://doi.org/10.1007/978-3-030-90238-4_30

With the rapid development of economy and society and the progress of science and technology, the inventions of appliances and handheld electronic products that computer, printer, tablet, humidifier, mobile phone etc. have taken a great change of the office means and the office environment, and people require a more appropriate extension socket to ensure the normal use of the products and daily charge. People have more and more demands and functions for extension socket, the appearance, usability and security of it have attracted people's attention. How to easily get electricity is the key issue that people consider at present in the limited office space. The requirements for the socket have changed from able to use to the fashionable appearance, good performance, safety, convenience and the others.

2 The Importance of User Experience Research in Design of the Office Socket

User experience (UX) is a design concept focusing on user's subjective feelings, which was proposed by American designer Donald Norman in the 1990s. User experience was first applied to human-machine interface and interaction design. With the development of user experience research and application, the User-Centered Design (UCD), namely User-Experience Design (UXD) has been highly valued and widely applied. In his book, The Psychology of Design, Norman said, "The user is not wrong, if the user has trouble using the product, it's because the design is imperfect" [2].

User research is the premise of design activities and plays a key role in the whole design process. Products can be designed that meet users' needs and promote social development only by taking users as the first research object and user research as the core design criterion [3]. The product should fit the user, not the other way around [4]. Product design is based on the designer's understanding of users, product design is no longer designer-centered, but really customer-centered. The concept of user experience enters the entire processes from the beginning of product development and continues throughout. A large part of the user experience is related to the user's psychological subconscious, which is completely based on the user's psychological and conscious cognition [5]. It is the starting point of the design to understand the characteristics of users and environmental needs, and then to master the characteristics of all aspects of the user needs.

As user needs change, people are not only satisfied with the functions of the product, but also put forward higher requirements for other aspects. User experience is particularly important in the design of the socket. The function of the socket is relatively simple, is a kind of auxiliary product but an indispensable product in our work, and it is not a simple wiring device or extension of the wire. If the design of office socket is purely considered from the aspect of electricity function, you can use the best conductive materials and the best insulating shell materials, overload protection, to ensure electricity safety. However, as a necessary office item, the office socket is not only to get electricity, but also to deal with the relationship between other items, and be in line with the user's habits and aesthetics. How to design a office socket with the users' popular that can meet the needs of today's office users, designers should apply user experience research theory and methods to find the existing problems of sockets and potential demands, and deal with the relationship between the socket and the user, office electrical appliances, non-electric products and environment. The pain points of users when using the sockets can be found, and they stimulate the creative inspiration of designers and make the product design more in line with the needs of consumers through design insight.

3 User Research on the Office Socket

User research is the first step of product design processes, the user's goals and needs can been analyzed through the user research. The user research of this subject includes two parts, the first is quantitative research, the results are analyzed and verified by mathematical methods through questionnaire distributed through the network. The second is qualitative research, users' needs can be obtained by exploring problems, understanding phenomenon, analyzing users' behavior and viewpoints.

3.1 User Research Processes of the Office Socket

(1) Quantitative research
The first stage is quantitative research, using questionnaire survey and data analysis methods. All kinds of information of the office socket have been mastered through the observation and analysis of the project team in the early stage. One online questionnaire has been designed according to the user information characteristics, the office environment, the current situation of the use of office socket and user behavior, including five aspects: the basic information of the respondents (gender, age, position, industry, number of people accommodated in the office); office environment (office environment, desk size, type of computer used in the office); office electricity related products (electrical appliances, Non-electric products); current situation of office socket(number of socket, number of jack, position of socket); user's habits and expectations (desired placement, power off habits, necessity of power off, necessity of setting independent switches, experience of socket after socket, wire length of socket). 21 objective questions have been set (single choice and multiple choice). The questionnaire has been issued and recycled through the questionnaire star platform and let users talk about those things about office electricity. The questionnaire survey has been completed online, collected 675 questionnaires, about 360 valid questionnaires among them.

(2) Qualitative research
The second stage is qualitative research. User needs have been obtained through design insight by observing and analyzing users' behavior through household survey and real user simulation. The research team consists of ten groups, each of which consists of five designers. Ten design companies were selected to observe and record that the office environment of the design company and the individual office environment. Employees of the design company have been questioned from users' information, consisting of the use of socket in daily life and their expectations for the socket in office. Real user simulation have been done to simulate the company's real office environment with a standard size of 1200 mm × 600 mm single desk, the user to layout the desktop, to rearrange the computer, electrical appliances, Non-electric products and simulate the working state, the whole processes of connecting

and unplugging according to their own habits, and the processes have been wholly recorded.

3.2 Results of User Research on the Office Socket

(1) Basic information about the respondents
Results of this questionnaire survey show that there is little difference between male and female users, 53% and 47% each. The age group of 18 to 25 years old is more(see Fig. 1). Among the office workers who participated in this survey, there are more grass-roots employees and more people engaged in manufacturing.

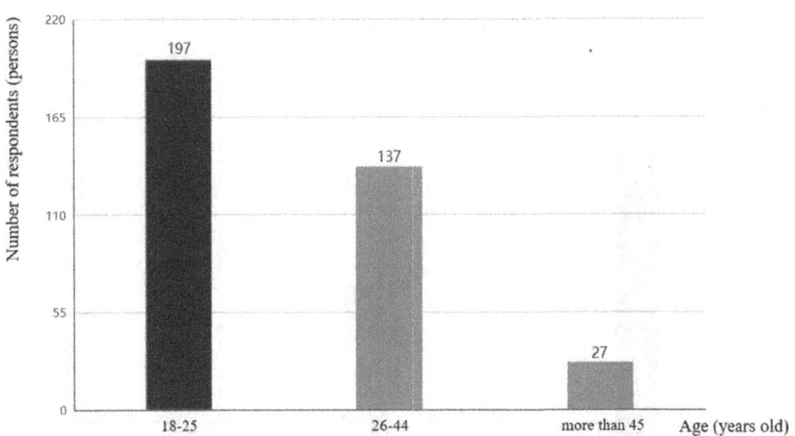

Fig. 1. The age distribution of the respondents

(2) Office environment
That more than 10 people working in one office is relatively more (see Fig. 2). Single office desk is maximum(see Fig. 3). The number of people using the 1200 mm × 600 mm desk size is more (see Fig. 4). Use desktop computers more often (see Fig. 5). There have electrical appliances such as sockets, telephone, Non-electric products as water cups on the desk.

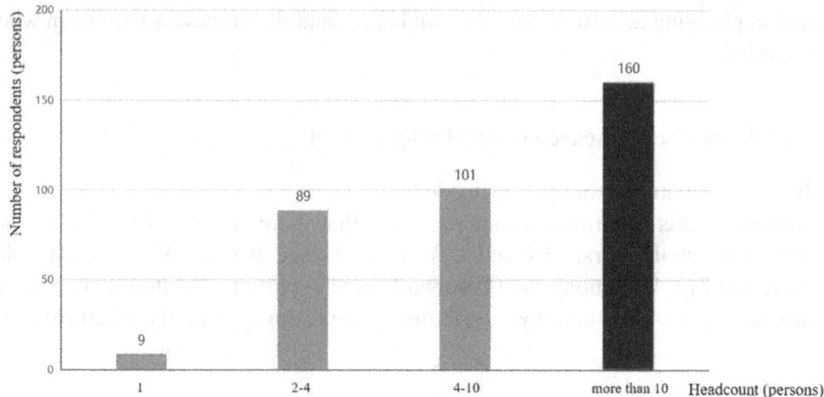

Fig. 2. Headcount in the office

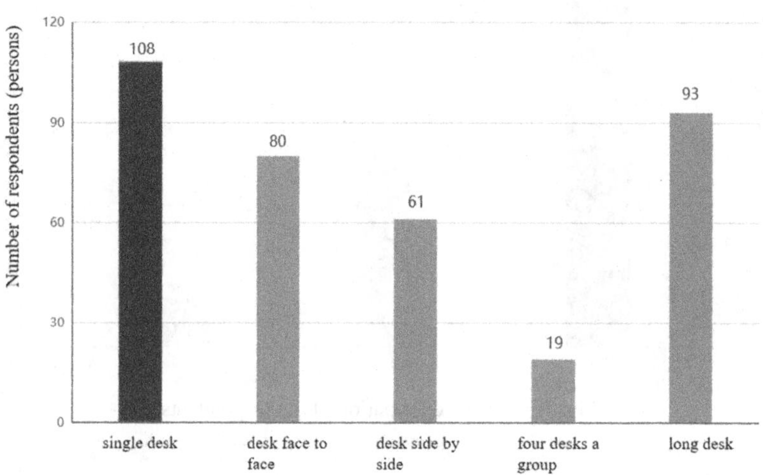

Fig. 3. Style of the office desk

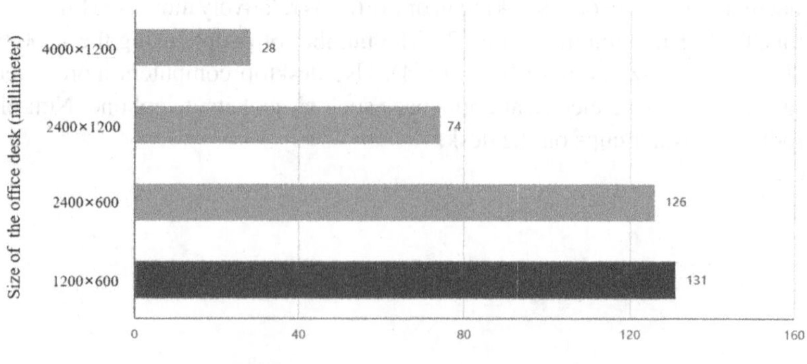

Fig. 4. Size of the office desk

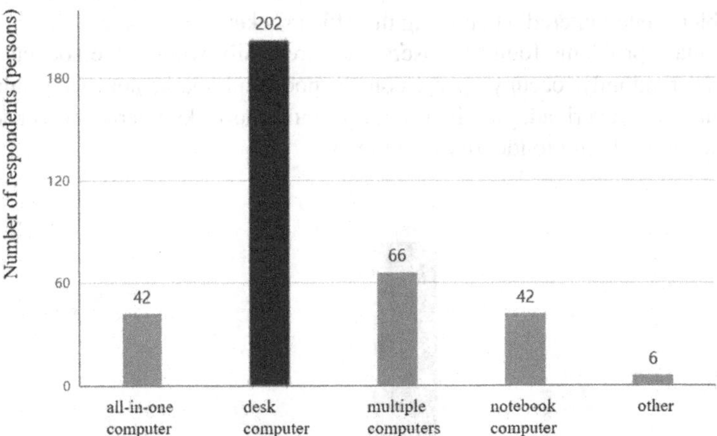

Fig. 5. Style of the office computer

(3) Current situation of office electrical products
Office employees who participated in this survey with more than 3 sockets in their personal area in the office is more (see Fig. 6). There are 5–7 jacks with one socket in the individual area(see Fig. 7). The position of the socket in the corner of the desktop is more. The number of workers is more hoped that the location of the socket buried in the ground.

(4) The Users' habits
73% of the office employees who participated in the survey are in the habit of turning off the power in an office environment. Many think it is necessary to turn off the power supply. More need independent switches, up to 65%. Up to 71% of those who have experience of socket and socket. Not enough and then socket and socket. Most of the Length of the socket wire is 1.2 m–1.8 m.

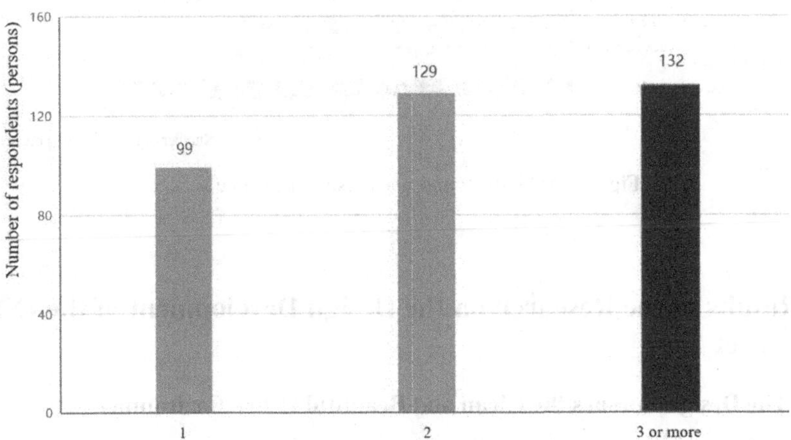

Fig. 6. Number of the office socket in personal area

(5) Problems encountered when using the office socket
The main problems found by users are wire easily wound, the sockets can't be placed randomly, occupy space, contact poor and spark, non watertight, worry about power overload, jack is difficult to pull, the jacks interact with each other, clutter and difficult to identify (see Fig. 8).

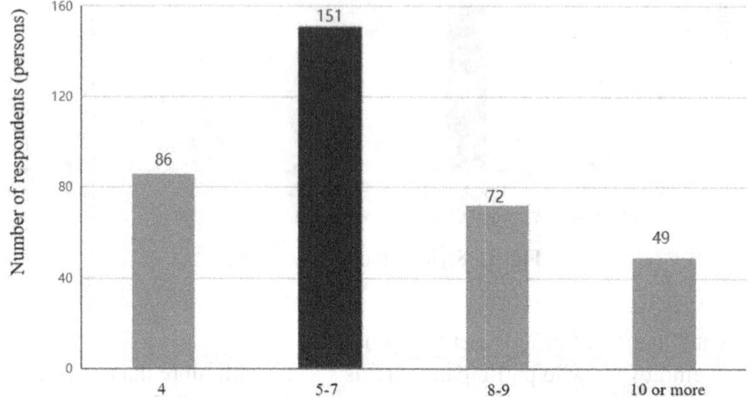

Fig. 7. Number of the jacks in the individual area

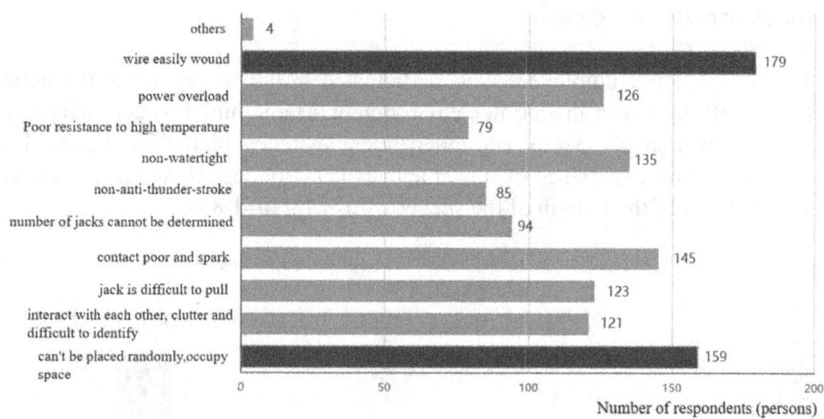

Fig. 8. Problems found when using the office socket

4 Results of the Research on the Design Development of the Office Socket

4.1 The Design Pursues the Clean and Beautiful Office Environment

Today, people need a holistic office environment, and the socket is an essential office product. It needs to be integrated with the electrical appliances and Non-electric products on the desktop, and coordinate and beautiful overall if placed on the desktop. Of course, it can be also placed in the place where the user can not see in the premise of not affecting the convenience of use, when not in use, people can not see the socket, when needed, and quickly meet the requirements of the users. Analysis and summary can be made mainly from the color, modeling, presentation methods. (1) There is no gender inclination demand for the socket. It can develop towards a neutral style, designed in black or white with fresh color. (2) The use of the socket will affect the appearance and cleanliness of the desktop. The socket shape can be designed more concise, easy to integrate with the environment. The color shape is simple and unified, simple and elegant, and the way of placing and fixing is diverse, which can be fixed on the wall. (3) Although the socket is small, but it will occupy the desktop space. We can design the smaller volume socket. (4) The number of jacks can not be determined, sometimes more and sometimes less. How to make the number of jacks humanized? We can do modular socket. The problem points, opportunities and design direction of the office socket have been found from the research and analysis of the appearance of the sockets.

4.2 The Design Pursues the Humanization of the Office Socket

The iterative innovation of socket is closely related to the pain points of users. The utility and effectiveness of function is the main consideration of most users when choosing a socket. (1) Lack of USB jack. Many consumer electronic products are using USB power supply or charging. Sockets with USB jack should become a standard configuration, and it can reduce the use of charging head, save costs, improve the use efficiency. (2) The jack number of the socket is not enough. The jack number of socket can be changed subjectively, and it can be artificial joined together. (3) Wire of the socket is messy and difficult to arrange. The wire of the socket can surround, shrink finishing. Design the socket with wire and storage box integration. (4) The jacks affect each other. The socket can be designed with misplaced jacks. (5) The plug is too difficult to pull out and requires both hands to operate. Plugging and pulling is the use way of the socket. The man-machine relationship of the socket is also extremely important. The user needs to fix the plug with one hand and hold the plug with the other hand. The user can pull out the plug with the mutual assistance of both hands. However, if the position is not reasonable, it will lead to inconvenience when the plug is inserted and unplugged. The socket can be designed in a variety of ways, can be designed to be inserted by adjusting the angle of the plug or by pressing the plug. (6) It is difficult to identify when pulling the plug, and it is often wrong to pull the plug due to the similar shape and color of the plug of the electrical appliances. How to identify the plug quickly? The plug can be marked manually or induced with the corresponding jack. (7) Problems of socket position and limits of use mode. How to deal with the placement of the socket, and the variety of placement. We can design the placement of the socket more stable, and the shape of the socket more adaptable to the environment. (8) The problems of long distance power, bending down to take power, easy to kick to trip to the socket. When the socket is placed under the table, the user needs to bend over to pull the plug. It is very inconvenient for ordinary users to operate, and is a very difficult thing for people with waist diseases to squat to plug it. It can indirectly lead to the plug loose and may even affect the safety of the user. How to deal with the distance between the the electrical appliances and the fixed power supply? we can replace the wire and make multiple forms of placement. (9) Problems of separation with modern forms of charging and smart socket. How to integrate with wireless charging, mobile internet operation, monitor AI socket, voice control operation, wireless charging design? (10) It can only be switched on and off manually. How to control the switch? use of voice control, time, remote switch, remote control and so on.

4.3 The Design Pursues the Security of the Office Socket

The socket is an auxiliary product, but it is one of the most indispensable products in our work. If there is a safety problem, the light can affect or damage the equipment worth thousands even tens of thousands of yuan, heavy related to the safety of people's lives. In recent years, accidents such as short circuit, electric leakage and even electrocution have occurred frequently due to unreasonable design of the socket. The sockets have become one of the biggest safety risks in the electrical age. The safety design of the socket is a very important design direction. (1) Plug easy to contact poor, unclear about the use status of the socket. We can design a prompt light to know whether the socket is running normally. (2) Socket power overload, but the user has no way of knowing. The power indicates at any time, there are corresponding prompts if the power is too large. (3) Security performance cannot be detected, and the user cannot know whether the plug is good or bad. Sockets can be designed with display related data, and has the alarm function. (4) High temperature resistance is poor, so design it can prompt when up to a certain temperature. (5) The water accidentally is spilled into the socket and causes a short circuit. Waterproof experience and design a waterproof socket. (6) Sockets is easy to catch dust and troublesome to clean up. We can design a socket with dust cover. (7) The users are not in the habit of turning off the power. We can design the sockets with multiple power off modes, regular power off or remote control can be implemented.

5 Discussion

With the improvement of people's living standard and the change of office environment, consumers have higher and higher requirements for the design and quality of the office socket, which promotes the development and progress of the design of the office socket. Through user research on the three stages of online questionnaire analysis, offline household research, real user simulation, the design development trend of appearance demand, practicability analysis and safety analysis of the office socket in today's environment has been study by analyzing the problem points of the products, then transformed into the opportunity points and innovation points of the design.

6 Conclusion

It is an effective means to practice user-oriented design to study the user experience of office electricity products, find out the problems of products, the market opportunities and innovation points of the product design. The office socket in the future should be redesigned good looking and easy to use combined with users' modern lifestyle and consumption concept updated and changed according to users' needs for the office socket from the aspects of beautiful and clean office environment, humanized use of placement and security.

References

1. Luo, B.J., Wu, Z.H., Li, Y.Q.: Man-machine safety design of socket. Ergonomics **17**(04), 65–68 (2011)
2. Norman, D.A.: Design Psychology. CITIC Press, Beijing (2010)
3. Zhao, Y., Jin, Y.Q.: Analysis of the design concept and method which take user research as the core. Popul. Liter. (17) 129 (2016)
4. Chen, G.Q.: Product Design Procedures and Methods. China Machine Press, Beijing (2011)
5. Hu, F.: User Insight: User Research Methods and Applications. China Building Industry Press, Beijing (2010)

Research on Improving Empathy Based on the Campus Barrier-Free Virtual Experience Game

Junyu Yang, Yawen Zheng, Tianjiao Zhao(✉), and Mu Zhang

Tianjin University, Tianjin, China
zhaotianjiao@tju.edu.cn

Abstract. As the world's population aging and chronic disease grow, the disability rate is also growing. Therefore, let more people pay attention to the disabled, and establishing the right accessibility concept is critical. However, barrier-free science popularization is not common at present. Most studies have shown that virtual reality can be moderately enhanced. Therefore, this research aims to explore the mechanism of empathy in the immersive virtual environment, and then form a set of design methods to be applied to empathy design.

According to three levels of emotional design, this research construct a design model includes the atmosphere layer, the interaction layer and the reflection layer. In this study, we tested the importance of visual, hearing, game form and game perspective on empathy of 34 participants. This experiment uses a method of combining qualitative and quantitative, and records the subject's eye data, ECG data, and subjective evaluation data. The results found that visual and game forms have a significant impact on empathy. Hearing is the strongest in people's subjective feelings, so it also has a certain influence on emotional arousal. The perspective of the game has the weakest influence on emotional arousal. On the whole, in a barrier-free virtual game environment in the form of low-saturated colors, intense music, and customs clearance form, the empathy for the disabled can be most stimulated. Therefore, the construction of subsequent barrier-free science games is recommended to adopt this principle for design, and it also provides a reference for the design of future empathy games.

Keywords: Empathy · Barrier-free · Virtual games · Immersion · Game design

1 Introduction

It is estimated that more than 1 billion people, equivalent to about 15% of the world population, suffer from some form of disability. Among people aged 15 and over, 110 million (2.2%) to 190 million (3.8%) have severe functional disorders. In addition, partly due to the aging of the population and the growth of chronic diseases, the disability rate is increasing [1]. At present, the total number of disabled people in China has reached 85 million, accounting for about 6% of the total population. Therefore, the society should pay more attention to the disabled groups from the physiological and psychological

C. Stephanidis et al. (Eds.): HCII 2021, LNCS 13094, pp. 434–450, 2021.
https://doi.org/10.1007/978-3-030-90238-4_31

needs of the disabled, and jointly promote the construction of a harmonious and equal society.

In terms of physiological needs, the concept of "barrier-free design" was introduced, which can be traced back to the 1930s. At that time, Denmark and other countries began to pay attention to the difficulties people with disabilities encountered in their lives and set a precedent for barrier-free design. Until today, there are still a large number of researches on barrier-free design at home and abroad, which are constantly extended and expanded, but its core has not changed. It is still to use modern technology to provide a safe and simple living environment for people with disabilities and provide them with equal opportunities to participate in society. While promoting the construction of barrier-free facilities, it is also very important to help the public establish a correct concept of barrier-free. Let more people pay attention to the voices and difficulties of people with disabilities, understand the barrier-free concept, and stop looking at them from a different perspective. However, the popular science of barrier-free design is not widespread at present.

With the continuous development of economy, science and technology and culture, various new devices have been developed and applied. Virtual reality technology is a new type of information transmission and simulation technology formed by the continuous improvement of computer, electronic information and other related technologies. Because of its strong adaptability, it can be widely used in many fields such as scientific research, daily information dissemination, and professional course teaching. The design of virtual reality has had a great impact on people's lives. People can use functional games to acquire the knowledge they want in the form of entertaining and teaching, and help others to subtly form some abilities and overcome some obstacles in their experiences. It has a very important influence and significance on people's life.

In the past ten years, the virtual human-computer interaction laboratory of Stanford University has been studying whether VR can really enhance empathy. With a few exceptions, most studies have shown that virtual reality can moderately enhance empathy. Ruixue Liu and Youqun Ren of East China Normal University explored the flow experience and empathy effect of learners in the process of learning scientific knowledge in immersive virtual environment and traditional teaching environment, and found that immersive virtual environment can enhance learners' flow experience and empathy ability in the process of learning scientific knowledge, promote learners' understanding of scientific knowledge and improve their academic performance [2]. Bachen et al., University of Santa Clara, USA, analyzed the effects of earthquake simulation games on empathy and learning interest from three angles of flow, existence and character recognition. Research shows that existence can significantly improve people's global empathy ability, and this empathy ability can enhance learners' learning interest [3]. Shin of Sun Yat-sen University in South Korea explored the application of immersive virtual reality technology in the process of experiencing stories. Research shows that the cognitive process of learners' learning experience, presence and flow experience will affect how individuals express VR stories and produce empathy [4]. Then, whether the interactive form of virtual reality can be applied to implement the barrier-free popular science experience and effectively improve people's empathy for the disabled will be the main problem of research.

This research aims to establish a certain empathy mechanism for barrier-free virtual experience games, so that people can subtly arouse empathy for the disabled in the process of experiencing games, establish correct barrier-free concepts, and achieve the effect of popular science in games.

This study chooses the old campus of Tianjin University as the scene of barrier-free virtual experience game, trying to explore the empathy mechanism in this game. According to the empathy component proposed by Morse [5], the cognitive level, the behavioral level, and the emotional level correspond to the three-level model of emotional design, thus constructing a three-level model of game design at the theoretical level, namely atmosphere level, interaction level and reflection level. Through this model, the game factors are refined, the design opportunities that can effectively empathize are analyzed and explored, and a set of empathy design methods is established, which is applied to the barrier-free virtual experience game, and also provides some reference for the design of empathy games in the future.

2 Research Method

2.1 Empathy Stimulation Experiment

Morse summarized the four key components of empathy through extensive review of empathy literature, such as Table 1. We found that the cognitive level, behavioral level and emotional level correspond to the three-level model of emotional design, namely instinct level, behavioral level and reflection level. Our research takes the barrier-free virtual experience game on campus as the original material, and constructs a new theoretical model for the game design by using the emotional design model. We divide it into atmosphere layer, interaction layer and reflection layer. Based on these three levels, the original materials are stimulated from the visual, auditory, game form and game perspective, combined with objective and subjective evaluation methods, to analyze the influence of the above elements on the empathy of the observer in the barrier-free virtual experience game. The whole research plan is shown in Fig. 1.

Table 1. Morse's components of empathy

Component	Definition
Emotive	The ability to subjectively experience and share in another's psychological state or intrinsic feelings
Moral	An internal altruistic force that motivates the practice of empathy
Cognitive	The helper's intellectual ability to identify and understand another person's feelings and perspective from an objective stance
Behavioural	Communicative response to convey understanding of another's perspective

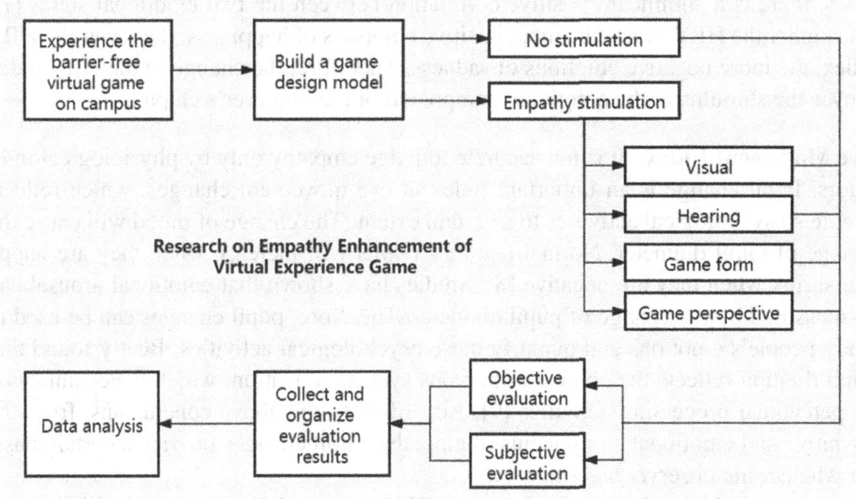

Fig. 1. Research process of empathy stimulation in virtual experience games.

2.2 Empathy Measurement Indicators

Scholars at home and abroad generally define empathy from its components or the process of showing empathy. Researchers who believe that empathy only contains cognitive components put forward that empathy is an individual's insight and understanding of others' inner feelings and behaviors in social life (Feshbach, 1969). Researchers who believe that empathy only contains emotional components put forward that empathy is the emotional resonance after individuals understand other people's experiences and emotional feelings (Eisenberg, 2000). Researchers who believe that empathy contains multiple dimensions have proposed that empathy can not only correctly understand each other's experiences and feelings, but also have certain emotional resonance with others (Hoffman, 2001). Therefore, to promote empathy, we need to stimulate cognition and emotion from its definition and composition.

ECG Indicators. A large number of studies on empathy have been conducted abroad since the late 1960s, but different studies have reached different conclusions due to the use of different empathy measurement indicators. Therefore, M.L. Hoffman summarized the measurement indicators of empathy and put forward three measurement indicators: physiological indicators, physical indicators and language indicators [6]. Among them, physiological indicators of empathy are widely used in adults, which provides an ideal basis for whether observers have empathy and the degree of arousal of empathy. One of the more commonly used physiological indicators is heart rate. According to Hoffman's analysis, the acceleration and deceleration of heart rate may be the essence of empathy arousal of observers.

According to relevant psychological research, different emotions experienced by humans also have certain differences in indicators such as heart rate variability. There is a significant difference between happiness and sadness in HRV index, and in most cases, there is a significant positive correlation between the two emotional states [7]. The higher the HRV index, the more positive emotions of happiness; the lower the HRV index, the more negative emotions of sadness. Therefore, the change in the HRV index proves the stimulus to the emotional component of the observer's empathy.

Eye Movement Index. It is not accurate to judge empathy only by physiological indicators. Pupil change is an important index in eye movement changes, which reflects people's psychological activities to a certain extent. The change of mood will cause the change of pupil diameter. Normal people's pupils will increase when they are happy and shrink when they are negative [8]. Studies have shown that emotional arousal can be measured by the change of pupil diameter. Therefore, pupil changes can be used to study people's emotions and quantify these psychological activities. Beatty found that pupil dilation reflects the change of nervous system activation, which is accompanied by perceptual processing activities [9]. According to the above conclusions, from the cognitive and emotional components of empathy, pupil changes provide a certain basis for whether the observer has empathy.

Therefore, this study uses the change of HRV index and the change of pupil diameter to analyze the stimulation from the cognitive and emotional components of the observer's empathy, and judge the effectiveness of the influence of different game environments on the observer's empathy.

2.3 Evaluation Method of Empathy Stimulation Effect of Barrier-Free Virtual Experience Game

In order to determine whether visual, auditory, game form and game perspective are effective in stimulating empathy, this paper intends to adopt a combination of quantitative and qualitative evaluation methods. Psychological research shows that when people are in a certain environment, human cognition and perception will be judged in different dimensions. The result of this judgment will be expressed through human physiological and psychological information. The purpose of this study is to evaluate the stimulation effect of materials by ECG measurement, eye movement measurement, subjective evaluation, in-depth interview and audio thinking. The evaluation process is shown in the following figure (Fig. 2).

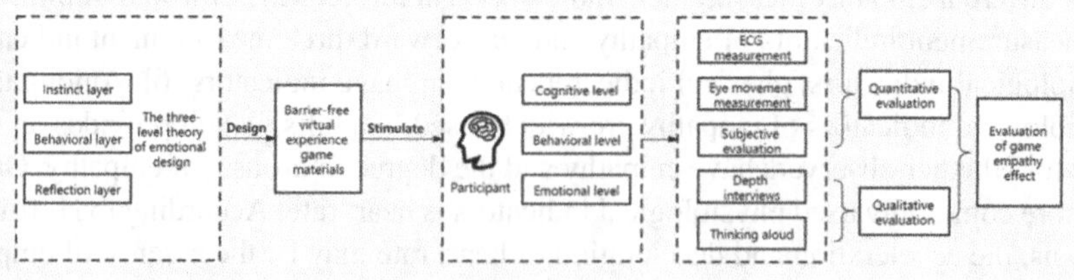

Fig. 2. Evaluation scheme of empathy effect of barrier-free virtual experience game.

Quantitative Evaluation. Quantitative evaluation is the use of mathematical methods, collecting and processing data, and judging the value of quantitative results to the evaluation object. Its logical reasoning is more rigorous, so it is more objective and scientific.

Subjective Data Evaluation. Semantic analysis is a method to study the meaning of things by using semantic difference scale, which was put forward by American social psychologist Osgood, C.E. and his colleagues in 1942. In psychological research, the semantic difference scale is widely used in the comparative study of culture, the comparative study of differences between individuals and groups, and the study of people's attitudes and views on the surrounding environment or things. The method is carried out in the form of paper and pen, and the subjects are required to evaluate a certain thing or concept on several semantic scales with 7 points (sometimes divided into 5 or 9) in order to understand the meaning and intensity of the thing or concept in various dimensions.

According to the purpose of this study, for barrier-free empathy games, a semantic difference scale was developed to evaluate the stimulation effect at cognitive, behavioral and emotional levels, with a five-level score consisting of eight pairs of antonyms. Let each subject evaluate the semantic difference scale after completing a set of experiments, and then interview it afterwards.

Objective Data Evaluation. In this paper, ECG data and eye movement data are combined, and the two data complement each other to establish a comprehensive evaluation model of empathy measurement. By collecting data from ECG equipment and eye tracking equipment, corresponding to the corresponding evaluation indexes, the weights of various factors are determined, and the evaluation indexes of empathy stimulation in different game environments are comprehensively obtained.

Qualitative Evaluation. Qualitative evaluation method can supplement quantitative evaluation results and increase the reliability of evaluation results. Audio thinking records reflect the thinking process of the observer's brain, and their thinking consciousness activities are completely orally expressed and recorded for further analysis. At the same time, in-depth interviews with observers at the end of the experiment will get the most direct and true information.

3 Establishment of Empathy Stimulation Experiment in Virtual Experience Game

3.1 Participants

Thirty-four students from Tianjin University volunteered to participate in this study, including 18 girls and 16 boys, aged between 20 and 24, and all the subjects had no cognitive impairment. The test environment is a quiet and bright indoor experimental environment, and each subject conducts experiments independently accompanied by researchers.

The use of the eye movement data, heart rate variability data, and subjective evaluation data recorded in this experiment all obtained the consent of the subjects.

3.2 Experimental Materials

Ergolab man-machine environment synchronization experimental platform, Tobii x2 eye tracker and single-lead ECG monitor are used as experimental equipment to capture and record the pupil diameter change data and heart rate variability (HRV) data of the subjects.

According to the atmosphere layer and interaction layer, the experiment is divided into two groups: A and B, each group has two factors, and each factor has two levels, as shown in Table 2.

The experiment material selected the Tianjin University campus barrier-free virtual experience game as the original material, intercepted 1 min 40 s in the game and recorded it as a video, and processed the video uniformly. Group A got 2 (visual: high saturation, low saturation) × 2 (hearing: soothing music, tension music) two-factor in-subject experimental materials, group B got 2 (game form: roaming, customs clearance) × 2 (game perspective: first-person, third-person) two-factor in-subject experimental materials. In order to eliminate the influence of the playing order of the stimulating materials on the experimental results, the materials of group A and group B were randomly presented in four different orders. At the same time, in order to ensure that the physiological data and eye movement data of the subjects have obvious changes, and to increase the contrast, the original material video was added before the official material was played to ensure the effective recording of the experimental data.

Table 2. Factors of empathy stimulation

	Factor	Level
Atmosphere layer	Visual	a1 high saturation color
		a2 low saturation color
	Hearing	b1 soothing music
		b2 tension music
Interaction layer	Game form	c1 roaming form
		c2 customs clearance form
	Game perspective	d1 first-person
		d2 third-person

3.3 Experimental Process

Before the start of the experiment, the subjects wear a single-lead ECG monitor and calibrate the Tobii x2 eye tracker. Without telling the subjects of the true purpose of the experiment, invite them to view the experimental materials. Through physiological signal monitoring and eye movement tracking, the physiological characteristics of the subjects were recorded. After the start of the formal experiment, the subjects were asked to fill in a questionnaire on the semantic scale after watching a piece of video material to record their opinions and feelings about the piece of material at the moment for subsequent analysis. After the experiment, each subject will be invited to do a short interview to understand their true feelings about the experiment. The purpose of the interview content is obtained with the consent of the subject. The specific experimental process is as follows (Fig. 3).

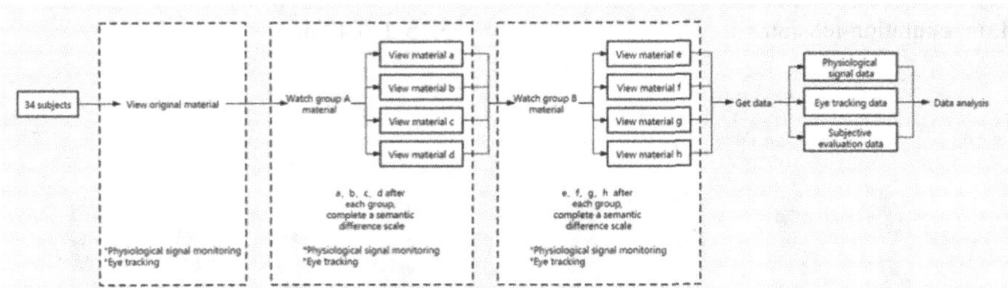

Fig. 3. Experiment process

4 Evaluation Results of Different Factors on Empathy Stimulus Effect in Virtual Experience Game Environment

4.1 Atmosphere Layer

Results of Behavioral Data. Table 3 and Table 4 show the change results of the corresponding heart rate variability (HRV) data and the statistical results of the change data of the corresponding pupil diameter when the subjects watch the experimental materials in different visual and auditory environments. Similarly, as shown in the Fig. 4, three core data detection results of the heart rate variability change of subjects when watching experimental materials in different visual and auditory environments are shown, namely, the standard deviation of all normal heartbeat intervals (SDNN), the root mean square of continuous difference (rMSSD), and the percentage of (R-R) intervals where the difference between adjacent NNs > 50 ms accounts for the number of sinus heartbeats (pNN50), and the size change of pupil diameter captured by eye tracker.

Table 3. Effects of different game environments on heart rate variability of subjects.

	SDNN	rMSSD	pNN50
High saturation-soothing	43.31 ± 9.27 ms	36.85 ± 10.90 ms	6.34% ± 5.25%
High saturation-tension	39.46 ± 11.80 ms	30.23 ± 9.49 ms	3.05% ± 3.23%
Low saturation-soothing	42.85 ± 10.69 ms	31.69 ± 11.32 ms	4.99% ± 5.78%
Low saturation- tension	46.08 ± 10.23ms	37.85 ± 12.81 ms	6.62% ± 5.15%

Table 4. Effects of different game environments on pupil size of subjects.

	Pupil diameter
High saturation-soothing	3.13 ± 0.42 mm
High saturation-tension	3.03 ± 0.36 mm
Low saturation-soothing	3.14 ± 0.42 mm
Low saturation-tension	3.15 ± 0.40 mm

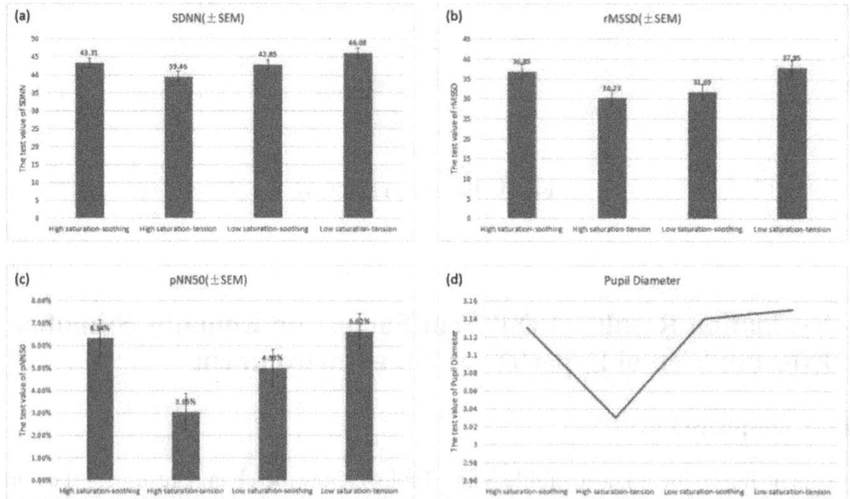

Fig. 4. Effects of different game environments on heart rate variability and pupil size. (a) SDNN (b) rMSSD (c) pNN50 (d) pupil size

It can be seen from the above experimental data that the experimental materials in different visual and auditory environments have obvious effects on the HRV (SDNN\rMSSD\pNN50) and pupil size of the subjects.

In the results of the heart rate variability HRV experimental data, the experimental data of low saturation-tension music environment and high saturation-tension music environment show a great contrast, and the HRV values of low saturation-tension music elements are relatively high, reaching SDNN 46.08 ± 10.23 ms, RMSSD 37.85 ± 12.81 ms and PNN50 6.62% ± 5.15%, respectively. The HRV values of high saturation-tension music elements are relatively low, reaching SDNN 39.46 ± 11.80 ms, RMSSD 30.23 ± 9.49 ms and PNN50 3.05% ± 3.23%, respectively. The data values of other game environment elements also showed corresponding fluctuations, which indicated that different visual elements and auditory elements caused different emotional states under the subjects' subconscious.

Similarly, in the experimental data of pupil diameter change, the high saturation-tension music element is quite different from other game environment elements, with the pupil diameter obviously smaller, reaching 3.03 ± 0.36 mm, while the low saturation-tension music element is also quite different from other game environment elements, with the pupil diameter larger, reaching 3.15 ± 0.40 mm. It shows that different visual elements and auditory elements cause corresponding changes in pupil diameter under the subject's subconscious, and high saturation-tension music elements and low saturation-tension music elements have the greatest influence on pupil diameter, showing opposite situations.

Table 5. Repeated measure ANOVA results of pupil size

Combination name	F	*P value*	Partial η2
Visual	15.847	0.000	0.353
Hearing	2.817	0.104	0.089
Visual*Hearing	8.304	0.007	0.223

Table 6. Repeated measure ANOVA results of heart rate variability

Combination name	Data value	HRV data		
		SDNN	rMSSD	pNN50
Visual	F	6.208	0.822	3.497
	P value	0.028	0.382	0.086
Hearing	F	0.012	0.027	1.856
	P value	0.915	0.871	0.198
Visual*Hearing	F	1.860	15.776	7.151
	P value	0.198	0.002	0.020

With pupil size as the dependent variable, the repeated measure ANOVA results of pupil size is shown in Table 5. The analysis shows that the main effect of visual is significant ($F = 15.847, p < 0.001$), and the pupil diameter in high saturation environment is significantly lower than that in low saturation environment. The main effect of hearing is not significant ($F = 2.817$, $p = 0.104$). The interaction between visual and hearing was significant ($F = 8.304$, $p = 0.007$).

With heart rate variability (HRV) as the dependent variable, the repeated measure ANOVA results of HRV is shown in Table 6. According to the analysis, the main effect of visual on SDNN data is significant ($F = 6.208$, $p = 0.028$), SDNN data in high saturation environment is significantly lower than that in low saturation environment. The main effect of hearing is not significant ($F = 0.012$, $p = 0.915$), and the interactive effect of visual and hearing is not significant ($F = 1.860$, $p = 0.198$). On rMSSD data, the main effect of visual is not significant ($F = 0.822$, $p = 0.382$), the main effect of hearing is not significant ($F = 0.027$, $p = 0.871$), and the interactive effect of visual and hearing is significant ($F = 15.776$, $p = 0.002$). On pNN50 data, the main effect of visual is not significant (F = 3.497, p = 0.086), the main effect of hearing is not significant (F = 1.856, p = 0.198), and the interactive effect of visual and hearing is significant (F = 7.151, p = 0.020).

Subjective Data Results. Evaluate subjective data, as shown in Table 7 and Table 8. The score in low saturation-tension music environment is the highest, reaching 27.59 ± 4.21, while the scores in high saturation-soothing and low saturation-soothing environments are the lowest, reaching 21.18 ± 5.37 and 21.29 ± 4.28 respectively. The repeated measure ANOVA results shows that the main effect of visual is not significant (F = 3.389, p = 0.075), the main effect of hearing is significant (F = 34.568, p < 0.001), and the interactive effect of visual and hearing is not significant (F = 3.205, p = 0.083).

Table 7. Subjective data results

	Score
High saturation-soothing	21.18 ± 5.37
High saturation-tension	25.38 ± 4.70
Low saturation-soothing	21.29 ± 4.28
Low saturation-tension	27.59 ± 4.21

Table 8. Repeated measure ANOVA results of subjective data

Combination name	F	*P value*	Partial η2
Visual	3.389	0.075	0.093
Hearing	34.568	0.000	0.512
Visual*Hearing	3.205	0.083	0.089

Discussion and Analysis. Behavioral data show that under four conditions, visual has a significant impact on emotion, and pupil diameter and SDNN data in low saturation environment are significantly higher than those in high saturation environment, which indicates that low saturation visual environment has the highest emotional arousal and is more conducive to empathy. Visual and hearing have a significant interaction with emotion, and the pupil diameter and HRV value in low saturation-tension music environment are obviously higher, which indicates that the game environment will promote empathy, which is consistent with subjective data.

However, according to subjective data analysis, hearing has a significant impact on the subjects' emotions, which is different from the results of behavioral data analysis, indicating that people are more sensitive to hearing subjectively, and under the subjects' subconscious, visual is also silently affecting people's emotions, even more than hearing.

4.2 Interaction Layer

Results of Behavioral Data. Table 9 and Table 10 show the change results of the corresponding HRV data and the statistical results of the change data of the corresponding pupil diameter when the subjects watch the experimental materials in different game forms and game perspectives. Similarly, as shown in the Fig. 5, it shows the three core data detection results of the heart rate variability changes (SDNN\rMSSD\pNN50) when the subjects watch the experimental materials under different game forms and game viewing angles, as well as the pupil diameter change data captured by the eye tracker.

Table 9. Effects of different game environments on heart rate variability of subjects.

	SDNN	rMSSD	pNN50
Roaming-first person	41.50 ± 12.42 ms	36.58 ± 12.06 ms	6.57% ± 5.92%
Roaming-third person	37.83 ± 13.18 ms	34.25 ± 10.22 ms	4.72% ± 4.02%
Customs clearance-first person	48.33 ± 18.38 ms	42.58 ± 12.80 ms	7.23% ± 3.84%
Customs clearance-third person	46.00 ± 10.11 ms	37.67 ± 16.43 ms	6.90% ± 7.50%

Table 10. Effects of different game environments on pupil size of subjects.

	Pupil diameter
Roaming-first person	3.07 ± 0.34 mm
Roaming-third person	3.20 ± 0.45 mm
Customs clearance-first person	3.30 ± 0.43 mm
Customs clearance-third person	3.23 ± 0.48 mm

From the above experimental data, it can be seen that the experimental materials in different game forms and game perspectives have obvious effects on the HRV (SDNN\rMSSD\pNN50) and pupil size of the subjects.

Combined with the heart rate variability HRV experimental data and pupil size experimental data, the customs clearance form-first person perspective element is quite different from other game environment elements, and HRV value and pupil diameter are obviously larger. They reached SDNN 48.33 ± 18.38 ms, RMSSD 42.58 ± 12.80 ms, PNN50 7.23% ± 3.84% and pupil 3.30 ± 0.43 mm, respectively, while the HRV value and pupil diameter in roaming form environment were obviously smaller, which indicated that different game forms and game perspective elements caused different emotional states of subjects under their subconscious.

With pupil size as dependent variable, the repeated measure ANOVA results of pupil size is shown in Table 11. The analysis shows that the main effect of game form is significant ($F = 18.808$, $p < 0.001$), and the pupil diameter in roaming form is significantly lower than that in customs clearance form. The main effect of game perspective

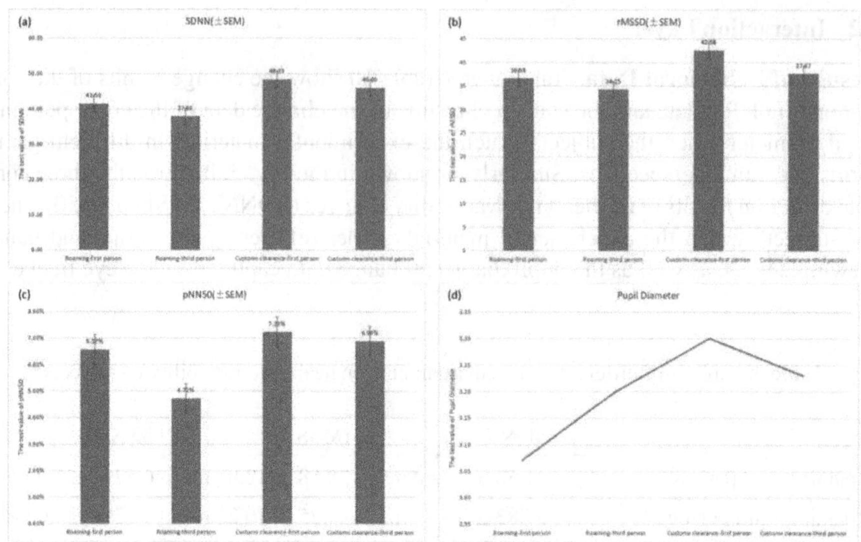

Fig. 5. Effects of different game environments on heart rate variability and pupil size. (a) SDNN (b) rMSSD (c) pNN50 (d) pupil size

Table 11. Repeated measure ANOVA results of pupil size

Combination name	F	*P value*	Partial η2
Game form	18.808	0.000	0.393
Game perspective	0.612	0.440	0.021
Game form*Game perspective	13.148	0.001	0.312

Table 12. Repeated measure ANOVA results of heart rate variability

Combination name	Data value	HRV data		
		SDNN	rMSSD	pNN50
Game form	F	6.885	3.882	4.759
	P value	0.024	0.074	0.052
Game perspective	F	1.585	1.936	0.160
	P value	0.234	0.192	0.696
Game form* Game perspective	F	0.034	0.175	1.746
	P value	0.856	0.684	0.213

is not significant (F = 0.612, p = 0.440). The interaction between game form and game perspective is significant (F = 13.148, p = 0.001).

With heart rate variability (HRV) as the dependent variable, the repeated measure ANOVA results of HRV is shown in Table 12. According to the analysis, the main effect of game form on SDNN data is significant (F = 6.885, p = 0.024), while the SDNN data in roaming form is significantly lower than that in customs clearance form, while the main effect of game perspective is not significant (F = 1.585, p = 0.234), and the interactive effect between game form and game perspective is not significant (F = 0.034, p = 0.856). According to rMSSD data, the main effect of game form is not significant (F = 3.882, p = 0.074), the main effect of game perspective is not significant (F = 1.936, p = 0.192), and the interactive effect between game form and game perspective is not significant (F = 0.175, p = 0.684). According to pNN50 data, the main effect of game form is not significant (F = 4.759, p = 0.052), the main effect of game perspective is not significant (F = 0.160, p = 0.696), and the interactive effect between game form and game perspective is significant (F = 1.746, p = 0.213).

Subjective Data Results. Evaluate subjective data, as shown in Table 13 and Table 14. The scores of customs clearance form-first person perspective and customs clearance form-third person perspective are significantly higher, reaching 27.26 ± 4.63 and 26.62 ± 6.32 respectively, while the scores of roaming form-third person perspective are the lowest, reaching 19.15 ± 5.00. The repeated measure ANOVA results shows that the main effect of game form is significant (F = 58.313, p < 0.001), the main effect of game perspective is not significant (F = 1.164, p = 0.288), and the interactive effect between game form and game perspective is not significant (F = 0.332, p = 0.568).

Table 13. Subjective data results

	Score
Roaming-first person	20.38 ± 6.01
Roaming-third person	19.15 ± 5.00
Customs clearance-first person	27.26 ± 4.63
Customs clearance-third person	26.62 ± 6.32

Table 14. Repeated measure ANOVA results of subjective data

Combination name	F	*P value*	Partial η2
Game form	58.313	0.000	0.639
Game perspective	1.164	0.288	0.034
Game form*Game perspective	0.332	0.568	0.010

Discussion and Analysis. Behavioral data show that under four conditions, the game form has a significant impact on emotion, and the pupil diameter and SDNN data in customs clearance environment are significantly higher than those in roaming form, which indicates that the game form of customs clearance has the highest emotional arousal and is more conducive to the generation of empathy. Game form and game perspective have significant interaction with emotion, and the pupil diameter and HRV value in the environment of customs clearance-first person are obviously higher, which indicates that the game environment will promote empathy. At the same time, the subjective data analysis shows that the game form has a significant impact on the subjects' emotions, and the score under the customs clearance form is significantly higher than that under the roaming form, which further shows that the customs clearance form will promote empathy.

4.3 Comprehensive Evaluation

Based on the above research results, it is found that visual and game forms have the highest emotional arousal, which is more conducive to promoting empathy. In comparison, the influence of hearing on emotion is far less obvious than that of visual, but it gives people the strongest subjective feeling, and on the whole, hearing also has certain influence. Another element, game perspective, has the weakest influence on emotional arousal, and the first-person perspective has higher emotional arousal than the third-person perspective, which is consistent with immersion theory [10]. However, after the interview, an interesting problem was found, most subjects said that the game environment from the third-person perspective can make him substitute into the barrier-free experience situation, thus generating empathy for the disabled. There are two reasons to summarize. The first is that the third-person perspective can make him substitute into the perspective of the disabled and experience the daily action situation of the disabled. The second is that the third-person perspective allows him to see from the perspective of a bystander or companion, feel the difficulties and inconveniences of the daily life of the disabled, understand their behavioral characteristics, and develop empathy.

5 Conclusion

This article takes the campus barrier-free virtual experience game as the research object. Through ECG experimental research, eye movement experimental research, and subjective scale test, starting from the atmosphere and interaction layer of game design, it has an understanding of visual, hearing, game form and game perspective. Each element was measured for empathy stimulation. The results show that the above four elements have different influences on people's empathy, which provides relevant theoretical support for the design of barrier-free popular science games, and also provides some reference for the design of empathy games in the future.

This study provides us with many useful discoveries:

Find a new opportunity for barrier-free construction, and carry out barrier-free science popularization in the form of virtual experience games, so as to promote the public's empathy for the disabled and call for attention to the disabled groups.

According to the three-level theoretical model of emotional design, a virtual experiential game design model of atmosphere-interaction-reflection is constructed, which stimulates empathy based on this theory.

Through the research, it is found that visual elements and game form elements can stimulate empathy best, and auditory elements are the strongest in people's subjective feelings. It is suggested that the subsequent campus barrier-free virtual experience games should be optimized by the rule of low saturation color-intense music-customs clearance form.

According to the immersion theory, the first-person game perspective has deeper immersion and can promote empathy, which is also proved by the experimental results, but the third-person perspective can stimulate empathy more subjectively. In view of this, it is suggested that the subsequent game design can be set with multiple perspectives, or set to a state where the disabled and bystander roles can be switched.

In this study, only college students aged 20–24 were selected as the research objects, which has certain limitations. Follow-up research can face more people, such as middle-aged people and children and other special groups. As a limited experimental study, this paper only studies four related elements of atmosphere layer and interaction layer, and does not deeply discuss the influence of reflection layer on empathy stimulation. Further research can be carried out for other game design elements.

Acknowledgments. We would like to express our gratitude to the participants and review experts for their contributions to the study reported here.

References

1. World Health Organization: The global burden of disease. Published by the Harvard School of Public Health on behalf of the World Health Organization and the World Bank. (2008)
2. Liu, R., Ren, Y.: Research on the flow experience and empathy effect in immersive virtual environment. Audio-Vis. Educ. Res. **40**(04), 99–105 (2019)
3. Bachen, C.M., et al.: How do presence, flow, and character identification affect players' empathy and interest in learning from a serious computer game? Comput. Human Behav. **64**(Nov), 77–87 (2016)
4. Shin, D.: Empathy and embodied experience in virtual environment: to what extent can virtual reality stimulate empathy and embodied experience? Comput. Human Behav. **78**, 64–73 (2018)
5. Mercer, S.W., Reynolds, W.J.: Empathy and quality of care. Br. J. Gener. Pract.: J. R. Coll. Gener. Pract. **52** (Suppl), S9–S12 (2002)
6. Hoffman, M.L., Guo, R.: Measurement index of empathy. Psychol. Dev. Educ. **7**(2), 30–32, 29 (1991)
7. Shi, H., et al.: Differences of heart rate variability between happiness and sadness emotion states: a pilot study. J. Med. Biol. Eng. **37**(4), 527–539 (2017). https://doi.org/10.1007/s40846-017-0238-0
8. Blackburn, K., Schirillo, J.: Emotive hemispheric differences measured in real-life portraits using pupil diameter and subjective aesthetic preferences. Exp. Brain Res. **219**(4), 447–455 (2012). https://doi.org/10.1007/s00221-012-3091-y

9. Beatty, D.D.: Visual pigments of the American eel Anguilla rostrata. Vision Res. **15**(7), 771–776 (1975)
10. Sherleena, B., et al.: Interprofessional empathy and communication competency development in healthcare professions' curriculum through immersive virtual reality experiences. J. Interprof. Educ. Pract. **15**(C), 127–130 (2019)

Trust and Automation: A Systematic Review and Bibliometric Analysis

Zhengming Zhang[1(✉)], Vincent G. Duffy[1], and Renran Tian[2]

[1] Purdue University, West Lafayette, IN 47906, USA
{zhan3988,duffy}@purdue.edu
[2] Indiana University Purdue University Indianapolis, Indianapolis, IN 46202, USA
rtian@iupui.edu

Abstract. With the development of machine learning, high-level automation in many aspects is promising or even fledgling. However, a challenge that appears at this moment is that how people trust in automation. This study is a systematic literature review of this topic using bibliometric analysis. Tools like Scopus, VOSviewer, AuthorMapper, MAXQDA, Vicinitas, Harzing's Publish or Perish, and Web of Science are used to conduct bibliometric analyses and content analyses. The results identify the most influential publications and trends of research topics. In addition, topics related to automated vehicles are found to be high-profile applications in the area.

Keywords: Trust · Automation · Bibliometric analysis

1 Introduction and Background

The corporation between human operator and automation have promoted higher working efficiency and accuracy. Though many automation products, such as highly automated vehicles, may be ready to walk into the markets and soon take the human operator's place. But to be entirely accepted by customs, many automation products still face significant challenges in dealing with acceptability. For example, according to Zmud and Sener's study from 2019, only 37% of Frisco residents are willing to purchase an SAE level 5 self-driving vehicle as their personally owned vehicles, and the biggest concern is that autonomous driving is too new [1]. Similar results from a survey study imply low acceptance of the high autonomy level's vehicles from users [2]. Since the decision-making process highly depends on machine learning models and machine learning models' mechanisms are too sophisticated for the public, users are likely to feel insecure when using the automation products, resulting in low acceptance.

The low acceptance rate will negatively impact the automation product's performance in the market, resulting in less automation research investment. Studies have proved that trust is a second factor (besides usefulness) affecting user's acceptance and intention-to-use attitudes towards advanced driver assistance systems, making the ease-of-use factor in the classical TAM (technology acceptance model) less critical [3, 4]. No matter the differences in applying trust as a direct factor or latent factor through usefulness and

C. Stephanidis et al. (Eds.): HCII 2021, LNCS 13094, pp. 451–464, 2021.
https://doi.org/10.1007/978-3-030-90238-4_32

perceived risk, these studies imply the importance of trust towards the system capability to increase the ADS utilization rates. Therefore, establishing a trust relationship between users and automation products is essential to boost the public's acceptability.

2 Purpose of Study

This study's main objective is to perform a systematic literature review of articles related to automation and trust. The results are helpful to identify the trending topics in the areas and those corresponding articles. Moreover, some potential undiscovered topics are discussed to enlighten the readers on this area's future work. With the increasing maturity of automation technology, a better understanding of user's trust in automation is essential to popularize and evolve the automation product. Most results from the bibliometric analysis are conducted through tools like Scopus, VOSviewer, AuthorMapper, MAXQDA, Vicinitas, Harzing's Publish or Perish, and Web of Science.

3 Data Collection and Methods

To collect a variety of articles in the area, We utilized several different academic databases. Except for Google Scholar, We used the search terms "trust" and "automation" for the other databases. Since Google Scholar returned an implausible number of results (646,000) use the search terms, We added the word "interaction" to narrow the searching scope. Also, each database provides different filters to users. We specified the filter used in Table 1, which allowed us to search related articles more precisely. In the rest of the study, we mainly used the results from Scopus and Web of Science because those two databases offer data exporting tools.

Table 1. Database search table with corresponding search terms, filter, and results

Database	Search term	Search type	Filter	Results
Scopus	"trust" AND "automation"	Article title, Abstract, Keywords	Time: 1990 – Now	2,022
AuthorMapper	"trust" AND "automation"	Unknown	Subject: User Interfaces and Computer Interaction	3,211
Web of Science	"trust" AND "automation"	Article title, Abstract, Keywords	Time: 1990 – Now	1,356
Google Scholar	"trust" AND "automation" AND "interaction"	Unknown	Time: 1995 – Now	135,000
Harzing's Publish or Perish	"trust" AND "automation"	Keywords	Maximum number of results: 400	400

We first presented trend analysis, which showed the upward or downward momentum of production and participation in the literature on the topic [5]. In other words, a trend analysis gives people a sense of the topic's popularity over time. Similar to the trend analysis, subject analysis demonstrates the diversity of the topic in academia. Furthermore, co-citation analysis is a well-known method to classify the existing literature [6], which is helpful to identify critical sub-topics and studies. Finally, the content analysis allowed us to monitor and outline research contents in the context of trust and automation [7]. Therefore, we review the topic of trust and automation from the shallower to the deeper using the abovementioned methods.

4 Results

4.1 Trend Analysis

Figure 1 is generated by the analysis tool of Web of Science. It shows the number of total publications in each year. There was a steady growth from 1994 to 2015, where the annual total number of publications is from 1 (1990) to 69 (2015). After 2015, the growth of publications in trust and automation is increasing rapidly. One potential reason for such a surge is the successful application of deep learning models. Before using machine learning models to develop automation technology, automation mainly assists human operators or independently finishes simple tasks. However, some successful applications of deep learning models (such as object detection using convolution neural network [8] and trajectory prediction using recurrent neural network [9], etc.) promote the next generation of automation and vastly enhance performance. Therefore, more researchers started to research to prepare the popularization of automation technology.

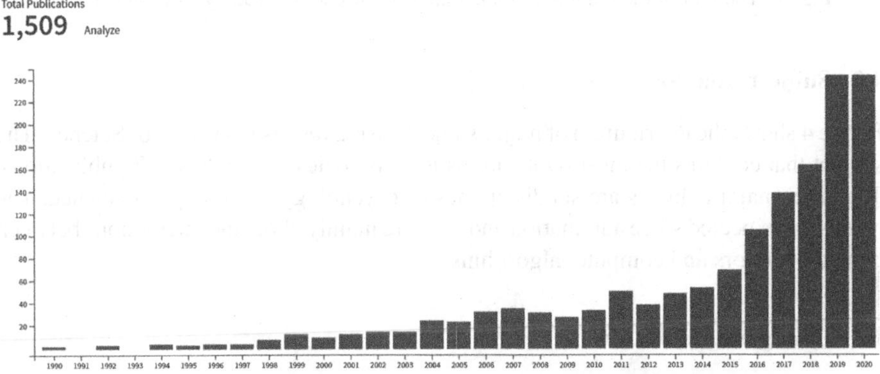

Fig. 1. The bar plot of the number of total publications in each year (1990–2020).

Moreover, the total citation trend per year (Fig. 2) and total author occurrence per year (Fig. 3) show the same abovementioned pattern. Note that all three trending figures peaked in 2020 except for the author occurrence, which showed a slight decrease in 2020.

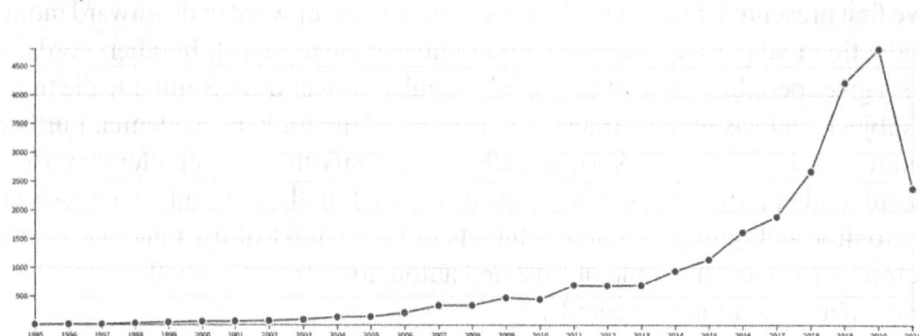

Fig. 2. The plot of the number of total citations in each year (1990–2020).

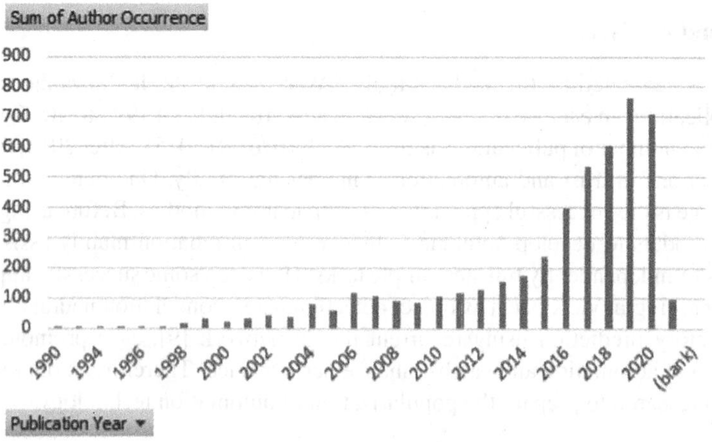

Fig. 3. The plot of the number of total author occurrences in each year (1990–2020).

4.2 Subject Analysis

Figure 4 shows the distribution of major subjects using results from Web of Science. The subject that contains the most publications is ergonomics, which has 347 publications. The other major subjects are subdisciplines of psychology and computer science. The results are expected since automation and trust are mainly about the interactions between human operators and computer algorithms.

4.3 Emergence Indicator

Figure 5 shows some analysis results from AuthorMapper. The trend of the number of publications is similar to the abovementioned trends, except that the peak is in 2020 instead of 2019. The part of keywords reported some frequently used terms in the sections of title, keyword, and abstract. The most frequently used term is "human factor," which addresses the importance of human operators' role in automation and trust. "Privacy" and "security" are two main concerns from the users, which many researchers are conducted to investigate how they influence trust.

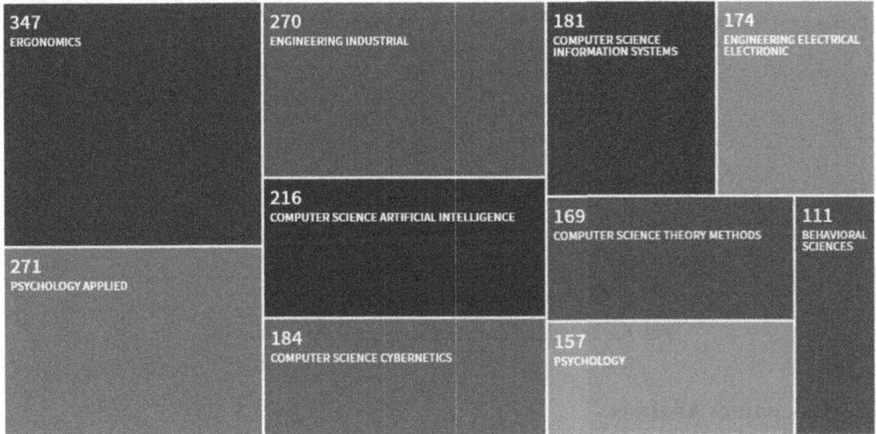

Fig. 4. The distribution of subjects using results from Web of Science.

SEARCH RESULTS

3211 ARTICLES | 8384 AUTHORS | 3164 INSTITUTIONS | 663 PUBLICATIONS

KEYWORDS

Human factors Trust
Privacy Security Automation
User experience Human-robot interaction Machine learning
Cloud computing
Artificial intelligence
Automated driving Internet of Things
Usability Human-systems integration
Situation awareness

YEAR PUBLISHED

1992 2021

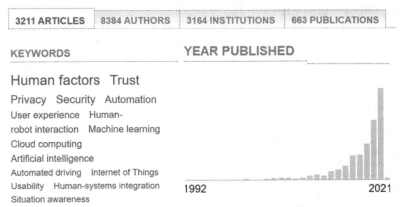

Fig. 5. Analysis results from AuthorMapper search

4.4 Engagement Measures

In addition to the emergence indicator, we calculate the engagement measures to show the public's attention to automation and trust. The analysis is conducted through Vicinitas, which helps track and analyze real-time and historical tweets of your social media campaigns and brands on Twitter [10]. Note the duration in Fig. 6 is relatively short. The sum of the followers of each user who posted related tweets is 312.8k, which is an indicator of the influence of the search terms. The steady impact (Fig. 6) might be affected by the other current affairs like the vaccine of COVID-19 and 2020 U.S. presidential election results.

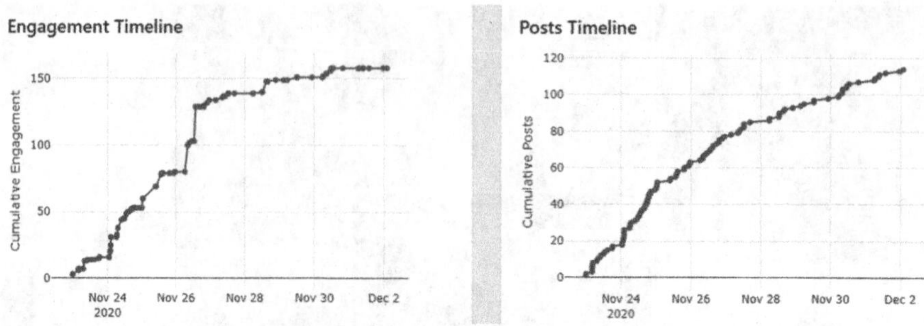

Fig. 6. Trends of cumulative engagement and post

4.5 Co-citation Analysis

The co-citation analyses are conducted in two parts. We used CiteSpace to analyze the data in the first part, while the second part is achieved through VOSviewer. The CiteSpace is a tool for visualizing and analyzing trends and patterns in scientific literature. It focuses on finding critical points in the development of a field or a domain, especially intellectual turning points and pivotal points [11].

References	Year	Strength	Begin	End	1990 - 2019
Lee JD, 2004, HUM FACTORS, V46, P50, DOI	2004	**23.61**	2005	2009	
Hoff KA, 2015, HUM FACTORS, V57, P407, DOI	2015	**15.99**	2016	2019	
Hancock PA, 2011, HUM FACTORS, V53, P517, DOI	2011	**9.68**	2012	2016	
Dzindolet MT, 2003, INT J HUM-COMPUT ST, V58, P697, DOI	2003	**9.5**	2006	2008	
Parasuraman R, 1997, HUM FACTORS, V39, P230, DOI	1997	**9.2**	1999	2002	
Dixon SR, 2006, HUM FACTORS, V48, P474, DOI	2006	**8.06**	2009	2011	
Dixon SR, 2007, HUM FACTORS, V49, P564, DOI	2007	**7.36**	2008	2011	
Merritt SM, 2008, HUM FACTORS, V50, P194, DOI	2008	**7.28**	2011	2012	
Beggiato M, 2013, TRANSPORT RES F-TRAF, V18, P47, DOI	2013	**6.74**	2015	2019	
Wiegmann Douglas, 2010, THEORETICAL ISSUES E, V2, P352, DOI	2010	**6.44**	2010	2015	
Onnasch L, 2014, HUM FACTORS, V56, P476, DOI	2014	**6.38**	2015	2019	
Parasuraman R, 2010, HUM FACTORS, V52, P381, DOI	2010	**5.96**	2011	2014	
Manzey D, 2012, J COGN ENG DECIS MAK, V6, P57, DOI	2012	**5.88**	2014	2017	
Wickens CD, 2007, THEOR ISS ERGON SCI, V8, P201, DOI	2007	**5.77**	2008	2011	
Moray N, 2000, J EXP PSYCHOL-APPL, V6, P44, DOI	2000	**5.75**	2003	2005	
Rovira E, 2007, HUM FACTORS, V49, P76, DOI	2007	**5.68**	2010	2012	
Verberne FMF, 2012, HUM FACTORS, V54, P799, DOI	2012	**5.59**	2015	2017	
Meyer J, 2004, HUM FACTORS, V46, P196, DOI	2004	**5.54**	2007	2009	
Beller J, 2013, HUM FACTORS, V55, P1130, DOI	2013	**5.31**	2015	2019	
Madhavan P, 2007, THEOR ISS ERGON SCI, V8, P277, DOI	2007	**5.27**	2008	2012	
Gold C, 2015, PROCEDIA MANUF, V3, P3025, DOI	2015	**4.99**	2017	2019	
Choi JK, 2015, INT J HUM-COMPUT INT, V31, P692, DOI	2015	**4.99**	2017	2019	
Hergeth S, 2016, HUM FACTORS, V58, P509, DOI	2016	**4.99**	2017	2019	
Dzindolet MT, 2002, HUM FACTORS, V44, P79, DOI	2002	**4.92**	2003	2007	
Bahner JE, 2008, INT J HUM-COMPUT ST, V66, P688, DOI	2008	**4.75**	2009	2012	

Fig. 7. Top 25 references with the strongest citation bursts (sorted by the intensity of the burst).

Figure 7 shows the top 25 references with the strongest citation bursts. Those publications could be considered as the most influential publications in areas of automation and trust. Note the first two articles have much more strength of citation bursts than the following articles. Both studies expressed an excellent literature review, but the difference is that Lee and See talked more about the challenges of trust in automation.

At the same time, Hoff and Bashir present a three-layered model which describes the relationship between some factors and trust [12, 13].

The specific citation bursts with clustered topics are shown in Fig. 8. The bottom area is more about automated vehicles. In this area, the interactions between driver and automation, such as the driver's take-over task, are essential. The right side of Fig. 8 focuses on the factors that influence users' trust related to the work of Hoff and Bashir.

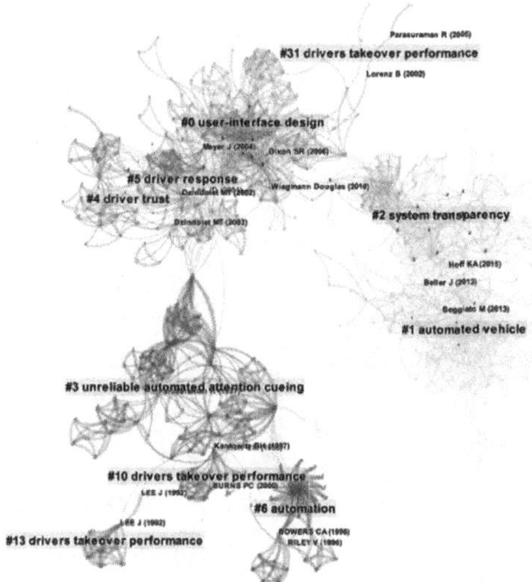

Fig. 8. Citation bursts with clustered topics using data from Web of Science

Create Map ×

Verify selected cited references

Selected	Cited reference	Citations	Total link strength
☑	parasuraman r, 1997, hum factors, v39, p230...	148	1042
☑	lee jd, 2004, hum factors, v46, p50, doi 10.15...	159	1030
☑	lee j, 1992, ergonomics, v35, p1243, doi 10.1...	107	777
☑	lee jd, 1994, int j hum-comput st, v40, p153, ...	95	740
☑	muir bm, 1996, ergonomics, v39, p429, doi 1...	77	558
☑	parasuraman r, 2000, ieee t syst man cy a, v3...	70	546
☑	muir bm, 1994, ergonomics, v37, p1905, doi ...	75	480
☑	dzindolet mt, 2003, int j hum-comput st, v58, ...	59	459
☑	parasuraman r., 1993, int j aviat psychol, v3, ...	49	415
☑	moray n, 2000, j exp psychol-appl, v6, p44, d...	39	314
☑	rempel jk, 1985, j pers soc psychol, v49, p95,...	34	300
☑	muir bm, 1987, int j man mach stud, v27, p52...	34	288
☑	merritt sm, 2008, hum factors, v50, p194, doi...	29	279
☑	lewandowsky s, 2000, j exp psychol-appl, v6,...	25	276
☑	madhavan p, 2006, hum factors, v48, p241, d...	25	253
☑	bainbridge l, 1983, automatica, v19, p775, d...	31	242
☑	dzindolet mt, 2002, hum factors, v44, p79, d...	24	234
☑	singh i. l., 1993, int j aviat psychol, v3, p111, ...	25	229

Fig. 9. The most co-cited articles using Web of Science data.

We used VOSviewer to conduct a co-citation analysis. Co-citation analysis aims to track the linkages among the intellectual works and map the evolutionary structure of scientific disciplines [14]. Figure 9 shows the most co-cited articles using Web of Science data sorted by total link strength. Some results are consistent with the citation burst analysis because the analysis methods are similar and used the same dataset. Figure 10 shows the relation between high co-cited articles. Moreover, We list the top six most cited publications in Table 2 to identify the most important articles [12, 15–19].

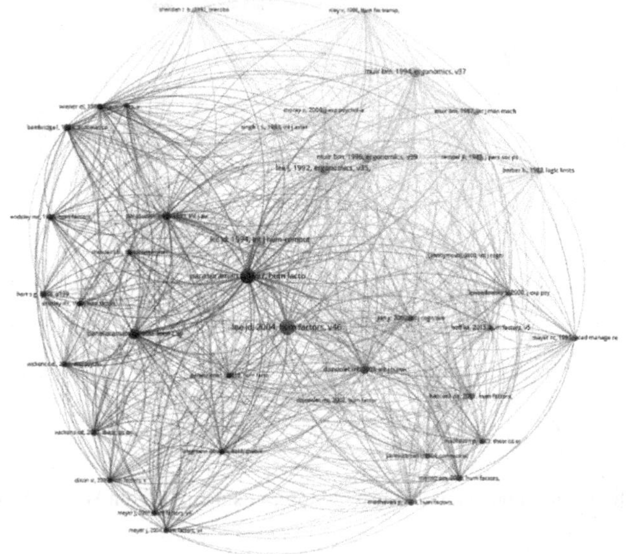

Fig. 10. VOSviewer content analysis diagram using web of science results.

4.6 Content Analysis

The results of content analysis are in two parts. We used AuthorMapper to generate the results for the first part. The first part identifies the leading authors and leading institutions and what is their contribution to the area of automation and trust. Understanding their focuses helps identify some important future work because they are the most active researchers or institutions in the area. According to the keywords from the leading authors' publications (Table 3), Ziefle and Boy paid much attention to understand what factors could influence the users' trust. At the same time, Bengler did more research on the interaction between human operators and automation.

Table 4 shows the leading institutions and their corresponding keywords. Note the autonomous driving is a trendy topic in the area. In other words, people may directly associate automation with automated vehicles. Numerous research in such topics addresses the importance of automated vehicles.

Table 2. Top 6 most co-cited publications

Authors	Year	Title	Co-citation
Lee, John D., and Katrina A. See	2004	"Trust in Automation: Designing for Appropriate Reliance"	159
Parasuraman, Raja, and Victor Riley	1997	"Humans and Automation: Use, Misuse, Disuse, Abuse"	148
Lee, John, and Neville Moray	1992	"Trust, control strategies and allocation of function in human-machine systems"	107
Lee, John D., and Neville Moray	1994	"Trust, self-confidence, and operators' adaptation to automation"	95
Muir, Bonnie M	1994	"Trust in automation: Part I. Theoretical issues in the study of trust and human intervention in automated systems"	77
Muir, Bonnie M., and Neville Moray	1996	"Trust in automation. Part II. Experimental studies of trust and human intervention in a process control simulation"	75

Table 3. Leading authors using AuthorMapper.

Authors	Years	Leading keywords	Count
Ziefle, Martina	2010; 2013–2021	Technology acceptance; Human factors; Privacy; User diversity; Trust in automation; Ambient assisted living; Autonomous driving	32
Bengler, Klaus	2014–2020	Conditional automation; Eye tracking; Mode awareness; Non-driving related task; Partial automation; Take-over performance	21
Boy, Guy André	2013; 2015; 2020	Experience Feedback; Human Spaceflight; Human-centered Design; Life-critical Systems (LCS); Technique For Human Error Rate Prediction (THERP)	18
Chen, Fang	2006; 2016–2020	Trust; Cognitive load; Predictive decision making; Uncertainty; Ambiguity Uncertainty; Automatic Quality Monitoring (AQM); Combined Point; Human-machine collaboration; Human-machine Trust	12
Philipsen, Ralf	2016–2020	Human factors; Autonomous driving; Business simulation game; Privacy; Technology acceptance; Trust in automation; Automation; Cyber-physical production systems; Decision support systems	12

Table 4. Leading institutions using AuthorMapper.

Institution	Years	Leading Keywords	Count
RWTH Aachen University	2010; 2012–2020	Technology acceptance; Human factors; Privacy Trust in automation; Ambient assisted living; Automated driving; Autonomous driving; User diversity	50
Delft University of Technology	2003; 2009; 2012; 2014; 2015; 2017–2020;	Automation; Autonomous vehicles; Explainable A.I.; Human-machine systems; Shared control; Trust; Workload	29
University of Central Florida	2011–2020	Human factors; Human-robot interaction; Trust; Assessment; Human-systems integration; military; Transparency; Active Indicator Probes; Adaptive associate systems	26
University of Southampton	2005; 2013–2014; 2016–2020;	Autonomous vehicles; Agency; Automation; Cognitive Work Analysis; EID; Handover; Human Factors methods; Human-computer interaction; Human-machine networks; Human-systems integration	22
Florida Institute of Technology	2013–2015; 2018; 2020	Design cognition; Design reasoning; Design support; Engineering design; Experience Feedback; Eye-tracking; Functional fixedness; Human Spaceflight; Human-centered Design	20

The result of the second part of content analysis is a word cloud. We used eight influential publications: the top three in Fig. 6 and all publications in Table 2 to generate the word cloud. Note there is one overlapping article [12] (Fig. 11).

Fig. 11. Word cloud using important publications

5 Discussion and Conclusion

Leading Articles: All the studies listed in Table 2 are typical human factor analyses on the issue of trust and automated systems, where they focus on the topic from different perspectives. Muir [18] proposed a model of human trust in automation, built upon interpersonal trust models. Expectations possessed by the users are the key factor to construct confidence in automation, which directly influences trust and behaviors. The study [15] illustrates the typical human trust situations in automation, which are overtrust, distrust, and appropriate trust and those corresponding results, abuse, disuse, and use. Lee and see [12] further developed a trust model integrating the prior two studies' results. They explained how trust is dynamically and mutually updated during the interaction between humans and automation. Experimental guidelines are given by Muir and Moray [19], and they further indicate some influential factors on trust, such as users' perception of the automation and understanding of the errors. Lee and Moray [17] study how the level of trust changes operators' strategy of use and how the trust is restored when an error from the automation systems happens.

Leading Authors: The leading authors in Table 2 were doing more wide research in trust and automation. For example, Ziefle and Philipsen studied trust in automation in order to promote a better user acceptance model for automation products. In contrast, the other three authors implemented advanced technologies such as eye-tracking devices and driving simulators to quantify the trust level empirically. Moreover, the automation in their context is mainly the automated vehicle, which is identified as the automation product receiving the most attention from the researchers.

Challenges: A recent large-scale survey [20] shows that more than 60% of people have initial negative trust in fully autonomous vehicles. And based on the framework of trust formation [21], the level of trust is likely to descend to the disuse level before achieving appropriate trust when the users' initial trust is low. Many users might deprecate the automation products because of distrust caused by the initial negative trust, in which the automation has no chance to earn trust from the users. Appropriate guidance for the public on automation is the key to calibrate the initial trust level.

Even though researchers understand the importance of trust, they are still facing two primary questions: what kind of trust we need to build (i.e., how to measure the level of trust implicitly and quantify it ?), and how could we promote such a trust relationship between users and automation? An overmuch trust will induce users to misuse the automation, while a lack of trust will make users disuse the automation [15]. In the context of autonomous driving, both misuse and disuse undermine traffic safety or even increase the possibility of fatal accidents. Therefore, we need a system that could calibrate the level of trust between users and automation products, which maintains the level of trust at an appropriate level. Recent studies have proven some empirical techniques for trust manipulation. For example, Helldin et al. show that people with uncertainty about the ADS possess less trust than those without [22], indicating uncertain information preventing users from overtrust. Also, Hock et al. provide evidence that reasoning feedback increases users' trust and acceptability [23].

Summary: Several significant findings are summarized below: 1. The research topic, trust and automation, has received more attention in both academia and industry than before due to automation breakthroughs. 2. Trust and automation are likely to be a trendy topic in the next several years. 2. Some important articles are identified. 2. The current most popular topic under automation and trust is the relationship between automated vehicles and trust. 3. Details about calibrating users' trust in automation to achieve appropriate trust might be a popular research direction soon.

6 Future Work

This study's results mostly came from the bibliometric analyses (trends and co-citation analysis, etc.) and content analyses (word cloud, leading keywords). Therefore, the results generally depicted the outline of automation and trust. Later, more detailed trending and challenges could be specified by an exhaustive literature review. Also, the study did not go beyond the scope of automation and trust; some emerging interdisciplinary topics might be identified in future work. Overall, understanding the relationship between automation and trust is the key to popularizing automation in the future.

References

1. Zmud, J., Sener, I.N.: Acceptance, trust and future use of self-driving vehicles (2019)
2. Hewitt, C., Politis, I., Amanatidis, T., Sarkar, A.: Assessing public perception of self-driving cars: the autonomous vehicle acceptance model. In: Proceedings of the 24th International Conference on Intelligent User Interfaces, pp. 518–527 (2019)
3. Ghazizadeh, M., Peng, Y., Lee, J.D., Boyle, L.N.: Augmenting the technology acceptance model with trust: commercial drivers' attitudes towards monitoring and feedback. In: Proceedings of the Human Factors and Ergonomics Society Annual Meeting, vol. 56, no. 1, pp. 2286–2290. Sage Publications, Los Angeles (2012)
4. Choi, J.K., Ji, Y.G.: Investigating the importance of trust on adopting an autonomous vehicle. Int. J. Human-Comput. Interact. **31**(10), 692–702 (2015)
5. Duffy, B.M., Duffy, V.G.: Data mining methodology in support of a systematic review of human aspects of cybersecurity. In: Duffy, V.G. (ed.) HCII 2020. LNCS, vol. 12199, pp. 242–253. Springer, Cham (2020). https://doi.org/10.1007/978-3-030-49907-5_17
6. Fahimnia, B., Sarkis, J., Davarzani, H.: Green supply chain management: a review and bibliometric analysis. Int. J. Prod. Econ. **162**, 101–114 (2015)
7. Guo, F., Lv, W., Liu, L., Wang, T., Duffy, V.G.: Bibliometric analysis of simulated driving research from 1997 to 2016. Traffic Inj. Prev. **20**(1), 64–71 (2019)
8. Ren, S., He, K., Girshick, R., Sun, J.: Faster R-CNN: towards real-time object detection with region proposal networks. In Advances in Neural Information Processing Systems, pp. 91–99 (2015)
9. Alahi, A., Goel, K., Ramanathan, V., Robicquet, A., Fei-Fei, L., Savarese, S.: Social LSTM: human trajectory prediction in crowded spaces. In: Proceedings of the IEEE Conference on Computer Vision and Pattern Recognition, pp. 961–971 (2016)
10. Twitter Analytics Tool for Tracking Hashtags, Keywords, and Accounts. Vicinitas. https://www.vicinitas.io/. Accessed 4 Dec 2020
11. Chen, C.: CiteSpace II: detecting and visualizing emerging trends and transient patterns in scientific literature. J. Am. Soc. Inform. Sci. Technol. **57**(3), 359–377 (2006)
12. Lee, J.D., See, K.A.: Trust in automation: designing for appropriate reliance. Hum. Factors **46**(1), 50–80 (2004)
13. Hoff, K.A., Bashir, M.: Trust in automation: integrating empirical evidence on factors that influence trust. Human Factors: J. Human Factors Ergon. Soc. **57**(3), 407–434 (2015). https://doi.org/10.1177/0018720814547570
14. Surwase, G., Sagar, A., Kademani, B.S., Bhanumurthy, K.: Co-citation analysis: an overview, pp. 179–185 (2011)
15. Parasuraman, R., Riley, V.: Humans and automation: use, misuse, disuse, abuse. Hum. Factors **39**(2), 230–253 (1997)
16. Lee, J., Moray, N.: Trust, control strategies and allocation of function in human-machine systems. Ergonomics **35**(10), 1243–1270 (1992)
17. Lee, J.D., Moray, N.: Trust, self-confidence, and operators' adaptation to automation. Int. J. Hum Comput Stud. **40**(1), 153–184 (1994)
18. Muir, B.M.: Trust in automation: Part I. Theoretical issues in the study of trust and human intervention in automated systems. Ergonomics **37**(11), 1905–1922 (1994)
19. Muir, B.M., Moray, N.: Trust in automation. Part II. Experimental studies of trust and human intervention in a process control simulation. Ergonomics **39**(3), 429–460 (1996)
20. Lee, J.D., Kolodge, K.: Exploring trust in self-driving vehicles through text analysis. Hum. Factors **62**(2), 260–277 (2020)
21. Ekman, F., Johansson, M., Sochor, J.: Creating appropriate trust in automated vehicle systems: a framework for HMI design. IEEE Trans. Human-Mach. Syst. **48**(1), 95–101 (2017)

22. Helldin, T., Falkman, G., Riveiro, M., Davidsson, S.: Presenting system uncertainty in automotive U.I.s for supporting trust calibration in autonomous driving. In: Proceedings of the 5th International Conference on Automotive User Interfaces and Interactive Vehicular Applications, pp. 210–217 (2013)
23. Hock, P., Kraus, J., Walch, M., Lang, N., Baumann, M.: Elaborating feedback strategies for maintaining automation in highly automated driving. In: Proceedings of the 8th International Conference on Automotive User Interfaces and Interactive Vehicular Applications, pp. 105–112 (2016)

Usability Assessment of Xiaomi Smart Band 4

Yiqing Zhou(✉), Jiaqi Tang(✉), Junchi Wu(✉), Jiayu Zeng(✉), and Marcelo M. Soares(✉)

School of Design, Hunan University, Changsha 410000, Hunan, People's Republic of China

Abstract. Since smartwatch might be a macrotrend in the future, a study is being conducted to evaluate the usability of Xiaomi Smart Band 4. In this research, several usability assessment methods are used to assess both the product itself and its website. Based on the consequence of the assessment, researchers concluded some suggestions to make the product more user-friendly, and made some improvement ourselves. This research aims to bring insights and inspirations for smartwatch market. It will serve as a support for smartwatch improve and further researches.

Keywords: Xiaomi Smart Band 4 · Homepage · Usability assessment · Design

1 Introduction

Recently, wearable devices have been trendy. According to new data [1], the sales of smartwatches are increased to 51%, and it is still rising. The rapid development of smart home and 5G technology might also boost smartwatch market. It is likely that smartwatches become a macrotrend in the near future. This study evaluated the usability of smart watches and found that there were some usability and user experience problems. Therefore it is undeniable that a more user-friendly smartwatch is needed.

Among all the smartwatch companies, Xiaomi, a Chinese electronics company, has won consumers affection for its high cost performance and became Top 1 wearable company with a market share of 10.7%. In the first quarter of 2019, Xiaomi sold 5.3 million smart band, ranking first in the world [2]. In order to further improve the usability of smartwatches, this research carried out a study to evaluate the usability of Xiaomi Smart Band 4. It used several analysis methodologies to assess both the product and its website, and made some improvement to the current design. It could provide insights and inspirations for smartwatch market.

2 Research Method and Process

The research contains several methods to evaluate the usability of both Xiaomi Smart Band 4 itself and its homepage. This study covers both empirical and unempirical methods, which were used to understand users' needs and analyze previous design. For the product, the research followed the steps incident diary, co-discovery, controlled experiment and property checklist. For the homepage, the research analyzed the tasks and carried out user tests, then a short interview and a questionnaire would come after the test.

C. Stephanidis et al. (Eds.): HCII 2021, LNCS 13094, pp. 465–477, 2021.
https://doi.org/10.1007/978-3-030-90238-4_33

2.1 Incident Diary

In order to get all the information of the user's experience, 3 participants were invited to use Xiaomi Smart Band 4 for a few days and record their activities, events and feelings in a form of diary. They were informed that the diary should include problems they met, their solutions to the problems and their stress rate (1–5).

2.2 Co-discovery

Four participants who have never used Xiaomi smart band before were invited to join the experiment in pairs. In order to open the topic faster, all the participants knew each other before. Xiaomi Smart band 4 were provided to each participants and the researcher helped them download the corresponding APP. Then participants were asked to explore the function by themselves. Certain hints would be given when necessary. Researchers also gave them Huawei band 4pro as a competitor.

After that, they were asked to tell their feelings and opinions about the smartwatches. For instance, which functions do you think are good? Is it easy to use? What confuse you the most? Is there any suggestions? (Fig. 1).

Users (Proportion corresponding to the marketing data):

Huawei & Xiaomi users are both included.

Age		Band Usage		Gender	
17-25	3	Novice	2	Male	4
25-35	2	Occasional	2	Female	2
35+	1	Expert	2		

Implementation

Recording device: iPad
Timer: Mobile Phone

Tasks

UI: Connect to Mobile; Online Payment via Mi Smart Band 4; Alarm; Workout Assistant;
Device: Put on the device; Charge the device

Measurement

When does movement start and end?

Ordering

of tasks and interface conditions

Hardware

Xiaomi Band 4 (NFC Edition)
vs
Huawei Band 4 Pro

Fig. 1. Four participants' background and tasks sample of the research.

2.3 Controlled Experiment

This method is used to analyze the performance of the product. It is effective to study specific design options through direct comparison. The study started with a testable hypothesis, then manipulated variables to support or refute the hypothesis. The test also contained a short interview to get ideas from users and discuss the mistakes made by users during the tasks. Experiment design.

2.4 Task Analysis

This research defined the main missions of the website and break it down into several small tasks to understand the structure, the flow of the website.

2.5 User Test

In order to find out the real users' needs and get to know how the users truly feel about the homepage, the study tested the interface with real users and collected data about users via questionnaires and interviews.

5 people were participated in this experiment, including 2 males and 3 females. 3 of them are novice users, who are familiar with online shopping process and have used this website before. Other participants are occasional users who have never seen this webpage.

Participants were asked to fulfil two tasks: sign up and sign in, and add one smart band to cart. In the mean time, how much time they took and error steps were recorded. After all the task were done, participants were asked to fill in a QUIS questionnaire (Appendix B) to evaluate users' satisfaction in 5 dimensions in a quantitative way [4]. Then it would have a short interview about the tasks and their opinions, such as how hard do they feel about accomplishing those tasks, is there anything confusing, and what do they think of the homepage in general, etc. When all the participants were finished, all the questionnaires were collected and were perform calculations to obtain the average score for each question.

3 Data Analysis and Discussion

3.1 Result of Incident Diary

Participant A, age 21, is a female new user. According to the users' diary (Appendix A), What bothers her most is that it always took a long time to get connected to the phone, and the function *Finding Band 4* didn't work at all. Participant A also found it hard to buckle up the strap since she often could not find the hole. Participant A had to turn the smartwatch around to solve that problem, which is a bit depressing.

Participant B, age 21, is a female novice user. Participant B claimed that it is hard to take of the main part to charge. The tester always got hurt when doing so and it was frustrating. The same problem also happened to participant C. Besides, the tester could not set the alarm through smartwatch, which is not considerate enough. The screen is also too small to her, which may cause misoperation.

Participant C, age 23, is a male novice user. The tester a bit allergic to the material, which bothers him the most. Other than that, he thought there were too many functions, especially some of which were useless, and the screen is too small for QR code detection. Participant C would put the smartwatch backwards sometimes, then The tester had to put it on again, which is a bit troublesome.

3.2 Result of Co-discovery

Participant A and Participant B thought the smartwatch could be better. They gave decent score on memorability and comfort. They agreed that the smartwatch could be more friendly to new users, since there was no guidance and some functions were too complicated.

Participant C and Participant D felt great about the smartwatch in general. They also thought it was too difficult and unfriendly for new users. They felt great about the memorability and comfort as well.

Here are some other negative comments mentioned by participants. For music function, they found it too difficult to use, and some felt they do not need this function at all. The sound adjustment button in the music interface is too small, which may lead to the inability to grasp the volume. For messages, some claimed that they could not find how to receive messages. For calls, the smartwatch can only can only hang up the phone instead of answering the phone, which is inconsiderate. Besides, the main part must be removed when charging and this part is very difficult to remove. Furthermore, It is very strange to find the band4 by controlling the vibration of the bracelet through the APP, because the band 4 vibration can not be felt when the band 4 cannot be found. The shape of the main body is too long and hard, which makes the wrist back uncomfortable, while the Huawei band 4pro has a little curvature, it will be a little more comfortable.

3.3 Result of Controlled Experiment

4 participants were asked to accomplish 5 tasks (3 UI-related tasks and 2 device-related tasks). We took the video of their whole experiment process to analyze their time consumption and task performance. For task A, which is connect to mobile, none of the participants make a mistake. Some said the process is too complicated. For task B, online payment, participants claimed that the light condition makes the code hard to identify, and they had difficulties getting familiar with the interaction actions since they could not remember previous settings. For task C, workout assistant, users were still not familiar with interaction actions. For task D, put on, participants often put the band and strap in wrong direction, and they found it hard to put it in. For task E, charging, participants found it hard to take it off, and they have to find the specific charger, which is troublesome and inconsiderate.

3.4 Result of Task Analysis

According to the homepage, there are 5 steps in purchasing a smart band, and these 5 steps can be broke down into small tasks. The detailed result are shown in Table 1.

3.5 Result of User Test

Most participants can complete all the tasks in a very short time without problem. Only one participant made 2 mistakes because of the confusing terminology. She mistakes *my order* for *my cart.* It reveals that the website is easy to use generally.

According to the questionnaire result (Appendix C), the study evaluate users' satisfaction using the average score of each questions. The result (Fig. 2 and Fig. 3) shows that the website in general might lack flexibility and fun. The consequence reveals that might because the homepage has kept the minimalism style to match their minimalist products. So the whole webpage is simple and clear without many decorations, in which case users might find it a bit dull and rigid. Besides, participants feel that the website is not that easy to use because of the terminology and lack of instructions.

Table 1. Table indicating 5 steps in purchasing a smart band

Step	Detailed Step	Action
1. Sign up	Contact information	Click *sign up* button on the top of the homepage
		Put your phone number in
	Verification	Click *create MI account*
		Enter code to verify
2. Sign in	Password log-in	Put MI ID and password in
		Click *sign in* button
	QR code log-in	scan QR code via mobile app
3. Select products	Choose product line	Pick one from the category navigator
		Click product information card
	Look at the product introduction	Get to know the product via banners
		Watch introduction videos
	Look at product details	Click *spec* button to check the details
		Click *comment* button to check the comments of previous users
4. Add to cart	Confirm product	Click *buy it now* button
		Choose size/color and fill in the address information
	Add to cart	Click *add to cart* button
		The product is added to your cart

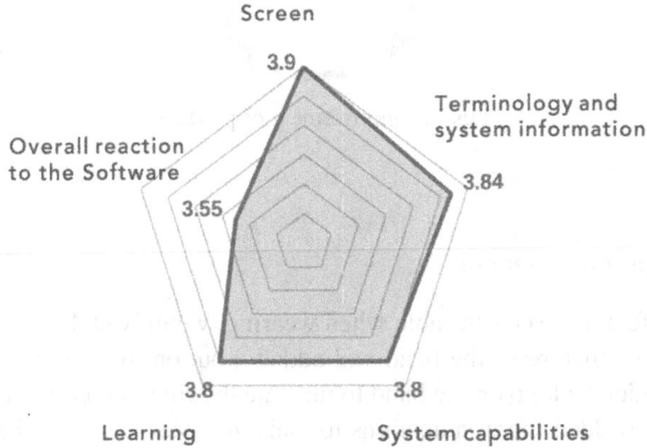

Fig. 2. Result of users' satisfaction questionnaire

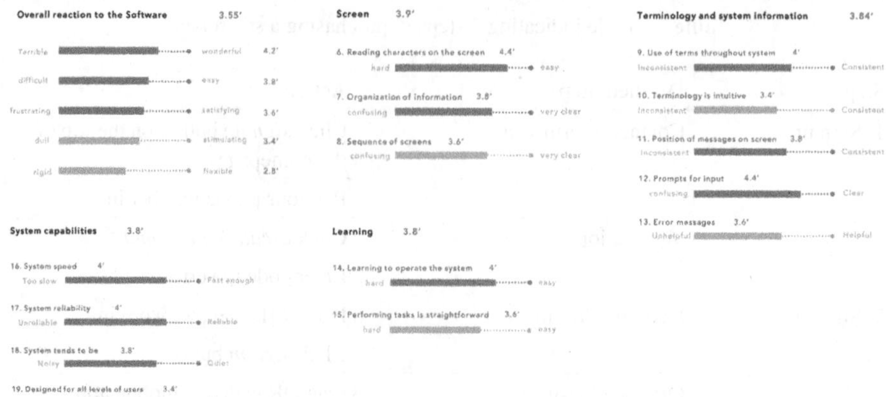

Fig. 3. Result of users' satisfaction questionnaire

4 Future Work and Redesign

4.1 Product Improvement

To make users feel more comfortable when wearing, the study moved the main part inwards. The study changed the structure of the band and added a button to take the main part down easier, and a logo is added on the band to indicate the right direction of wearing it. For the interface, the study added more indications to make it easier to use, and a guidance will be shown when new users power on the smartwatch for the first time. Besides, the study also make the buttons bigger to prevent error operation (Fig. 4).

Fig. 4. Improvement of product

4.2 Homepage Improvement

To make users feel more comfortable when wearing, we moved the main part inwards. We changed the structure of the band and added a button to take the main part down easier, and we added a logo on the band to indicate the right direction of wearing it. For the interface, we added more indications to make it easier to use, and a guidance will be shown when new users power on the smartwatch for the first time. Besides, we also make the buttons bigger to prevent error operation.

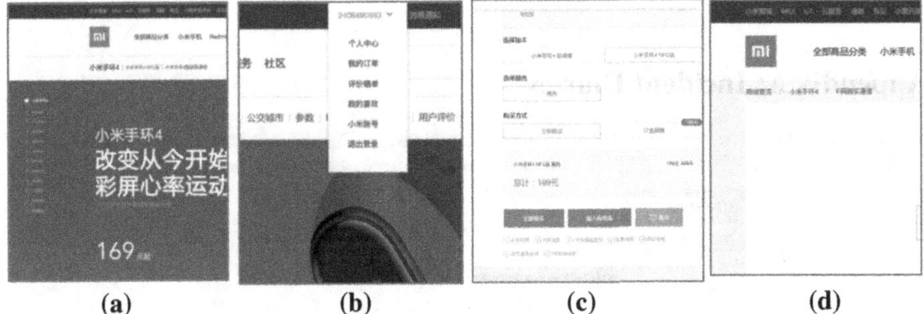

Fig. 5. Improvement of current website

To solve the problem of unclear main points, a navigation bar (Fig. 5a) is added on the left side of the webpage. It also works as a feature list. In this case users can have a clear idea of the features of the smart band. Users can also jump to whatever section they would like to know by clicking the list, and they can go back to the top easily. Besides, My Order button is moved into the pull-down menu (Fig. 5b) to prevent users from mistaking it for My Cart button.

In the *F code* page, the study added the introduction page to the navigator (Fig. 5c) and adjusted the workflow, in which case users can either go back to the introduction page or jump to the *Mi Online Shop* page using the navigator. A *check out now* button (Fig. 5d) is added right beside *add to cart* button to simplify the purchase process. Users do not have to add the product to the cart first, then go to another webpage to check out.

5 Conclusion

Though Xiaomi Smart Band won a huge success as the best seller among Chinese smartwatch market, there is still room for improvement. For the product, here are some suggestions: Firstly, it is too difficult for new users since there is no guidance. Secondly, there are too many complicated functions while some functions are less used. Thirdly, The screen is so small that it may cause error operation. Besides, the buckle makes it uncomfortable to wear sometimes. For the website, though banners make the whole webpage simple and chic. Besides, terms should be chosen more carefully. Some of the terms are not straightforward enough, which might cause misunderstanding. The flow of the website can be improved as well. It is confusing at some point, and it can also be simplified.

This research also reveals that smart band has many uncomfortable user experience and it need more improvement both in the product and the website. Other than that, this study are relatively limited. In the future, the follow-up work is necessary to expand more different kind of smart watches, choose more different type of users, and extend the test time. To make sure the research is comprehensive and accurate.

Appendix

Appendix A: Incident Diaries

Date	Time	Event	Solution	Stress Rate
5. 26	8:20am	It's not easy to buckle up the strap, because it's hard to find the hole.	Turn the buttoned part to the back of the wrist to see the hole better.	2
	9:00am	If you do n' t link to your phone, you won' t see the weather	Use your phone to watch the weather.	1
	9:00am	It ' s slow to connect to the phone and it takes a long time	No solution, just wait for a long time.	5
	14:00pm	The material is not breathable and it gets a little hot after wearing for a long time	No solution, just keep still.	1
5. 27	10:00am	Can't open the music function, there is no certain guidance	Give up the music function	1
5. 28	9. 30am	Although there is a function to find Band 4 on the APP, the Band 4 cannot be found by vibrating	Look for Band 4 for long time.	5
Date	Time	Event	Solution	Stress Rate
5. 29	7: 30am	You can't adjust the alarm clock directly through Band 4, but through your mobile phone	Set many alarms at a time, choose on the Band 4, but it is very troublesome	3
	2:00pm	When testing your heart rate, you can't determine whether your heart rate is normal, but you can refer to it on other smart band	Find solutions on Internet.	1
5. 30.	8. 00am	When charging, take off the main part, because the strap is very tight, it is difficult to take off, and the hand is very painful	Try to pull the strap to both sides first and then push down the main part.	3
	11. 20am			
	11. 20am	Can only hang up the phone through the Band 4, can not answer the phone	No solution	2
	1. 30pm	The screen is very small, some functions will be touched by mistake	Turn pages carefully	2
Date	Time	Event	Solution	Stress Rate
5. 25	11. 30am	Fins that the system has too many default functions, some are useless	I deleted some Band 4 functions on the APP myself	4
	1. 00pm	Wearing Xiaomi Band 4 will be uncomfortable when sleeping, but it needs to be worn to detect sleep quality	No solution	2
	5. 20pm	When scanning the Alipay code through the Xiaomi smart Band 4, it is difficult to recognize the QR code because the screen is too small.	Reach into the scanner	3
5. 26	7. 00am	when I wear the product, I find that the screen is reversed	Wear it again	1
	12. 00am	Wear the band for a long time outside, it will be hot, and it will become dirty.	Clean it	3
	1. 00pm	It's hard to take off the theme and charge	No good solution	4
5. 30.	2. 30pm	I found that wearing the Band 4 makes me allergic, and there are many small acne around my wrist	Hand product to others	5

Appendix B: User Test Questionnaire

用户满意度问卷

USER SATISFACTION QUESTIONNAIRE

请您结合使用小米官网的感受，回答一下问题。
Please rate your satisfaction with the system according to your experience of using the official website of xiaomi.

1. 总体印象 Overall reaction to the Software

非常不满意 Terrible					非常满意 wonderful
使用困难 difficult					使用简单 easy
令人沮丧的 frustrating					令人满足的 satisfying
无趣的 dull					有趣的 stimulating
死板的 rigid					灵活的 flexible

Screen

2. 阅读屏幕上的信息 Reading characters on the screen

困难 hard					简单 easy

3. 页面信息结构 Organization of information

非常混乱 confusing					非常清晰 Very clear

4. 页面跳转顺序 Sequence of screens

非常混乱 confusing					非常清晰 Very clear

Terminology and system information

5. 系统术语使用 Use of terms throughout system

不一致的 Inconsistent					一致的 Consistent

6. 术语使用是直观的 Terminology is intuitive

非常不同意 Inconsistent					非常同意 Consistent

7. 元素在页面中的位置 Position of messages on screen

不一致的 Inconsistent						一致的 Consistent

8. 输入提示 Prompts for input

混乱的 Confusing						清晰的 Clear

9. 错误提示 Error messages

没有帮助的 Unhelpful						有帮助的 Helpful

Learning

10. 学习如何操作系统 Learning to operate the system

困难 hard						简单 easy

11. 执行任务很简单 Performing tasks is straightforward

非常不同意 hard						非常同意 easy

System capabilities

12. 系统响应速度 System speed

迟钝的 Too slow						迅速的 Fast enough

13. 系统的可靠性 System reliability

非常不可靠 Unreliable						非常可靠 Reliable

14. 系统偏向于 System tends to be

吵闹的 Noisy						安静的 Quiet

15. 所有用户都能轻松使用该网站 Designed for all levels of users

非常不同意 Never						非常同意 Always

Appendix C: User Test Result

Participant 01: female, age 21, notive user										
Task result		Questionnaire result								
time	error	Q1	Q2	Q3	Q4	Q5	Q6	Q7	Q8	Q9
2m	0	4	5	4	3	3	5	5	4	4
Questionnaire result										
Q10	Q11	Q12	Q13	Q14	Q15	Q16	Q17	Q18	Q19	
4	4	5	5	4	4	5	4	5	4	

Participant 02: female, age 49, occational user										
Task result		Questionnaire result								
time	error	Q1	Q2	Q3	Q4	Q5	Q6	Q7	Q8	Q9
4m	2	4	3	4	4	3	3	3	4	3
Questionnaire result										
Q10	Q11	Q12	Q13	Q14	Q15	Q16	Q17	Q18	Q19	
3	4	4	3	3	3	4	4	4	3	

Participant 03: male, age 21, notive user										
Task result		Questionnaire result								
time	error	Q1	Q2	Q3	Q4	Q5	Q6	Q7	Q8	Q9
2m	0	4	2	2	2	2	4	4	4	4
Questionnaire result										
Q10	Q11	Q12	Q13	Q14	Q15	Q16	Q17	Q18	Q19	
3	4	4	4	4	3	3	4	3	4	

Participant 04: male, age 22, occational user										
Task result		Questionnaire result								
time	error	Q1	Q2	Q3	Q4	Q5	Q6	Q7	Q8	Q9
3m	0	5	5	4	4	4	5	5	5	5
Questionnaire result										
Q10	Q11	Q12	Q13	Q14	Q15	Q16	Q17	Q18	Q19	
4	5	4	4	5	5	5	5	4	4	

Participant 05: female, age 21, notive user										
Task result		Questionnaire result								
time	error	Q1	Q2	Q3	Q4	Q5	Q6	Q7	Q8	Q9
2m	0	4	4	4	4	2	5	2	4	4
Questionnaire result										
Q10	Q11	Q12	Q13	Q14	Q15	Q16	Q17	Q18	Q19	
3	2	5	2	4	3	3	4	2	2	

Appendix D: Controlled Experiment Recording

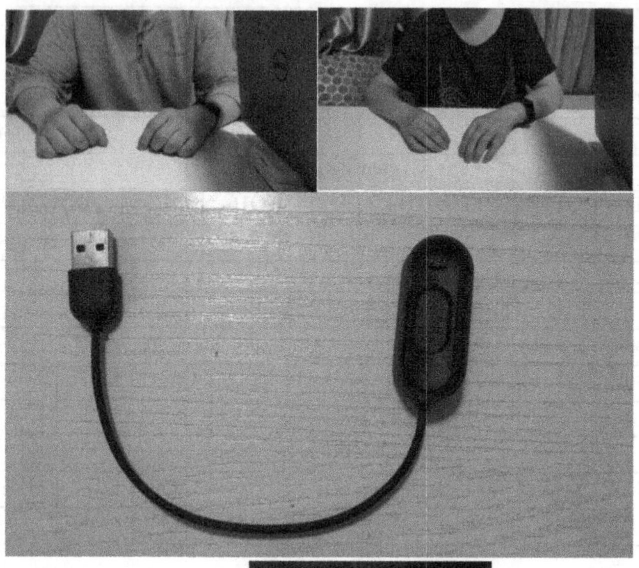

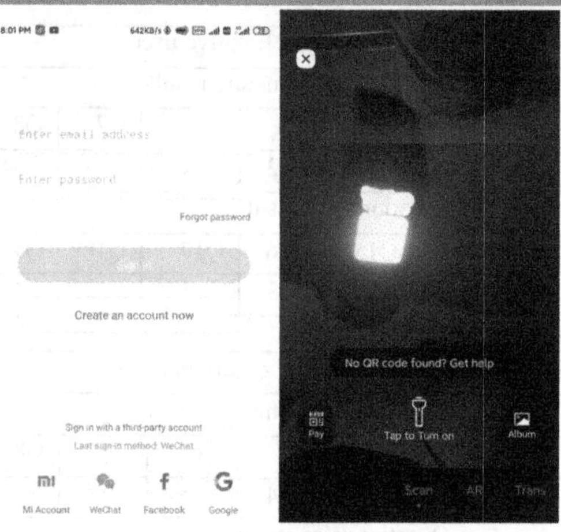

References

1. McCourt, D.: Growing sales proves smartwatches are a slow-burning trend, NEXTPIT (2019)
2. Chin, J.P., Diehl, V.A., Norman, K.L.: Development of a tool measuring user satisfaction of the human-computer interface. Int. J. Geriatr. Psychiatry (1993)
3. Harris, L.S.: U.S. department of health and human services. J. Hosp. Palliat. Nurs. **4**(4), 206–207 (2010)
4. Valentim, N.M.C., Rabelo, J., Oran, A.C., Conte, T., Marczak, S.: A controlled experiment with Usability Inspection Techniques applied to Use Case Specifications: comparing the MIT 1 and the UCE techniques. In: ACM/IEEE International Conference on Model Driven Engineering Languages & Systems. ACM (2015)

5. Author, F.: Contribution title. In: 9th International Proceedings on Proceedings, pp. 1–2. Publisher, Location (2010)
6. LNCS Homepage. http://www.springer.com/lncs. Accessed 21 Nov 2016

Cross-Cultural Design

Factors Affecting e-Commerce Satisfaction in Qatar: A Cross-Cultural Comparison

Muth Mary Abraham(✉) and Pilsung Choe(✉)

Department of Mechanical and Industrial Engineering, Qatar University, Doha, Qatar
{ma1901095,pchoe}@qu.edu.qa

Abstract. Expanded internet penetration, per capita income and changes in consumer spending trends in Qatar have turned e-Commerce into a preferred choice for consumers, enterprises, and online service providers. An initial pilot survey has been conducted in a randomly selected 45 people to examine the factors influencing logistics quality and interface quality. The collected data has been analyzed for internal consistency using Cronbach's alpha test of reliability. The results from Cronbach's alpha test ($\alpha = 0.983$) have shown good internal consistency in the construct's reliability. Mann - Whitney U test was used to compare the means of independent groups to determine whether there is a statistically significant difference between gender, age category, nationality, and language in the ratings of e-commerce satisfaction variables. This study concludes that logistics service and interface quality have a strong relation in customer satisfaction and user experience of the e-Commerce sector in each of the primary constructs within the cultural context of Qatar.

Keywords: e-Commerce · User experience · Mann - Whitney U-test

1 Introduction

A significant variation in the shopping patterns of consumers in Qatar during Pandemic is that the change from physical shopping to e-Commerce platforms. E-Commerce growth is expected to be QR 12 billion ($3.2billion) by 2022 as per the statistics of the Ministry of transportation and communication (MOTC) [1]. Last-mile delivery is the most successful order fulfillment from the retailer's perspective, whereas curbside delivery, click and collect from the retailer are the other available options in Qatar. An extensive focus on delivery quality and interface quality is increasingly crucial for e-Commerce companies and customers insist their online purchases have to be delivered rapidly. Delivery quality is a comprehensive fulfillment in meeting the customer demand at the lowest possible cost and managing the material flow and information in every stage till the product reaches the customer with order accuracy. Interface quality, like delivery quality, is a measure of service quality in the e-Commerce context and can be explained as the user engagement with the e-Commerce interface [2].

C. Stephanidis et al. (Eds.): HCII 2021, LNCS 13094, pp. 481–494, 2021.
https://doi.org/10.1007/978-3-030-90238-4_34

Online platforms offer customers a selection of services and products dependent on their preferences; because of this reason, an enormous population is getting attracted to online retailing in the current pandemic situation. Considerable advantages of web-based shopping for a regular consumer are convenience, simple product assessments, and immediate accessibility. Even though the internet penetration in the region is high compared to the other middle eastern countries, e-Commerce usage is still seen in the infancy stage. Several e-Commerce startups have begun operations with innovative methodologies, which contrasts based on what was spearheaded by established internet business organizations. The digital adaptation of Small and Medium Enterprises, which structures under Qatar's Ministry of Transport and Communications program, not only representing the country's digital scenario but also demonstrating SMEs on the most proficient method to benefit from boosting and advancing their organizations in operational expense and assets.

The objective of this study is to investigate and develop a model for the factors affecting e-Commerce customer satisfaction in the State of Qatar to identify and analyze the factors of logistics service and interface quality that affects customer and user experience in the cross-cultural context.

2 Literature Review

Zeithaml et al., (2001, 2002) [5] recommended the e-SERVQUAL as a measure of website e-service quality and characterized as scale on which electronic media encourages the simplicity of efficient shopping, buying, and transportation of items and services. This approach with seven different variables using a three-phase procedure comprising two sets of empirical and exploratory data analysis [6]. Parasuraman et al. (2005) [7] discussed that service quality is dominated by services administered by individuals in offline settings, whereas in online settings it is replaced by consumer to website interactions. Flavián et al. (2006) [8] proposed that the trust of a website is the foundation of user satisfaction which arises from previous experience. Website familiarity and usability have been investigated and concluded that user experience has an impact on satisfaction. Corporate competitiveness [9] can be increased through the enhancement of service and material flows along the supply chain; logistics service providers play an eminent role in organizations. Yang et al., (2009) [10] emphasize the importance of efficient logistics for organizations in the transportation context and explain the relationship with a theoretical foundation. Website Interactivity, online completeness, ease of use, and content clarity have a substantial impact on information quality, according to data analysis using WebQual™ Scale [11].

Gounaris et al., (2010) [12] suggested a model which shows that service quality pertains a beneficial impact on e-satisfaction, as well as influencing the consumer's behavioral intentions, such as site revisit, word-of-mouth, and repeat purchase, both directly and indirectly. The findings show that cognitive assessments come before emotional reactions and satisfaction is strongly influenced by consistency. Fang et al. (2014) suggested a model in which website quality elements, such as usability and technological adequacy positively influence perceived usefulness and satisfaction, regardless of cultural differences [3]. E-service quality, which includes response time, successful interaction, and website customization, was identified by Arya and Srivastava as the most important contributor to e-loyalty for a service website [8].

Blut et al. (2015) conducted a hierarchical meta-study approach with sixteen attributes, four dimensions for e-service quality, and identified outcomes are customer satisfaction, word of mouth, and repurchase intentions. In terms of predictive potential, the more common scales are E-S-Qual, eTailQ, and E-RecS-QUAL. E-S-QUAL scale is a 22-item, four-dimensional scale that measures performance, fulfillment, system availability, and privacy [13]. The E-RecS-QUAL scale, which includes 11 items in three dimensions: responsiveness, reward, and interaction, is only relevant to customers who had nonroutine interactions with the sites [14]. Rita et al. (2019) found that e-service quality has a significant effect on customer trust and customer satisfaction of online shopping sites. Web design is essential for achieving website satisfaction or increasing customer satisfaction and repurchase intention [15]. The findings demonstrated that in e-Commerce, customer service is unrelated to e-service quality. The most important website factors for increasing e-SQ, according to Zhao et al. 2019 [16], are website design, information, and technological support, which have been investigated in the telecom industry. Website security is an important aspect for e-Commerce customers, and service provider has to ensure the security of data. Customers will take website security for granted in most cases.

Customer satisfaction on logistics facilities, particularly in the E-Commerce business, has a significant impact on their efficiency. Y. K. Huang et al. (2009) examined logistics efficiency in e-Commerce systems in comparison with Taiwan and other countries in retailing. The delivery process involved in online shopping is a physical process for products and the activities have to be involved efficiently in the supply chain process [16]. Firm logistics and third-party logistics (3PL) are the different modes of logistics adopted by e-Commerce firms and are sustainable advantages of a firm's performance [17].

In addition to the fundamental components of availability, timeliness, and condition in logistics efficiency, Xing et al. (2011) incorporated return in an e-PDSQ assessment scale to identify how the merchant deals with damaged, undesired, or faulty items [18]. Researchers suggested any shipping charges could have a significant negative effect on the purchase decisions of customers, and free shipping has a positive effect on customers' purchase intention but has a negative impact on delivery timeliness when compared with most e-Commerce firms. Customer satisfaction is significantly decreased by delay or non-arrival of a product, order accuracy, and product damage. In this regard, Yu et al. (2014) suggested that it is important to emphasize not only the delivery time but also the quality of the delivered product while discussing delivery capability. Implementing technologies such as the delivery tracking method requires better infrastructure in the physical distribution of products aiming at flexibility [19]. Post-sales services received less attention in an e-Commerce context, Cao et al., 2018 [20], examined the drivers, practices, and outcomes of electronic service in an organizational context.

This study uniquely adds to the literature because it focuses specifically on the analysis of satisfaction with e-Commerce websites and logistics services. This study tests an established website quality scale, WebQual™, in the context of general interaction with e-Commerce, and examines the relationships among website layout, navigation, security, information quality, and aesthetics. Logistics efficiency of service providers will be assessed based on SERVQUAL scale in terms of delivered product condition, shipping charges, timeliness of delivery, and return handling which are the post-purchase factors.

3 Proposed Framework

Several researchers explain the e-service quality and logistics quality separately using WebQual™ and SERVQUAL scale, but in this study, a combination of those two scales has been considered to describe the online and offline transactions in the E-Commerce environment. In this section, a discussion of hypothesis development for the model which is proposed has been accomplished. The important factors which were supposed to be essential to attain satisfaction were investigated following the extensive review of literature, and these factors were studied as the important reason for attaining e-Commerce success in Qatar. The conceptual frameworks of Blut et al. (2016) [21] and Cao et al. (2018) [20] are the basis of the recommended framework (see Fig. 1) for this study.

3.1 Hypothesis Development

The proposed constructs such as website layout, navigation, security, information quality, aesthetics, product condition, delivery charges, delivery timeliness, returns handling have been discussed in this section to develop the hypothesis for the analysis of interface quality and logistics quality.

Interface Quality. Website layout differs in mobile devices and desktop versions. Layouts are available in various styles and perceptions also may vary among cultural preferences. As a result, different perceptions will emerge among users as layout design determines the ease of interaction among users with the interface. According to Gregory et al. (2010) [22], website design is associated with consumer perception of the organization, and layouts tend to improve consumer interaction with the interface. H1: Website layout has a positive influence on interface quality to achieve user experience.

Salehi et al. (2012) [22] described the website's e-convenience dimensions of which navigation is one of the important dimensions for interface quality. H2: Navigation has a positive influence on interface quality to achieve user experience.

Perceived security in online browsing content is the state of being confident, the security of important information, or risk-free while browsing online content. According to Khalid et al. (2018) [23], the primary factor influencing customer satisfaction with e-Commerce systems is the quality of e-Commerce services, particularly security [4] and payment methods. H3: Perceived security has a positive influence on interface quality to achieve user experience.

Many researchers, including Montoya-Weiss et al. [24], Koo et al. [25], and Hsu et al. [26], exposed that having a high level of relative and fundamental Information quality influences user satisfaction when using e-services websites. H4: Information quality has a positive influence on interface quality to achieve user experience.

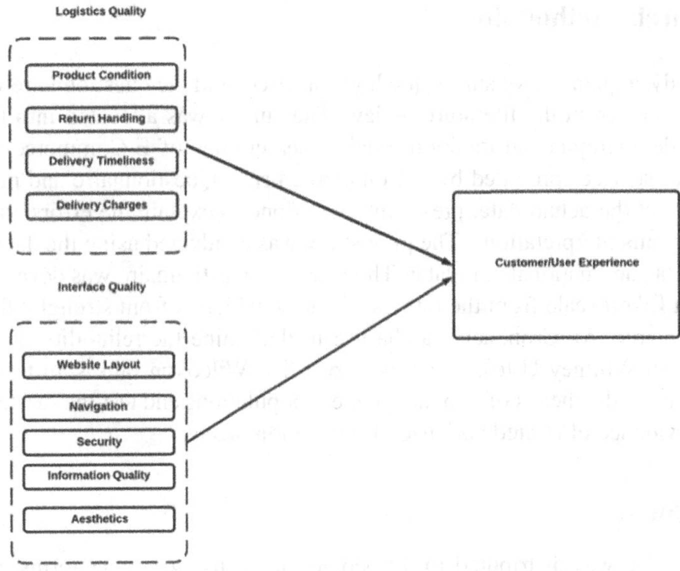

Fig. 1. A conceptual framework of e-Commerce customer satisfaction and user experience

Aesthetics affects user's regular visits to the website, and every user has his or her perceptions about the presentation of content in the interface. The impact of website color treatments on user trust, satisfaction, and e-loyalty differed across cultures, with findings indicating that color appeal significantly increases user trust and experience across cultural groups. H5: Perceived aesthetics has a positive influence on interface quality to achieve user experience.

Logistics Quality. Last-mile delivery is the last leg of the supply chain, is a vital aspect of logistics in E-Commerce. Third-party logistics or E-Commerce firms ensure that products acquired into a warehouse or hub are properly picked, packaged, and delivered to the customer in a good condition and within the agreed time frame.

H6: Product condition after delivery has a positive influence on logistics quality to achieve customer satisfaction
H7: Delivery Charge has a positive influence on logistics quality to achieve customer satisfaction
H8: Delivery Timeliness after delivery has a positive influence on logistics quality to achieve customer satisfaction
H9: Returns handling has a positive influence on logistics quality to achieve customer satisfaction.

4 Research Methodology

For this study, a quantitative survey method was used, and the questionnaire was developed using data from the literature review. The survey was allocated into three parts which are demographic information, quality perceptions of E-Commerce interfaces, and logistics services provided by E-Commerce firms. Questionnaire and pilot review were to collect the actual data, pre-testing was done to explain the errors and to order to avoid any misinterpretations. The pilot study was conducted using the data collected from residents and nationals of Qatar. The research questionnaire was developed using a five-point Likert scale from the ratings of 1 to 5, which is from strongly disagree (1) to strongly agree (5). Cronbach's alpha test to determine the reliability of constructs and the Mann-Whitney U test which is also called Wilcoxon rank-sum test has been used to compare the means of two independent populations and test the existence of the statistical evidence of related variance of population means.

4.1 Pilot Study

A questionnaire was distributed to 40 people online using google forms among the citizens and residents in Qatar over one month from October 2020 to November 2020. Out of 45 questionnaires, 35 were responded, 2 were invalid and 33 were acceptable.

5 Reliability Analysis and Mann-Whitney U Test

Descriptive statistics for the respondents based on the questionnaire items are discussed in this section. The questionnaire includes 6 items of interface quality and five items for customer satisfaction. The questions related to all variables are integrated in a 5-point Likert scale. The result outcomes of means and standard deviations (SD) indicated that for the construct of interface quality and logistics quality, respondents managed to perceive levels of agreement on the measurement items with mean scores over 3.67 and 4.15 respectively. The findings show that most respondents had a medium levels of user experience, with most mean scores of approximately 4.03 and customer satisfaction of 3.85. These findings imply that respondents are not highly satisfied with the quality of websites and logistics quality of e-Commerce companies.

Table 1. Overall reliability statistics

Factors	Mean	SD	Cronbach's α
Customer/User Exp.	3.94	0.754	0.890
Logistics Quality	3.67	0.971	0.822
Interface Quality	4.15	0.838	0.877

The Cronbach's alpha test is used to determine the construct's reliability and alpha value higher than 0.7 is considered that the instrument is internally consistent. The overall Cronbach's alpha value of the model proposed in this study is of 0.890 and the construct-wise Cronbach's alpha value is summarized in Table 1. Cronbach's alpha for web interface quality is 0.877, which is shown in Table 2. Logistics consistency has a Cronbach's alpha value of 0.822, which is also good. Since the questionnaire's reliability indicates that the items of the constructs are well connected, actual data for the main study can be collected.

Table 2. Item reliability statistics.

Factors	Mean	SD	Cronbach's α
Product Condition	3.85	1.121	0.882
Delivery Charges	3.36	1.270	0.883
Delivery Timeliness	3.91	1.128	0.880
Return handling	3.58	1.275	0.893
Customer satisfaction	3.85	1.004	0.886
Website Layout	3.97	1.262	0.880
Navigation	4.27	1.039	0.869
Security	4.21	0.927	0.875
Information Quality	4.23	0.902	0.876
Aesthetics	4.05	0.947	0.885
User Experience	4.03	1.045	0.880

Mann-Whitney U test has been conducted to compare logistics and interface quality factors in e-Commerce satisfaction under the categories such as gender, age group nationality, and language. A significance level of 0.1 was used in consideration of a small sample size as a pilot study. Table 3 *shown below summarizes mean, standard deviation, and median of factors based on gender and age group.*

Table 3. Mean, SD and median of the samples based on gender and age group.

Factors		Gender		Age group	
	Mean & SD	Male (N = 10)	Female (N = 23)	Below 35 (N = 14)	Above 35 (N = 19)
Product Condition	Mean	4.100	3.740	4.000	3.740
	SD	0.876	1.214	1.240	1.046
	Median	4.000	4.000	4.000	4.000
Delivery Charges	Mean	3.300	3.390	3.790	3.050
	SD	1.337	1.270	0.975	1.393
	Median	3.500	4.000	4.000	3.000
Del. Timeliness	Mean	3.700	4.000	4.210	3.680
	SD	1.252	1.087	1.122	1.108
	Median	4.000	4.000	4.500	4.000
Return Handling	Mean	3.900	3.430	3.790	3.420
	SD	1.197	1.308	1.188	1.346
	Median	4.000	3.000	4.000	3.000
Website Layout	Mean	3.900	4.000	3.930	4.000
	SD	1.449	1.206	1.439	1.155
	Median	4.500	4.000	5.000	4.000
Navigation	Mean	4.400	4.220	4.070	4.420
	SD	0.843	1.126	1.207	0.902
	Median	4.220	5.000	5.000	5.000
Security	Mean	4.600	4.040	4.000	4.370
	SD	0.516	1.022	1.038	0.831
	Median	5.000	5.000	4.000	5.000
Information Qua	Mean	4.450	4.130	4.140	4.290
	SD	0.685	0.980	1.008	0.839
	Median	4.750	4.500	4.250	4.500
Aesthetics	Mean	4.500	3.850	3.750	4.260
	SD	0.527	1.027	0.956	0.903
	Median	4.500	4.000	4.000	4.500

Table 4 *summarizes mean, standard deviation, and median of factors based on language and nationality.*

Table 4. Mean, SD and median of the samples based on language and nationality.

Factors		Language		Nationality	
	Mean & SD	Non-Arab (N = 12)	Arab (N = 21)	Qatari (N = 8)	Non Qatari (N = 25)
Product Condition	Mean	3.670	3.950	3.750	3.880
	SD	0.985	1.203	1.165	1.130
	Median	4.000	4.000	4.000	4.000
Delivery Charges	Mean	3.080	3.520	2.880	3.520
	SD	1.311	1.250	1.356	1.229
	Median	3.500	4.000	2.500	4.000
Del. Timeliness	Mean	3.580	4.100	3.750	3.960
	SD	1.621	0.700	0.707	1.241
	Median	4.000	4.000	4.000	4.000
Return Handling	Mean	3.580	3.570	3.500	3.600
	SD	1.505	1.165	1.069	1.354
	Median	4.000	3.000	3.000	4.000
Website Layout	Mean	3.420	4.290	4.630	3.760
	SD	1.379	1.102	0.518	1.363
	Median	3.000	5.000	5.000	4.000
Navigation	Mean	3.750	4.570	5.000	4.040
	SD	1.215	0.811	0.000	1.098
	Median	3.500	5.000	5.000	4.000
Security	Mean	3.580	4.570	4.750	4.040
	SD	1.084	0.598	0.463	0.978
	Median	3.500	5.000	5.000	4.000
Information Qua	Mean	3.630	4.570	4.630	4.100

(*continued*)

Table 4. *(continued)*

Factors		Language		Nationality	
	Mean & SD	Non-Arab (N = 12)	Arab (N = 21)	Qatari (N = 8)	Non Qatari (N = 25)
	SD	1.025	0.618	0.694	0.935
	Median	4.000	5.000	5.000	4.500
Aesthetics	Mean	4.000	4.070	4.630	3.860
	SD	1.022	0.926	0.443	0.995
	Median	4.000	4.000	4.750	4.000

The Mann-Whitney U test results are summarized in Table 5 and Table 6. The test revealed a significant difference in the preference of Interface aesthetics among male (*Median* = *4.5, n* = *10)* and female (*Median* = *4, n* = *23), U* = *73, z* = *−1.697, p* = *0.090* and *r* = *0.295*. There is a significant difference in the preference of Interface aesthetics among people who aged below 35 years old (*Median* = *4, n* = *14)* and above 35 years old (*Median* = *4.5, n* = *19), U* = *85.5, z* = *−1.784, p* = *0.074* and *r* = *0.311*.A significant difference is showed in the delivery timeliness among people who aged below 35 years old (*Median* = *4.5, n* = *14)* and above 35 years old (*Median* = *4, n* = *19), U* = *89.5, z* = *−1.684, p* = *0.092* and *r* = *0.293.*

Table 5. Mann-Whitney U test results on gender and age group.

Factors	Gender				Age group			
	(Male and Female)				(Below & Above 35)			
	U	Z	p	r	U	Z	p	r
Prod. Condition	97.000	−0.746	0.455	0.130	105.500	−1.060	0.289	0.185
Del. Charges	110.000	−0.204	0.838	0.036	93.500	−1.499	0.134	0.261
Del. Time	96.500	−0.770	0.441	0.134	89.500	−1.684	**0.092***	0.293
Ret. Hand	91.500	−0.951	0.342	0.166	111.000	−0.828	0.408	0.144
Web. Layout	114.000	−0.042	0.967	0.007	127.500	−0.214	0.830	0.037
Navigation	110.000	−0.223	0.823	0.039	116.000	−0.706	0.480	0.123
Security	82.000	−1.398	0.162	0.243	107.000	−1.024	0.306	0.178
Info. Quality	96.500	−0.759	0.448	0.132	122.000	−0.419	0.675	0.073
Aesthetics	73.000	−1.697	**0.090***	0.295	85.500	−1.784	**0.074***	0.311

*Significant at $\alpha = 0.1$

The test revealed a significant difference in the satisfaction of delivery timeliness among Arab (*Median* = *4, n* = *12)* and non-Arabic speakers (*Median* = *4, n* = *21), U* = *118.5, z* = *−1.684, p* = *0.092* and *r* = *0.293*. There is a significant difference in navigation experience among Qatari (*Median* = *5*, *n* = *8*) and non- Qatari (*Median* = *4*, *n* = *25*), $U = 48$, $z = -2.489$, $p = 0.013$ and $r = 0.433$. A significant difference is showed in the security perceived among Qatari (*Median* = *5*, *n* = *8*) and non - Qatari (*Median* = *4*, *n* = *25*), $U = 58$, $z = -1.908$, $p = 0.056$ and $r = 0.332$. A significant difference is showed in the experience website information quality among Qatari (*Median* = *5*, *n* = *8*) and non - Qatari (*Median* = *4.5*, *n* = *25*), $U = 62$, $z = -1.671$, $p = 0.095$ and $r = 0.291$. A significant difference is showed in the aesthetics among Qatari (*Median* = *4.75*, *n* = *8*) and non - Qatari (*Median* = *4*, *n* = *25*), $U = 53$, $z = -2.036$, $p = 0.042$ and $r = 0.354$.

Table 6. Mann- Whitney U test results on language and nationality samples.

Factors	Language (Arab and Non-Arab)				Nationality (Qatari and Non-Qatari)			
	U	Z	p	r	U	Z	p	r
Prod. Condition	99.000	−1.060	0.289	0.185	93.000	−0.311	0.756	0.054
Del. Charges	100.500	−1.499	0.134	0.261	72.000	−1.225	0.220	0.213
Del. Time	118.500	−1.684	**0.092***	0.293	73.500	−1.183	0.237	0.206
Ret. Hand	118.000	−0.828	0.408	0.144	89.500	−0.456	0.649	0.079
Web. Layout	79.500	−0.214	0.830	0.037	66.500	−1.506	0.132	0.262
Navigation	78.500	−0.706	0.480	0.123	48.000	−2.489	**0.013***	0.433
Security	59.000	−1.024	0.306	0.178	58.000	−1.908	**0.056***	0.332
Info. Quality	54.500	−0.419	0.675	0.073	62.000	−1.671	**0.095***	0.291
Aesthetics	123.000	−1.784	**0.074***	0.311	53.000	−2.036	**0.042***	0.354

*Significant at $\alpha = 0.1$

6 Results and Discussions

After investigating through an extensive literature review on the theoretical foundation of WebQualTM AND SERVQUAL scale, which includes user/customer experience as independent variables, interface quality comprising website layout, navigation, security, information quality, and aesthetics and logistics quality comprising return handling, product condition, delivery charges and delivery timeliness as dependent variables. The conceptual model, along with the hypothesis, is being proposed in this paper for a further detailed study. Due to small sample size, this study has been considered significance level (α) as 0.10 and the results of pilot study are acceptable. The sample size will be increased in the detailed study to investigate more valid significance. The questionnaire used in the pilot study was a combination of open-end and closed-end questions to identify more variables related to satisfaction.

The significant difference in the aesthetic quality of web interface is higher in female as they tend to choose more visually aesthetics aspects than men. The older people prefer more aesthetics feature than younger people to acquire an emotional attitude of easiness in usability whereas it is seen lesser in younger people as they prefer to switch from multiple sites. Aesthetic features of web interface should be configured according to the gender and age of customers and their respective product inclinations, which can improve the dissatisfaction of aesthetics to an extent. Delivery timeliness is also a crucial factor for people above age 35 years because this category mostly depends on these platforms for essential household items and easily perishable goods. A significant difference is noticed in the delivery timeliness and aesthetics among Arab and non-Arab speakers and can be explained due to the diversity in the cultures. Delivery timeliness is highly related to the supply chain process and infrastructure of e-retailer and improving the delivery strategy and infrastructure can enhance the delivery timeliness satisfaction among non-Arabs. Website quality attributes such as navigation, security, information quality, and aesthetics have significant differences among Qatari and non-Qatari due to the quality comparison of other efficient e-Commerce platforms in their native countries. Online retailers can provide value addition to Navigation and information quality by delivering online support at the time interface interaction and also provide security to the data privacy of consumers.

7 Conclusion

In Qatar, e-Commerce is still in its developing stage and this paper discusses the primary variables influencing user experience and customer satisfaction. This study is to construct a preliminary model, and more detailed research may be undertaken by further testing the variables. e- Commerce is an excellent alternative for retailers and consumers due to its convenience. This study identifies that there are several possibilities to optimize e-Commerce from the consumer perspective.

References

1. MOTC Homepage. https://tdv.motc.gov.qa/e-Commerce-in-qatar. Accessed 07 June 2021
2. Chang, H.H., Chen, S.W.: The impact of customer interface quality, satisfaction and switching costs on e-loyalty: internet experience as a moderator. Comput. Hum. Behav. (2008). https://doi.org/10.1016/j.chb.2008.04.014
3. Fang, Y., Qureshi, I., Sun, H., McCole, P., Ramsey, E., Lim, K.H.: Trust, satisfaction, and online repurchase intention: the moderating role of perceived effectiveness of E-commerce institutional mechanisms. MIS Q. **38** (2014). https://doi.org/10.25300/MISQ/2014/38.2.04
4. Chiu, C.-M., Hsu, M.-H., Lai, H., Chang, C.-M.: Re-examining the influence of trust on online repeat purchase intention: the moderating role of habit and its antecedents. Decis. Support Syst. **53** (2012). https://doi.org/10.1016/j.dss.2012.05.021
5. Zeithaml, V.A., Parasuraman, A., Malhotra, A.: A Conceptual Framework for Understanding e-Service Quality: Implications for Future Research and Managerial Practice, Cambridge, MA (2000)
6. Zeithaml, V.A., Parasuraman, A., Malhotra, A.: Service quality delivery through web sites: a critical review of extant knowledge. J. Acad. Mark. Sci. **30** (2002). https://doi.org/10.1177/009207002236911

7. Parasuraman, A., Zeithaml, V.A., Malhotra, A.: E-S-QUAL. J. Serv. Res. **7** (2005). https://doi.org/10.1177/1094670504271156
8. Arya, S., Srivastava, S.: Effects of user's primary need on relationship between e-loyalty and its antecedents. Decision **42**(4), 419–449 (2015). https://doi.org/10.1007/s40622-015-0103-3
9. Green, K.W., Whitten, D., Inman, R.A.: The impact of logistics performance on organizational performance in a supply chain context. Supply Chain Manag.: Int. J. **13** (2008). https://doi.org/10.1108/13598540810882206
10. Yang, C.-C., Marlow, P.B., Lu, C.-S.: Assessing resources, logistics service capabilities, innovation capabilities and the performance of container shipping services in Taiwan. Int. J. Prod. Econ. **122** (2009). https://doi.org/10.1016/j.ijpe.2009.03.016
11. Kim, H., Niehm, L.S.: The impact of website quality on information quality, value, and loyalty intentions in apparel retailing. J. Interact. Mark. **23** (2009). https://doi.org/10.1016/j.intmar.2009.04.009
12. Gounaris, S., Dimitriadis, S., Stathakopoulos, V.: An examination of the effects of service quality and satisfaction on customers' behavioral intentions in e-shopping. J. Serv. Mark. **24** (2010). https://doi.org/10.1108/08876041011031118
13. Blut, M.: E-service quality: development of a hierarchical model. J. Retail. **92** (2016). https://doi.org/10.1016/j.jretai.2016.09.002
14. Akinci, S., Atilgan-Inan, E., Aksoy, S.: Re-assessment of E-S-Qual and E-RecS-Qual in a pure service setting. J. Bus. Res. **63** (2010). https://doi.org/10.1016/j.jbusres.2009.02.018
15. Rita, P., Oliveira, T., Farisa, A.: The impact of e-service quality and customer satisfaction on customer behavior in online shopping. Heliyon **5** (2019). https://doi.org/10.1016/j.heliyon.2019.e02690
16. Agatz, N.A.H., Fleischmann, M., van Nunen, J.A.E.E.: E-fulfillment and multi-channel distribution – a review. Eur. J. Oper. Res. **187** (2008). https://doi.org/10.1016/j.ejor.2007.04.024
17. Xianglian, C., Hua, L.: Research on e-commerce logistics system informationization in chain. Procedia Soc. Behav. Sci. **96** (2013). https://doi.org/10.1016/j.sbspro.2013.08.095
18. Xing, Y., Grant, D.B., McKinnon, A.C., Fernie, J.: The interface between retailers and logistics service providers in the online market. Eur. J. Mark. **45** (2011). https://doi.org/10.1108/03090561111107221
19. Yu, J., Subramanian, N., Ning, K., Edwards, D.: Product delivery service provider selection and customer satisfaction in the era of internet of things: a Chinese e-retailers' perspective. Int. J. Prod. Econ. **159** (2015). https://doi.org/10.1016/j.ijpe.2014.09.031
20. Cao, Y., Ajjan, H., Hong, P.: Post-purchase shipping and customer service experiences in online shopping and their impact on customer satisfaction. Asia Pac. J. Mark. Logist. **30** (2018). https://doi.org/10.1108/APJML-04-2017-0071
21. Blut, M., Chowdhry, N., Mittal, V., Brock, C.: E-service quality: a meta-analytic review. J. Retail. **91** (2015). https://doi.org/10.1016/j.jretai.2015.05.004
22. Gregory, A., Wang, Y.(Raymond), DiPietro, R.B.: Towards a functional model of website evaluation: a case study of casual dining restaurants. Worldw. Hosp. Tour. Themes **2** (2010). https://doi.org/10.1108/17554211011012603
23. Salehi, F., Abdollahbeigi, B., Langroudi, A.C., Salehi, F.: The impact of website information convenience on E-commerce success of companies. Procedia Soc. Behav. Sci. **57** (2012). https://doi.org/10.1016/j.sbspro.2012.09.1201
24. Khalid, A., Lee, O., Choi, M., Ahn, J.: The effects of customer satisfaction with e-commerce system. J. Theor. Appl. Inf. Technol. **96**, 481–491 (2018)
25. Montoya-Weiss, M.M., Voss, G.B., Grewal, D.: Determinants of online channel use and overall satisfaction with a relational, multichannel service provider. J. Acad. Mark. Sci. **31** (2003). https://doi.org/10.1177/0092070303254408

26. Koo, C., Wati, Y., Park, K., Lim, M.K.: Website quality, expectation, confirmation, and end user satisfaction: the knowledge-intensive website of the Korean National Cancer Information Center. J. Med. Internet Res. **13** (2011). https://doi.org/10.2196/jmir.1574
27. Hsu, C.-L., Chang, K.-C., Chen, M.-C.: The impact of website quality on customer satisfaction and purchase intention: perceived playfulness and perceived flow as mediators. Inf. Syst. e-Bus. Manag. **10** (2012). https://doi.org/10.1007/s10257-011-0181-5

Factors Influencing Trust in WhatsApp: A Cross-Cultural Study

Gabriela Beltrão[1(✉)] and Sonia Sousa[1,2]

[1] School of Digital Technologies, Tallinn University, Narva Rd 25, 10120 Tallinn, Estonia
{gbeltrao,scs}@tlu.ee

[2] University of Trás-os-Montes e Alto Douro (UTAD), Vila Real, Portugal

Abstract. This research presents a cross-cultural investigation of users' trust in technology to map how different factors can influence their propensity to trust. The study focuses on three Portuguese-speaking countries, Brazil, Mozambique, and Portugal, aiming to provide results that are not tied to a specific context while also focusing on underexplored populations. The study consisted of a survey using the Human-Computer Trust Model, in which the participants (n = 91) were asked to assess their trust in the application WhatsApp. Their responses were analyzed by country, gender, and generation to identify the effects of these aspects on the respondents' trust levels. The results indicate that these factors' influence on trust is manifold. The effect of culture is tied to the context of use and local particularities and not to intrinsic cultural characteristics. Differences across genders varied according to the users' generation, and in this case, were only significant among older respondents. Distinct demographic groups also had differences in privacy concerns, which showed to be correlated to trust. This investigation highlights the importance of considering users' differences in trust and generates insights into addressing their concerns during the design. To that end, this study presents a framework to assist practitioners in including trust as an aspect of personas.

Keywords: Cultural differences and HCI · Gender and HCI design · Heuristics and guidelines for design · User survey · Trust in technology

1 Introduction

The vertiginous increase in technology adoption in the past years has been essential for most segments of society and even more crucial during the COVID-19 pandemic. However, while keeping people together, technology also allowed the growth of surveillance and misinformation. Moreover, it increased existing social inequalities, changing the logic of our economic system to what is explained as "surveillance capitalism" [1]. The coming to light of these events have undermined individuals' trust at a generalized level, contributing to a broader social phenomenon conceptualized as the post-truth era [2]; that is, individuals can no longer distinguish truth from false information. Although these events permeate most layers of society, they have technology at its center, trusting less in it and through it.

C. Stephanidis et al. (Eds.): HCII 2021, LNCS 13094, pp. 495–508, 2021.
https://doi.org/10.1007/978-3-030-90238-4_35

Findings from different spheres reinforce the multifaceted character of this issue [3, 4] and reinforce the importance of these multidisciplinary approaches to solve it. Lewandowsky et al. [2] propose the concept of "Technocognition", a response to post-truth based on the use of technological solutions and psychological principles, an approach that is also following value-sensitive design (VSD) principles [5], which accounts for human values through the design process.

In this unsettled context, trust plays a decisive role in HCI: it is at the basis of the user's relationship with technology [6], affecting its acceptance and uptake [7] and determining whether a system will be adopted or not. However, while progress has been made towards understanding trust in technology, there are still gaps in understanding how this mechanism operates for different individuals.

Designing trustworthy systems is not only a way for companies to reach better quantifiable results (i.e., downloads of an application) but is also necessary to build sustainable relationships with users. Designers must have the knowledge and the resources to design technology that is trustworthy and is recognized as such by its users. For that, practitioners need to understand both the trust mechanisms and how the users perceive them. Still, trust is considered generically in the design, impairing systems' adoption and usage because users do not see their needs addressed. This study focuses on broadening the knowledge about users' differences in trust and proposes ways to translate it into the design practice, following paths suggested by Technocognition and VSD principles.

1.1 Measuring Trust

Progress has been made towards understanding trust in the context of HCI, but there is still little consensus about how to measure it [6], and the user, in most cases, is considered a generic character. To date, the most adopted models use either constructs from the literature in interpersonal trust or information technology (IT) [8, 9], making them effective when this division is clearly defined [10] but problematic when used in more ambiguous contexts.

This study adopts the Human-Computer Trust Model (HCTM), a socio-technical model that measures trust in technology considering both system and human-like aspects [11]. For the trust assessment, it employs the Human-Computer Trust Scale (HCTS) [11], an instrument with twelve items covering three attributes: (1) Risk Perception (RP), (2) Competence (COM), and (3) Benevolence (BEN), and three items referring directly to Trust (TR). The Trust Score is calculated based on the average results of all the constructs.

1.2 Differences in User Trust

As the importance of trust in technology has become more recognized, its relationship with other elements, intrinsic and extrinsic to the users, has also become more evident. Nevertheless, there is still little research focused on the specific effects of essential factors, as culture, age, and gender, on trust.

Culture is commonly investigated based on the cultural dimensions model [12], according to which culture works at a collective level, as a mental software that guides patterns of thinking, feeling, and acting. The model allows the analysis of national cultures based on the quantification of six dimensions, making it suitable for investigating cross-cultural differences. Although no dimension refers directly to trust, evidence from other studies demonstrates that these characteristics can affect propensity trust among individuals [13, 14].

Differences in trust between genders are suggested in the literature. Evidence from studies in economy indicates that men are more likely to take risks [15] and present a more instrumental trust behavior: they tend to trust more because they expect more in return [16]. In HCI, it was found that women tend to trust less in autonomous agents [17], are in general more critical in aesthetic judgments [18], and tend to have more negative expectations regarding the acceptance of new technology [19]. Findings mostly point towards a similar direction, that women tend to trust less, although the size of the gender gap may vary according to context and age level [15]. However, it is noteworthy that gender differences in trust may result from a social hierarchy in society [15, 16, 19], as explained by the social role theory [20], and thus not the root cause of the contrasts.

Differences in trust across age ranges are similarly often attested but lacking conclusions about each generation's specific truss characteristics. Therefore, this study adopts the concept and terminology of generations [21], identifying patterns in values and social behaviors across the age ranges. The terminology is widely adopted in multiple disciplines, with studies that attest to differences in why and how each generation uses technology [22–26], but with little reference to trust.

In sum, there are indications that the aspects above listed can affect trust, but without consolidated findings of how they, together, influence users' propensity to trust in technology. Similarly, there is still a gap in the translation of theoretical conclusions into the design practice. Addressing both of these issues, this study proposes an exploratory inquiry and a framework to be used by practitioners.

2 Research Methodology

Primarily, this research aims to identify the effects of culture, gender, and generation on trust. To achieve this goal, this study is guided by a central research question: can the Human-Computer Trust Model (HCTM) be used to map trust behaviors across different cultures? Along with five hypotheses defined based on the literature:

- H1: The country of origin will influence users' trust in WhatsApp.
- H2: Portugal will have the lowest levels of trust in WhatsApp.
- H3: Brazil will have the highest levels of trust in WhatsApp.
- H4: Males will have higher levels of trust in WhatsApp across the countries.
- H5: Users' generation will influence their trust in WhatsApp.

2.1 Procedure

The research focused on mapping trust behaviors across three countries, Brazil, Portugal, and Mozambique, by assessing their trust in a single object, WhatsApp. It used an online survey consisting of the assessed Portuguese version of the HCTS [27] and complementary questions referring to users' country, age range, gender, and WhatsApp other messaging applications usage. In addition, two instruments were included to evaluate other aspects related to trust and enrich the investigation: the Domain-Specific Innovativeness scale (DSI) [28] to measure users' adoption categories [29]; and Privacy Concerns, evaluated through five constructs [30], namely (1) Awareness to Privacy Issues, (2) Privacy Concerns, (3) Privacy Intrusion, (4) Perceived Effectiveness of Privacy Policy, and (5) Privacy control.

WhatsApp was chosen as the object of research due to its current relevance to the theme. The application is the most popular messaging application worldwide [31]; however, it has been the target of controversies since its acquisition by Facebook in 2016, and most recently due to changes in the application's privacy terms, which implied usage of the users' data by the social media [32]. Although Facebook's use of WhatsApp's metadata has been noted previously [33], the discontent caused by the recent changes was boosted by other current issues involving the platforms, such as phishing and misinformation, which raised concerns about the risks of the application.

The study participants were restricted to the educational context to ensure coherence between the data of the three countries while still ensuring the representativeness of the investigated users' groups. Before starting the survey, a scenario of usage of the application was presented, exploring the notion of digital dwelling enabled by WhatsApp [34] and the risks intrinsic to its interactions. The scenario was introduced in the form of mock-ups of the application and supplementary texts.

The survey had a total of 31 questions, divided into five sections: (1) demographic information, (2) usage patterns, (3) DSI, used in full, (4) Privacy Concerns used partially, and the (5) HCTS, used in full.

2.2 Participants

The data collection occurred during March and April 2021, with 102 participants enrolled in the study. Of those, 11 did not complete the survey and were discarded, resulting in 91 valid responses. The majority of the respondents were from Portugal, with 58.2% (n = 53) responses, followed by 23.1% (n = 21) from Brazil and 18.7% (n = 17) from Mozambique.

There were 59.3% (n = 54) females, 34.6% (n = 36) males and 1.1% (n = 1) non-binary respondent, which was not considered due to its low representativeness. The age ranges were grouped according to the generations ranges proposed by Parker & Igielnik [26], consisting of 62.6% (n = 57) from Generation Z (up to 24 years old), 25.3% (n = 23) Millennials (between 25 and 40 years old), 10,9% (n = 10) from Generation X (41 to 56 years old), and 1.1% (n = 1) Baby Boomer (57 or more years old), which was also not considered.

3 Results

First, the reliability of the trust results was assessed through the calculation of Cronbach's alpha coefficient, which should be above 0.7 [35]. This condition was satisfied in this study, with Cronbach's alpha = .86.

Next, the results were weighted by country to overcome the differences in sample sizes [36], ensuring that each had an equivalent representation of approximately one-third. Finally, the weight was calculated based on the total sample (n = 91) and executed through the built-in SPSS procedure.

3.1 Cross-Cultural Trust Assessment

The cross-cultural Trust Score was 3.24 (SD = .79), corresponding to a trust level of 64.74%, considered low and unacceptable according to the instrument's threshold. The lowest score was of Risk Perception (RP) and the highest for Competence (COM), indicating that users perceive the application as risky but competent. Table 1 presents the complete results.

Table 1. Cross-cultural trust assessment results

	N	Mean	Std. Deviation	CV	Level (%)
Trust Score	91	3.24	.79	.24	64.74
RP	91	2.77	1.09	.39	
BEN	91	3.06	1.16	.37	
COM	91	4.00	.96	.24	
TR	91	3.12	1.00	.32	

Influence of Culture

The results of the trust assessment per country revealed differences in trust per country. Portugal had the highest Trust Score (M = 3.50, SD = .53), followed by Brazil (M = 3.12, SD = .64) and Mozambique with the lowest. The significance of the differences was tested using independent samples t-tests to reduce the risk of error, as samples were small and equal variances were not assumed in all cases [37].

As shown in Table 2, the difference was only significant in the comparison between Portugal and Brazil, for which the first had a significantly higher Trust Score, $t(58) = -2.51$, p = .015.

Thus, this sample accepts H1, as the country of origin influences users' trust in WhatsApp; Rejects H2: Portugal does not have the lowest level of trust in WhatsApp and rejects H3: Brazil does not have the highest scores of trust in WhatsApp.

Influence of Gender

In comparing genders, the Trust Score was slightly higher for females (M = 3.33, SD = .74) than males (M = 3.17, SD = .83). However, independent sample t-tests attested that the difference between the genders was not significant, $t(87) = 0.95$, p = .34, as shown in Table 3.

This result does not demonstrate the effect of gender on trust and thus rejects H4, as males do not have higher levels of trust on WhatsApp. Further differences across genders are discussed in the following sections.

Table 2. Sample results for *t-test* results comparing countries

				t-test			
Variables	N	M	SD	t	df	Sig. (2-Tailed)	Decision
Brazil	21	3.12	.64	.12	48.29	.91	Not significant
Mozambique	17	3.10	1.06				
Brazil	21	3.12	.64	−2.51	58	.01	Significant
Portugal	53	3.50	.52				
Portugal	53	3.50	.52	−1.89	43.20	.06	Not significant
Mozambique	17	3.10	1.06				

Table 3. Sample results for *t-test* comparing genders

				t-test			
Variables	N	M	SD	t	df	Sig. (2-Tailed)	Decision
Female	54	3.41	.64	.95	87	.34	Not significant
Male	36	3.06	.78				

Influence of Generation

The comparison of results per generation revealed that Generation Z had the highest Trust Scores (M = 3.51, SD = .61), followed by Generation X (M = 3.09, SD = .88) and Millennials (M = 2.99, SD = .71). However, as presented in Table 4, the difference was only meaningful in one case, with Generation Z having significantly higher Trust Scores than Millennials', $t(73) = 3.66$, $p < .01$.

These results are enough to demonstrate the existence of differences in trust scores between the groups, thus accepting H5: generations influence trust results.

Influence of Gender per Generation

It is suggested in the literature that the influence of gender on trust may be interrelated to individuals' age [15]. Therefore, to further investigate these differences, these two variables were analyzed together, and Trust Scores were compared between genders within each generation.

As presented in Table 5, the differences in results varied according to generations, being smaller across the younger respondents, although not significant in all cases. Generation X had the biggest difference and was the only case in which females had the lowest trust. Generation X females' Trust Score (M = 2.61, SD = .99) was significantly lower males' (M = 3.67, SD = .46), $t(12) = -2.54$, $p = 0.03$; the other differences were not statistically significant.

Table 4. Sample results for *t-test* results comparing generations

				t-tests			
Groups	N	M	SD	t	df	Sig. (2-Tailed)	Decision
Generation Z	57	3.52	.61	3.66	73	< .01	Significant
Millennial	23	2.99	.71				
Generation Z	57	3.52	.61	1.73	56	.09	Not Significant
Generation X	10	3.09	.88				
Generation X	10	3.09	.88	-.83	43	.41	Not Significant
Millennial	23	2.99	.71				

Table 5. Sample results for *t-test* results comparing genders within generations

					t-tests			
Variable	Groups	N	M	SD	t	df	Sig. (2-Tailed)	Decision
Generation Z	Female	28	3.54	.49	1.24	27	.22	Not significant
	Male	16	3.45	.98				
Millennial	Female	6	3.23	.90	.345	19.26	.73	Not Significant
	Male	21	2.82	.69				
Generation X	Female	8	2.61	.99	−2.64	9.56	.03	Significant
	Male	7	3.67	.46				

The results supported the assumption from the literature, indicating that there are, in fact, differences in trust between genders but that they are subject to generation.

3.2 Privacy Concerns

Privacy concerns were analyzed by construct, with Awareness to Privacy receiving the highest and most homogeneous score (M = 5.49, SD = 1.64), followed by Privacy Concerns (M = 4.90, SD = 2.19), Privacy Intrusion (M = 4.45, SD = 2.24), Perceived Effectiveness of Privacy Policy (M = 3.73, SD = 1.97) and lastly, Privacy Control (M = 3.40, SD = 2.07). Next, results were analyzed by groups to identify critical differences in the results.

Differences in Privacy Concerns

Millennials attributed the lowest scores for Privacy Control (PCTL) (M = 3.17, SD = 2.07) and Perceived Effectiveness of Privacy Policy (POLICY) (M = 3.32, SD = 1.99), and the highest for Privacy Concerns (PCON) (M = 5.39, SD = 2.28) and Privacy Intrusion (INTRU) (M = 5.05, SD = 2.09), characterizing the most concerned generation. On the other hand, Generation Z had the highest scores for POLICY (M = 4.15, SD = 1.87) and the lowest for INTRU (M = 3.98, SD = 2.26), consequently the least concerned.

The comparison between countries revealed that Brazil had the highest PCON (M = 5.05, SD = 2.27) and lowest PCTL scores (M = 2.71, SD = 1.71), Portugal, the highest POLICY score (M = 4.00, SD = 1.60), and Mozambique the highest INTRU (M = 4.00, SD = 2.44).

Correlation Between Privacy Concerns and Trust
Pearson's correlation coefficient was used to assess if the relationship between privacy concerns and Trust Scores was meaningful. The correlation was significant in all cases and was the strongest for POLICY, positively correlated to trust, $r(91) = .62$, $p < .01$. Positive effects were also observed for PCTL, $r(91) = .48$, $p < .01$; and for AWARE, $r(91) = .26$, $p = .01$. On the other hand, negative correlations were observed for PCON, $r(91) = -.47$, $p < .01$, and for INTRU, $r(91) = -.38$, $p < .01$.

4 Discussion

Based on the sample in this study, culture, gender, and generation influence trust in technology, but it happens in ways that differ from the literature suggestions on the topic.

First, the study demonstrated that culture affects users' trust, but that does not happen as per the cultural dimensions model [12]. In fact, the results were inverse from what should be expected based on the model, with Portugal with the highest and Brazil with the lowest Trust Score.

Nonetheless, the low trust in Brazil may be explained by the local issues involving WhatsApp in recent years and its relationship with the country's political unrest [38]. At the same time, Portugal is part of a region with more regulations regarding privacy and data usage. These findings indicate that contextual factors regarding the use of the specific object have a more substantial influence on users' trust than general cultural characteristics. Therefore, other models should be considered for the investigation of differences in trust in technology across cultures.

The results for gender reveal that trust differences between males and females vary according to the age range, as pointed out by [15]. While the results opposed the expectations that women would trust less at a general level, the assumption was valid within Generation X, where women had the lowest trust. Besides, the diminishing gender gap among younger generations indicates that the difference in trust between genders is changing over time. Such a trend would highlight the fact that gender differences are dynamic, following the views from social role theory [20]. These findings demonstrate that gender generalizations may be problematic and reinforces the importance of considering gender in relation to users' generation.

Without considering gender, differences in trust between generations were attested only between Generation Z and Millennials; however, the results may have been affected by this study's sample size and the higher participation of younger generations. Nevertheless, it was enough to verify that users' generation influences their trust in technology.

Additionally, Millennials, the intermediate age group, had the lowest Trust Score, discarding a linear relationship between age and trust; that is, it is not merely decreasing or increasing over time.

Privacy Concerns

The results of privacy concerns align with findings from the trust assessment. Millennials' lowest trust was reflected in their highest scores in most privacy concerns; Generation Z's highest trust was in agreement with their lowest Privacy Intrusion and highest Perceived Effectiveness of Privacy Policy.

The differences in the Privacy Concerns between the countries reinforce the premise that contextual factors significantly influence trust. For example, Portugal is under stricter privacy regulations, which likely affected its highest score for the Perceived Effectiveness of Privacy Policy. The opposite is observed in Brazil and Mozambique, which have the lowest Privacy Control and the highest Privacy Intrusion, respectively.

The analysis of correlations corroborated the relationship between privacy concerns and trust, which was significant in all cases. Furthermore, it revealed that the strongest correlation is with Perceived Effectiveness of Privacy Policy, which positively affects trust, and the second-strongest is with Privacy Control. These findings are relevant as they indicate that enhancing (effective) privacy policies and users' control are essential measures to increase users' trust.

Implications for the Design Practice

Overall, the results demonstrate that the HCTM is suitable for mapping trust behaviors across cultures, answers this study's research question, and emphasizing this tool's potential for investigating differences in users' trust in technology. Although this research relied mainly on the Trust Scores, comparisons of the results of the trust attributes can provide more detailed insights into how different variables affect users' trust.

5 Framework: Including Trust in the Personas Creation

Aiming to bridge the gap between research and practice regarding trust in technology, the authors proposed a tool that designers and other practitioners could use to facilitate the inclusion of trust in the initial phases of the design during the creation of personas. It indicates the essential factors that influence users' trust and insights about users' differences to support the designers' inquiry, as per this study.

The initial framework went under a round of expert reviews, consisting of the evaluation and feedbacks from 5 designers from 4 different countries (Brazil, Estonia, Iran, and Russia). According to the designers, trust is only included in the personas when customers indicate trust issues, which happens infrequently. Moreover, they claimed that it is still unclear when and how to include trust during the design process and that so far, they lack the tools to do so. For that reason, all the experts considered the tool relevant and valuable for their practice.

The reviews were summarized using a feedback matrix [39] and improved according to the suggestions. The model is presented in the appendix and is intended to assist the design practice. It highlights factors that should be considered to include "trust" as personas' attributes and brings the findings from this specific study to encourage the reflection about other segments of technology. Finally, the framework demonstrates that it is possible to transpose research insights to the design practice, and the HCTM can also be used to this end.

6 Conclusion

This investigation made evident that there is a gap in understanding how different factors influence users' trust in technology. Nevertheless, it is possible to work on this issue. This study demonstrated that trust is dependent on various factors, some of which can be anticipated and considered from the initial stages of the design. Considering aspects that influence trust within specific groups is a way to guide design decisions towards more effective solutions.

This research aimed to investigate and map how factors as culture, gender, and age influence users' trust in technology. For that, the HCTM, initially designed to assess propensity to trust, showed to be also an effective tool.

Factors that are easily considered, such as culture, gender, and generation, can provide insights into creating trustworthy technology. However, the findings and framework presented here are just the initial efforts towards this end, and its insights should be used to guide further investigations.

Limitations

The main limitation of this research was its sample size, which may have affected the soundness of the results. Although the procedures adopted aimed at reducing the chance of error, the number of responses was below the recommended threshold in some cases.

Recommendations for Future Research

The approach adopted by this research was efficient for generating insights for HCI literature and practice. Hence, the first recommendation for future research is to develop cross-cultural studies using the HCTS on a larger scale and target other cultures.

Besides that, there is an overall lack of information about specific populations in the HCI literature. Namely, regions such as Africa and Latin America are generally underexplored and should be more often considered. Similarly, minority groups, as non-binary populations, are often disregarded. The HCI field is dominated by a relatively homogeneous demographic group, which can fail to address the needs of other social segments. Research should increasingly include minorities and adopt an inclusive standpoint.

Finally, more efforts should be put towards creating tools to assist the development of trustworthy systems and bridge the gap between research and practice regarding trust in technology.

Appendix

See Fig. 1.

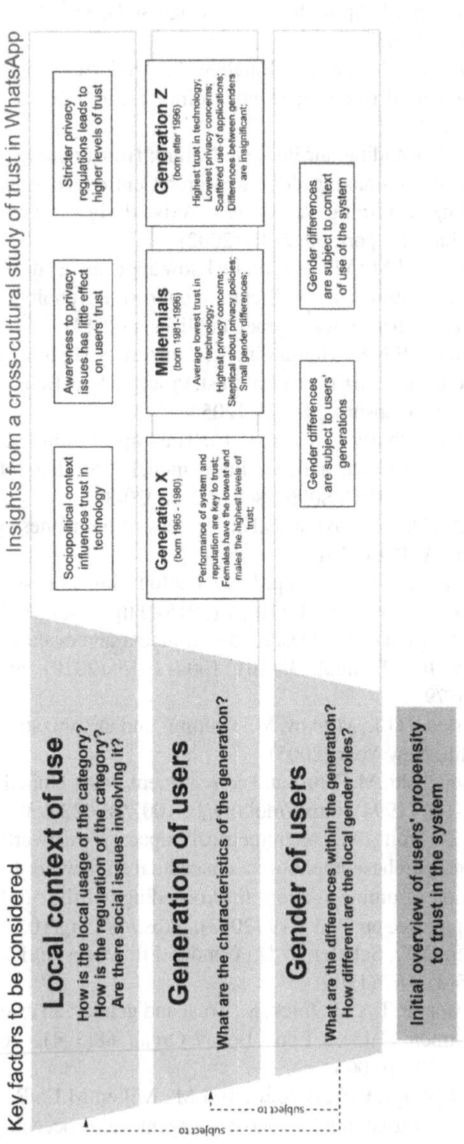

Fig. 1. Framework: Creating Personas for Trustworthy Design.

References

1. Zuboff, S.: Big other: surveillance capitalism and the prospects of an information civilization. J. Inf. Technol. **30**(1), 75–89 (2015). https://doi.org/10.1057/jit.2015.5
2. Lewandowsky, S., Ecker, U.K., Cook, J.: Beyond misinformation: understanding and coping with the "post-truth" era. J. Appl. Res. Mem. Cogn. **6**(4), 353–369 (2017). https://doi.org/10.1016/j.jarmac.2017.07.008
3. Podavini, N., Flore, M., Verile, M., Balahur, A.: Understanding citizens' vulnerability to disinformation and data-driven propaganda. Joint Research Centre (European Commission). EU, Luxemburgo (2019)
4. Mair, D., et al.: Understanding our Political Nature: How to put knowledge and reason at the heart of political decision-making. Joint Research Centre (European Commission) (2019)
5. Friedman, B., Kahn, P., Borning, A.: Value sensitive design: theory and methods. University of Washington technical report, pp. 2–12 (2002)
6. Söllner, M., Leimeister, J.M.: What we really know about antecedents of trust: a critical review of the empirical information systems literature on trust. Psychology of Trust: New Research, D. Gefen, Verlag/Publisher: Nova Science Publishers (2013)
7. Kassim, E.S., Jailani, S.F.A.K., Hairuddin, H., Zamzuri, N.H.: Information system acceptance and user satisfaction: the mediating role of trust. Procedia Soc. Behav. Sci. **57**, 412–418 (2012). https://doi.org/10.1016/j.sbspro.2012.09.1205
8. McKnight, D.H., Choudhury, V., Kacmar, C.: The impact of initial consumer trust on intentions to transact with a web site: a trust building model. J. Strateg. Inf. Syst. **11**(3–4), 297–323 (2002). https://doi.org/10.1016/S0963-8687(02)00020-3
9. Tripp, J., McKnight, D.H., Lankton, N.K.: Degrees of humanness in technology: what type of trust matters? In: AMCIS (2011)
10. Lankton, N.K., McKnight, D.H., Tripp, J.: Technology, humanness, and trust: rethinking trust in technology. J. Assoc. Inf. Syst. **16**(10), 1 (2015). https://doi.org/10.17705/1jais.00411
11. Gulati, S., Sousa, S., Lamas, D.: Design, development and evaluation of a human-computer trust scale. Behav. Inf. Technol. **38**(10), 1004–1015 (2019). https://doi.org/10.1080/0144929X.2019.1656779
12. Hofstede, G., Hofstede, G.J., Minkov, M.: Cultures and Organizations: Software of the Mind, vol. 2. Mcgraw-Hill, New York (2005)
13. Yamagishi, T., Yamagishi, M.: Trust and commitment in the United States and Japan. Motiv. Emot. **18**(2), 129–166 (1994). https://doi.org/10.1007/BF02249397
14. Chong, B., Yang, Z., Wong, M.: Asymmetrical impact of trustworthiness attributes on trust, perceived value and purchase intention: a conceptual framework for cross-cultural study on consumer perception of online auction. In Proceedings of the 5th International Conference on Electronic Commerce, pp. 213–219 (2003). https://doi.org/10.1145/948005.948033
15. Byrnes, J.P., Miller, D.C., Schafer, W.D.: Gender differences in risk taking: a meta-analysis. Psychol. Bull. **125**(3), 367 (1999)
16. Buchan, N.R., Croson, R.T.A., Solnick, S.: Trust and gender: an examination of behavior and beliefs in the Investment Game. J. Econ. Behav. Organ. **68**(3–4), 466–476 (2008). https://doi.org/10.1016/j.jebo.2007.10.006
17. Hillesheim, A.J., Rusnock, C.F., Bindewald, J.M., Miller, M.E.: Relationships between user demographics and user trust in an autonomous agent. In: Proceedings of the Human Factors and Ergonomics Society Annual Meeting, vol. 61, no. 1, pp. 314–318. SAGE Publications, Sage CA (2017). https://doi.org/10.1177/1541931213601560
18. Oyibo, K., Ali, Y.S., Vassileva, J.: Gender difference in the credibility perception of mobile websites: a mixed method approach. In: Proceedings of the 2016 Conference on User Modeling Adaptation and Personalization, pp. 75–84 (2016). https://doi.org/10.1145/2930238.2930245

19. Venkatesh, V., Morris, M.G., Davis, G.B., Davis, F.D.: User acceptance of information technology: toward a unified view. MIS Q., 425–478 (2003). https://doi.org/10.2307/30036540
20. Eagly, A.H., Wood, W., Diekman, A.B.: Social Role Theory of Sex Differences and Similarities: A Current Appraisal. Lawrence Erlbaum Associates Publisher (2000)
21. Strauss, W., Howe, N.: The Fourth Turning: What the Cycles of History Tell us About America's Next Rendezvous with Destiny. Crown, New York (2009)
22. Howe, N., Strauss, W.: The Next 20 Years: How Customer and Workforce Attitudes Will Evolve. Harvard Business Review (2014). https://hbr.org/2007/07/the-next-20-years-how-customer-and-workforce-attitudes-will-evolve. Accessed 04 Feb 2021
23. Obal, M., Kunz, W.: Trust development in e-services: a cohort analysis of millennials and baby boomers. J. Serv. Manag. (2013). https://doi.org/10.1108/09564231311304189
24. Noah, B., Sethumadhavan, A.: Generational differences in trust in digital assistants. In: Proceedings of the Human Factors and Ergonomics Society Annual Meeting, vol. 63, no. 1, pp. 206–210. Sage CA, Los Angeles. SAGE Publications (2019). https://doi.org/10.1177/1071181319631029
25. Calvo-Porral, C., Pesqueira-Sanchez, R.: Generational differences in technology behaviour: comparing millennials and Generation X. Kybernetes (2019). https://doi.org/10.1108/K-09-2019-0598
26. Parker, K., Igielnik, R.: On the cusp of adulthood and facing an uncertain future: what we know about Gen Z so far. Pew Research Center's Social & Demographic Trends Project (2020). https://www.pewresearch.org/social-trends/2020/05/14/on-the-cusp-of-adulthood-and-facing-an-uncertain-future-what-we-know-about-gen-z-so-far-2/. Accessed 17 Mar 2021
27. Pinto, A., Sousa, S., Silva, C., Coelho, P.: Adaptation and validation of the HCTM Scale into Human-robot interaction Portuguese context: a study of measuring trust in human-robot interactions. In: Proceedings of the 11th Nordic Conference on Human-Computer Interaction: Shaping Experiences, Shaping Society, pp. 1–4 (2020). https://doi.org/10.1145/3419249.3420087
28. Goldsmith, R.E., Freiden, J.B., Eastman, J.K.: The generality/specificity issue in consumer innovativeness research. Technovation **15**(10), 601–612 (1995). https://doi.org/10.1016/0166-4972(95)99328-d
29. Rogers, E.: Diffusion of Innovations. Free Press, New York (2003)
30. Xu, H., Dinev, T., Smith, H.J., Hart, P.: Examining the formation of individual's privacy concerns: toward an integrative view. In: ICIS 2008 Proceedings, no. 6 (2008). https://aisel.aisnet.org/icis2008/6
31. Deloitte Global Mobile Consumer Survey Brasil, 10 May 2020 (2019). https://pesquisas.lp.deloittecomunicacao.com.br/global-mobile-consumer-19. Accessed 23 Nov 2020
32. Doffman, Z.: WhatsApp Beaten By Apple's New iMessage Privacy Update. Forbes (2021). https://www.forbes.com/sites/zakdoffman/2021/01/03/whatsapp-beaten-by-apples-new-imessage-update-for-iphone-users/?sh=32113a023623. Accessed 21 Mar 2021
33. Rastogi, N., Hendler, J.: WhatsApp security and role of metadata in preserving privacy. arXiv Prepr. arXiv1701, 6817, 269–275 (2017)
34. O'Hara, K.P., Massimi, M., Harper, R., Rubens, S., Morris, J.: Everyday dwelling with WhatsApp. In: Proceedings of the 17th ACM Conference on Computer Supported Cooperative Work & Social Computing, pp. 1131–1143 (2014). https://doi.org/10.1145/2531602.2531679
35. Tavakol, M., Dennick, R.: Making sense of Cronbach's alpha. Int. J. Med. Educ. **2**, 53 (2011). https://doi.org/10.5116/ijme.4dfb.8dfd
36. Landau, S.: A handbook of statistical analyses using SPSS. Chapman & Hall (2019).
37. De Winter, J.C.: Using the Student's t-test with extremely small sample sizes. Pract. Assess. Res. Eval. **18**(1), 10 (2013). https://doi.org/10.7275/e4r6-dj05

38. Avelar, D.: WhatsApp fake news during Brazil election ‘favoured Bolsonaro’. The Guardian (2019). https://www.theguardian.com/world/2019/oct/30/whatsapp-fake-news-brazil-election-favoured-jair-bolsonaro-analysis-suggests. Accessed 19 Dec 2020
39. Doorley, S., Holcomb, S., Klebahn, P., Segovia, K., Utley, J.: Design Thinking Bootleg. Institute of Design at Stanford (2018). https://static1.squarespace.com/static/57c6b79629687fde090a0fdd/t/5b19b2f2aa4a99e99b26b6bb/1528410876119/dschool_bootleg_deck_2018_final_sm+%282%29.pdf. Accessed 28 March 28/

The Research on the User Experience of Consultation Designed by China's Medical Mobile Media Platforms Under the Background of COVID-19

Lingxi Chen, Yuxuan Xiao(✉), and Linda Huang

Changsha University of Science and Technology, Changsha, China
huanglinda@csust.edu.cn

Abstract. At the beginning of 2020, the outbreak of COVID-19, a black swan event, brought a huge impact and change to people's normal life. Online consultation has become the best choice for users to consult and seek medical treatment. This paper starts from the problem of the experience of User Consultation of the Medical Mobile Media Platforms under the Backdrop of COVID-19, method based on an online survey from of 167 valid questionnaires, we studied the influence and role of medical APP on the society and the rigid medical demand. Based on the research purposes, the content of the Mobile Medical App User Consultation, the Use Effect of Mobile Medical App User Consultation are studied on field research. Research shows that the consultation design is not perfect, the application effect varies due to the differences in age, region, condition, cost and other aspects of the patients. But we can be predicted that in addition to its unique diagnostic and therapeutic advantages in the face of epidemic diseases such as COVID-19, it can also play a role in the diagnosis and treatment of common diseases and chronic diseases.

Keywords: Medical mobile media platforms · User experience of consultation · App interface design

As the popularization of IT and wide use of Internet, medical service, a common thing in modern life, has undergone profound changes in its service model. In China, with the double blessing of big data and AI technology, online consultation Apps like Ping An Good Doctor, Hao Dai Fu Online, Dr. Clove and Tencent Doctorwork are springing up, delivering great convenience for people to seek medical advice. At the beginning of 2020, the outbreak of Covid-19, a black swan event, disturbed people's normal life. For the purpose of reducing the risk of being infected from direct human interaction and communication, online consultation has become the best choice for users to consult and seek medical treatment.

At present, the functions of China's mainstream mobile medical App have been consistent, most of them undertaking the basic functions of registration, drug usage inquiry, access to health information, disease self-examination and self-diagnosis and online consultation. The regular epidemic prevention and control enables mobile medical

C. Stephanidis et al. (Eds.): HCII 2021, LNCS 13094, pp. 509–521, 2021.
https://doi.org/10.1007/978-3-030-90238-4_36

Apps to be the top priority for many users to seek for diagnosis and treatment under the backdrop of the raging COVID-19.

1 Research Objects and Methods

1.1 Research Objects

Research on user consultation experience of mobile medical App: For the purpose of creating a better user experience, this paper collects and analyzes data from the aspects of users' behavior, preference and use feeling, exploring potential factors that influence user experience of mobile medical App. Thus, developers are able to be equipped with relevant management strategies and suggestions to enhance the consultation experience of mobile medical App users and improve the medical service.

1.2 Research Methods

Questionnaire. Designed on the basis of user experience, the questionnaire covers 200 online consultation service users from 6 provinces and cities on the platform of Wenjuanxing mainly through Wechat, QQ and many other diversified online communication channels. In total, 187 questionnaires were collected with a collection rate of 93.5%, 167 of them valid, with a validity rate of 83.5%. There were 51 males and 100 females in the surveyed, and 16 were unwilling to disclose their gender. The ratio of male to female stood as 1:1.96. Most of the respondents were young and middle-aged people aged from 18–35, accounting for 88.63%. This paper will sort out and analyze the questionnaires collected, in order to have a better understanding about the usage and experience of users and to draw relevant conclusions.

Interview Method. The corresponding interview outline were drawn up in advance, and in-depth communication were conducted with consultation service users from Hunan, Jiangxi, Guangdong, Beijing and Shanxi Provinces via video link or telephone. Related interviews were recorded and progress was made to make in-depth understanding about related issues. Moreover, the interview materials were used as a supplement to the questionnaires.

Mathematical Statistic Method. Statistics software is Applied to achieve corresponding data results and the obtained questionnaire, interview content and the search data are summarized and analyzed.

2 Core Concepts of the Study

2.1 User Experience

Originated from human-computer interaction technology, "User Experience" was first proposed by Donald Nor-man and Appeared in people's view in the mid-1990s. According to the 3-layers theory, the universal experience of human is divided into 3 different layers, namely instinct layer, behavior layer and reflection layer. As for user experience, a standardized definition has not taken shape. The ISO9241–210 defines user experience as: it emphasizes the completion of tasks and goals and cognitive information processing; focuses on the psychological, physical and emotional perception and reaction in the process of interaction between users and products. However, Hassenzahl M is prone to believe that user experience is the result of interaction between user's psychological and cognitive states as expectation, demand, motivation and the complexity, usability and functionality.

For sure, with the updated design, expanded content, extended structure and widespread Application of products, the connotation of user experience is continuously expanding. In terms of the elements of user experience, some influential theories, such as User Engagement and Flow Experience formed up. The current user experience researches mainly focus on guiding the design and system evaluation, thus paying less attention to users. This paper concentrates on the consultation service of mobile medical App, aiming at figuring out factors that exert influence on the evaluation and assessment of users' choices and experience against the backdrop of public health emergency from the most basic functional level.

2.2 Mobile Medical Applications

APP, the abbreviation of Application, is designed for and Applied to mobile intelligent terminal on the basis of Internet. On the basis of its definition and some scholars' understanding about similar software in medical field, mobile medical App mainly provides services as medical consultation, inquiry, registration, medical product purchase and professional information checking. Characterized by diversity, pluralism and easy-operating, more than 2000 mobile medical Apps now stand in the Chinese market, most of them share similar functions despite of differences among some of them.

3 Research Result and Analysis

3.1 Content Analysis of Mobile Medical App User Consultation

Selection of Consultation Platform. Because of the large number of mobile medical Apps, users' choices on different platforms are stepwise and are more likely to favor those Apps that they have used before. As it is showed in Fig. 1, the Apps in the first-tier are either the extension of the PC end in the mobile end, such as Hao Da Fu Online and Dr. Clove; or the entrepreneurs from IT field, the first batch of pioneers in the mobile medical field, such as Chunyu Doctor; or group enterprises reluctant to miss the potential value of mobile medical service and relying on strong capital background to build the star products in Internet medical field, such as Ping An Good Doctor.

Analysis of Consultation Demand. Under normal circumstances, the users' consultation content is relatively balanced on the mobile medical App, mostly involving the related content in general medical consultation, such as consultation to doctor, conventional disease symptoms inquiry and access to health information. However, as the COVID-19 rages, users' concern and need for consultation related to the epidemic

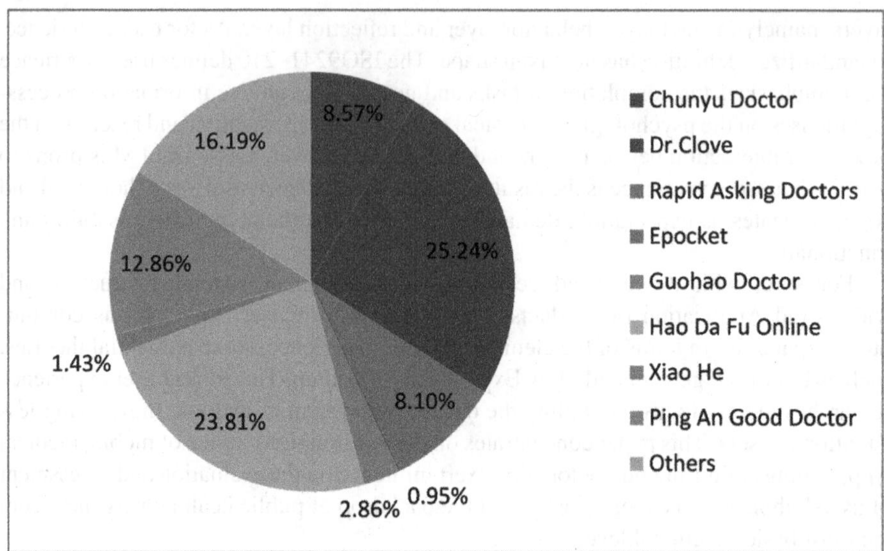

Fig. 1. What kinds of medical consultation Apps or Applets have you used?

increased significantly, and the proportion of medical consultation for internal medicine climbed sharply, and mostly for symptoms related to COVID-19, such as cold, headache, diarrhea and fever, accounting for 35.47%. Compared with the usual circumstances, other kinds of consultation content did not show too much fluctuation and remained centered in conventional chronical diseases, such as cardiovascular diseases and dermatosis.

It is worth mentioning that women accounted for more than 50% of the people who used medical Apps to ask for consultations during the Covid-19 epidemic; men were only 36.67%. After in-depth interviews, it was found that the consultation of female users was not limited to themselves, but the conditions of other members of their family, which indicts that some male users completed the APP consultation in an indirect way, and the proportion of them was 46% of the confirmed interviewed female users (Table 1).

Analysis of the Purpose of User Consultation. The data shows that during the COVID-19 epidemic, there is a big difference between the content of the interviewed users' usual consultations and those during the epidemic, and there is a certain contrast between the real needs and purposes of users' consultations. 63.47% of users (106 people) have inquired about Covid-19 during the epidemic, and only 36.53% of users have inquired about non-Covid-19. In the former, 10 people were randomly sampled and learned through interviews that they were all in low-risk infection areas and had no history of exposure to Covid-19. They all actively consulted about covid-19 during routine chronic diseases to eliminate their doubts.

Compared with the usual user consultation, the purpose is to be able to obtain the doctor's professional guidance in an easy and quick way, and advice on existing and conventional diseases. During the epidemic, more users hope to complete two main diagnosis and treatment goals through APP consultation: on the one hand, the APP can

Table 1. What kind of diseases are consulted on medical consultation APPs?

Consultation Content	Rate(%)
Internal Medicine (cold, headache, diarrhea, fever)	34.57%
Orthopedics (ligament injury, lumbar disc herniation, Cervical spondylosis,arthritis,etc.)	13.17%
Otolaryngology (rhinitis, pharyngitis, tonsillitis, otitis media, tinnitus,etc.)	16.87%
Dermatology (eczema, acne, urticaria)	15.23%
Obstetrics and gynecology	4.53%
Pediatric	6.17%
Stomatology	6.58%
Others	2.88%

answer their own diseases that are inconvenient to go to the hospital due to epidemic prevention and control. On the other hand, it can help to eliminate the possibility of contracting Covid-19 and gain a sense of psychological security. Therefore, during the

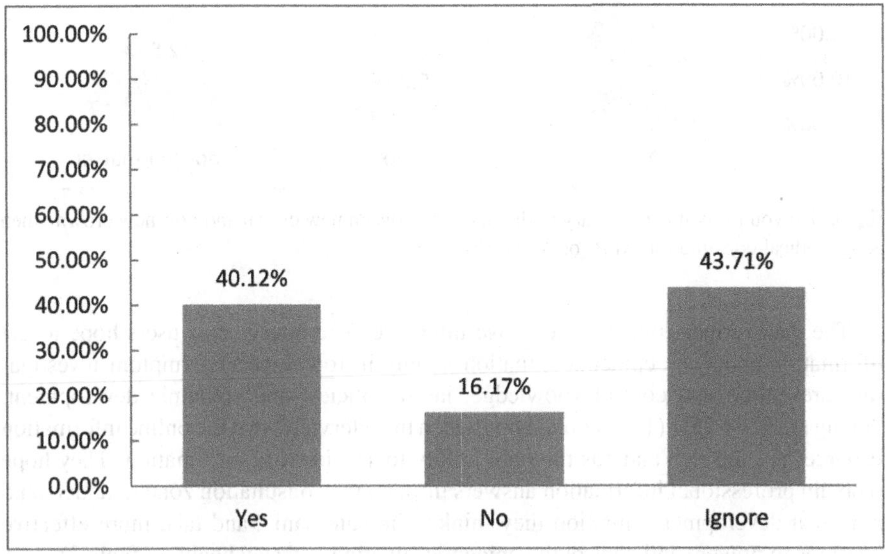

Fig. 2. When using the medical consultation APP or Applet, do you observe any special interface about the Covid-19?

consultation process, 40.12% of users will take the initiative to find out whether there is a special interface for the Covid-19 on the APP, and only 16.17% of users don't care about it (Fig. 2).

It needs to be particularly pointed out that in Table 3, telephone interviews were conducted among a sample (8 people) of 43.71% of users who did not notice whether there is a special interface for the Covid-19 in the App, and of which 7 people, accounting for 87.5% of the sample ratio, emphasized or agreed that it was necessary to distinguish between Covid-19 and non-Covid-19 consultations in medical consultation Apps so that patients can get targeted consultations. This ratio basically coincides with the 82.04% of the feedback obtained for the same questions raised by all users in the questionnaire, indicating that users still have a clearer understanding and judgment of the severity of the Covid-19 epidemic, and have a better understanding of the information concerning epidemic situation (Fig. 3).

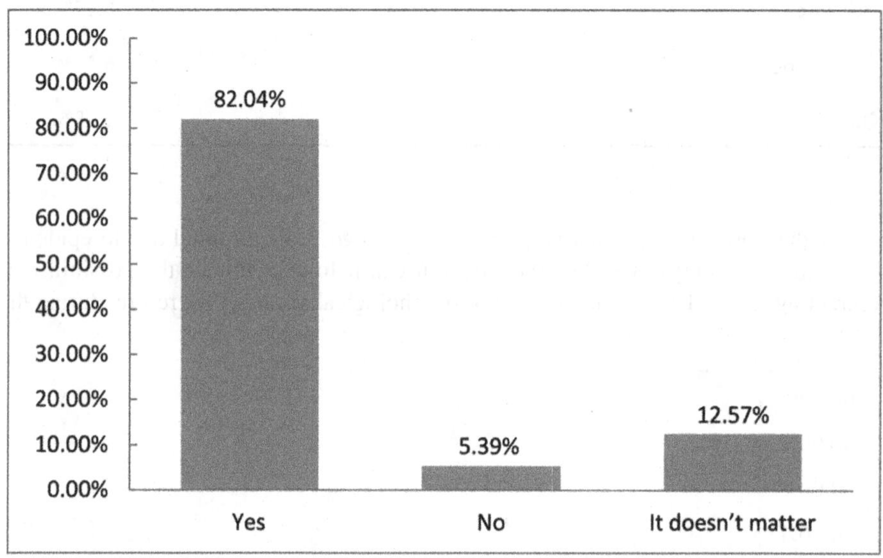

Fig. 3. Do you think it is necessary to distinguish between new crown and non-new crown when using medical consultation APPs or Applets?

The data further confirms the above inference. The interviewed users hope to get information about the epidemic situation mainly in four aspects: symptom investigation, prevention and control knowledge, latest policies, and epidemic development. Among them, 74.25% (124 people) admitted in the interviews that the online information resources are too rich and has the poor ability to self-identify information. They hope to obtain professional information answers through the consultation zone to understand more real development situation they think, eliminate panic, and take more effective measures to prevent and control the epidemic situation and stay healthy (Table 2).

Table 2. What modules can be added to therapeutic APPs or small programs during the period of the epidemic?

Modules	Rate(%)
Covid-19 popular science	28.84%
Covid-19 prevention knowledge	20.22%
Covid-19 symptoms self-examination	22.91%
Covid-19 doctor consultation	13.48%
More detailed map of Covid-19	14.02%
Others	0.54%

3.2 Analysis of the Use Effect of Mobile Medical App User Consultation

On the whole, users' satisfaction with the consultation effect of mobile medical Apps is in the upper middle position. Different Apps have different levels of user satisfaction with their online services. Those with higher satisfaction are all platforms in the first echelon sequence in the previous analysis, which have simple operation, high efficiency of consultation, safety and convenience, and the professionalism of hospitals and doctors. However, dissatisfaction is mainly concentrated in three aspects: general consultation results, difficult operation, and low consultation efficiency.

Interface Operation. The interface designs of all major medical APPs are similar, and the functions are almost the same. Health self-examination, registration Appointment, medical payment, medical consultation and other medical services are all available. The important service entrances for Appointment registration, online consultation, and health self-diagnosis are all in the default service section. The search bar and grid navigation settings also provide users with an intuitive and quick search entrance, making it as convenient as possible for users to get the service what they want with the least number of clicks.

Therefore, the survey shows that the highest degree of satisfaction is mainly due to the "easy-operating" of the consultation, accounting for 37.87%. Through data analysis, the age factor in this option highlights the key role.

Among users who think APP is "easy-operating", people under 18 years old and 56–75 years old accounted for only 7.96% and 7.08% respectively. People under 18–35 years old accounted for 45.13%, 36-55year-old people for 39.82%. Due to using smart terminals for a long time during their studies, work, and daily life, young and middle-aged people are relatively familiar with APP design and operation. Too many operational obstacles are not existed in the use of this professional medical APP for consultation, while minors and elderly groups are obviously familiar with the interface content. Numerous mobile medical Apps show inadaptability (Fig. 4).

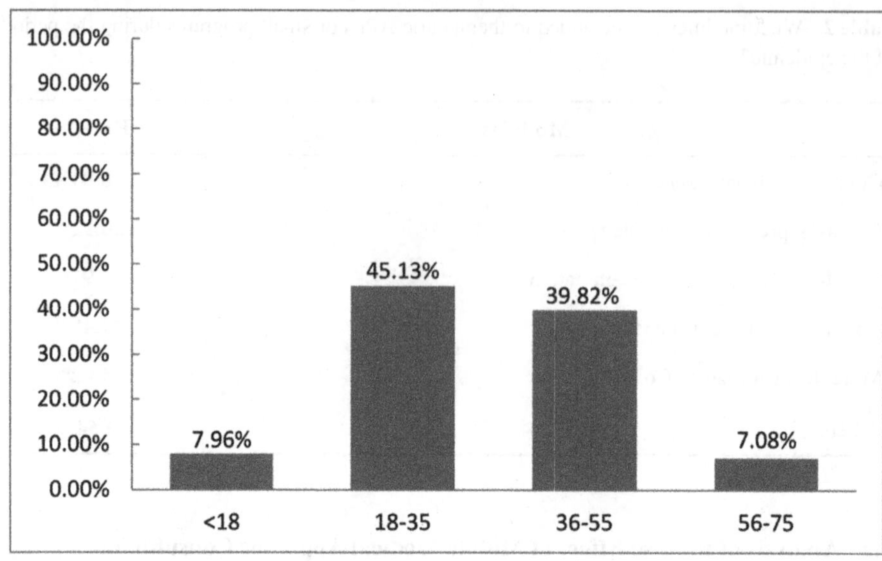

Fig. 4. People who think medical consultation Apps are "easy-operating".

Refined analysis of the data, and it was further found that the gender factor shows great differences in different age groups, especially in people aged 56–75. In the aforementioned user demand analysis, it was mentioned that 46% of the purpose of female users consultation was not limited to their health, but also others' health conditions in the consultation. Among them, 21people were 56–75 years old, accounting for 45.6%. Added 6 females of the same age who only consulted their own health, the total of 28 accounted for 28% of the total surveyed female users, which is much higher than the 5 males surveyed at the same age, which account for 9.8%. It shows that among the elderly, women are more frequently to use mobile medical APP and are more familiar with the interface.

In particular, 27.85% of the interviewed users showed that they knew or used the free clinic for Covid-19, while 72.15% did not hear about it at all, including 51 people who used APP for consultation during the epidemic, accounting for 53.69% of APP consultation users. After the outbreak, several mainstream medical Apps quickly put up anti-epidemic online consultation zones on their interfaces, and established an online consultation and prevention mechanism. Although with the integration of health committees, the CDC and the central media's epidemic information, medical Apps such as "Dr. Clove" can use their own powerful IP to upload and release policies, and spread protection knowledge protection so that the public has a clearer and comprehensive understanding of the epidemic. However, the professional medical APP port linkage is not satisfactory.

Consultation Efficiency. People who are satisfied with the "high efficiency of consultation" are mainly young adults aged 18–35, accounting for 59.18%; those under 18 account for 8.16%; those aged 36–55 account for 30.61%; and those aged 56–75 are the least, only 2.04%. Among the people who think that Apps can provided the "low consultation efficiency", the 36–55 years old accounted for 57.69%. After further investigation and analysis, feedback time and cost have become the two major influencing factors for users to judge the efficiency of consultation (Fig. 5).

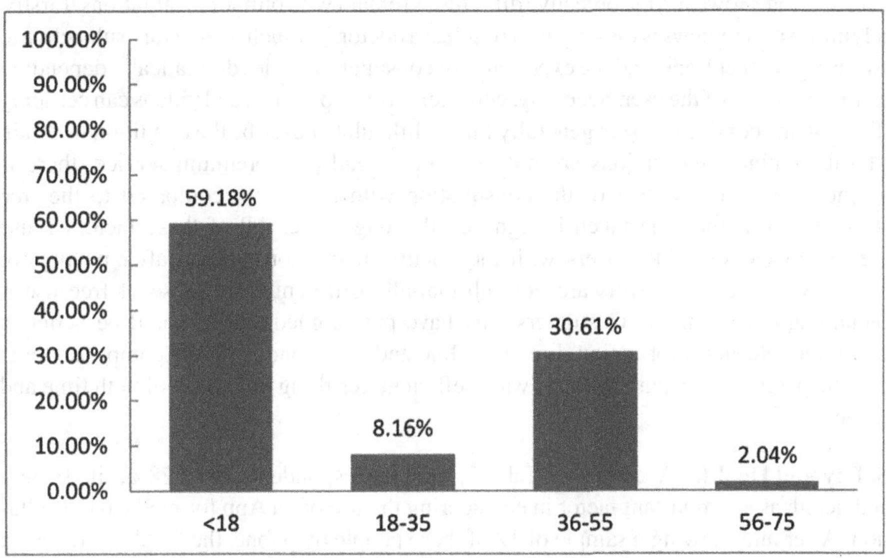

Fig. 5. People who don't think medical consultation Apps are "easy-operating".

At present, the major medical APPs can provide consultation services of different payment levels and service levels, including free consultation, Free clinic zone, online expert consultation, famous doctor direct access, etc. Both free and paid consultation services are the most commonly used. In the free consultation, users generally need to click the label to quickly consult the condition and describe the symptoms in detail. They can also add pictures to upload the condition and check the pictures, so as to facilitate the doctor's confirmation. After asking questions, the doctor replies to the consultation questions in the form of an online dialogue. Since this service only provides a simple analysis of the condition, the doctor advises on the treatment. For users with non-serious diseases, the information feedback time is more satisfactory. 18–35 years old age group user are mostly daily discomfort, such as light microscopic disease or common diseases, some of the patients aged 36–55 years old have common chronic diseases, but not serious diseases. The above two types of free consultation can meet their basic medical consultation needs and save the time of offline hospital diagnosis and treatment. Therefore, the efficiency of consultation is Approved by the users.

Meanwhile, the proportion of people aged 36–55 who considered the consultation to be inefficient stands at 57.69%, which is significantly higher than other age groups. Upon investigation, most of these users do have specific information needs for the condition they are consulting about and normally use the paid consultation function to purchase different premium service. The premium service basically includes pre-consultation assessment, graphic consultation, case collection, case analysis, on-demand interview, treatment plan, out-of-hospital care, inpatient and surgical arrangements, etc. and the price ranges from 20 to 2000 yuan. The price of consultation varies from 5 to more than 200 yuan depending on the doctor's professional title and the hospital's rank.

As a premium service, users have significantly higher expectations when asking for a consultation. However, the objective difficulties remain with online consultations. Firstly, relying on past reviews is the only way to select a doctor, but such reviews are subjective to past users' perceptions, and the experience of consultation varies dramatically depending on the condition of the user. Secondly, while text, voice, pictures, and videos can certainly illustrate the condition, users generally find it difficult to describe the condition in detail. Thirdly, online consultations are not one-to-one, and as a premium service, there is an increase in the content of the consultation with the doctor compared to the free ones, however, the time taken is significantly longer, etc. All of these factors cause the dissatisfaction of the users with the quality of the consultation after paying for it, believing that the results are not substantially different from those of free users, remaining quite general. Customers who have paid argued that it would be better to check the relevant information directly online and then conduct a self-comparison and that the purchase of a paid package was inefficient, resulting in a waste of both time and money.

Safety and Health. According to Table 3, 52 of the respondents, or 17.28%, cited safety and health as an important factor in considering the use of an App for medical consultation. After interviewing a sample of 12 of these people by phone, the feedback received was that the APP consultation was the best initiative to reduce visits to crowded places like hospitals, given the impact of the COVID-19 pandemic. It can reduce the risk of pandemic spreading, but also promote the prevention and control of the COVID-19. Among the respondents who chose the safe and healthy option, young and middle-aged people accounted for the highest proportion with 39 in total or 75%, indicating that the awareness of epidemic prevention and control among the group as the backbone of society.

Table 3. Reasons for being satisfied with the medical consultation App during COVID-19 pandemic.

Reasons for being satisfied with the medical consultation App	Rate (%)
Easy to operate	37.87%
High efficiency	16.61%
Professionalism of hospitals and doctors	11.96%
Reasonable charges	5.65%
Safe and convenient	17.28%
Good experience with UI	9.30%
Others	1.33%

Degree of Professionalism. The degree of professionalism of hospitals and doctors is an important indicator for users to judge when they make an App consultation. The data on this dimension shows consistency with the data on users' access to the APP. Those who chose the degree of professionalism of APP consultation were concentrated in two age groups, 18–35 and 36–55 years old, with 34 people, accounting for 35.78% of those who had used the APP consultation. Older people are still skeptical about the professionalism of online consultations. Young people make up half of the identification group, indicating that they are relatively more receptive to online consultation as a form of Internet healthcare. Among them, 77.78% of the respondents under 18 years old considered that the hospitals and doctors of the App consultation were not professional sufficient, which was the main reason why their age group was less likely to use the APP consultation.

In-depth interviews with users under the age of 18 revealed that they were mostly concerned with physical or psychological symptoms during puberty, and were less willing to communicate and consult on private issues through online consultation, believing that there were loopholes in the protection of privacy on the Internet and the possibility of being spied on and discussed by others. Some of the respondents who had experienced the App consultation found that hospitals and doctors' responses were not targeted enough, and there were even interdisciplinary consultations.

4 Conclusion and Recommendations

4.1 Conclusion

1. The COVID-19 Pandemic promoted the use of medical APPs for consultation. 62 of the respondents, or 86.11% of those who had never used a medical App, said that they were already using or planning to have access to a medical App for consultation services. Given the normalization of epidemic prevention and control, safety and health, avoiding cross infection and mitigating risk are the main factors in their consideration of this service.
2. Based on the premise that the COVID-19 Pandemic is likely to co-exist with humans in the long term, the functional design of medical Apps needs to echo such infectious diseases. With 157 people, or 94.01% of the total number of respondents, believing that there is a requirement to improve or further develop the consultation module or other functional modules related to the COVID-19 pandemic, so that people can better access the corresponding knowledge, enter the team to compare themselves and report the information, which will play an essential function in the prevention and control of the epidemic.
3. Due to the pandemic, the unique advantages of Internet hospitals and Internet diagnosis and treatment were highlighted, the public expects to access this service for breaking the geographical limitation of medical resources and to receive more professional and stable medical services. 52.17% of respondents indicated the highest concern for professional health information, and 74.77% of users were willing to pay for online consultation on the premise of obtaining professional and targeted treatment.

4. As mobile healthcare enters a phase of rapid commercial development, more potential problems have been exposed in the industry. 98 of the respondents, or 58.68%, expressed concerns about the reliability of the online consultation. The potential for health literacy gaps between doctors and patients, the lack of industry standards, inadequate regulatory constraints, and the inability to ensure the accuracy of data security may all affect the stability and safety of the mobile healthcare service model, ultimately creating problems with the quality and safety of healthcare services.

4.2 Recommendations

1. For the psychological and operational requirements of people in different age groups and consultation groups, the UI design of the medical App should be emphasized. The layout of the interface should be simple as possible and adopt a color palette that matches the subjective impression of the medical product, such as white, green, blue, and other light colors for a clean, comfortable and relaxed feel. The interface could be added with voice control, font enlargement, and other functional buttons to improve the operating experience for special groups such as those with poor eyesight and older people.

   Since there are various types of online consultations nowadays, the homepage content should not be placed excessively, and the essential content should be unified in one module to simplify the complexity of the product's functional structure. For example, all different types of consultation functions could be placed on one functional page for consultation and then distinguished into different areas in the secondary pages.
2. Special function module for COVID-19 should be implemented. Modules could be differentiated in terms of colors, icons, and fonts to facilitate user recognition. The consultation process should be rationalized to distinguish COVID-19 from other conditions and to make the consultation more targeted.
3. The professionalism of the medical App should be focused on improving the efficiency of the consultation. First, rigorous quality audits should be conducted for doctors and hospitals entering the platform. Secondly, the user's anamnesis and results should be easily and quickly identified, recorded, and accessed in the App to more accurately determine the user's types of diseases. Thirdly, a discussion platform for complex cases should be provided for remote sharing, backup, and consultation of cases, sharing of files, technology and information, and other functions, thereby enhancing the professionalism of APP consultations and offering better quality healthcare services.
4. The platform should be strengthened to protect patient privacy and personalization settings should be added for the user. The functions could be designed to meet the user's requirements for accessing their own medical information and doctor's updates, while in this way the user could choose the topics they are concerned about, and the APP would provide relevant health literacy knowledge and information targeted to enhance and develop the user's health literacy and behaviors. Meanwhile, information relating to both the users and members of their family who have difficulties operating the App should be able to be added to help them complete the App consultation. The quality and security of medical services should be improved to

ensure the privacy of users' consultation information, which will reduce the bounce rate and boost user stickiness.

The limitation of this paper lies in the fact that in researching the user experience of medical App consultation, the impacts are considered more from the patient's perspective and the other side of the APP connection, i.e., doctors and hospitals, is not researched as a user. Therefore, the relationship between the influence of doctors and hospitals on user experience is yet to be further explored and verified in subsequent researches.

References

1. Norman, Miller, J., et al.: What You see, some of what's in the future, and how we go about doing it: HI at Apple computer. In: Conference Companion on Human Factors in Computing Systems (1995)
2. Daniel, L.: Understanding user experience. Web Tech. **5**(8), 42–43 (2000)
3. Ergonomics of Human-system Interaction—Part 210: Human-Centred Design for Interactive Systems, 15 March 2010. http://www.doc88.com/p-9902136715348.html
4. Hassenzahl, M.: User experience-a research agenda. Behav. Inf. Technol. **25**(2), 91–97 (2006)
5. Liu, R., Zhang, Y., Yu, J.: Research on user experience of mobile medical apps from the perspective of health literacy. J. Mod. Inf. **10**(40), 62–71 (2020)
6. Yao, Z.: Research on user experience of sports dance APP. Sichuan Sports Sci. **5**(39), 117–122 (2020)
7. Gudigantala, N., Song, J., Jones, D.: User satisfaction with web-based DSS: the role of cognitive antecedents. Int. J. Inf. Manag. **31**(4), 0–338 (2011)
8. Baldwin, J.L., Singh, H., Sittig, D.F., et al.: Patient portals and health apps: pitfalls, promises, and what one might learn from the other. Healthcare **5**(3), 81 (2016)

Research on the Attractive Factors and Design of Cultural Derivative Commodities Under Cultural Sustainability

Kuo-Liang Huang[1(✉)], Na Xu[1], Hsuan Lin[2], and Jinchen Jiang[1]

[1] Department of Industrial Design, Design Academy, Sichuan Fine Arts Institute, Chongqing, China
{shashi,2018028,jiangjinchen}@scfai.edu.cn

[2] Department of Product Design, Tainan University of Technology, Tainan, Taiwan (R.O.C.)
te0038@mail.tut.edu.tw

Abstract. "Cultural commodity" refers to playing a positive role in the background of cultural sustainability. Nevertheless, cultural commodities have significantly different characteristics from general commodities. It faces multiple challenges regarding design and development. This study attempts to extract attractive factors from cultural commodities that people highly appreciate and analyze their correlations with design attributes. The research design integrates qualitative and quantitative methods. Firstly, derivative commodities of cultural relics that represent the National Palace Museum of China were chosen as research objects. Secondly, attractive factors and commodity attribute features were extracted through the evaluation grid method (EGM) in Miryoku Engineering. Finally, correlations between the "attractive factors" and "design attributes" were analyzed through the "Quantitative Theory I." This study concluded an "evaluation grid diagram of cultural commodity" and the "design rule" of cultural commodities.

Keywords: Cultural heritage · Cultural commodity · Kansei Engineering · Miryoku Engineering · Product design · Cultural sustainability · Design evaluation · Attractive factors

1 Introduction

People began to think and pursue higher spiritual and cultural significance and cultural identity after they gained satisfying conditions of material life following changes in economic patterns and social structures in the 21st Century [1]. However, the traditional cultures of countries across the world are challenged by cultural homogenization in the current globalization trend. Consequently, sustainability's relation to inheritance and

C. Stephanidis et al. (Eds.): HCII 2021, LNCS 13094, pp. 522–538, 2021.
https://doi.org/10.1007/978-3-030-90238-4_37

the development of traditional cultures has become increasingly important [2–4]. Cultural commodities are a new form of cultural factors in cultural relics that are designed according to the modern lifestyle. A cultural commodity adds values and is designed for the pursuit of spiritual satisfaction [5]. Therefore, commodities rich in cultural connotation can awaken people's memory of figures, events, and things in a certain time and space, thus replenishing their souls and bringing increased pleasure to their life [6, 7]. Then, what are the differences between cultural commodities and general commodities? Throsby [6] pointed out in Cultural Economics that cultural commodities' characteristics not only vest immortal and cultural significance to their internal significance and symbol characteristics. They also can generate significant economic benefits. Therefore, a cultural commodity can promote social and economic development and improve the influences of local cultures in the global market from the perspective of cultural sustainability [8–10]. Therefore, the key challenge in the design and development of cultural commodities is how to transform cultural connotations and vest features of cultural commodities cleverly and achieve resonance with consumers.

The challenge that designers encounter among cultural connotations, customer cognitions, and product features is similar to the challenge of the complicated and unpredictable black box [11]. Customers' "expectation" of the interesting cultural commodities and "specific conditions and features" of cultural commodities can be used as references for designers to make correct and decent cultural commodities designs. From this perspective, existing relevant studies mainly focus on general commodities. Therefore, further support of the design and development of derivative commodities under cultural sustainability is needed [12].

To address these problems, this study attempted to establish the "attractive factor" of cultural commodities to customers and used it to recognise essential "features" of commodity design and relations between the "attractive factor" and these "features." This study focuses on two problems: (1) What is the "attractive factor" of the cultural commodity that customers are interested in? (2) How is the "attractive factor" related to the "design conditions and features" of the cultural commodity? Based on the study of these two problems, a design rule regarding the relationship between the attractive factor of cultural commodities and relevant design elements was proposed. This was conducive to ensuring the design idea's effectiveness during the design and development of cultural commodities.

2 Literature Review

2.1 Culture and Cultural Commodity

Culture is a product of the civilization process of humankind [13]. Differences in the definition of culture by scholars from different disciplines are manifested using the concept [2]. Williams [14] proposed three generalized interpretations of "culture." First, culture refers to a universal process of knowledge, spirit and aesthetic development. Second, culture refers to a nation, time, or specific lifestyle of a group. Third, culture is the finished product and practices of wisdom, especially referring to art activities. Additionally, Yang Yufu [15] argued that culture could be divided into three levels: (1) culture at the metaphysical level (thinking activity and language), (2) culture at the form level (communication and interaction system), and (3) culture at the physical level (tools or specific forms that people use). Cultures in all forms possess considerable cultural, economic, artistic, and educational values [16]. It is the irreplaceable source of people's life and inspiration [17]. "Cultural heritage" with the concept of cultural property includes not only the perspective of art and historical value, but also the interactive ability of the cultural carrying of objects and memory in its cultural value [17].

A cultural commodity is a creative production activity, through cultural sustainable economic activities [18], with intellectual properties and some social significance [8]. The United Nations Educational, Scientific, and Cultural Organization (UNESCO) notes that cultural derivative commodity has both economic and cultural properties. It becomes the carrier of cultural features, values, and significance through expression and inheritance, arousing resonance of customers, and passing on cultures [17]. Therefore, "cultural commodity" is an artistic and practical life commodity that is designed from cultural topics based on the creative added value of design. This is done to strengthen the attraction and emotional connection of the culture and thereby promote cultural sustainability.

2.2 Design and Image of Cultural Commodity

People and the use of commodities have always been influenced by culture, and design drives cultural changes, transforming each other [13]. The "Design of a cultural commodity" is to transform "culture" into "creativity" and manifest it in a specific commodity with the added value of "design." Design can be divided into three stages: (1) the extracting style and characteristics of a culture, (2) forming a conceptual mode of the design, and (3) finishing the design of a cultural commodity [19]. Leong and Clark [20] highlight the importance of studying the design structure of a cultural commodity, including its shape, psychological ideology, behavior, rites, and customs. The researchers proposed a design philosophy for cultural integration. Based on this design structure, Lin [6] further proposed three levels of cultural commodities, including external, middle, and internal levels. The design of a cultural commodity is a process that reflects or checks cultural features before redefining cultural characteristics. This is done to create a new commodity that can adapt to modern society and meet customers' cultural and aesthetic needs. Based on the opinion of product semantics, Krippendorff [21] once proposed that an artifact is composed of two major levels, including its "form" and "meaning." Users understand the usage of an artifact through its form and meaning. Therefore, a designer should pay attention to and consider users' context in consuming products to form connections between designers and users. This communication mode has great enlightenment for designers. The sustainability in the green design and cultural sustainability of a cultural commodity cannot be ignored [22] (Fig. 1).

An image represents some object or event, and it transfers relevant information. An image has distinctive sensation features, including the cognitive evaluation and affective evaluation [23]. An image of a commodity refers to the distinctive sensation features that customers developed from the delivering information of the commodity. Consequently, it can arouse inner resonance and the consumption motivation of customers [24]. However, a value judgment of customers' hobbies covers perception, psychological decision-making, social background, and art evaluation. It is, therefore, a fuzzy and

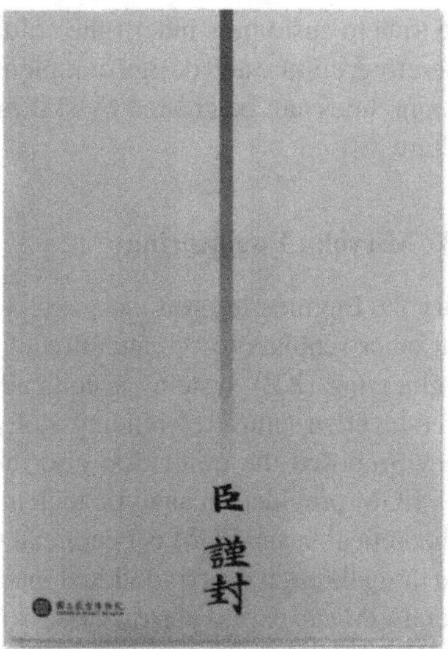

Fig. 1. File Folder Confidential Letter: Left is front; right is reverse Source: National Palace Museum Shop

complicated concept [25]. For general commodities, Jiao, Zhang, and Helander [26] believed that it was absolutely an essential means to use the emotion or sensation of customers for the different designs of commodities. This was done to search design elements corresponding to the commodities' image and transform the sensation demands (attractive factor) of the customers into the design features of the target commodities. Clausing [27] pointed out that the mapping problem between customer demands and product features was a primary topic. Further, the customer-oriented design philosophy could assure that the designed commodities would resonate with the desired functions and images of customers, thus improving the success rate of product development.

Beyond the requirements of general commodities, cultural commodities have special cultural connotations, symbols, and recognitions. They provide people with a different consumption option (except for daily necessities) and become a special symbol of an individual's lifestyle and personal style. During the design of cultural commodities, it is necessary to explore and understand the background and knowledge of the culture. We can, then, establish a design development and new design direction of commodities by extracting and transforming cultural connotations [20]. Moreover, the design of cultural commodities should be combined with the concerned psychology and perceived stimuli of customers. This will ensure that users develop emotional resonance through the attractive factor of the commodity [28]. A successful cultural commodity must have a good, interesting design that can deliver cultural contents in an easy, cognitive way. It must also represent the background of the origin [29]. Therefore, cultural commodities can be recognised in global marketing by combining successful cultural and commodity design, thus enabling them to further spread their values [6]. Based on the above studies, cultural commodities' design and development require not only to inherit cultural assets and aesthetic forms with cultural characteristics. It also requires integration with daily life to help customers inherit the cultural features, thus creating a delicate life culture. Therefore, culture and design complement each other. The irreplaceable value of cultural commodities can be created by skillfully combining and transforming the images of a culture.

2.3 Miryoku Engineering

Miryoku Engineering was a research technology proposed by Masato Ujigawa and several other scholars to "create attractive products and space." It is a link of the Kansei Engineering (KE) system. Specifically, Junichiro Sanui and Masao Inui improved a three-layer structure referencing Kelly's [30] *The Psychology of Personal Constructs.* They proposed the evaluation grid method (EGM) for Miryoku Engineering [25, 31, 32]. EGM provides an analytic technique to attractive factors of commodities that have a theoretical basis. EGM is beneficial to mastering people's psychological cognition on one thing through deep individual interviews with highly involved ethnic groups. Moreover, EGM can export abstract assessment items and objective specific conditions. They can also export features, organise perceptible evaluation constructs of interviewees to the commodity, and analyze the attractive commodity factor to customers [33, 34]. EGM has been widely applied in relevant studies.

This study analyzed the relations of the attractive factor with objective specific conditions and features. That is, the linear or nonlinear perceptual design mode between perceptual images and design elements [35, 36] is often applied in Miryoku Engineering. Quantitative Theory Type I was developed by a Japanese statistician Chikio Hayashi and is a categorical multiple regression analysis method [35]. It is equivalent to using categorical parameters as dummy variables to test the influence intensity of explanatory variables on criterion variables [36]. Each item is composed of several categories. The constructed regression formula can be used to predict the variability of data and events. The function is expressed as:

$$y = \Sigma \beta x + e \tag{1}$$

where y is the predicted value, β is the weight of a category, x is the items, and e is a random number.

The analysis results and reference degree of the above function can be referred to by the following numerical values. (1) Partial correlation coefficient: the higher the numerical value of a partial correlation coefficient, the stronger the impact of categories on target items. (2) Multiple correlation coefficient (R): the higher numerical value of R refers to the higher reliability of analysis results. Among them, R value 0.40–0.69 is a strong correlation; 0.70–1.00 is a highly correlated [37].

3 Method

3.1 Research Design

According to the research objective and problems, this study combined qualitative and quantitative analyses based on the methodology of Miryoku Engineering. Firstly, the attractive factors of cultural derivative commodities to individual respondents were constructed through EGM. The relations of the attractive factors with objective specific design conditions and features were analyzed. Subsequently, assessments of different respondents were reviewed by an affinity diagram, and a common EGM which were combined. Thirdly, the Involvement Degree Scale and Association Scale of Attractive Factor were designed, and relevant data were collected through a questionnaire survey. Finally, the relations between attractive factors and design elements were investigated through the Quantitative Theory Type I.

3.2 Subjects

Research objects were chosen as Lifestyles on the official marketing websites (National Palace Museum Shop, www.npmshops.com) of cultural derivative commodities of the National Palace Museum. This institution has inherited valuable Chinese culture for thousands of years, possesses nearly 700,000 cultural relics, and receives more than 6.14 million of visitors every year.

3.3 Materials

This study assured the integrity and representativeness of the research objects and attractive cultural commodities to customers. The top three categories of cultural derivative commodities concerning marketing volumes in the past twelve months were selected as research objects from thirteen categories. These categories consisted of home décor, kitchen & tableware, tea sets, mugs/thermos, travels, personal accessories, computer accessories, puzzles/toys/crafts, keychains, fans, stamps, food/drink, and beauty & personal care. Finally, a total of 39 cultural commodities were chosen. Later, the pictures, names, brief descriptions, and specifications of the preliminarily selected cultural commodities were collected from the official website of the National Palace Museum Shop. Thirty-nine pieces of 10 cm × 16 cm of physical pictures were created (Fig. 2.).

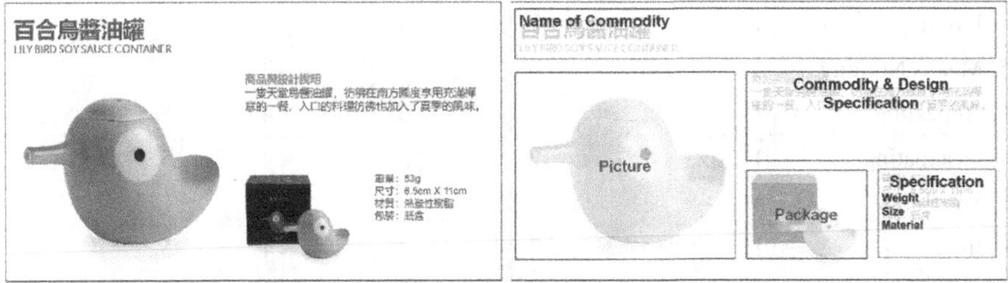

Fig. 2. Picture of cultural commodity (left) and layout of relevant information (right)

3.4 Participants

The highly-involved participants were recruited according to the attractive factors of cultural commodities. They included three cultural popularization workers of the museum, three designers of cultural commodities, three salesmen of cultural commodities, and three fans of cultural commodities. A total of twelve participants were selected, and the EGM extracted attractive factors. Respondents were recruited randomly from visitors in the exhibition and marketing regions of cultural derivative commodities in the National Palace Museum. They analyzed the relation between attractive factors and design elements.

3.5 Instruments

This study confirmed the recruited respondents' cognitive ability, who analysed the relationship between attractive factors and design elements with cultural commodities. To do so, the revised personal involvement inventory (RPII) (Zaichkowsky 1994) was applied to measure the involvement degree. The RPII had 10 questions and was measured with a 5-point Likert scale. The total score was 50, and only respondents who achieved higher than 35 points were included for the measurement of the Association Scale of Attractive Factor. The Association Scale of Attractive Factor was compiled according to common EGM results. Please refer to the appendix for details (Table 1).

Table 1. RPII Scale

What's you opinion on cultural commodities?	Strongly disagree	Disagree	Moderate	Agree	Strongly agree
1. Important	1	2	3	4	5
2. Boring	5	4	3	2	1
3. Relevant	1	2	3	4	5
4. Exciting	1	2	3	4	5
5. Means Nothing	5	4	3	2	1
6. Appealing	1	2	3	4	5
7. Fascinating	1	2	3	4	5
8. Worthless	5	4	3	2	1
9. Involving	1	2	3	4	5
10. Not Needed	5	4	3	2	1

Data source: Zaichkowsky, J.L. The personal involvement inventory: Reduction, revision, and application to advertising. Journal of advertising 1994, 23, 59–70.

3.6 Data Collection Procedure

The data acquisition in the present study was divided into two parts. The process is shown in Fig. 3. (1) Extraction of attractive factor: an individual EGM was performed on all 12 respondents through individual interviews. Next, three designers of cultural commodities were invited to review the individual EGM results according to an affinity diagram and integrate them into a common EGM. (2) Analyzing the relationship between attractive factors and design elements: firstly, a scale was designed and manufactured. Secondly, respondents were collected randomly from visitors in the exhibition and marketing areas of cultural commodities in the National Palace Museum. They were asked to fill in the RPII. Respondents who gained a score of over 35 were guided to the next stage. Meanwhile, the rest of the respondents finished tasks in the questionnaire survey. Finally, data were analyzed based on Quantitative Theory Type I.

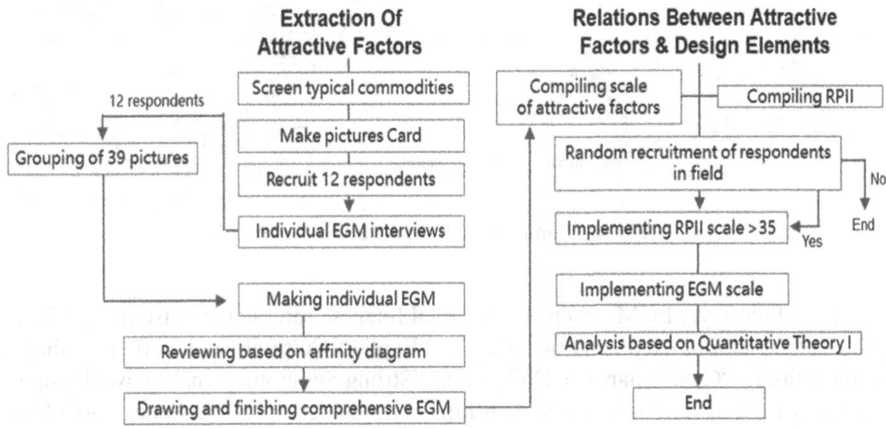

Fig. 3. Workflow of the questionnaire survey

Materials and Methods should be described with sufficient details to allow others to replicate and build on published results. Please note that publication of your manuscript implicates that you must make all materials, data, computer code, and protocols associated with the publication available to readers. Please disclose at the submission stage any restrictions on the availability of materials or information. New methods and protocols should be described in detail while well-established methods can be briefly described and appropriately cited. Research manuscripts reporting large datasets that are deposited in a publicly available database should specify where the data have been deposited and provide the relevant accession numbers. If the accession numbers have not yet been obtained at the time of submission, please state that they will be provided during review. They must be provided prior to publication. Interventionary studies involving animals or humans, and other studies require ethical approval must list the authority that provided approval and the corresponding ethical approval code.

4 Results

4.1 Construction of EGM for Cultural Commodities

Individual EGM was constructed to 12 respondents and reviewed according to the affinity diagram. On this basis, the final common EGM was formed Fig. 4.

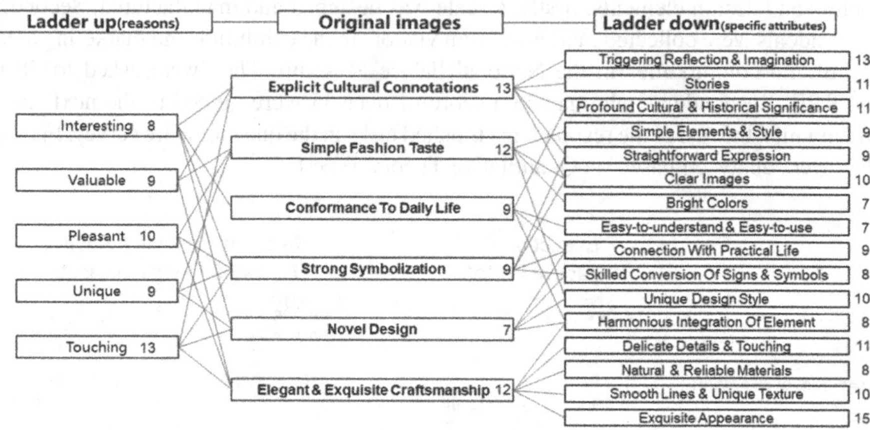

Fig. 4. Common EGM of attractive factors

In the three-layer EGM structure, "Original Images" refer to the outstanding "Properties" of the commodity. This includes the "Explicit Cultural Connotation," "Simple Fashion Taste," "Conformance to Daily Life," "Strong Symbolization," "Novel Design," and "Elegant and Exquisite Craftsmanship." The ladder up is the upper part of the structure, which represents the "Abstract Reasons" for the outstanding "Properties" of cultural commodities. There are five reasons for this, consisting of "Interesting," "Valuable," "Pleasant," "Unique," and "Touching." Ladder down refers to the lower part of the structure, which represents the main "Specific Attributes" for the outstanding "Properties" of cultural commodities. There are 16 specific attributes. Besides, the numbers behind the boxes refer to the number of subsets in the affinity diagram, whereas straight lines imply a correlation between two boxes.

4.2 Relations Between Attractive Factors and Design Elements

To disclose the relations between the abstract reasons in the attractive factors of cultural commodities and design elements, the official investigation began after the primary assessment of respondents by the RPII. Later, data were analysed via the Quantitative Theory Type I. A total of 261 participants were investigated, and 107 respondents became valid samples according to the RPII. Among the valid samples, there were 41 males (38%) and 66 females (62%). Concerning age groups, there were 3 respondents younger than 20 years old (2.8%), 35 respondents between 21–30 years old (32.7%), 25 respondents between 31–40 years old (23.4%), 22 respondents between 41–50 years old (20.5%),

15 respondents between 51–60 years old (14%), and 7 respondents over 61 years old (6.5%).

In the EGM of cultural commodities, "Interesting" in the ladder up covers four original images, including "Explicit Cultural Connotation," "Conformance to Daily Life," "Strong Symbolization," and "Novel Design." According to the analysis results in Table 2, the determination coefficient was R2 = 0.780. This demonstrates the strong correlation between "Interesting" and design elements.

Table 2. Relations between "Interesting" and design elements

Items	Partial Correlation Coefficients	Category	Categories Score
Novel Design	0.625	Harmonious integration of elements	0.682
		Unique design style	0.050
		Connection with practical life	−0.018
		Skilled conversion of signs and symbols	−0.292
Strong Symbolization	0.507	Straightforward expression	0.206
		Connection with practical life	0.173
		Skilled conversion of signs and symbols	−0.215
		Clear images	−0.294
Conformance to Daily Life	0.460	Connection with practical life	0.188
		Profound cultural and historical significance	−0.099
		Harmonious integration of elements	−0.120
		Easy-to-understand and easy-to-use	−0.304
Explicit Cultural Connotation	0.185	Straightforward expression	0.297
		Clear images	0.065
		Stories	0.007
		Profound cultural and historical significance	−0.011

(continued)

Table 2. (*continued*)

Items	Partial Correlation Coefficients	Category	Categories Score
		Triggering reflection and imagination	−0.063
C = 6.79	R = 0.883	$R^2 = 0.780$	

Note. N = 107

Among the four original images, the "Novel Design" is the primary influencing factor of "Interesting," followed by "Strong Symbolization," "Conformance to Daily Life," and "Explicit Cultural Connotation" successively. This is according to the partial correlation coefficients. In the individual items, the value of categories score (regression coefficient) reflects the specific condition and feature, making the biggest contribution to "Interesting" in each item. Based on the results, we recognise that "Harmonious integration of elements" was the most prominent one in "Novel Design." Meanwhile, "Skilled conversion of signs and symbols" shows the highest negative value. This reflects that, while "Skilled conversion of signs and symbols" is one of the important components in "Novel Design," it is useless to "Interesting." The rest can be done in the same way. The order of the columns in Table 2 re-ranks the results after our analysis, which is for better reading. Table 3, 4, 5 and Table 6 are in the same processing in the following text.

Table 3. Relationship between "Valuable" and design elements

Items	Partial Correlation Coefficients	Category	Categories Score
Elegant and Exquisite Craftsmanship	0.60	Delicate details and touching	0.528
		Natural and reliable materials	0.258
		Smooth lines and unique textures	−0.010
		Exquisite appearance	−0.017
		Harmonious integration of elements	−0.305
Simple fashion taste	0.42	Simple elements and styles	0.146
		Unique design style	0.086
		Bright colors	0.052
		Clear images	−0.274

(*continued*)

Table 3. (*continued*)

Items	Partial Correlation Coefficients	Category	Categories Score
Explicit Cultural Connotations	0.30	Clear images	0.184
		Triggering reflection and imagination	0.182
		Straightforward expression	−0.076
		Profound cultural and historical significance	−0.079
		Stories	−0.178
C = 6.084	R = 0.776	$R^2 = 0.601$	

Note. N = 107

Table 4. Relations between "Pleasant" and design elements

Items	Partial Correlation Coefficients	Category	Categories Score
Elegant and Exquisite Craftsmanship	0.625	Natural and reliable materials	0.173
		Exquisite appearance	0.122
		Harmonious integration of elements	−0.029
		Delicate details and touching	−0.138
		Smooth lines and unique texture	−0.248
Strong symbolization	0.279	Skilled conversion of signs and symbols	0.076
		Connection with practical life	0.000
		Straightforward expression	−0.022
		Clear images	−0.041
Simple fashion taste	0.216	Simple elements and styles	0.073
		Unique design style	0.040
		Bright colors	0.030
		Clear images	−0.020
C = 5.536	R = 0.822	$R^2 = 0.675$	

Note. N = 107

Table 5. Relationship between "Unique" and design elements

Items	Partial Correlation Coefficients	Category	Categories Score
Strong symbolization	0.791	Skilled conversion of signs and symbols	0.245
		Clear images	−0.011
		Connection with practical life	−0.094
		Straightforward expression	−0.150
Simple fashion taste	0.740	Simple elements and style	0.270
		Unique design style	0.053
		Clear images	−0.071
		Bright colors	−0.053
Novel design	0.534	Unique design style	0.059
		Harmonious integration of elements	0.033
		Skilled conversion of signs and symbols	−0.076
		Connection with practical life	−0.101
C = 5.725	R = 0.858	$R^2 = 0.737$	

Note. N = 107

Table 6. Relationship between "Touching" and design elements

Items	Partial Correlation Coefficients	Category	Categories Score
Novel design	0.566	Skilled conversion of signs and symbols	0.151
		Harmonious integration of elements	−0.006

(*continued*)

Table 6. (*continued*)

Items	Partial Correlation Coefficients	Category	Categories Score
		Unique design style	−0.043
		Connection with practical life	−0.095
Explicit cultural connotations	0.560	Stories	0.038
		Straightforward expression	−0.012
		Profound cultural and historical significance	−0.098
		Triggering reflection and imagination	−0.098
		Clear images	−0.222
Elegant and Exquisite Craftsmanship	0.418	Delicate details and touching	0.059
		Exquisite appearance	0.019
		Harmonious integration of elements	0.016
		Natural and reliable materials	0.005
		Smooth lines and unique texture	−0.098
Strong symbolization	0.390	Connection with practical life	0.031
		Clear images	0.017
		straightforward expression	−0.191
		Skilled conversion of signs and symbols	−0.191
C = 6.119	R = 0.775	$R^2 = 0.601$	

Note. N = 107

5 Conclusions and Suggestions

This study constructs an EGM of cultural derivative commodities in the National Palace Museum based on attractive factors (Fig. 4), including six original images, five ladder ups (reasons) and 16 ladder downs (specific attributes) of the original "Properties." The relations between attractive factors and design elements of cultural commodities are disclosed on this basis, and a deduction model of images of cultural commodities is constructed effectively. Research conclusions can provide substantial benefits for inheriting and developing traditional cultures through cultural commodities under cultural sustainability in facing the cultural homogenization challenge.

As previously mentioned, a cultural commodity has significantly different essences from a general commodity. Therefore, this study focuses on cultural commodities. Based on relevant academic studies, the attractive factors of cultural commodities to customers are extracted, and their relations with design elements are determined. This study fills the holes in relevant research. Practicing, designing, and developing cultural commodities is also different from that of general commodities. The constructed EGM for cultural commodities and the relations between attractive factors and specific conditions and features can provide references for the design and development of cultural commodities. Five tables (Table 3, 4, 5 and Table 6) can be used as the design rule of cultural commodities. They can also help design ideas to master the expectation of customers to cultural commodities effectively. Finally, we believe that cultural commodities are different from general commodity designs. They not only carry the cultural connotations of the past, but also echo and embody current life; and cultural commodity designers act as indirect cultural communicators, shouldering the task of transmitting and spreading cultural values.

Although this study has some theoretical and practical significance, it is limited with tangible derivative commodities from Chinese culture but excludes other intangible cultures. Future studies can further deepen the design of cultural commodities based on existing research conclusions.

Acknowledgments. Supported by the Science and Technology Research Program of Chongqing Municipal Education Commission (Grant No. KJZD-K201901001, and Grant No. 20SKGH148).

References

1. Hełdak, M., Kurt Konakoğlu, S.S., Kurtyka-Marcak, I., Raszka, B., Kurdoğlu, Ç.B.: Visitors' perceptions towards traditional and regional products in Trabzon (Turkey) and Podhale (Poland). Sustainability **12**, 2362 (2020)
2. Soini, K., Dessein, J.: Culture-sustainability relation: towards a conceptual framework. Sustainability **8**, 167 (2016)
3. Schoormans, J.: Routledge handbook of sustainable product design. Des. J. **21**, 553–557 (2018)
4. Grevstad-Nordbrock, T., Vojnovic, I.: Heritage-fueled gentrification: a cautionary tale from Chicago. J. Cult. Herit. **38**, 261–270 (2019)
5. Lin, R.-T.: Design of creative cultural products: from perspectives of perception technology, humanization design and cultural creativity. Humanit. Soc. Sci. Newsl. Q. **11**, 32–42 (2009)

6. Lin, R.-T.: Transforming Taiwan aboriginal cultural features into modern product design: a case study of a cross-cultural product design model. Int. J. Des. **1**, 45–53 (2007)
7. Dalmas, L., Geronimi, V., Noël, J.-F., Tsang King Sang, J.: Economic evaluation of urban heritage: an inclusive approach under a sustainability perspective. J. Cult. Herit. **16**, 681–687 (2015)
8. Throsby, D.: The Economics of Cultural Policy. Cambridge University Press, Cambridge (2010)
9. Yair, K., Press, M., Tomes, A.: Crafting competitive advantage: crafts knowledge as a strategic resource. Des. Stud. **22**, 377–394 (2001)
10. Tu, H.-M.: The attractiveness of adaptive heritage reuse: a theoretical framework. Sustainability **12**, 2372 (2020)
11. De Souza, M., Dejean, P.: Cultures and product relationship in a globalized environment. In: Anais do P&D Design, pp. 513–522 (1998)
12. Chai, C., Shen, D., Bao, D., Sun, L.: Cultural product design with the doctrine of the mean in confucian philosophy. Des. J. **21**, 371–393 (2018)
13. Çakmakçıoğlu, B.A.: Effect of digital age on the transmission of cultural values in product design. Des. J. **20**, S3824–S3836 (2017)
14. Williams, R.: Keywords: A Vocabulary of Culture and Society. Oxford University Press, Oxford (2014)
15. Yufu, Y.: Cultural basis of design: design, symbol and communication. Asiapac, Taibei (1998)
16. Tang, C., Zheng, Q., Ng, P.: A study on the coordinative green development of tourist experience and commercialization of tourism at cultural heritage sites. Sustainability **11**, 4732 (2019)
17. Vecco, M.: A definition of cultural heritage: From the tangible to the intangible. J. Cult. Herit. **11**, 321–324 (2010)
18. Moubarak, L.M., Qassem, E.W.: Creative eco crafts and sustainability of interior design: schools in Aswan, Egypt as a case study. Des. J. **21**, 835–854 (2018)
19. Lin, R.-T.: Essence and research on cultural creative industry. J. Des. **16** (2011)
20. Leong, B.D., Clark, H.: Culture-based knowledge towards new design thinking and practice—a dialogue. Des. Issues **19**, 48–58 (2003)
21. Krippendorff, K.: On the essential contexts of artifacts or on the proposition that "design is making sense (of things)." Des. Issues **5**, 9–39 (1989)
22. Bai, Z., Mu, L., Lin, H.-C.: Green product design based on the BioTRIZ multi-contradiction resolution method. Sustainability **12**, 4276 (2020)
23. Gartner, W.C.: Image formation process. J. Travel Tour. Mark. **2**, 191–216 (1994)
24. Chuang, M.C., Chang, C.C., Hsu, S.H.: Perceptual factors underlying user preferences toward product form of mobile phones. Int. J. Ind. Ergon. **27**, 247–258 (2001)
25. Ishihara, S., Ishihara, K., Nagamachi, M., Matsubara, Y.: An automatic builder for a Kansei Engineering expert system using self-organizing neural networks. Int. J. Ind. Ergon. **15**, 13–24 (1995)
26. Jiao, J., Zhang, Y., Helander, M.: A Kansei mining system for affective design. Expert Syst. Appl. **30**, 658–673 (2006)
27. Clausing, D.: Total Quality Development: A Step-by-step Guide to World Class Concurrent Engineering. ASME Press, New York (1994)
28. Desmet, P.: Three levels of product emotion. In: Proceedings of the International Conference on Kansei Engineering and Emotion Research, pp. 236–246 (2010)
29. Dal Palù, D., Lerma, B., Bozzola, M., De Giorgi, C.: Merchandising as a strategic tool to enhance and spread intangible values of cultural resources. Sustainability **10**, 2122 (2018)
30. Kelly, G.: The Psychology of Personal Constructs: A Theory of Personality, vol. 1. W.W. Norton and Company, New York (1995)

31. Minyuan, M.: Discussion on perception of new product development in Japan. J. Ind. Mater. **280**, 160–172 (2010)
32. Nagamachi, M.: Kansei engineering as a powerful consumer-oriented technology for product development. Appl. Ergon. **33**, 289–294 (2002)
33. Sanui, J., Tolman, F., Kalay, Y.E., Shou-qian, S., Yun-he, P.: Visualization of users' requirements: Introduction of the Evaluation Grid Method. In: Proceedings of the 3rd Design & Decision Support Systems in Architecture & Urban Planning Conference, pp. 365–374
34. Ujigawa, M.: The evolution of preference-based design. Res. Dev. Inst. **46**, 1–10 (2000)
35. Hayashi, C.: On the quantification of qualitative data from the mathematico-statistical point of view. Ann. Inst. Stat. Math. **2**, 35–47 (1950)
36. Tsuchiya, T., Maeda, T., Matsubara, Y., Nagamachi, M.: A fuzzy rule induction method using genetic algorithm. Int. J. Ind. Ergon. **18**, 135–145 (1996)
37. Sugiyama, K., Nouel, K.: The Basic for Survey and Analysis by Excel. Kaibundo Publishing, Tokyo (1996)

Cross-Cultural Design in Consumer Vehicles to Improve Safety: A Systematic Literature Review

Priyanka Koratpallikar(✉) and Vincent G. Duffy

Purdue University, West Lafayette, IN 47906, USA
pkoratpa@purdue.edu

Abstract. In our interconnected global culture, now more than ever products need to be designed with a multicultural audience in mind. One of the most global products is consumer vehicles, such as cars, minivans, SUVs, etc. used for personal transport, such as for commuting, traveling, or daily activities. Consumer vehicles are becoming more and more of "computers on wheels" rather than merely a transportation device. Therefore, as the understanding of human perception while driving increases, there will be the additional need to study how the human-computer interaction affects driving cross-culturally and how automakers can take these into consideration when designing safe to use interfaces. As access to consumer vehicles increases exponentially around the world, it will be important to ensure that vehicles meet the anthropometric and cultural needs of consumers around the world. (Rhim 2020) This paper demonstrates a systematic literature review of articles from credible, peer-reviewed journals on the topic of Cross-Cultural design in the automotive industry, especially for consumer vehicles. The methodology for this literature review began with retrieving relevant articles from the reliable databases of Google Scholar, Springerlink, and Web of Science. The keywords of "Cross-Cultural design", "vehicles", and "safety" were used in several relationship analyses. The analyses were conducted by utilizing bibliometric Softwares including CiteSpace, VOSViewer, and Harzing. A takeaway from this review is that vehicle usage is trending upward on a global scale and that Cross-Cultural design will need to be incorporated to ensure safe usage by a global, multicultural society.

Keywords: Cross-cultural design · Vehicles · Safety

1 Introduction and Background

1.1 Cross-Cultural Design and Vehicles

Cross-cultural design is designing products and services to be compatible with any given local culture to ensure high usability and satisfying user experience (Plocher 2012). Since global users cross all international boundaries, it is imperative that products intended to be used on a global scale take into account the variability of local cultures. The framework for Cross-Cultural psychology was determined by Hoftstede,

C. Stephanidis et al. (Eds.): HCII 2021, LNCS 13094, pp. 539–553, 2021.
https://doi.org/10.1007/978-3-030-90238-4_38

who outlined five dimensions for studying cultures: Power Distance (how much less powerful members of society are willing to accept that power is distributed unequally in their society), Uncertainty Avoidance (extent by which members are threatened by uncertainty), Individualism-Collectivism (a spectrum on whether the individuals in the society identify as individuals or as members of a group), Masculinity-Femininity (a measure of how distinct gender roles are defined in the society), and Long-Term vs Short-Term Orientation (how future oriented the society is). (Plocher 2012).

Background. The history of consumer vehicles can be traced back to Henry Ford's Model T in 1908. *(Ford,* n.d) A vehicle meant for the everyday "common man", it was revolutionary not due to the vehicle itself but due to its ability to be mass produced via an assembly line during manufacturing. While the Model T may have been initially made for American audiences, today all major automakers consider some level of Cross-Cultural design in vehicle design (Khan 2016).

Relevant Prior Writing. As will be discussed throughout this systematic literature review, many relevant prior writings exist. Two textbooks will be referenced in this systematic literature review: Occupational Safety and Health for Technologists, Engineers, and Managers by David Goesth and Safety and Health for Engineers by Roger Brauer. The current leading author on the topic of Cross-Cultural design in vehicles is Andre Dietsch, who has authored several research articles in the field such as "Analyses on the heterogeneity of car-following behaviour: Evidence from a Cross-Cultural driving simulator study" (Cheng, Xiaobei, Wuhong, Dietrich, et al. 2019). Additional research is being conducted on Cross-Cultural studies of autonomous vehicles as seen in "Human moral reasoning types in autonomous vehicle moral dilemma: a Cross-Cultural comparison of Korea and Canada." (Rhim 2020).

Uniqueness of Topic. While Cross-Cultural design has existed in some capacity for as long as society has been globalized, the specific applications to vehicles, in particular autonomous vehicles, for improved safety is a decidedly unique topic. Prior research does exist for Cross-Cultural studies of cross cultural reasoning, but it did so from a moral standpoint rather than a safety perspective. For instance, "Human moral reasoning types in autonomous vehicle moral dilemma: a Cross-Cultural comparison of Korea and Canada." (Rhim 2020). Since one of the major benefits of autonomous vehicles will be increased safety, ensuring that an increasingly globalized society is able to safely operate autonomous vehicles will be of the utmost importance. As previously mentioned in the Background sub-section, all major automakers consider some level of cross-cultural design because vehicles are a global product. However, the research relating to Cross-Cultural design as it applies to ***safety*** of the user in vehicles is a research area that has not been explored extensively.

Applications Justifications. Cross-cultural design is important in vehicle design because consumer vehicles are a global product. A 2021 study showed that more than 91% of American households have access to at least one vehicle (Peterson 2021) and another study estimates the current global number of cars, trucks, and buses has exceeded 1.4 billion (Chesterton 2018). With so many vehicles available all around the globe, it is vital for the safety of users for vehicles to have Cross-Cultural design.

Human-Computer Interaction. The field of Human-Computer Interaction (HCI) is focused on designing computer and other technology interfaces with the interaction of humans with the interface in mind. (Interaction Design Foundation, n.d.) Consumer Vehicles are a perfect application of this topic because of the increased technology use in vehicles. Computer interfaces within consumer vehicles include GPS navigation system, internal controls, audio controls, etc. Since users should be focusing on the road while driving, it is important for these controls to not be distracting and easy to use at a glance.

1.2 Systematic Review Procedure

The Softwares used for this study were MAXQDA, Harzing, and VOS Viewer. Leading articles were retrieved from Google Scholar, ResearchGate, and Springerlink. Papers involving both Cross-Cultural Design and Automotive industry (or Vehicles) from Google Scholar were used in the creation of trend graphs. Additionally, Google Scholar was used to retrieve metadata in Harzing. Consequently, the metadata was utilized to create cluster analysis and co-citation analysis in VOS Viewer. Lastly, content analysis was also conducted on the relevant articles.

2 Purpose of Study

The aim of this systematic literature review is to study articles about Cross-Cultural design and vehicles. The study focuses primarily on highlighting the importance of Cross-Cultural design in vehicles, both from an anthropometric perspective and a cultural reasoning perspective. The consequences of a vehicle designed without taking into account the diverse physical dimensions and cognitive abilities of a global population can lead to deadly results. For instance, despite providing overall more protection for women than men, short women are more likely to be killed by airbags due to being outside the protection range (*Cummings,* et. al.). Additionally, pregnant women are also especially at risk for injury from airbags (*Björnstig,* et al.). This provides an opportunity for Cross-Cultural design to be applied in the automotive industry to improve safety for diverse drivers.

3 Methodologies

3.1 Data Collection

The credible research databases of Web of Science and Google Scholar were used for the keyword search to retrieve data. The keywords used were "Cross-Cultural design" AND "vehicles" (the operator AND was used to ensure results included both keywords). Data such as titles, keywords, abstract, etc. were gathered from the database of Web of Science. However, the Web of Science database had fewer results than Google Scholar. The software Harzing's Publish or Perish was used to find relevant data from a variety of databases. Due to the vastness of the search results, Harzing's Publish or Perish is considered to be the best search resource for this systematic literature review. The search using the keywords "Cross-Cultural design" AND "vehicles" resulted in 28 articles in Web of Science and 1000 articles (the max allowed) in Harzing's Publish or Perish. When deemed appropriate, SpringerLink database was utilized. However, the now defunct AuthorMapper which utilized the SpringerLink database was not available for this systematic literature review, but its publications were available via Google Scholar.

3.2 Trend Analysis

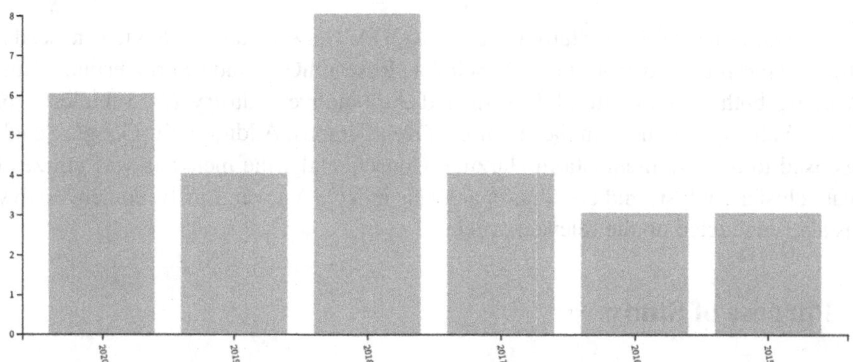

Fig. 1. By utilizing the Web of Science analytics tool a trend analysis was conducted on a total of 28 articles on the topic of Cross-Cultural design and vehicles from 2020 to 2015. The X-axis shows the years from 2015–2020 and the Y-axis shows the count. (Web of Science, n.d.).

The number of articles in 2018 increased from 3 in 2015 and 2016 to 8 in 2018. The number of articles written in 2020, according to Web of Science, was 6. Therefore, the field of Cross-Cultural design in vehicles is a growing, emerging field with many research study potentials (Fig. 1).

Emergence Indicators

Table 1. Emergence Indicator (Web of Science, n.d)

Year	Number of articles
2015	3
2016	3
2017	4
2018*	8
2019	4
2020	6

* Year of maximum published articles in the past 6 years therefore indicating emergence.

Field: Source Titles	Record Count	% of 28
SUSTAINABILITY	2	7.143 %
8TH INTERNATIONAL CONFERENCE ON AMBIENT SYSTEMS NETWORKS AND TECHNOLOGIES ANT 2017 AND THE 7TH INTERNATIONAL CONFERENCE ON SUSTAINABLE ENERGY INFORMATION TECHNOLOGY SEIT 2017	1	3.571 %
9TH INTERNATIONAL CONFERENCE ON EDUCATION AND NEW LEARNING TECHNOLOGIES EDULEARN17	1	3.571 %
ACCIDENT ANALYSIS AND PREVENTION	1	3.571 %
AMERICAN JOURNAL OF SEMIOTICS	1	3.571 %
APPLIED ACOUSTICS	1	3.571 %
CHI 20 EXTENDED ABSTRACTS OF THE 2020 CHI CONFERENCE ON HUMAN FACTORS IN COMPUTING SYSTEMS	1	3.571 %
COGNITION TECHNOLOGY WORK	1	3.571 %
COMPUTERS IN HUMAN BEHAVIOR	1	3.571 %
CROSS CULTURAL DESIGN APPLICATIONS IN MOBILE INTERACTION EDUCATION HEALTH TRANSPORT AND CULTURAL HERITAGE CCD 2015 PT II	1	3.571 %

Fig. 2. Analysis of source title for Cross-Cultural design for vehicles. (Web of Science, n.d.)

The below Fig. 3 demonstrates the frequency of authorship on the topic of Cross-Cultural Design and Vehicles. The author with the most number of articles in this field is Hofstede with 9 articles. The years of analysis were 1980 to present (Fig. 2 and Table 1).

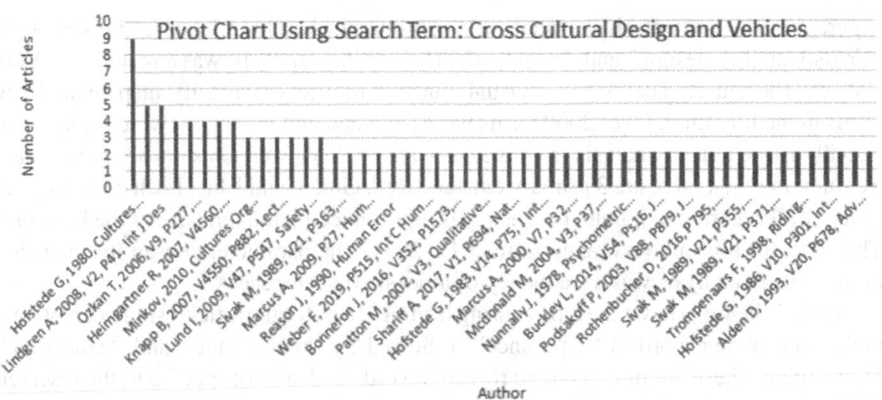

Fig. 3. Analysis of article frequency for Cross-Cultural design and vehicles. (Web of Science, n.d.)

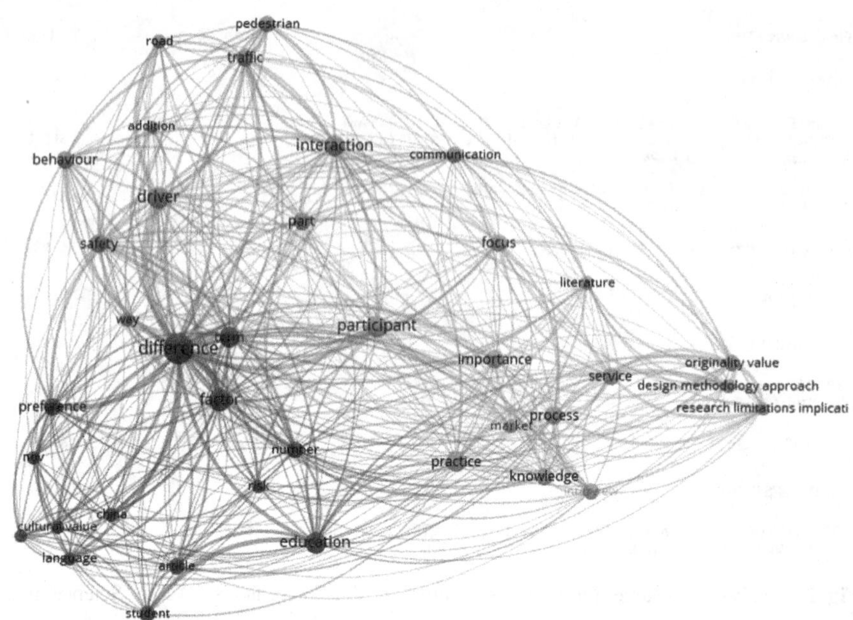

Fig. 4. Cluster analysis for Cross-Cultural design and vehicles *(VOSviewer, n.d.)*

4 Results

4.1 Content Analysis of Leading Terms via VOSViewer

Figure 4 is the content cluster analysis for bibliometric data related to the keywords "Cross-Cultural design" and "vehicles". The cluster analysis was created via VOS Viewer. The cluster analysis is a visual analysis tool to graphically map related key terms in the literature. Over 2000 terms were analyzed and the top 40 were represented visually in the cluster analysis from the software VOS Viewer. Clusters are represented by color. In Fig. 4 there are 3 primary clusters: Red, Green, and Blue. Each cluster represents a different research field. Frequency is represented by the font size of the keyword. The years for this analysis range from 2012–2021. The parameter chosen for inclusion in this cluster analysis was a frequency of at least 10 occurrences.

Table 2 (below) shows the top occuring leading terms for the articles studied. Notably, the top occurring word is "difference", followed by "participant", and "education". Interestingly, these are more generic research words and are not specific to the research field of Cross-Cultural design in vehicles. The top keywords in the table do not represent this research field. In order to find relevant keywords, an in depth study of the cluster analysis would be most beneficial.

This cluster analysis in Fig. 4 shows three clusters of keywords based on frequency and are represented by the colors Red, Green, and Blue. The largest cluster is the Green cluster (on terms related to "participant") closely followed by Blue (on terms related to "driver") and lastly the smallest cluster is Red (on terms related to "education").

The most abundant cluster is the Green cluster. The top keywords are "participant", "importance", and "focus". These keywords are not specific to the field of Cross-Cultural design or vehicles and therefore may not be of much significance for researchers looking for keywords in this field.

The most relevant cluster to the field of vehicles is the Blue cluster. The top terms in this cluster include "driver", "safety", "road", "pedestrian" and "traffic". Therefore, these keywords likely will yield better results for the topic of Cross-Cultural design in vehicles than the keywords from the Green cluster.

The Red cluster is likely most related to the field of Cross-Cultural design due to the keywords "cultural value" and "china". Interestingly, the most prominent keyword in this cluster is "difference" perhaps hinting at the use of this term when discussing cultural differences.

Table 2. Table of Leading Terms for "Cross-Cultural design" AND "vehicles"(*VOSviewer,* n.d.)

Term	Occurrences
Difference	93
Participant	72
Education	55
Interaction	47
Knowledge	38

Above in Table 2 is a leading table that was compiled on the keywords for the search constraints of "Cross-Cultural design" AND "vehicles". The top five most frequently occurring terms were "difference", "participant", "education", "interaction", and "knowledge". These terms are prominently displayed in the cluster analysis in Fig. 4. These keywords are likely to produce the most relevant results when researching the topic of Cross-Cultural design in vehicles.

4.2 Co-citation Analysis via VOSViewer

In Fig. 5 the results from the VOS Viewer co-citation analysis on 54 articles on the topic of Cross-Cultural design and vehicles can be seen. The frequency of authors referencing each other in the reference section can be seen through VOS Viewer's co-citation analysis. The thickness of the line indicates the frequency of co-citation between authors in the reference section or co-authoring of papers. Additionally, the colors indicate researchers focused on differing fields. Only authors with at least three co-citation were included. The Web of Science and Harzing database were utilized to generate the Fig. 5 co-citation analysis. The figure shows that the field of Cross-Cultural design in vehicles is heavily influenced by the author Andre Dietrich from Technische Universität München in Germany. Dietrich has authored papers such as "Implicit communication of automated vehicles in urban scenarios: Effects of pitch and deceleration on pedestrian crossing behavior." (2019).

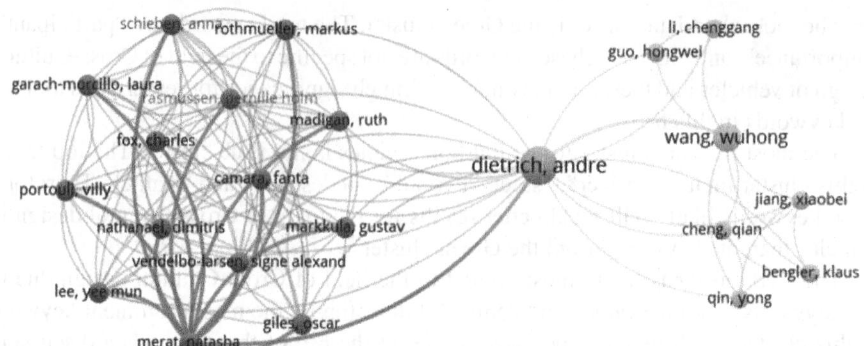

Fig. 5. Results from co-citation analysis of 54 articles on the topic of Cross-Cultural design and vehicles. (VOSViewer, n.d.)

Table 3 shows that all authors were actively publishing on the topic since at least 2011. However, the majority of the authors began after 2016, which leads to the conclusion that the field of Cross-Cultural design in vehicles was not a major field of research until at least 2016. This could be due to a variety of factors, but the rapid globalization and interconnectedness of the world in the 2010s could play a role. Therefore, Cross-Cultural design in vehicles is still an emerging field with room for further exploration in the future. As traditional vehicle technology transitions into autonomous vehicle technology, accounting for Cross-Cultural differences will be more crucial than ever before.

In addition, Table 4 shows that the leading universities and institutions for the research field of Cross-Cultural design in vehicles are in Germany, China, and the United Kingdom. This includes the university for the leading researcher, Dietrich, who hails from Technische Universität München in Germany. Interestingly, all the universities listed in Table 4 are focused in the field of engineering and technology, but have a wide variety of application industries, such as manufacturing, robotics, and autonomous vehicles.

4.3 Content Analysis from MAXQDA

A word cloud is a useful visual tool for quick analysis because the size of the word corresponds to frequency. The software MAXQDA (premium free trial) was used to get keywords from 7 key articles which were chosen through co-citation analysis and other relevant methods. A word cloud, as seen in Fig. 6, was generated after importing the articles to MAXQDA. This visual highlights the important keywords for this paper.

Terms that were not relevant to the research field such as the prepositions "the", "and", etc. were removed to reduce the number of unnecessary terms. As can be seen visually in Fig. 6, the three largest words are "moral", "vehicle" and "AV". For the research area of Cross-Cultural design in vehicles, these keywords are appropriate. Notably, researchers interested in pursuing this topic should note that "AV" is short for "Autonomous Vehicles".

Table 5 outlines the leading articles as discovered through VOS Viewer from the bibliometric data from Web of Science and Harzings "Publish or Perish" site. The earliest

Table 3. The table shows leading authors and their leading keywords.

Author	Years	Leading Keywords	Count
Camara, Fanta	2018–2021	Game Theory, Autonomous Vehicles, People, Pedestrian, Driver, Human Behavior	24
Cheng, Qian	2018–2020	Technology, Electronics, Vehicles	11
Dietrich, Andre	2016–2021	Ergonomics, Cross-Cultural, Transportation, Vehicles	35
Merat, Natasha	2012–2020	Driver, Behaviour, Vehicle, In-Vehicle, Driving, Automated, Automation	25
Portouli, Villy	2018–2021	Automated Vehicles, Driver, Traffic, Road, Users, Pedestrian	11
Wang, Wuhong	2011–2020	Mechanical Engineering, Technology, Design Engineering, Cross-Cultural Design	29

Table 4. The table shows leading institutions and their keywords.

Institution	Country	Leading Keywords	Count
Technische Universität München· Institute of Ergonomics Diplom	Germany	Engineering, Automotive, Transportation Cross-Cultural Design, Technology	11
Beijing Institute of Technology School of Mechanical Engineering	China	Technology, Mechanical Engineering, Cross-Cultural Design, Automotive	34
University of Oxford	United Kingdom	Engineering Science, Morality, Ethics, Behavior	12
University of Leeds	United Kingdom	Transport, Economics, Econometrics, Ergonomics, Vehicles	15

published article was 2009 and the latest was 2020. The most common phrase in the article titles is "Cross-Cultural differences". Table 6 outlines the leading keywords from the word cloud analysis.

Keyword Breakdown. A detailed discussion of the importance of the keywords from Table 6 is outlined below.

Moral - Moral is an important keyword in the context of Cross-Cultural design for vehicles because there are many moral and ethical problems that arise from utilizing cultures that designers are not familiar with. For instance, cultural appropriation that further alienates the population from using the technology (Brake 2020).

Fig.6. Utilizing MAXQDA software the above Word Cloud for key terms in the leading articles was generated. (MAXQDA, n.d.)

Table 5. Information for The Leading Articles used for MAXQDA

Author(s)	Title and Publication Info	Year
Brauer, Roger L	"Designing for the Workforce" in Safety and Health for Engineers. (Third ed., pg 1097o–1097z)	2016
Du, Na, Dawn Tilbury, Lionel Robert, X. Jessie Yang, and Anuj Pradhan	"A Cross-cultural study of trust building in autonomous vehicles."	2018
Goetsch, David L	"Safety, Health, and Competition in the Global Marketplace" in Occupational Safety and Health for Technologists, Engineers, and Managers. (Eighth ed, pg 97–104)	2015
Jeon, Myounghoon, Andreas Riener, Ju-Hwan Lee, Jonathan Schuett, and Bruce N. Walker	"Cross-cultural differences in the use of in-vehicle technologies and vehicle area network services: Austria, USA, and South Korea." In Proceedings of the 4th International Conference on Automotive User Interfaces and Interactive Vehicular Applications, pp. 163–170	2012

(*continued*)

Table 5. (*continued*)

Author(s)	Title and Publication Info	Year
Khan, Tawhid, Matthew Pitts, and Mark A. Williams	"Cross-cultural differences in automotive HMI design: a comparative study between UK and Indian users' design preferences." Journal of Usability Studies 11, no. 2	2016
Petiot, Jean-François, Cécile Salvo, Ilkin Hossoy, Panos Y. Papalambros, and Richard Gonzalez	A Cross-Cultural study of users' craftsmanship perceptions in vehicle interior design." International Journal of Product Development 7, no. 1–2	2009
Rhim, Jimin, Gi-bbeum Lee, and Ji-Hyun Lee	"Human moral reasoning types in autonomous vehicle moral dilemma: a Cross-Cultural comparison of Korea and Canada."	2020

Table 6. This table summarizes the keyword frequency.

Keyword	Occurences
moral	301
vehicle	169
AV	148
people	147
design	139

Vehicle - Vehicle is an important keyword for this topic because it represents the practical industry application of the theoretical field of Cross-Cultural design.

AV - AV stands for Autonomous Vehicle. AVs will be the future of vehicles and therefore is an important aspect to study when considering Cross-Cultural design of vehicles. Cross-Cultural understanding of technology acceptance and compliance will be important when studying public perception of AVs.

People - People will ultimately be the users of vehicles - whether human-driven or autonomously driven. Cross-Cultural design is an instance of human-centered design and therefore including "people" or "human" among the keywords for this topic is imperative.

Design - Design represents the method for applying Cross-Cultural principles.

4.4 Content Analysis via CiteSpace

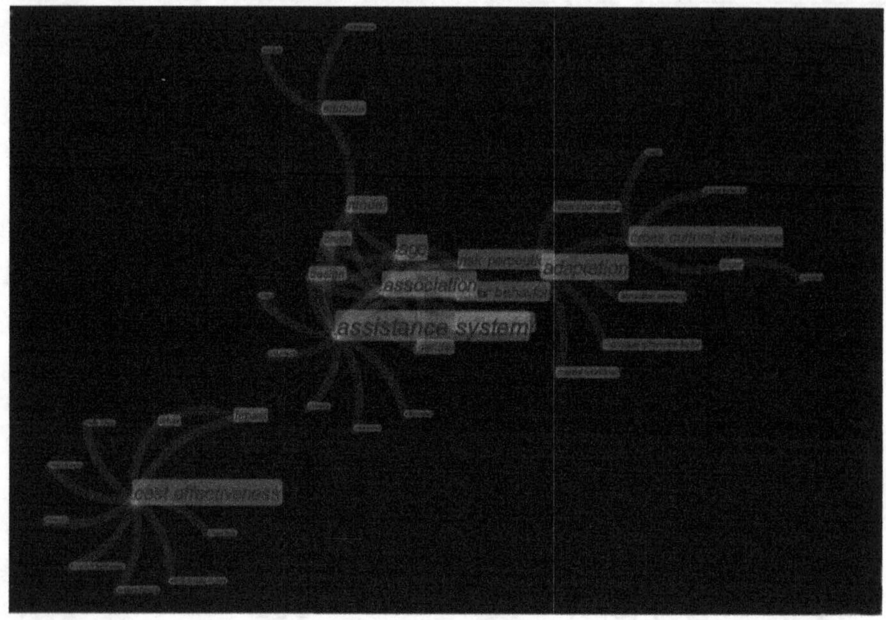

Fig.7. Cluster analysis via CiteSpace for Cross-Cultural design and vehicles (*Citespace,* n.d.)

A content analysis of the leading articles on the topic of Cross-Cultural design and vehicles was performed using the software CiteSpace. Three distinctive clusters are visible. One cluster is related to the term "cost effectiveness". Relevant terms in this cluster include "impact" and "driver". Another cluster is related to the term "assistance system" and includes the terms "association", "design" and "age". The last cluster is related to the term "adaptation" and includes the keywords "Cross-Cultural Difference" and "mental workload". These keywords differ from the VOS Viewer cluster analysis in Fig. 4 (Fig. 7).

4.5 Engagement Indicator

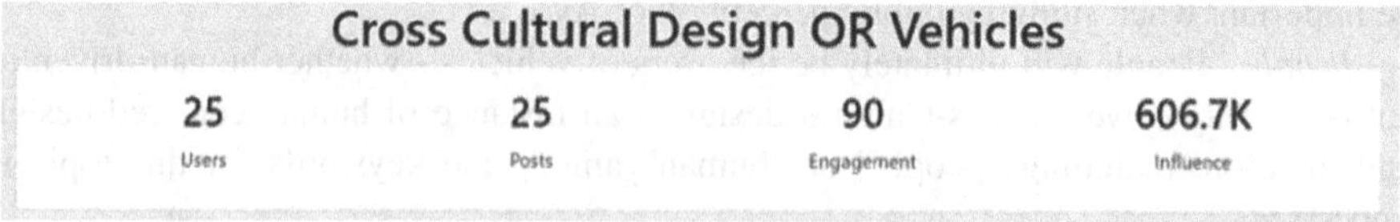

Cross Cultural Design OR Vehicles

25	25	90	606.7K
Users	Posts	Engagement	Influence

Fig. 8. Engagement Analytics by Vicinitas: Twitter Analytics Tool (Vicinitas, n.d.)

Twitter is one of the top social media platforms today, with 187 million active users (Statista 2021). The Twitter Analytics Tool, Vincinitas, was used to see public engagement on the topic of "Cross-Cultural design" OR "vehicles" (see Fig. 8). The operator "OR" was used rather than "AND" to yield more results for analysis. The engagement result was 90 with only 25 users. Therefore, the conclusion to be drawn from this analysis is that the topic of Cross-Cultural design in terms of vehicle design is not a topic popular in the public at the moment.

5 Discussion

5.1 Reappraisal

New insights gained through this systematic literature review of Cross-Cultural Design of Vehicles include an understanding of Hofstede's principles, Cross-Cultural perceptions of autonomous vehicles, and difference between global product design and Cross-Cultural Design.

5.2 Cross-Cultural Design in Vehicles as an Emerging Area

Understanding of Cross-Cultural Design as it applies to enhancing safety is described as one of the objectives of this course. In addition, Cross-Cultural Design is an explicit chapter in the Handbook of Human Factors and Ergonomics by Gabriel Salvendy. Other textbooks used for this course that have relevant chapters are Brauer's Safety and Health for Engineer (Chapter: Designing for Global Workforce pg 1097o–1097z) and Goetsch's Safety and Health for Technologists, Engineers, and Managers (Chapter: Safety, Health, and Competition in the Global Marketplace pg 97–104).

6 Conclusion

As the world moves towards increased globalization, Cross-Cultural Design will be more necessary than ever. Consumer vehicles will be one among many products that will need to have Cross-Cultural Design considerations in order to be effectively and safely operated by humans around the world. The results from the analyses show that Cross-Cultural Design is a growing, emerging field. The leading authors in this research field, such as Dietrich, understand the need for Cross-Cultural Design in Vehicles, as seen in the co-citation analysis and Dietrich's leading paper.

7 Future Work

The National Science Foundation in the United States funds research initiatives, including for the field of Cross-Cultural design in vehicles. Three examples of current research grants by the National Science Foundation in the United States will be highlighted to show areas where future research is being conducted.

The first award is "Modeling, Measuring and Controlling Human Comfort in Human-Autonomous-Machine Interaction (HaMI)" (direct weblink: https://www.nsf.gov/awardsearch/showAward?AWD_ID=1845779&HistoricalAwards=false) at Clemson University in South Carolina and began in 2019 and is expected to conclude in 2024. The aim of this research is to understand human interaction and perceptions when dealing with autonomous machines, and how the machines can be designed to increase confidence and comfortability with autonomous machines. Since humans will likely work closely with autonomous machines in the future whether it be autonomous vehicles or surgical robots, it will be important to understand how to effectively design these systems to interact with humans.

The second award is "Summer Institutes for Cross-Cultural Anthropological Research" which began research in 2020 and anticipates ending the study in 2023. The study is being conducted in New Haven, Connecticut. The purpose of this study is to study Cross-Cultural anthropology as it relates to design. The results of this finding will be vital in the future of the field of Cross-Cultural design. An increased understanding in Cross-Cultural anthropology will be provide designers with better resources for Cross-Cultural design.

Lastly, the third award is "IRES Track One: Involving Undergraduates in Research on Design and Cross-Cultural Perceptions of Cuteness in Robotic Gadgets". The research began in 2019 and is planning to continue until 2022. The purpose of this study is to understand Cross-Cultural perception of robotic technology and monitor the users' trust. It is assumed that "cuteness" correlates with a user's acceptance of robotic technology.

References

Ashta, A., Stokes, P.J., Smith, S.M., Hughes, P.: Japanese CEOs cross-cultural management of customer value orientation in india. Manag. Decis. (2021). https://doi.org/10.1108/MD-06-2020-0776

Brake, N.A., et al.: Cross-cultural engineering skill development at an international engineering summer boot camp. Paper presented at the ASEE Annual Conference and Exposition, Conference Proceedings. www.scopus.com. Accessed June 2020

Brauer, R.L.: Designing for the Workforce in Safety and Health for Engineers, (Third ed., pp. 1097o–1097z)

Bjornstig, U., Haraldsson, P.O., Polland, W., Sandstrom, T.: Awareness of the risk of air bag-associated injuries essential. Lakartidningen, **99**(28–29), 3022–3026 (2002)

Chan, W.T., Li, W.: Assessing professional cultural differences between airline pilots and air traffic controllers. https://doi.org/10.1007/978-3-030-49183-3_19, www.scopus.com

Cheng, Q., Jiang, X., Wang, W., Dietrich, A., Bengler, K., Qin, Y.: Analyses on the heterogeneity of car-following behaviour: evidence from a cross-cultural driving simulator study. IET Intel. Transport. Syst. **14**(8), 834–841 (2019)

Chesterton, Andrew. How many cars in the world? CarsGuide (2018). https://www.valuepenguin.com/auto-insurance/car-ownership-statistics. Accessed 23 April 2021

Cummings, P., McKnight, B., Rivara, F.P., Grossman, D.C.: Association of driver air bags with driver fatality: a matched cohort study. BMJ **324**(7346), 1119–1122 (2002)

DePauw University (2019) NSF Grant for Japan 'Cuteness' in Robots. https://www.depauw.edu/news-media/latest-news/details/34340/. Accessed 22 April 2021

Dietrich, A., Maruhn, P., Schwarze, L., Bengler, K.: Implicit communication of automated vehicles in urban scenarios: effects of pitch and deceleration on pedestrian crossing behavior. In: Ahram, T., Karwowski, W., Pickl, S., Taiar, R. (eds.) IHSED 2019. AISC, vol. 1026, pp. 176–181. Springer, Cham (2020). https://doi.org/10.1007/978-3-030-27928-8_27

Du, N., Tilbury, D., Robert, L., Yang, X.J., Pradhan, A.: A Cross-cultural study of trust building in autonomous vehicles (2018)

Ford Motor Company (n.d.) The Model T.https://corporate.ford.com/articles/history/the-model-t.html. Accessed 22 April 2021

Goetsch, D.L.: Safety, health, and competition in the global marketplace. In: Occupational Safety and Health for Technologists, Engineers, and Managers, Eighth ed, pg 97–104 (2015)

Interaction Design Foundation (n.d.). https://www.interaction-design.org/literature/topics/human-computer-interaction. Accessed 10 June 2021

Jeon, M., Riener, A., Lee, J.-H., Schuett, J., Walker, B.N.: Cross-cultural differences in the use of in-vehicle technologies and vehicle area network services: Austria, USA, and South Korea. In: Proceedings of the 4th International Conference on Automotive User Interfaces and Interactive Vehicular Applications, pp. 163–170 (2012)

Harzing's Publish or Perish. (n.d.). https://harzing.com/resources/publish-or-perish. Accessed 22 April 2021

Khan, T., Pitts, M., Williams, M.A.: Cross-cultural differences in automotive HMI design: a comparative study between UK and Indian users' design preferences. J. Usability Stud. **11**(2), 45–65 (2016)

MAXQDA (n.d.). https://www.maxqda.com/. Accessed 22 April 2021

McIlroy, R.C., et al.: Who is responsible for global road safety? a cross-cultural comparison of actor maps. Accid. Anal. Prev. **122**, 8–18 (2019)

National Science Foundation (n.d.) "Awards Simple Search". https://www.nsf.gov/awardsearch/. Accessed 22 April 2021

Plocher, T., Rau, P.-L.P., Choong, Y.-Y.: Cross-cultural design. Handb. Hum. Fact. Ergon. **162**, 252–279 (2012)

Petiot, J.-F., Salvo, C., Hossoy, I., Papalambros, P.Y., Gonzalez, R.: A cross-cultural study of users' craftsmanship perceptions in vehicle interior design. Int. J. Prod. Dev. **7**(1–2), 28–46 (2009)

Peterson, B.: Car Ownership Statistics 2021 Report (2021). https://www.valuepenguin.com/auto-insurance/car-ownership-statistics. Accessed 23 April 2021

Rau, P.-L.P., Plocher, T., Choong, Y.-Y.: Cross-cultural Design for IT Products and Services. CRC Press (2012)

Rhim, J., Lee, G.-B., Lee, J.-H.: Human moral reasoning types in autonomous vehicle moral dilemma: a cross-cultural comparison of Korea and Canada. Comput. Hum. Behav. **102**, 39–56 (2020)

Statista (2021). Twitter Usage. Accessed 22 April 2021. https://www.statista.com/statistics/272014/global-social-networks-ranked-by-number-of-users/

VOSviewer (n.d.). https://www.vosviewer.com/. Accessed 22 April 2021

Web of Science (n.d.). https://apps-webofknowledge-com.ezproxy.lib.purdue.edu/WOS_GeneralSearch_input.do?product=WOS&search_mode=GeneralSearch&SID=7EKUw7yEVcCGVrUwOlu&preferencesSaved. Accessed 22 April 2021

Cross-Cultural Differences of Designing Mobile Health Applications for Africans

Helina Oladapo(✉) and Joyram Chakraborty(✉)

Department of Computer and Information Sciences, Towson University, 7800 York Road, Towson, MD 21252, USA
{holadapo,jchakraborty}@towson.edu

Abstract. With the increasing use of mobile applications in Western countries, more emphasis should be taken when designing them for the African community to increase usage. In this paper, a pilot study was conducted with 30 participants to investigate the cultural differences in the design of a mobile health application for the African community residing in Maryland, United States. It highlights the cultural factors that need to be considered when designing a mobile application for the African community. The findings revealed the differences with adoption by gender and age. Differences in age and gender were measured using cultural dimensions. Health management along with cultural markers were examined to better understand factors that affect behavioral changes in the design of a mobile health application.

Keywords: Africans · Cross-cultural design · Mobile health application

1 Introduction

Studies have found that individuals from different backgrounds do not exhibit the same values, beliefs, customs, and perspectives towards Human Computer Interaction (HCI) systems and culture plays a critical role in interface design [1]. Research has shown that incorporating the localization of the user interface should be compatible with the cultural characteristics of the specific country [2]. Integrating the cultural aspects of the African community into the interface design influences the behavior and increases the adoption of mobile health applications. It was reported that technological solutions fail because the local cultural factors of the participants' native countries were neglected [3]. To avoid cultural issues in interface design, interface designers must be aware of the cross-cultural elements of typeface, number, date and time formats, images, symbols, colors, flow, and functionality [4] when designing for different cultures that use the user interface in several ways. Mobile health applications are promising tools. Therefore, when specifically designed with Africans' culture and health in mind, they become useful for behavioral adoption. A self-developed intervention tool, called AfriBP, was developed on a smartphone to monitor blood pressure (BP).

C. Stephanidis et al. (Eds.): HCII 2021, LNCS 13094, pp. 554–563, 2021.
https://doi.org/10.1007/978-3-030-90238-4_39

2 Research Background

National Health and Nutritional Examination Survey (NHANES) data highlighted that uncontrolled blood pressure, a major health concern in the United States, is more common in Blacks than whites and in males than females [5]. Hypertension (HTN) is predominantly experienced by persons of African origin during adulthood [6]. With respect to cross-cultural differences in the design of the AfriBP for the African communities, emphasis is placed on the prevalence of high BP/HTN, which is expected to increase to 68% (125.5 million) of the population of sub-Saharan Africa by 2025 [4]. AfriBP would be used to address these problems within the African communities. Due to the rapid growth and use of smartphones and mobile devices, including expansion into the healthcare industry, millions of people are currently utilizing more than 100,000 health-related behavior apps to manage chronic health diseases [7, 8]. Mobile health technologies are self-management behavior tools for promoting the care of HTN, among other illnesses [9]. Specifically, mobile health applications for HTN with the goal of achieving a healthy diet and exercise are known to be effective in instituting behavior lifestyle changes [10].

Africans' behavior patterns in physical inactivity and unhealthy eating habits can significantly impede health. In Sub-Saharan Africa, risk factors such as rapid urbanization, aging, and diet can contribute to lifestyle changes, resulting in metabolic conditions like HTN [11, 12]. Nutrition transition is the change in dietary patterns that occur with low- and middle-income countries as they move from a traditional to a westernized diet [13]. The dietary habits of the African migrants often change within a short period of time upon migration, thereby contributing to unhealthy diets [14]. Food habits are challenging to change for the West African community due to sociocultural beliefs. However, Africans consume large portions of carbohydrates, which are starchy. In West Africa, daily energy supply (DES) is derived mainly from carbohydrates rather than from animal fat. On the contrary, DES in North America and the European Union is derived from animal fat. Therefore, in West Africa, carbohydrates, not animal fats, are a major source of DES [15]. Physical inactivity is a global concern contributing to the risk factors of premature death and impeded quality of life [16]. Africans are physically inactive due to cultural and individual reasons. Healthy lifestyles of the African community would lead to better health and prolonged quality of life. The aim of this study is to better understand the cross-cultural differences experienced with the African communities in the United States in the design of the mobile health applications to improve and manage their health.

3 Culture

Cross-cultural studies identify the similarities and differences of user experiences within culture and provide a better understanding of the relationship between culture and mobile health applications [17, 18]. Hofstede defined culture as "the collective programming of the mind which distinguishes the members of one group or category of people from another" [19]. This implies that people from different cultures have distinctive patterns of thinking, feeling, and potentially acting which are learned throughout their lifetime. These cultural differences are manifested in rituals, symbols, heroes/heroines, and values [19], thereby influencing individuals' perceived choice, preference, and pattern of behavior in the design of the interface of a mobile health application. Hofstede's dimensions of culture have been applied to user interface design to understand cultural differences [20] Furthermore, Hofstede is mostly researched in HCI literature [21], proposing six dimensions of culture from his work including power distance, individualism-collectivism, masculinity-femininity, uncertainty avoidance, long-short term orientation, and indulgence-restraint [22].

4 Methodology

AfriBP incorporated cultural and health features into the design. The mobile application was designed using Flutter framework, a Google cross-platform framework for iOS and Android. Flutter used the Dart programming language to create the AfriBP application. The AfriBP application securely stored data in the Google Firebase system, Cloud Firestore (Figs. 1 and 2).

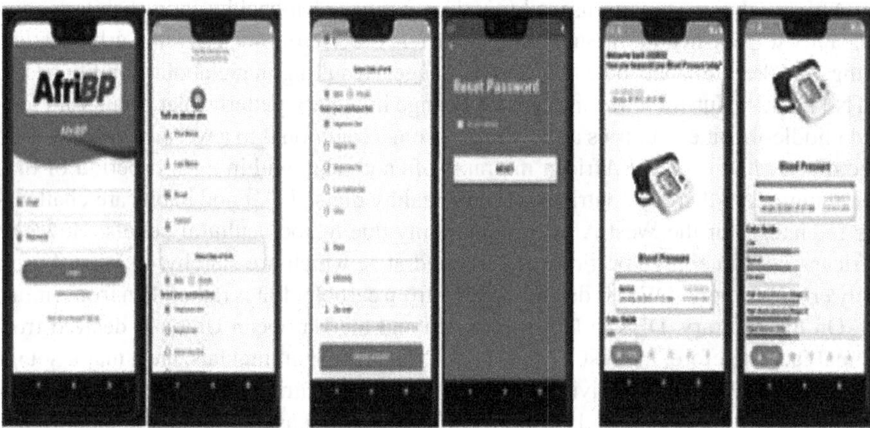

Fig. 1. Home screens

Cultural features are provided to enhance participants' health and lifestyle management. The recommended cultural features included the Exercise fitness and Eat Healthy Diet tabs based on the BP category, readings, and body mass index (BMI) for normal, elevated, or high blood pressure in stage 1 and 2 for the users. Furthermore, the Exercise fitness tab included pick-up soccer, basketball leagues, volleyball leagues, soccer leagues, and tennis, as well as African dance fitness. The Eat Healthy Diet tab is a food recommendation for Nigerian and Ghanaian users that is used to improve the health management of BP. Likewise, users were provided with clickable links to check for the nearest healthy shop within a geolocation to provide recommendations based on where the user lived (geolocation). The health-related features of the AfriBP included the demographic, nutritional diet, and geolocation recommendations based on where the user lives for account registration. In addition, the health-related features enabled users to manually enter BP measurements, and allowed for calculation of BMI, generation of

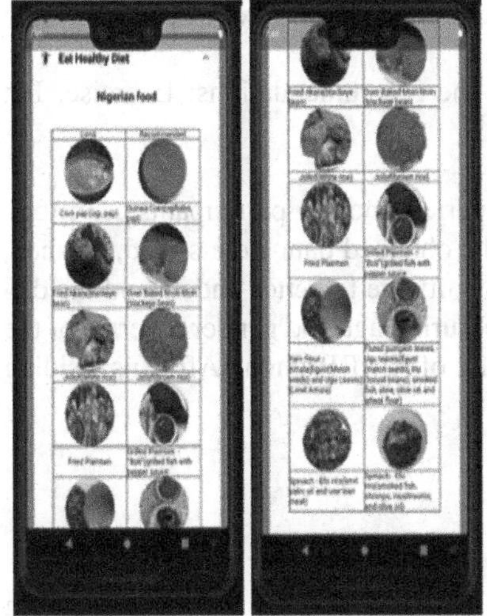

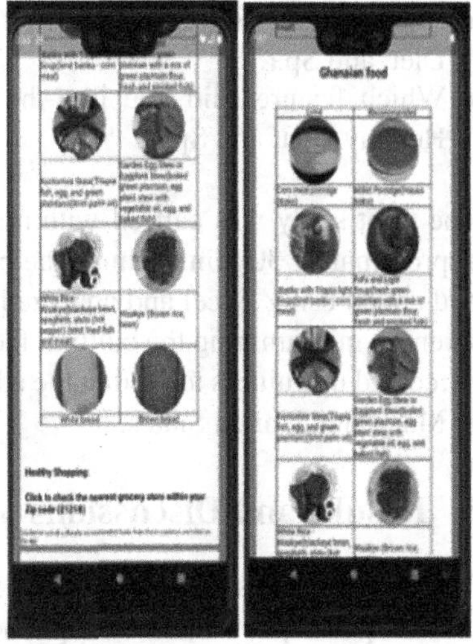

Fig. 2. Ghanaian and Nigerian diet recommendation screens

health history and bar chart, indication of cultural elements, exportation of health data and BMI in PDF and printable format, notification reminders, and display of user profile.

A qualitative statistical design was used for the data collection using the Zoom platform and Qualtrics Survey software. A random sampling approach was utilized. Recruitment of participants in Maryland, United States was carried out through recommendations and word of mouth solicitations. The sample included 30 Nigerian and Ghanaian participants, aged 18–69 years. Half (n = 15) of the individuals were male; the other 15 were female.

The study acquired the approval of the Institutional Review Board to conduct the online pilot usability testing experiment. The participants were invited to the Zoom platform via email. An Excel spreadsheet was used by participants to sign-in using the pre-provisioned credentials identifier. Before completing the consent form, the goal of the pilot study was explained to the participants. Participants were informed that participation is voluntary, and they have the right to withdraw or discontinue their involvement in the study at any time with the assurance of the confidentiality of data. The participants must also agree to have their BP recorded in the AfriBP application tool. Upon agreeing to sign the consent form, the participants could join the study. A qualitative questionnaire was developed with open-ended and closed questions. Participants were informed to complete the pre-questions about demographics during the usability testing. Then, they proceeded to navigate briefly through the user interface and complete some actual tasks to test the AfriBP application. After which, the participants were requested to complete the post-questionnaires based on the following questions (measuring differences in age and gender):

- If this personalized app is available on Google Play or Apple app store would you consider using this to manage your BP health?

- What features did you like the best about the recommendations: Exercise, Eat Healthy Diet, and Spa?
- Which features did you like the least about the recommendations: Exercise, Eat Healthy Diet, and Spa?

The pilot study was carried within a 12-month period and the participants each took approximately 30 min to finish the procedure. The obtained data were stored, cleaned, and coded using Excel and analyzed using SPSS software to better understand the differences in examining the cultural factors and measuring age and gender in terms of the perceived usefulness to predict the adoption and use of AfriBP among Africans residing in Maryland, United States.

5 Results and Discussions

The research revealed the following findings:

All participants had smartphones with 66.7% using iPhones and 33.3% using Androids. The percentage of male and female participants was equal (50%). Ages of the participants ranged from 18–19 with 6.7%, 20–29 with 30%, 30–39 with 26.7%, 40–49 with 23.3%, 50–59 with 6.7%, and 60–69 years with 6.7%.

The findings regarding age revealed that an equal percentage of 28.0% of those aged 20–29 and 30–39, and 24.0% of those aged 40–49 years reported that they would consider using the Google Play or Apple store to manage their BP health.

Table 1. Age if personalized app was available to manage BP health

Age		If this personalized app is available on Google play or Apple app store, would you consider using this to manage your BP health? Why?		Total
		Yes	No	
	18–19	1(4.0)	1(20.0)	2(6.7)
	20–29	7(28.0)	2(40.0)	9(30.0)
	30–39	7(28.0)	1(20.0)	8(26.7)
	40–49	6(24.0)	1(20.0)	7(23.3)
	50–59	2(8.0)	0(0.0)	2(6.7)
	60–69	2(8.0)	0(0.0)	2(6.7)
	Total	25(100.0)	5(100.0)	30(100.0)

Gender results highlighted that 52.0% of males and 48.0% of females reported they would consider using the Google play store or Apple store to manage their BP health.

Findings for the gender component revealed that 80.0% of males and 80.0% of females reported perceived usefulness of cultural dimensions as the best feature of the recommendations leading to the adoption of AfriBP.

Table 2. Gender if personalized app was available to manage BP health

		If this personalized app is available on Google play or Apple app store, would you consider using this to manage your BP health? Why?		Total
		Yes	No	
Gender	Male	13(52.0)	2(40.0)	15(50.0)
	Female	12(48.0)	3(60.0)	15(50.0)
	Total	25(100.0)	5(100.0)	30(100.0)

Perceived usefulness in health management (Gender). 86.7% of participants were interested in the recommended features for the health management provided, while 13.3% responded "no." However, both male and female participants equally shared perceived usefulness for health management.

Perceived usefulness in cultural dimensions (Age). Participants within the ages of 20–29, 30–39, and 40–49 years equally shared (25%) perceived usefulness in the cultural dimensions, unlike other age groups.

Perceived usefulness in health management (Age). There were significant age differences between age groups within 20–29 (34.6%), 30–39 (19.2%), and 40–49 years (23.1%) perceived usefulness in health management.

The resultant implications from the research predicted some differences for Africans in the adoption and use of mobile applications for measuring BP. In Table 1, implications from this result indicated that there would be an increase in uptake for individuals within the ages 20–29 and 30–39 years. This is because these groups were technologically savvy and adept with technology within these ages. This was observed during the usability testing experiment between ages 40–49 years, indicating differences in the way participants would consider using the app if it was available in Google Play or the Apple store to manage their BP health. In addition, one participant within the above age groups stated that "yes, because of the recommendations and information that are specific to my user group," they would use the personalized app. Another individual stated that "yes, I am in the middle and my health needs better monitoring." Overall, the AfriBP was a better way of improving health with "the information handy it helps to know the next steps to take according to my readings."

Table 2, as it relates to gender, showed that more men were adopting mobile health technology than females [23]. One of the male participants stated that "I would consider using it because of its sufficient features that aligns with traditional BP monitors" Likewise, one of the female participants highlighted the importance of using the.

AfriBP for BP health: "yes, the AfriBP has relevant information on diets and exercise which are key to BP improvements." "I would consider using the app because it directly offers specific recommendations for my cultural food I consume while other apps would offer very broad recommendations that can be difficult to apply to the entirety of my diet."

When comparing perceived usefulness in cultural dimensions and health management for gender, there seemed to be a higher percentage of participant preferences for the recommended features for health management than cultural dimensions. Both males and females were equally adopting the recommended features.

The health management for health history of the participants from the BP readings, BMI, health history, reminders, and the geolocation (zip code) were useful information needed for the Africans to stay healthy. One participant stated, "I liked how you can find close leagues and direct links to the nearest stores to improve your health just by providing the Zip code." As stated by a participant, the AfriBP ability to download and share data (currently print) to exchange with caregivers or relatives with needs to know" on how to effectively care for health. Another benefit emphasized was the reminder system "it is very useful as well as it will keep you on track of your health."

Regarding cultural dimensions, Africans liked to play soccer, which is a form of Exercise fitness. Likewise, Africans generally love dancing, which is also a form of Exercise fitness. Eat Healthy Diet are recommended food images for Africans to improve eating habits. One participant stated, "The Exercise features are better, especially the part of giving recommendations on different types of sports and locations. I like the recommendation of sport leagues." In addition, the Eat Healthy Diet culturally recommended foods were options provided to the participants. One participant stated that, "I like how the healthy diet section had the specific cultural African foods lined up so that the regular food to the alternatives/limited food were right beside each other." Some participants stated that, "When the right diets are eaten then one can be strong to add exercises with it." "It is relative and considerate of my cultural cuisine," and "It has recommendations for healthy African food." One participant emphasized that "the use of pictures for the recommended meals and the direct links to the nearest shops." This highlights the cultural element with "The Eat Healthy Diet portion of the app is the best part. It has a lot of potential. I think it is best to show good and bad ways to prepare each food. (Ex. Plantain can be boiled rather than fried in oil.)".

Implications from the results highlighted that both genders and all ages were interested in the recommendation for the perceived usefulness in cultural dimensions. However, the percentage of interest for the cultural dimensions was much higher for gender than ages within 20–29, 30–39, and 40–49 years. This indicated that the AfriBP needed to be better designed to fit the needs of the participants within the 18–19, 50–59, and 60–69 year-old age groups. The best features about the recommendations for Exercise, Eat Healthy Diet, and Spa were the Exercise fitness for ages between the years of 20–29 and 30–39. The best features, in terms of gender, were the recommendations for Exercise, Eat Healthy Diet, and Spa. The males had a higher percentage (66.7%) of the best recommended feature (Exercise fitness) "I like that the exercise section offered specific and access links to sports organizations in my area." While females with the percentage (53.8%) for the Eat Healthy Diet feature. "I liked the pictures that came with the food recommendations and the links attached to the recommended locations." "I think this feature should be highlighted as the niche for this app."

The Spa was the least favored feature the participants highlighted because "it was not cultural and is also expensive" "I knew nothing about spa, it is not part of my regime." Spa is a stress management service offered to Africans. With Africans' physical inactivity and unhealthy eating habits it was necessary to identify the perceived need of the participants to increase usage in the design of the AfriBP.

When comparing the perceived usefulness in cultural dimensions as related to age with the perceived usefulness in health management, this implies that participants for the perceived usefulness in health management adopted and considered it useful by different ages than in cultural dimensions that equally consider its usefulness. When designing for Africans, cultural factors of the participants' needs should be taken into utmost consideration to increase their satisfaction.

Participants in the age groups between 20–29 (34.6%), 30–39 (19.2%), and 40–49 years (23.1%) adopted perceived usefulness in health management differently. Meanwhile, the participants in these age groups adopted the perceived usefulness in cultural dimensions equally.

With Africans' physical inactivity and unhealthy eating habits it was necessary to identify the perceived need of the participants to increase usage in the design of the AfriBP.

6 Conclusion

This paper provides an understanding of cross-cultural differences from the preliminary findings regarding the way this group of people perceived such tools. The participants showed some significant increases and differences in ages within 20–29 and 30–39 than 40–49 years of the Africans behavioral adoption. If the personalized app were available on Google Play or the Apple app store, participants would be willing to use the application to manage their BP health. In terms of gender, the findings showed significant differences within men adopting the mobile health technology than females due to their cultural and health features. Although, with gender, there were significant differences and increases in the percentage of participants adopting the recommended features for the health management than cultural dimensions as the best feature for its usefulness as both adopted the AfriBP equally. Moreover, both genders were equally interested in cultural dimensions while age groups 20–29, 30–39, and 40–49 years had varying preferences in the recommendations for cultural dimensions. This means that investigations about issues within these age groups need to create a better understanding and design to meet their needs. Participants equally shared and adopted the cultural dimensions within the ages 20–29, 30–39, and 40–49 years than differently for the health management considering its usefulness. The participants adopted the best features for the recommendations for Exercise fitness, Eat Healthy Diet, and Spa differently for the age and gender. Age groups within 20–29 and 30–39 years preferred Exercise fitness. Males had a higher percentage of adoption of the Exercise fitness than females for the Eat Healthy Diet. Incorporating the cultural factors of the Africans' habits of playing soccer and dancing as a form of Exercise fitness and Eat Healthy Diet are recommended food images for Africans to improve healthy behaviour lifestyle. In addition, the health management such as the BP readings, BMI, health history, reminder notification, and the geolocation (zip code) will improve the health of Africans. However, the Spa was the least preferred feature the participant wanted because it was not considered a cultural feature.

Generally, Africans have a poor physical inactivity and unhealthy eating habits so embracing their perceived usage would help improve the design of the AfriBP in designing a successful user interface. The limitation was the number of participants and variation in age range. Future work for this study will include using a larger group of participants to get more insight to effectively design for the youngest, middle-age and elderly Africans. This will potentially contribute to the knowledge of researchers and developers in designing mobile applications based on cross-cultural findings. The AfriBP tool will also provide awareness and help improve the behavior of Africans in the United States in managing their healthcare concerns.

References

1. Kyriakoullis, L., Zaphiris, P.: Culture and HCI: a review of recent cultural studies in HCI and social networks. Univ. Access Inf. Soc. **15**(4), 629–642 (2015). https://doi.org/10.1007/s10209-015-0445-9
2. Dong, Y., Lee, K.P.: A cross-cultural comparative study of users' perceptions of a webpage: with a focus on the cognitive styles of Chinese, Koreans and Americans. Int. J. Des. **2**, 2 (2008)
3. Kim, H., Gupta, S.: Investigating customer resistance to change in transaction relationship with an internet vendor. Psychol. Mark. **29**(4), 257–269 (2012)
4. Oladapo, H., Owusu, E., Chakraborty, J.: Effects of culturally tailored user interface design. In: Ahram, T.Z., Falcão, C.S. (eds.) AHFE 2021. LNNS, vol. 275, pp. 845–853. Springer, Cham (2021). https://doi.org/10.1007/978-3-030-80091-8_100
5. Yoon, S.S., Fryar, C.D., Carroll, M.D.: Hypertension prevalence and control among adults: United States, 2011–2014, pp. 1–8. Hyattsville, MD, USA: US Department of Health and Human Services, Centers for Disease Control and Prevention, National Center for Health Statistics (2015)
6. Modesti, P.A., et al.: Cardiovascular health in migrants: current status and issues for prevention. A collaborative multidisciplinary task force report. J. Cardiovasc. Med. **15**(9), 683–692 (2014)
7. Edwards, E.A., et al.: Gamification for health promotion: systematic review of behaviour change techniques in smartphone apps. BMJ Open **6**(10), e012447 (2016)
8. Istepanian, R.S., Woodward, B.: M-health: Fundamentals and Applications. John, Hoboken (2016)
9. Logan, A.G.: Transforming hypertension management using mobile health technology for telemonitoring and self-care support. Can. J. Cardiol. **29**(5), 579–585 (2013)
10. Mann, D.M., Kudesia, V., Reddy, S., Weng, M., Imler, D., Quintiliani, L.: Development of DASH Mobile: a mHealth lifestyle change intervention for the management of hypertension. Stud. Health Technol. Inform. **192**, 973 (2013)
11. Agyei-Mensah, S., de-Graft Aikins, A.: Epidemiological transition and the double burden of disease in Accra, Ghana. J. Urban Health **87**(5), 879–897 (2010). https://doi.org/10.1007/s11524-010-9492-y
12. Mbanya, J.C.N., Motala, A.A., Sobngwi, E., Assah, F.K., Enoru, S.T.: Diabetes in sub-Saharan Africa. Lancet **375**(9733), 2254–2266 (2010)
13. Galbete, C., et al.: Food consumption, nutrient intake, and dietary patterns in Ghanaian migrants in Europe and their compatriots in Ghana. Food Nutr. Res. **61**(1), 1341809 (2017)
14. Satia-Abouta, J., Patterson, R.E., Neuhouser, M.L., Elder, J.: Dietary acculturation: applications to nutrition research and dietetics. J. Am. Diet. Assoc. **102**(8), 1105–1118 (2002)

15. Bosu, W.K.: An overview of the nutrition transition in West Africa: implications for non-communicable diseases. Proc. Nutr. Soc. **74**(4), 466–477 (2015)
16. Ding, D., Kolbe-Alexander, T., Nguyen, B., Katzmarzyk, P.T., Pratt, M., Lawson, K.D.: The economic burden of physical inactivity: a systematic review and critical appraisal. Br. J. Sports Med. **51**(19), 1392–1409 (2017)
17. Marcus, A.: Cross-cultural user-experience design. In: Barker-Plummer, D., Cox, R., Swoboda, N. (eds.) Diagrams 2006. LNCS (LNAI), vol. 4045, pp. 16–24. Springer, Heidelberg (2006). https://doi.org/10.1007/11783183_4
18. Papayiannis, S., Anastassiou-Hadjicharalambous, X.: Cross-cultural studies. In: Goldstein, S., Naglieri, J.A. (eds.) Encyclopedia of Child Behavior and Development, pp. 438–440. Springer US, Boston, MA (2011). https://doi.org/10.1007/978-0-387-79061-9_738
19. Hofstede, G.: Cultures and Organizations: Software of the Mind. 125. McGraw-Hill, New York (1991)
20. Marcus, A.: Cross-cultural user-interface design. In: Smith, M.J.S.G. (ed.) Proceedings of the Human-Computer Interface Internat. (HCII), vol. 2, pp. 502–505. Lawrence Erlbaum Associates, Mahwah (2001)
21. Gasparini, I., Pimenta, M.S., Palazzo M. de Oliveira, J.: Vive la différence!: a survey of cultural-aware issues in HCI. In: X Brazilian Symposium on Human Factors in Computer Systems (IHC 2011), pp. 13–22 (2011)
22. Hofstede, G.: Dimensionalizing cultures: The Hofstede model in context. Online Readings Psychol. Cult. **2**(1), 2307–2919 (2011)
23. Peprah, P., et al.: Knowledge, attitude, and use of mHealth technology among students in Ghana: a university-based survey. BMC Med. Inform. Decis. Mak. **19**(1), 1–11 (2019)

"Tell Me Your Story, I'll Tell You What Makes It Meaningful": Characterization of Meaningful Social Interactions Between Intercultural Strangers and Design Considerations for Promoting Them

María Laura Ramírez Galleguillos[1(✉)], Aya Eloiriachi[2], Büşra Serdar[3], and Aykut Coşkun[1,3]

[1] Koç University Arçelik Research Centre for Creative İndustries, Koç University, Istanbul 34450, Turkey
{mgalleguillos18,aykutcoskun}@ku.edu.tr
[2] Sociology Department, Koç University, Istanbul 34450, Turkey
[3] Media and Visual Arts Department, Koç University, Istanbul 34450, Turkey

Abstract. Positive meaningful interactions are encounters that promote positive attitudes and learning about others, which are needed to develop healthy social fabrics and cultural diversity. However, individuals tend to interact more with people like themselves often avoiding encounters with others that seem to be different, for example, with intercultural strangers. Though previous HCI work has been concerned with exploring meaningful experiences with products and technologies as a way of promoting product attachment, the field lacks studies exploring how design could facilitate intercultural MSI. Designing interventions to support intercultural MSI requires i) understanding what characteristics make these interactions meaningful and ii) how these characteristics can be addressed through design. In this study, we contribute to the literature by producing knowledge on these aspects. Based on an analysis of 56 real-life stories about intercultural MSI and an idea generation session with designers, we characterize intercultural MSI with four dimensions (outcomes, feelings, context, and elements) and we identify four design considerations to be taken into account when designing interventions to support intercultural MSI. Hence, our contribution is to formulate this knowledge while highlighting how the characteristics and perceptions of intercultural MSI can be applied to design new technologies that promote this kind of interaction.

Keywords: Meaningful interactions · Collocated interactions · Migration

1 Introduction

Positive and meaningful interactions are encounters that promote a positive attitude and learning about other social groups. These interactions are needed to develop healthy social fabrics and promote diversity in society [11, 40]. Also, interacting with new people can promote a sense of trust and community building with others, supporting the

C. Stephanidis et al. (Eds.): HCII 2021, LNCS 13094, pp. 564–583, 2021.
https://doi.org/10.1007/978-3-030-90238-4_40

integration of different communities as part of the same social system [8, 50], fostering collaboration among different communities [18, 26, 46]. However, individuals tend to prefer interactions with people who are similar to themselves [21, 24], often avoiding interactions with people perceived as different, such as strangers or social and cultural outgroups[1] [42, 48]. Hence, intercultural interactions (i.e., interactions between immigrants and locals) might be avoided not only because of language differences, but also due to biases and prejudices against other social or cultural groups.

The design and HCI fields are concerned with turning current situations into preferred ones [38, 49]. Hence, we believe technological products, services, and spaces that influence individuals' behavior [51] might also be used to encourage meaningful social interactions (MSI) between intercultural groups [13, 30]. Though previous HCI work has previously explored meaningful experiences with products and technologies [9, 16, 19] as a way of promoting product attachment [34], the field has been less concerned with exploring how design could facilitate intercultural MSI between individuals, which face additional barriers for interactions (i.e., avoidance, biases, language).

Proposing technological interventions that support intercultural MSI requires i) understanding which characteristics make these interactions meaningful and ii) how these characteristics can be addressed through design. Hence, we explored these questions by collecting and analyzing real-life stories about intercultural MSI experienced by 56 participants. Based on an analysis of these stories and an idea generation session with designers, we characterize intercultural MSI with four dimensions (outcomes, feelings, context, and elements) and we present a set of design considerations that can guide practitioners and researchers when aiming to propose design solutions that facilitate MSI between intercultural strangers.

2 Related Work

In this section, we review previous work related to meaningful interactions in HCI, collocated interactions, and migration-related projects in HCI.

2.1 Meaningfulness in HCI

HCI researchers have been interested in exploring positive and meaningful experiences mostly with technologies and products as a way of promoting attachment towards such technologies. This is the case of previous work proposing frames for meaningful experiences in HCI. For example, [34] suggested five aspects of the experience of meaning: connectedness, purpose, coherence, resonance, and significance. Also, [16] studied how to apply Slow Design theory to the design of an electric fruit juicer establishing more mindful usage of products. Other designs have also been proposed to promote meaningful experiences, for example, Fibo [9], a pregnancy wearable that allows a partner to feel the movements of their unborn child through jewelry, emphasizing the need to enable meaningful experiences through thoughtful design. However, these studies have

[1] Ingroup refers to a social group an individual identifies as part of, while outgroup is a social group individuals do not feel identified with.

been dedicated to promoting meaningful experiences with technology rather than exploring meaningful human-to-human social interactions and ways in which technology can facilitate these interactions.

We have found only 3 prior works focusing on MSI [12, 25, 27]. For example, [27] studied MSI in the context of social media. The authors explained that meaningful interactions are those with emotional, informational, or tangible impact people believe enhance their lives, the lives of their interaction partners, or their relationships. They propose that attributes most likely to facilitate MSI include strong (e.g., friends and family) and community ties (e.g., neighbors). Further, they state that MSI are normally planned in advance. Concerning meaningful interactions that present close ties, [25] studied grandparents and grandchild MSI for cultural exchange activities. The authors suggested that technology-mediated support of intergenerational immigrant cultural exchange must shift the perspective from "barriers" to consider how they might foster further engagement. Finally, [12] explored children and teenagers' perceptions of MSI by studying if and when meaningful social interaction occurred during gameplay, how it occurred, and with which impact. The authors proposed design recommendations to support serious games for MSI. For instance, they indicated that designing for MSI requires taking into consideration participants' preferences, needs, and requirements to support interactions that are both desired and meaningful to those interacting. Hence, suggesting that the MSI field requires building guidelines that are specific to the kind of MSI.

In HCI, social interactions with nearby strangers have been studied under the scope of collocated interactions, which are interactions held with people located in the same or nearby places [28, 30]. Such explorations have focused mostly on fostering encounters with unacquainted others through interactions that promote the feeling of serendipity and unexpectedness to start a new interaction with strangers. Examples of the latter are the explorations of serendipitous interactions [7], playful interactions [29], opportunistic interactions [35]; and emergent interactions [36]. We observed that this field has mostly concentrated on how interactions are started more than on how they develop to be positive or meaningful. Furthermore, how to overcome the avoidance of such interactions due to intercultural or intergroup conflicts has not been deeply studied.

With all the above, the present study has key differences from prior work. First, concerning the object of the study, because unlike them this study is concerned with exploring interactions between humans rather than with products. Second, regarding the kind of human-to-human MSI. MSI with intercultural strangers present further barriers to be developed because of the absence of strong community ties, the lack of planning of encounters, and the individual's tendency to avoid such interactions due to biases. Hence, the field lacks design guidelines and design knowledge that is specific to MSI between intercultural strangers which can guide practitioners and researchers when developing HCI solutions to promote this kind of MSI.

2.2 Migration in HCI

HCI researchers have been previously concerned with migration issues, specifically exploring how migrants can be supported in their integration process to a new country. For example, developing technological projects for migrants to keep in contact with their significant others [14, 37], to learn a language and develop professionally [1, 2], to understand laws and regulation in the new country [17, 39], mobility in their new city [32], and to support their digital inclusion [20]. Likewise, there have been works supporting community building, for example through storytelling for the empowerment of migrant women in Finland [4] and Germany [54], of Palestinian youth in the West Bank and East Jerusalem refugee camp [47], and of Syrian refugees at Za'atari camp [31, 55].

These works exemplify the broad spectrum of migration works in HCI. However, we observe there has been less focus on promoting social integration through exploring direct contact between migrant groups and locals or even between migrants and migrants. Furthermore, the studies that have directly focused on developing positive contact between different social or cultural groups, have mostly pursued this aim through in-person activities like workshops. In these situations, optimal conditions for positive contact [43, 44] can be ensured by conducting activities that ignite equal status among participants, shared goals, besides authority sanction [3]. However, in public places of the city, which is where we find most people from different cultural and social groups, optimal conditions are not necessarily met.

In summary, how to support MSI between intercultural strangers through technology has not been deeply explored in HCI. Hence, as a first step towards addressing this gap, we characterize this kind of interaction by studying real-life stories to extract knowledge that can guide practitioners and researchers when designing to promote intercultural MSI.

3 Methodology

3.1 Research Context

This study was developed in Istanbul. This is a cosmopolitan and multicultural city, that historically has represented a top destination for internal and external migration [23, 53] even before the Ottoman Empire [33]. Even though there is an important migrant population, since the war in Syria started, there has been a rise in violence and social exclusion towards immigrants [22]. All these reasons make Istanbul a rich case to explore interactions between locals and migrants living in this city.

3.2 Research Procedure

We created and distributed an online form to collect the stories. Our goal was to understand which interactions are categorized as meaningful by the participants and identify characteristics of intercultural MSI. Hence, we relied on storytelling to capture events, participants' perceptions, expectations, and thoughts concerning their previous experiences of intercultural social interactions [15]. Then, having listed these characteristics, we aimed to create implications of these stories for design by showcasing spaces of interventions that can make new interactions meaningful.

We used Jot Form to collect the stories as at the moment of this study COVID-19 regulations were restricting in-person activities. Further, the written format of storytelling seemed to be suitable to reveal events that explain participant's perceptions, expectations, and thoughts about intercultural MSI and inform design [15]. With this collection, we expected to 1) understand the significance of these kinds of interactions according to participants and, 2) extract lessons from previous memorable intercultural interactions to guide designers when creating solutions that promote these interactions. The form had 4 stages:

1. We first presented a broad definition of MSI to the participants, i.e., encounters promoting a positive attitude and learning about an intercultural stranger.
2. Then, we asked them to share a real-life story of a moment on which they had such interactions in a collocated fashion.
3. Third, we asked them to signal where the interaction was developed and what made it meaningful.
4. Finally, they filled a table selecting and describing 3 key moments of the interaction and attaching to each moment 3 adjectives from a given list.

With this format, we were able to access key information about previous experiences, places where interactions took place, what made interactions meaningful, and how participants perceived these interactions through the adjective's selection. Participants were also asked to voluntarily fill demographic questions about their nationality, age, gender, migration status, and languages they spoke. The survey was available in Turkish and English, and the form could be filled either by text or by audio to broaden its accessibility. The survey was distributed through social media, self-organized WhatsApp groups of both migrants living in Istanbul and locals, and the networks of some organizations working to promote social inclusion in Istanbul as KUSIF and IMECE. It was open for 2 months. We collected 56 stories of people currently living in Istanbul, 19 were in Turkish, 37 in English.

3.3 Analysis

The collected stories were analyzed using reflexive thematic analysis [5, 6, 10]. We used mixed coding as we were expecting to find characteristics that make a specific social interaction meaningful, which could be transferred into design considerations to create technological solutions. We brainstormed characteristics of interactions that were used as initial codes. These were people involved, duration of interaction, facilitator of interactions, and meaningfulness. Later, as we performed a round of open coding, we were flexible to add new codes (e.g., feelings provoked by such interactions). Finally, we extracted 5 main themes from those codes which frame the characteristics of intercultural MSI: Elements involved in intercultural MSI, places where these interactions took place, feelings provoked, the significance of these interactions for participants, and facilitators of the interactions.

In parallel, the tables with key moments and attached adjectives were analyzed to understand participants' perception of MSI between intercultural strangers and the development of key-events of the story which could signal points where design and technology could intervene to promote this kind of interaction. We did this with 2 different kinds of analysis. First, we analyzed the most frequently used adjectives to describe the MSI as an overall experience and key moments during this experience [45, 52], which, in turn, gave us a sense of a general perception about the interactions. Second, we analyzed the word selection concerning the key moments they were attached to and the progressions of events of each story.

4 Findings

Through the analysis of the stories, we have been able to identify different characteristics and perceptions of intercultural MSI. In this section, we specify and explain these findings and their relevance.

4.1 Characteristics of MSI

What Made the Interaction Meaningful?
We have found 3 aspects that can make an intercultural interaction a meaningful one. First, experiencing kindness through helping each other or doing small favors. For example, S6[2] expressed that what made the interaction meaningful was "*that even though the driver was a stranger, he was ready to help me and not leave me half of the way*". Therefore, experiencing something in which a person is in a vulnerable position and is being helped by others can make an interaction meaningful. Second way of making an interaction meaningful is having the opportunity to develop bonds among people by finding similarities and common. As an example, S12 mentioned that "(people) *always think they will find people with similar interests or to click with, but it's very rare. We are still friends* (with the person of the interaction) *and talk about these topics*". The third aspect that can make an intercultural interaction meaningful is learning something new either about the "other" by exploring each other's life, experiences, or cultures. In S30, the participant and a group of their friends take a trip to explore Şanlıurfa, a city in Southeast Turkey. One of the locals they encountered offered them to be their guide. They ended up spending the holiday together and the local person gave all kinds of cultural details and historical information. He spontaneously invited them to his cousin's wedding, and they ended up sharing a personal event while also learning about the details regarding a different culture's wedding rituals.

Finally, the stories also showed that these different ways of building meaningfulness might promote future intercultural interactions. For instance, half of the stories revealed participants being engaged in new intercultural interactions after the one reported either with the same or different person. This point seems to indicate that having a meaningful experience might lead to people being more open to interacting with others in the future. This is relevant considering the bigger purpose of fighting avoidance of interactions by promoting positive intercultural interactions between strangers.

What Feelings are Related to Intercultural MSI?
We have found that MSI have various ways of impacting participants emotionally either

[2] From now one references to stories will use Sn as the format where S refers to Story and n to the number.

while the interaction is taking place or after it ends. It could be argued that it is natural for interactions to ignite some kind of feeling, which can be the case of interactions in contexts of strong or community ties. However, it is relevant to highlight the emotional quality of intercultural MSI as they are promoted between unacquainted "others" who present cultural differences and with a high tendency to be avoided.

Further, the feelings expressed in the stories are interesting because they present a broader range of feelings in comparison to collocated interactions, which usually promote feelings of unexpectedness and serendipity. The feelings participants associated with intercultural MSI were fun, belonging, special connection, vulnerability, trust, gratefulness, or oneness. Moreover, we have found that, in most stories, participants reported that feelings can be mutually felt (i.e., between themselves and the strangers they were interacting with), as in the case of feeling a special connection, closeness, or oneness. This idea portrays the possibility of increasing meaningfulness when feelings are mutually and collectively experienced. However, there are still feelings individually felt, which can take place at the moment of interaction as in the case of vulnerability and belonging or, even after the interaction is finished, as in the case of gratefulness.

What Contexts and Situations Facilitate MSI?

We have found that intercultural MSI normally take place in public places. This is understandable since these are the spaces where we encounter strangers and social and cultural differences, in comparison to private places as home. Consequently, these interactions take place while participants were in a public place to do something else. For instance, an individual goes to the supermarket intending to do groceries, or to a migration office because they need to deal with some kind of procedure, and they happened to be engaged in an intercultural MSI meanwhile. Hence, these interactions are not planned, and the participants of such interactions do not present the initial intention of interacting with others. For this reason, we analyzed the context of the stories that we will discuss concerning the mobility level (i.e., if someone is transiting or staying in the place) and specific situations that enable MSI between intercultural strangers.

Mobility Level: public places are spaces where people can freely move around or stay for some time. In these places, interactions can be held while participants are moving from one place to another or staying in that place. This determines the level of mobility, which ranges from transitory (i.e., when individuals are passing by to go somewhere else) to staying (i.e., when people stay in the place), meaning that the place where the interaction is held is the final destination. Examples of transitory levels are the intercultural MSI developed while people are walking on the streets or using public transportation. Examples of staying level are the interactions held while people are sitting in coffee shops and parks. From our collected stories, 31 took place in transitory levels while 25 occurred in staying level. These levels are relevant because they seem to connect to specific situations where intercultural MSI can take place. Also, most of the interactions happened at transitory level, revealing that most of the interactions were not long, however, they were meaningful for participants.

Situations that Enable MSI: By analyzing the stories we have found different situations that can enable intercultural MSI. These are exploring cultures, connecting with nature, and transporting.

1. Exploring cultures: in these situations, a participant goes somewhere to learn about cultures, history, cities, among others. Consequently, these interactions are normally held in cultural places. For example, we received 3 different meaningful interaction stories taking place in museums and their surroundings. In S37, the interaction started in the queue to enter the museum, and the participants shared many facets of themselves and their cultures. In fact, the participant stated that they bonded with each other so much that they ended up going to the museum's cafeteria together. Therefore, when individuals are exploring cultures and history, they seem more open to learning about others, their life, and culture, which, in turn, enables intercultural MSI.
2. Connecting with nature: in these situations, participants visit natural and open places to enjoy nature in lakes, parks, forests, and others. For example, in S2 the participant met a couple on a hitchhiking trail with whom he spent the next 7 h conversing, sharing life stories and lessons while exchanging food and drinks. For the participant, this experience was so relevant that they stated, "*that random conversation gave me the best life review session I've ever had in a while*". Hence, in these situations, people might have a more open attitude towards others and enjoy a time together to bond.
3. Transporting: these are the situations in which people are moving from one place to the other either inside a city, within cities, or between countries. Therefore, they have as a setting different means of transportations like planes, buses, or taxis. For example, in S5, the participant meets with a stranger on an airplane, and they end up talking about their hobbies. The participant mentioned, "*we started talking about our hobbies. She made me realize how people have a variety of interests and how each individual in society is different from the other*". Further, when asked why it was meaningful, the participant explained that "*a totally random person filled her 3 h of loneliness with an enjoyable conversation while on that trip*". Hence, while transporting to other places, participants seem to have time available to be open to interacting with others in a meaningful way.

In all these situations we can observe that the time available plays a key role for interactions to start in the first place. Therefore, these interactions are normally connected to situations in which people must wait for something, for example, buying a ticket to enter the museum, waiting for their bus, waiting to arrive at their destination, or having free time such as when sitting in a park. This point is relevant as it can draw lines to new spaces, situations and mobility levels to promote intercultural MSI.

What Elements are Present in Intercultural MSI?

We found 5 elements that configure different intercultural MSI (Table 1) concerning their internal aspects. These are:

Group Size. This element refers to the number of people involved in a given MSI. According to our stories, we have identified intercultural MSI in pairs (1 and 1), between a person and a group of people, between 2 or more groups of people, and between 1 person and a crowd of people. For example, interactions held between a person and a crowd refers to the experience of being part of a massive event as it was portrayed in S30 referring to an intercultural MSI while attending a 2-day seminar. The experience frames an event that the attending crowd will have in common in the future, in addition to representing a space where many interactions in smaller group sizes can take place. Hence, not all MSI are developed in pairs, conceived as a stereotypical interaction.

Physical Proximity. This element refers to how physically close people are when the interaction starts. This element presents 3 levels: First, not being physically close, for example, while experiencing a virtual interaction. Second, being physically distant, as in S27 where the participant and their parents were fishing had a meaningful interaction with an old lady watching them from her balcony. And third, being physically close on a meter range.

Kind of Communication. This element refers to the way by which participants communicate with each other. The stories refer to interactions developed through different communication channels. For example, there are MSI developed through oral communication and physical communication (body language). In S4, the interaction took place on a public bus where the participant had an intercultural MSI with a stranger without verbally communicating. The participant was offered one of the strangers' ear-pads to listen to music, while the participant offered the stranger gum in return. The participant stated that they shared music for around fifteen minutes in silence until the young stranger left the bus while waving to the participant. Even though physical communication is not a new phenomenon, this finding is relevant because meaningful interactions are generally associated with deep talks, for example, having a lengthy conversation about life experiences as was mentioned in S2 "life review" session. Hence, this finding seems to explain that people actually can create meaningful moments even without directly speaking.

Duration of the Interaction. This element refers to the length of the interaction. Hence, there is a range of duration, starting from very brief to longer interactions. For example, in some of the collected stories, brief interactions lasted less than 15 min, therefore indicating that MSI does not necessarily require a long period to be experienced as meaningful. This was the case of S19, a story in which the participant notices two tourists having a hard time understanding the transportation options in Kadıköy, İstanbul. The participant approaches them and helps them, they talk for some minutes, then, they simply thank the participant and continue with their journey. However short, for the participant, this was an intercultural MSI through which they got to briefly help others. Longer interactions could last hours or even days. For example, in S1, the participant accidentally meets with a pair of people, and they end up spending three days together, exploring the car business and the environment while sharing stories. The participant mentioned, "*I remember this one scene where I was walking behind them in a market, it felt like having two guardian angels or two fathers. They brought a lot of peace to my heart*".

Meaningfulness. This element refers to two different ways in which meaningfulness can be built. However, this does not necessarily mean they are the only ways of building meaningfulness in social interactions. According to the stories collected, meaningfulness can be achieved through an exchange of intercultural facet or through sharing human connections and experiences beyond the cultural divide. Hence, activities related to exchanging involved a tangible asset that is exchanged between the participants of the interactions. Also, turns of exchanging those assets can be identified. Examples of the latter are exchanging directions, words, or artifacts. Another way of building meaningfulness was through sharing, which we identified as something that is mutually and synchronously experienced. Thus, no turns of giving and taking were observed. Also, the act of sharing in intercultural MSI was more related to intangible things. For example, sharing a smile, a glance, a song, or feelings and experiences which are perceived as mutual. An example of the latter is S3, where the participant shared an experience in which they felt understood simply by looking in the eyes of the stranger they were interacting with and stating that "*I smiled in relief that I wasn't the only one feeling the same way*".

Table 1. Summary of the characteristics of intercultural MSI

Characteristics	Overview
Outcomes of interactions that make them meaningful	Promoting kindness Developing bonds Learning
Feelings ignited by interaction	Mutually felt as compatibility, closeness, similarity Individually felt as belonging, trust, gratefulness
Contextual characteristics of the interaction	Level of mobility Situations that can facilitate intercultural MSI (exploring cultures, connecting with nature, transporting)
Elements of the interaction	Group size Physical proximity Kind of communication Duration of interaction Meaningfulness

4.2 Perceptions of Intercultural MSI

By analyzing the adjectives which were used to describe MSI and the key moments in these interactions, we were able to identify perceptions of intercultural MSI. We found that intercultural MSI are mainly evaluated as unexpected, emotional, and genuine encounters. Further, our analysis of the adjectives also revealed that participants normally highlighted 4 different phases of development of the stories. These were igniting, meaningful moment, closure, and self-reflection. The first three phases happen during an interaction. Igniting stage is the first key moment when the interaction is started. The meaningful moment is the second key stage when something is shared or exchanged between participants. The closure is the third key moment referring to how the interaction ended. On the other hand, in some cases, participants selected an after-interaction moment as a fourth moment, which refers to a reflective moment in which they think again about the interaction.

As we asked participants to select adjectives concerning such key moments (Table 2), we observed that the igniting moment is generally perceived as unexpected, awkward, and/or fun. Even though perceived as unexpected, in the collected stories the participants were able to overcome the awkwardness of the igniting stage and advance to the meaningful moment, which was perceived as emotional, fun, and judgment-free. This is the moment that makes a regular interaction a meaningful one. Finally, the third key moment, which referred to the closure of the interaction or to the reflective process ignited after the interaction was finished, was perceived as pleasant, emotional, and genuine.

These findings showcase that intercultural MSI are interactions that, due to their characteristics and events development can evolve from unexpected and awkward to a pleasant and honest experience. Also, with this analysis, we have found stages of progression which can highlight different key moments where technologies can intervene when aiming to promote intercultural MSI. Further, our findings indicate that these intervention moments involve further stages rather than just the direct interaction (i.e., thinking and reflecting after the interaction is finished). Furthermore, according to the stories, interactions that involve reflecting about participants' experiences, life, beliefs seem to promote openness to future interactions as well. This finding is relevant as most collocated interactions focus on the interaction per se, while what happens after can be key in promoting meaningfulness and influencing future interactions.

Table 2. Summary of most selected words overall and per key moment.

Overall	First	Second	Third
Unexpected	Unexpected	Emotional	Pleasant
Emotional	Awkward	Fun	Emotional
Genuine	Fun	Judgement-free	Genuine

5 Applying the Findings in the Design Process

In this study, we aimed to identify characteristics and perceptions of intercultural MSI that could inform the design of interventions to promote this kind of interaction between intercultural strangers. Exploring how the findings of this study could contribute to this aim, we conducted an idea generation session with four designers to transfer the presented findings into designing interventions. The First 3 authors of this study participated as facilitators, leading the activities and discussions of each group. In this session, we created 3 groups and gave each group a design challenge to ideate on. Each group had 45 min to create a design solution. This session was conducted online, through Zoom and Miro. We created a Miro board for each team, stating their challenge and all the characteristics introduced previously. Participants performed 2 idea generation activities. The first was individual brainstorming and the second, group brainstorming. Then, participants created a journey map (Table 3) to detail their idea concerning each key moment we presented in the previous section. Finally, at the end of the brainstorm session, the participants were asked to further explain how the characteristics and the perceptions could support their design process.

Challenge 1: How can we Design Technologies that Facilitate Intercultural MSI Through Bonding Experiences While in a Park? During the brainstorming session, Group 1 ideated a digital environment that suggests interactions by identifying common grounds based on similar location journeys. The group argued that people often need to find and pursue excuses to start an interaction. Therefore, they were inspired by a situation in which someone drops a napkin and then another person picks it up and gives it back, hence igniting an interaction. Consequently, people can "drop" traces of their journey, which others can collect digitally according to their location to find similarities and start an interaction. Hence, even though the challenge was located in a park, these interactions can be started practically anywhere with geolocation. Further, this environment could create interaction possibilities by removing the need to be simultaneously physically present. Therefore, the digital environment would act as a common ground detector which enables people to leave traces about themselves and their journey to further match them with the people who have a common ground (e.g., common interests, activities, hobbies), and finally, suggest new activities for them to experience together (e.g., a concert which they both might be into).

The group explained that this tool can facilitate both 1:1 or group size interactions. They also suggested it could promote feelings like bonding, feeling attached, and motivation, thus allowing for new friendships to be made. Finally, we provided them the park as a location, yet they produced an idea where the park itself was an excuse to have a meaningful interaction, not the meeting location. This point is interesting as we observed that while ideating, geographical locations can act as inspiration for creating ideas about other places as well.

Challenge 2: How can we Design Technologies that Facilitate Intercultural MSI Through Kind Experiences While Using the Bus? During the brainstorming session, Group 2 redesign the bus trip experience to promote intercultural MSI. This was done through interventions that motivate people to give and take favors from others. These favors could range from helping with homework, sharing advice, helping with small daily tasks like giving directions, recommendations, or simply spending time with them during the bus trip. By doing so, the bus is redesigned as a space to casually volunteer, be kind, help, share knowledge or spend meaningful small moments while on the bus trip. Hence, it takes advantage of waiting time to do something kind and meaningful.

The group proposed 3 interventions on the bus. First, a reserved seat sector for people who would like to give or receive a favor. Second, public announcements to share needs and to ask for volunteers to provide help. Third, a button that counts the amount of MSI developed by either giving or receiving kindness. With this, kindness givers and takers could reflect on their interactions as well as encourage others to be kind in their turn. According to this Group, not only does the intervention focus on igniting the interaction by making the announcements and providing the reserved seats, but it also intervenes in the meaningfulness stage by ensuring the presence of kind acts. Besides, it might allow for reflection to be started through the meaningfulness button which materializes both the end and the realization of the meaningfulness of the interaction.

Challenge 3: How can we Design Technologies that Facilitate Intercultural MSI Through Learning While in the Museum? During the brainstorming session, Group 3 ideated an intervention for facilitating intercultural MSI using the idea of the museum as a setting that presents its attendants with stories of the past. In that sense, this group took that idea and extrapolated it to people from different cultures who also have different experiences and stories. Hence, through learning those stories the group thought that interactions could be facilitated. Moreover, reflections about others and their experiences could be ignited. Also, they discussed that people go to the museum to learn, and once they learn something their perspectives change as well. The general idea was to have a hybrid "museum of everything", in which attendants to the museum bring a "piece" of themselves to share in the museum. This piece would be exhibited in the digital archive of the museum as well as in a physical space of the museum through digital projections. Each piece would also have a small description to which people that go to the museum can reply and leave comments. Participants can also decide to send a digital postcard with the piece to someone else they would like to share it with. The group called it the museum of everything because this hybrid idea can be applied to connect different museums in the world. For example, someone in Istanbul can access pieces from people attending museums in other cities and countries as well. With this, they aimed to connect people from different cultures through their experiences and pieces of their lives. After uploading and commenting on a piece, they can meet the author of the pieces they liked if they wish to, hence, creating a network of people learning about each other.

Participants' Reflections about the Findings and Activity.
With the brainstorm session we conducted and related comments about designers' experience, we have initially identified different uses of the findings in different design tasks (i.e., ideating, detailing, and diversifying solutions). We have observed that when aiming to ideate new solutions, the findings seem to be effective to signal new ways of approaching intercultural MSI. The latter is exposed by the ideas created during the brainstorming. However, it is relevant to note as well that not all the characteristics and findings presented might be used at the same time to ideate because they might restrict too much the space of interventions. Therefore, after discussions with the designers who took part in the brainstorming sessions, we suggest to other practitioners to select at most 3 of the presented characteristics and integrate them into a design challenge. This because the participants agreed that integrating more characteristics could be constraining for creativity. Concerning place selection, it seems to be quite relevant for ideation because participants during the activity were being inspired by connecting the place their challenge was located in, with activities that can be developed in that place.

We also observed that the 4 key moments journey map (Table 3) helped highlight crucial actions to develop the interaction and creating ideas with a holistic approach. This was showcased by the fact that the 3 ideas created focused on promoting intercultural MSI

Table 3. Journey map of each idea.

	During the interaction			After
Idea	Ignite	Meaningful moment	Closure	Reflect
Museum of everything	Invitation to upload or share a piece of yourself	Learning about others and reflecting on our own experiences	Invitation to meet the author of the piece or send a postcard of the piece to a significant other	While deciding what to share of themselves and when replying to a piece, sending the postcards
Bus kind	Seating in a reserved space Announcements asking to give or take favors	Promoting kindness through receiving or giving help or a favor Sharing experiences, listening to others' advice, etc	People can offer future help or just leave the bus at their stop	Clicking on the "meaningfulness button"
Leaving Traces	People leave digital traces of their journey, so the common ground detector identifies it	Matching with people who have similarities	Invitations to meet, spend time together about their similarities	Not included

by going beyond starting an interaction, promoting other feelings than unexpectedness and serendipity, and integrating considerations to open up meaningfulness, closure, and reflection stages.

However, we also received comments about other ways of applying these findings to other design tasks. For example, when the aim is to detail a drafted design solution, it was observed that the complete list of characteristics could be of use to determine these elements of the solution. In the same way, when aiming to diversify existing solutions these findings might also present some benefits. For example, practitioners could ideate on how the existing solution would be concerning different elements. For example, Next2you [41] is a proximity-based social mobile application that encourages interaction between people who are regularly within close proximity of each other. For redesigning this app from an MSI perspective, we could think how it would instead of promoting interpersonal communication (i.e., 1 to 1 group size), promote group to group, physically close meetings through learning about each other. We understand that the characteristics, and specifically the elements of intercultural MSI, might look like constraints for the design process. However, by presenting and relocating existing products within new contexts and requirements, new features can come to life that directly addresses ways of promoting intercultural MSI, for example, giving the option for group interactions rather than just pairs of people.

6 Design Considerations

With the characteristics and perceptions introduced above, we contribute to the HCI and design literature by highlighting intervention spaces for new technologies as well as informing decisions during the design process of such solutions. Therefore, we have created 4 design considerations based on the findings of 56 real-life stories about intercultural MSI and the brainstorming session using those findings to ideate technological solutions. In this section, we discuss those design considerations.

6.1 Identify Different Elements' Configurations to Go Beyond Stereotypical Interactions

Through this study, we have laid the possibility and need for technological interventions to go beyond the igniting stage to promote intercultural MSI. This is even more relevant in the case of intercultural interactions which tend to be avoided. Through this study, we have understood that interactions to be meaningful need to pass from the igniting stage to a meaningful moment, and technologies aiming to promote intercultural MSI would focus on how to frame that meaningful moment.

We have presented an outline that can be helpful to determine different ways in which technologies can intervene to promote meaningfulness in each stage of interactions. Additionally, we have presented characteristics that can support practitioners to go beyond stereotypical interactions (i.e., interactions in pairs through oral communication). These findings can be used to think about other frontiers of interactions as they highlight spaces to create new solutions that promote intercultural MSI using different configurations of its elements. Consequently, we suggest practitioners explore different configurations of the presented characteristics, and specifically of the elements presented, to propose solutions that persist beyond the igniting stage and are grounded in different ways of promoting meaningfulness. Hence, interesting cases to further explore are how groups of people can be involved in intercultural MSI in different public places, with different kinds of impacts (i.e., kindness, bonding, and learning) of intercultural MSI and different element configurations. For example, how technology can support groups of people bonding, briefly, in bus stops; or, how can technology support crowds of people meaningfully meeting thorugh kind actions while transiting on streets. The aforementioned examples are just cases to exemplify how the presented elements can signal new configurations of intercultural MSI to go beyond the typical interactions previously explored in the collocated interactions field (i.e., 2 people interaciton through oral comunication). Hence, the recommendation here is to take the list of elements and analyze them to evaluate new alternatives for new or existing technological solutions.

6.2 Carefully Selecting Context of Interaction for Linking Places with Meaningfulness

Through the analysis of the stories, we have grasped the relevance of understanding the context of interaction to promote encounters that fit with it. People are generally in a place not only with an aim, time availability, and also with a specific level of mobility. These can influence the engagement or avoidance of interactions. Additionally, most of the stories collected happen while people are going somewhere (i.e., transporting), connecting with nature, or trying to explore a culture. Consequently, these findings signal ways of connecting people in places that might be considered as appropriate situations for promoting intercultural MSI.

Therefore, the suggestion here is for practitioners and researchers to carefully consider the context of the interaction to make a better fit between places, mobility levels, and situations of interactions. For example, future studies could explore how we can use the waiting time to design technologies for public places to promote intercultural meaningfulness by sharing or exchanging kind actions or even physical objects; also, how can we design for people to meaningfully connect while commuting to different places of the city or even different cities. Further, participants' responses can be an indirect way to receive feedback on the practicality of spaces to engage meaningful interactions based on a relevant experience. Hence, future research could explore how different public places and services could promote intercultural MSI by analyzing its current context according to the characteristics presented.

6.3 Pair Aspects that Make MSI Meaningful with Related Activities

In this study we have presented 3 actions that can make intercultural interactions meaningful, namely MSI through bonding, kindness, and learning. These impacts also signal specific activities that can be related to specific kinds of MSI. For example, bonding can be facilitated through different activities that highlight similarities between people. At the same time, design and technology can be of use when exploring how to highlight those similarities by proposing new designs of places, services, or technologies. Hence, practitioners and researchers might benefit from considering these activities when proposing technological interventions that promote intercultural MSI. We believe that the findings of this study might inform more situated interventions, concerning activities, places that allow such activities, and elements of interactions associated with those activities. For example, one such case could be, as expressed on the brainstormed ideas, museums facilitating learning, not just about history, but about other humans and their experiences. This example signals how practitioners and researchers might make use of linking specific actions to promote a specific kind of MSI in a suitable context or place.

6.4 Exert Feelings to Promote Meaningfulness

Another interesting aspect of our findings was the associated feelings of intercultural MSI. It is interesting for us, and we believe for our field, that even though these interactions were developed with intercultural strangers (i.e., people that present cultural differences besides being strangers), these interactions have a high emotional component. Hence, we believe, HCI can make use of promoting feelings when aiming to encourage these interactions in public places. Consequently, future work can explore how spatial design, technologies, and services can induce the associated feelings in people; and how regular activities developed in specific places can also exert these feelings. For example, many stories related to feelings such as sharing and intimacy which relate to MSI facilitate the connection of people through expressing individuals' feelings and thoughts, even to strangers. These conversations seem to facilitate intercultural MSI. Hence, the question here could be how we can, through design and technology, facilitate mutual feelings of vulnerability and intimacy safely. At the same time, how we can share more of ourselves, our cultures, and our identities to connect with others. Making a case, we could take the idea developed in challenge 1 (i.e., digital environment that leave traces about participants and their journeys to match them with people who share a common ground) and make use of these traces to connect people through places where they have felt such feelings and from these, start interactions with one another.

6.5 Limitations and Future Work

This is an initial study, and it presents different limitations. One of them is the method followed to collect stories. As this study was conducted while experiencing high restrictions to in-person contact due to the COVID-19 pandemic, we took an online approach. This, even though it facilitated the collection of several stories, also constrains the public that could access the activity. Besides, this fact limited the opportunities for researchers to make further questions to the participants about their stories which could complement the data collected. Finally, another limitation concerns the representation of different migrant communities and local people that participated in this study, as we could not discriminate the number of participants from different cultures. However, as it has been explained previously, this work is more focused on unpacking characteristics and perceptions of MSI that can inform the design of technologies rather than providing data that is representative of all migrant communities. Hence, there might be other characteristics we have not found yet, which other researchers can contribute to identifying as well. Future work in line with this study could refer to applying the same method in another context, another country, or another group, also sharing the stories collected here and using them as props for further discussions about intercultural MSI. Finally, using the perceptions collected as design inspiration to promote intercultural MSI can be another road for future work in this field.

7 Conclusion

In this paper, we elaborated on the characteristics, perceptions, and phases of intercultural MSI along with four design considerations for designing interventions to support these interactions. We believe our work would inspire HCI researchers and practitioners in their efforts to study and support intercultural MSI by signaling new spaces for technological interventions as well as supporting ideating, detailing, and diversifying solutions. Previous work from HCI has explored meaningful experiences (for product attachment) and meaningful interactions (in the media landscape) in very different contexts than the one we have unpacked through this study (for positive intercultural contact). Therefore, the findings we have presented and discussed open up a space to create new technological interventions aiming to promote more meaningful social interactions between intercultural strangers. Hence, facilitating positive human-to-human contact among people that might normally avoid interacting. In the future, we will continue with implementing and testing the intervention ideas in real-life settings besides further formulating guidelines that frame intervention spaces for and locals.

References

1. Abujarour, S., Krasnova, H., Hoffmeier, F.: ICT as an enabler : understanding the role of online communication in the social inclusion of Syrian refugees in Germany, p. 17, January 2018
2. Atif, J., Alghunaim, A.: Tarjimly (2019)
3. Bargal, D., Bar, H.: A Lewinian approach to intergroup workshops for Arab-Palestinian and Jewish youth. J. Soc. Issues. **48**(2), 139–154 (1992)
4. Bengs, A., Hägglund, S., Wiklund-engblom, A., Majors, J., Ashfaq, A.: Designing for social inclusion of immigrant women : the case of TeaTime, 1610 (2018). DOI:https://doi.org/10.1080/13511610.2017.1348931.
5. Braun, V., Clarke, V.: One size fits all? what counts as quality practice in (reflexive) thematic analysis? Qual. Res. Psychol. **2020**, 1–25 (2020)
6. Braun, V., Clarke, V.: Reflecting on reflexive thematic analysis. Qual. Res. Sport, Exerc. Health **11**(4), 589–597 (2019)
7. Brown, C., Efstratiou, C., Leontiadis, I., Quercia, D., Mascolo, C.: Tracking serendipitous interactions: how individual cultures shape the office. In: Proceedings of the 17th ACM Conference on Computer Supported Cooperative Work and Social Computing, pp. 1072–1081 (2014)
8. Brown, R., Hewstone, M.: An integrative theory of intergroup contact. Adv. Exp. Soc. Psychol. **37**, 255–343 (2005). https://doi.org/10.1016/S0065-2601(05)37005-5
9. Carpenter, V.J., Overholt, D.: Designing for meaningfulness: a case study of a pregnancy wearable for men. In: DIS 2017 Companion - Proceedings of the 2017 ACM Conference on Designing Interactive Systems, 95–100 (2017). https://doi.org/10.1145/3064857.3079126
10. Clarke, V., Braun, V.: Thematic analysis. In: Teo T. (eds) Encyclopedia of Critical Psychology, pp. 1947–1952https://doi.org/10.1007/978-1-4614-5583-7_311
11. Fonseca, X., Lukosch, S., Brazier, F.: Social cohesion revisited: a new definition and how to characterize it. Innov. Eur. J. Soc. Sci. Res. **32**(2), 231–253 (2019)
12. Fonseca, X., Slingerland, G., Lukosch, S., Brazier, F.: Designing for meaningful social interaction in digital serious games. Entertainment Comput. **36**, 100385 (2021). https://doi.org/10.1016/j.entcom.2020.100385
13. Gaver, W.W.: Affordances for Interaction : the Social is Material for Design. September (1996). https://doi.org/10.1207/s15326969eco0802
14. Gifford, S.M., Wilding, R.: Digital escapes? ICTs, settlement and belonging among karen youth in Melbourne Australia. J. Refugee Stud. **26**(4), 558–575 (2013). https://doi.org/10.1093/jrs/fet020
15. Golsteijn, C., Wright, S.: Using narrative research and portraiture to inform design research. IFIP Conf. Hum.-Comput. Interact. **2013**, 298–315 (2013)
16. Grosse-Hering, B., Mason, J., Aliakseyeu, D., Bakker, C. Desmet, P.: Slow design for meaningful interactions. In: Proceedings of the SIGCHI Conference on Human Factors in Computing Systems, New York, NY, USA, 3431–3440 (2013)
17. Harney, N.: Precarity, affect and problem solving with mobile phones by asylum seekers, refugees and migrants in Naples, Italy. J. Refugee Stud. **26**(4), 541–557 (2013). https://doi.org/10.1093/jrs/fet017
18. Harrison, S., Dourish, P.: Re-place-ing space: the roles of place and space in collaborative systems. In: Proceedings of the 1996 ACM Conference on Computer Supported Cooperative Work, pp. 67–76 (1996)
19. Hassenzahl, M., Eckoldt, K., Diefenbach, S., Laschke, M., Lenz, E., Kim, J.: Designing moments of meaning and pleasure . Experience Des. Happiness Underst. Experiences, **7**(3), 21–31 (2013)

20. Hespanhol, L., Davis, H., Fredericks, J., Caldwell, G., Hoggenmüller, M.: The digital fringe and social participation through interaction design. J. Commun. Inf. **14**(1), 4–16 (2018). SE-Special Issue: Designing Participation for the Digital Fringe
21. Ingram, P., Morris, M.W.: Do people mix at mixers? structure, homophily, and the "life of the party." Adm. Sci. Q. **52**(4), 558–585 (2007)
22. International Crisis Group 2018. Turkeys Syrian refugees: defusing metropolitan tensions. Eur. Report. **248** (2018)
23. International organization for migration 2018. World Migration Report (2018)
24. Karimi, F., Génois, M., Wagner, C., Singer, P., Strohmaier, M.: Homophily influences ranking of minorities in social networks. Sci. Rep. **8**(1), 1–12 (2018)
25. Liaqat, A., Axtell, B., Munteanu, C., Cultural, I., In, B., Families, I.: Participatory design for intergenerational culture exchange in immigrant families : how collaborative narration and creation fosters democratic engagement, **5** (2021)
26. Light, A., Howland, K., Hamilton, T., Harley, D.A.: The meaning of place in supporting sociality. In: DIS 2017 - Proceedings of the 2017 ACM Conference on Designing Interactive Systems, vol. 2, pp. 1141–1152 (2017). https://doi.org/10.1145/3064663.3064728
27. Litt, E., Zhao, S., Kraut, R., Burke, M.: What are meaningful social interactions in today's media landscape? a cross-cultural survey. Soc. Media + Soc. **6**(3), 205630512094288 (2020). https://doi.org/10.1177/2056305120942888
28. Lucero, A., et al.: Interaction between nearby strangers: Serendipity and playfulness. ACM Int. Conf. Proc. Ser. **1**, 521–532 (2018). https://doi.org/10.1145/3297716
29. Lucero, A., Holopainen, J., Ollila, E., Suomela, R., Karapanos, E.: The Playful Experiences (PLEX) framework as a guide for expert evaluation. In: Proceedings of the 6th International Conference on Designing Pleasurable Products and Interfaces, New York, NY, USA, 221–230 (2013)
30. Lundgren, S., Fischer, J.E., Reeves, S., Torgersson, O.: Designing mobile experiences for collocated interaction. In: CSCW 2015 - Proceedings of the 2015 ACM International Conference on Computer-Supported Cooperative Work and Social Computing, pp. 496–507 (2015). https://doi.org/10.1145/2675133.2675171
31. Maitland, C., Tomaszewski, B.: Promoting participatory community building in refugee camps with mapping. Technology (2015). https://doi.org/10.1145/2737856.2737883
32. Maseltov App (2015). http://www.maseltov.eu/
33. McIlwaine, C.: Super-diversity, multiculturalism, and integration: an overview of the Latin American population in London, UK. Cross-Border Migr. among Lat. Am. **2011**, 93–117 (2011)
34. Mekler, E.D., Hornbæk, K.: A framework for the experience of meaning in human-computer interaction. Conf. Hum. Factors Comput. Syst. - Proc. **2019**, 1–15 (2019). https://doi.org/10.1145/3290605.3300455
35. Monastero, B., Lucero, A., Takala, T., Olsson, T., Jacucci, G., Mitchell, R.: Multimedia ubiquitous technology for opportunistic social interactions. ACM Int. Conf. Proc. Ser. **2018**, 545–550 (2018). https://doi.org/10.1145/3282894.3286058
36. Mondada, L.: Emergent focused interactions in public places : a systematic analysis of the multimodal achievement of a common interactional space. **41**, 1977–1997 (2009). https://doi.org/10.1016/j.pragma.2008.09.019
37. Navarrete, C., Huerta, E.: A bridge home: the use of the internet by transnational communities of immigrants. In: Proceedings of the Annual Hawaii International Conference on System Sciences. 6, February 2006. https://doi.org/10.1109/HICSS.2006.3
38. Nelson, H.G., Stolterman, E.: The Design Way (2019)
39. NETT.WERKZEUG (2017). http://www.nett-werkzeug.de/en/. Accessed 09 Feb 2019
40. Novy, A., Swiatek, D.C., Moulaert, F.: Social cohesion: a conceptual and political elucidation. Urban Stud. **49**(9), 1873–1889 (2012)

41. Paasovaara, S., Olshannikova, E., Jarusriboonchai, P., Malapaschas, A., Olsson, T.: Next2You: a proximity-based social application aiming to encourage interaction between nearby people. ACM Int. Conf. Proc. Ser. **2016**, 81–90 (2016). https://doi.org/10.1145/3012709.3012742
42. Paluck, E.L., Green, D.P.: Prejudice reduction: what works? a review and assessment of research and practice. Annu. Rev. Psychol. **60**(2009), 339–367 (2009). https://doi.org/10.1146/annurev.psych.60.110707.163607
43. Pettigrew, T., Tropp, L.: How does intergroup contact reduce prejudice? meta-analytic tests of three mediators. Eur. J. Soc. Psychol. **38**(2008), 922–934 (2008). https://doi.org/10.1002/ejsp.504
44. Pettigrew, T.F.: Intergroup contact theory. *Ann.* Rev. Psychol. **49**(1), 65–85 (1998). https://doi.org/10.1146/annurev.psych.49.1.65
45. Poria, S., Cambria, E., Winterstein, G., Huang, G.B.: Sentic patterns: dependency-based rules for concept-level sentiment analysis. Knowl.-Based Syst. **69**(1), 45–63 (2014). https://doi.org/10.1016/j.knosys.2014.05.005
46. Putnam, R.D.: Tuning in, tuning out: the strange disappearance of social capital in America. PS: Polit. Sci. Polit. 28(4), 664–684 (1995)
47. Sawhney, N.: Voices beyond walls: the role of digital storytelling for empowering marginalized youth in refugee camps. In: Proceedings of the 8th International Conference on Interaction Design and Children, pp. 302–305 (2009)
48. Sherman, D.K., Brookfield, J., Ortosky, L.: Intergroup conflict and barriers to common ground : a self - affirmation perspective, 1–13 (2017). https://doi.org/10.1111/spc3.12364
49. Simon, H.A.: The Sciences of the Artificial (2019)
50. Tausch, N., et al.: Secondary transfer effects of intergroup contact: alternative accounts and underlying processes. J. Pers. Soc. Psychol. **99**(2), 282–302 (2020). https://doi.org/10.1037/a0018553
51. Tromp, N., Hekkert, P., Verbeek, P.-P.: Design for socially responsible behavior: a classification of influence based on intended user experience. Des. Issues. **27**(3), 3–19 (2011)
52. Tsai, A.C.-R., Wu, C.-E., Tsai, R.T.-H., Hsu, J.Y.: Building a concept-level sentiment dictionary based on commonsense knowledge. IEEE Intell. Syst. **28**(2), 22–30 (2013)
53. Turkey: A Transformation from Emigration to Immigration (2003). https://www.migrationpolicy.org/article/turkey-transformation-emigration-immigration. Accessed 03 Feb 2019
54. Weibert, A., Ribeiro, N.O., Aal, K., Wulf, V.: "This is my story···": Storytelling with tangible artifacts among migrant women in Germany. In: DIS 2017 Companion - Proceedings of the 2017 ACM Conference on Designing Interactive Systems, pp. 144–149 (2017). https://doi.org/10.1145/3064857.3079135
55. Xu, Y., Maitland, C.: Participatory data collection and management in low-resource contexts: a field trial with urban refugees. In: Proceedings of the Tenth International Conference on Information and Communication Technologies and Development, pp. 18 (2019)

Intercultural HMIs in Automotive: Do We Need Them? – An Analysis

Peter Rössger(✉)

Beyond HMI, Hohe Street 4, 71032 Böblingen, Germany
peter.roessger@beyond-hmi.de

Abstract. The automotive business is a global industry, vehicles are developed globally for a global market. Due to cost, the number of variants limited. This leads to a delicate balance the vehicle industry needs to solve constantly. Users in different cultures have different preferences, experiences, use cases, and contexts of use. Technological standards differ from region to region. Localized user interface solutions are useful to fulfill the different needs. On the other hand, automotive OEMs try to limit the number of vehicle variants to keep cost and complexity of vehicle development and production low. Cars often carry a brand related image; they relate to specific user experiences. For example, Porsche cars are recognized as typical German and carry the respective image.

Keywords: Automotive industry · Intercultural HMI · Usability · User experience

1 Introduction

This paper describes cross cultural differences in perception and rating of human-machine-interfaces (HMIs). The focus will be on automotive driver information systems, but the generic approach allows a transfer to other domains. Focusing on driver information systems also adds the highly relevant aspect of safety and driver distraction. A row of studies indicates differences in use and cognition of HMIs. German user for example appreciated a high value look & feel of the interfaces, ease of use and strict use of German as the HMI language. US Americans valued an easy interaction process. Japanese accepted complex interaction procedures if a high-tech look & feel was provided.

Studies from other domains support the assumption, that cultural aspects are highly relevant for the success of a product. On the other hand, the use of smartphones, PCs, and other devices with one HMI solution may equal intercultural differences.

This paper analyzes background, parameters, situation, and perspective of intercultural HMI designs in cars. Vehicle specific HMI parameters will be discussed. The situation of today will be analyzed. Possible solutions will be presented.

1.1 Background

Writing about culture and intercultural differences is always a dangerous endeavor. It is easy to step into stereotypes, to overgeneralize, to mis-interpret, and to mis-understand.

C. Stephanidis et al. (Eds.): HCII 2021, LNCS 13094, pp. 584–596, 2021.
https://doi.org/10.1007/978-3-030-90238-4_41

If you talk about "The Germans", "The Asians" or "The US-Americans" you will never meet everyone, there will always be exceptions, and of course it may be inadequate to categorize humans in such a way. On the other hand, it makes sense to think about intercultural differences to maximize the value of technology.

Technology is not a value, just by its sheer existence. It becomes valuable when it is used, applied, makes a difference. Technology has a value when it makes people's lives better. Better by means of faster, safer, easier, or only just funnier. Many technologies are developed, produced, and sold in a 'one-size-fits-all' mentality. Independent from user preferences, needs, wishes, and dreams, independent from use cases and contexts.

Over the years a certain awareness on cultural differences was established. Humans have different experiences, backgrounds, ideas, mind sets depending on the cultural they grew up in depending on where they life on the globe.

This paper reflects on the influences of intercultural differences on HMI (human machine interfaces) of cars. Vehicles are developed in a comparably small number of countries and cultures. The developers reflect their ideas, their thinking, and their values into the technological solutions. Cars and trucks are sold and used globally. We face a user group that belongs to a different cultural group than the developers. This may lead to misunderstanding, confusion, and sometimes rejection of technologies [1].

The HMI design of infotainment systems (the highly integrated technologies representing functions like navigation, media, radio, phone, internet access) is critical under three aspects:

1. Safety: the complex systems shall be used safely. A high level of intuitiveness allows an interaction with low mental capacities. Since the interaction with infotainment HMIs in vehicles is a secondary or tertiary task, sufficient mental resources shall remain for driving, which is the primary task [1, 2].
2. Efficiency: the technologies cost money, they eat up resources, they take room on dashboards and in the brains of their users. So, a certain price is paid. An HMI shall allow efficient access to useful functions. It shall be easy, fast, and ideally fun [1].
3. Branding: the HMI is part of the interior of a vehicle. Automotive brands differ more and more in design since technology gets continuously standardized and optimized.

It makes sense to have a closer look at intercultural differences [4–6, 20], to analyze the situation in automotive technology, and to discuss the interrelated influences between consumer devise like smartphone, tablets, and PCs on one side, and the vehicle world on the other [3]. Culture influences the expectations of users in and interaction with technological systems [26].

1.2 Definition Culture

Early sources define culture as a complex whole including knowledge, believes, moral values, art, law, and customs [7]. Later, culture was defined as the creator of identity of groups and group processes. Today culture is defined as the mental programming of humans that belong to a group. Opposed to earlier definitions of culture, focusing on high culture this also includes ordinary objects, acts, or values. Cultural values are learned, not genetically determined [9].

Figure 1 shows the relationship between personality, culture, and human nature. Human nature is genetically determined, all humans have it in common. This very often refers to physiological abilities, how far can she reach, how much can he lift. Also, visual parameters, sensing in general, the needs for food, basic human relationships, the need for safety are in this category. On the other end is the personality of humans. Each and every one of us is unique, has unique abilities, thoughts, talents, targets, an individual life. Even in very homogeneous groups like police, businesspersons, or medical staff, individual differences are ubiquitous. This is determined by both: learning and genetics.

The middle layer represents culture. It defines groups, groups have certain culturally defined parameters in common. Culture is learned. Most of it before the age of 12. We can adapt to other culture at a later age, we can learn and understand it. But the core measure we use is what we learned as a child [8, 9].

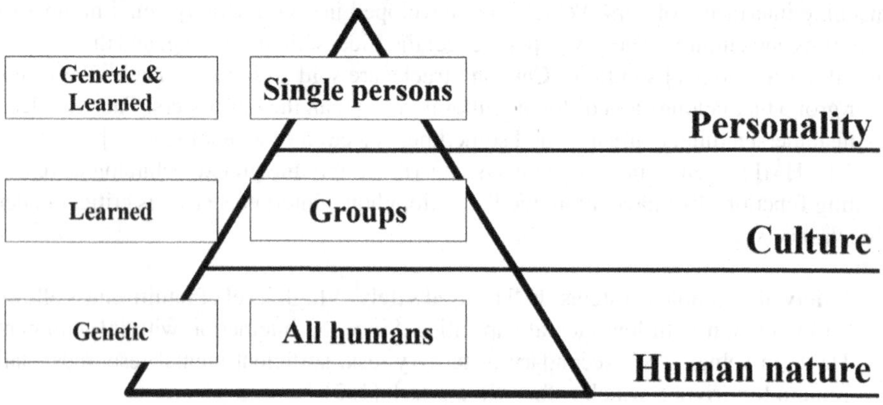

Fig. 1. Personality, culture, and human nature.

Core of all definitions is a common set of values, acts, views etc. that distinguishes one group from another. It allows the communication between group members, influences behavior, learning, and cognitive processing [20].

Often this is determined by geographical differences. But also, with-in one region cultural differences occur, for example steered by personal preferences or education. So, a typical biker culture exists, or a common culture in groups of sports fans.

Culture has explicit components, in explicit, and unaware. Only the explicit components are visible. Conflicts occur on the other levels [20]. For this paper, if not mentioned differently, the geographical culture is in focus, the country, the region users, developers, vehicle producer are located in.

2 Parameters of Intercultural Design

This chapter discusses a few typical parameters that differ between different cultures. They are analyzed under the aspect of implementation in automotive HMI solutions. The rules are generic, they apply for other technologies and use cases as well. This paper focusses on in-vehicle applications. The mentioned parameters serve as examples only.

2.1 Fonts

Cultures are highly determined and represented by the language [10]. Depending on the language different sets of characters are used in written texts. Most European languages use the Latin character set, with language dependent special characters. During HMI implementation it is important that the used fonts provide the respective characters.

With-in Europe, some languages use Cyrillic characters, Greek and Georgian have own sets. Asian characters provide, opposed to European fonts, more content with-in a single sign, like for example syllables. In addition, Chinese for example has 20.000 characters [11].

Fig. 2. Selected fonts used in different parts of the world.

Figure 2 shows some examples of fonts used in different languages and different regions of the world. They all have specifics and characteristics that provide challenges for the development of HMIs.

- Left column from top to bottom:
  - Latin: the font used in most Western languages like English, German, French, Spanish, Polish etc. Each of these languages (besides English) has a few additional language specific letters. Examples are the German ö, the Swedish å, or the Spanish ñ. The letters are separated, the number depending on the language is between 24 and about 35. Can be typed with typewriters using a 1:1 philosophy, meaning, every letter has one key.
  - Greek: used in Greece. Handled in the same way as Latin characters.
  - Chinese: based on icon style graphics the Chinese letters where introduced. Different versions are available, the overall number exceed 20,000. Hard to type with a typewriter, usually phono graphics in English is used. This means that a word is typed using the English alphabet and then turned into a Chinese font.

- Georgian: just an example of an exotic and outside Georgia hardly known font.

- Middle column top to bottom:
  - Japanese: two different fonts are known in Japan; one is like Chinese.
  - Thai: unique font, not used in other languages.

- Right column top to down:
  - Cyrillic: used for Russian and a few other Eastern European languages. Separate letters like in Latin fonts, 33 letters in the Russian version.
  - Korean: unique font used for Korean only. Each group consists of a consonant-vowel-consonant combination.
  - Hebrew: written right to left, which makes a large difference to other languages mentioned.
  - Arabic: connected letters, written right to left.

The use of different fonts in different languages has clear consequences for the implementation of HMIs. The respective fonts and letters shall be provided. Keyboards shall be implemented. For the use of non-Latin, Greek, or Cyrillic character different means of input are required. A specific challenge is the use of Hebrew and Arabic since an input from right to left is required.

Another problem is the length of words. German or even more Finnish use extremely long words, English is clearly shorter. Chinese, Korean, and Japanese require even less space.

2.2 Units and Formats

For units two different systems are used globally: the metric system and the imperial system. The imperial system is used in the USA, Liberia, the Virgin Island, and the Northern Mariana Islands. It is partially used in the UK, Thailand, North Korea, and some smaller former British overseas areas. The metric system is used in the rest of the world.

The metric system is a closed system, different kinds of units like distance, areas, and volumes are based on each other. Examples:

- 10 mm add up to a centimeter.
- 1000 mm add up to a meter.
- 1000 m add up to a kilometer.
- 1 m by one meter make a square meter (apartments for example are measured in square meters, mine has $95m^2$).
- A volume of 10 cm by 10 cm by 10 cm makes a liter.

The imperial system uses independent units with one kind of units, like distances, and between different kinds of units. The is no natural relationship between the units. Examples:

- 12 inch make a foot.
- 3 foot make a yard.
- 1760 yards make a mile.
- One acre is 43,650 square foot.
- One square foot is 1,728 cubic inches.

The transcription of number, dates, and values differs between different cultures. Examples:

- 10/11/12 is the 12th of November 2010 in many parts of Asia, the 11th of October 2012 in the USA, and, if written this way, the 10th of November 2012 in Europe. In Germany the date is written in the format dd.mm.yy (or yyyy) [13].
- 12,000 is twelve in Germany and twelve thousand in the USA, 12.000 vice versa.
- Temperatures in the US are degrees Fahrenheit, in Europe degrees Celsius.

These differences need to be considered when designing HMI solutions for different markets. The correct use of units and formats is essential for usability and user experience of devices. Not following the respective rules will miss acceptance and market success.

2.3 Colors

Colors are an excellent example of intercultural differences. Opposed to characters and measure an emotional component comes along.

In HMI design colors play a vital role. Besides the sheer esthetic aspect, colors offer carry additional information. Green usually indicates "good", "ok", or "uncritical". Red means the opposite. Other connotations between colors and meanings are known, like white for innocence or purity and black for grief and elegance. This attribution is culturally determined and differs from culture to culture. In Asia red is not necessarily connected with critical conditions, and in some parts of China white is the color of grieving [14, 15].

Table 1 shows a selection of colors and countries and the respective meanings colors have (based on [15]). The most obvious issue is the meaning of red in Asian cultures. Seen as an indicator for danger in the Western world, red does not carry that meaning in China or India. For an automotive HMI development this may cause problems when red is used as a carrier of information only.

The use of colors is emotionally loaded. This may lead to inadequate emotions of the user. In Germany dark, black dominated HMIs are elegant, for Asians they may look sad and dull. On the other hand, the Asian colorful websites or infotainment systems look childish and inappropriate to European users.

2.4 Gestures and Icons

Gestures have an extremely culture driven context [16, 17]. Some, like the stiff middle-finger work globally, but most have totally different meanings in different regions of the world. A positive or neutral gesture in one country may be deeply insulting in others.

Also, the use of icons carries the chance of misinterpretation. Icons are often simplified representations of real objects. Here two issues may occur:

Table 1. Meaning of colors in selected countries.

Country	Color	Meaning
USA	Black	Authority, death, grieving, eternity, sin, style
	White	Heaven, luxury, wedding, purity, truth
	Yellow	Cowardice, energy, fun, peace, fighting the evil
	Red	Danger, anger courage, excitement, heat, love
	Green	Luck, growth, envy, nature, safety
Germany	Black	Death, grieving, luxury, style, design
	White	Purity, innocence, cleanness, wedding, goodness
	Yellow	Envy, lie, fun, summer, sun
	Red	Danger, anger, heat, love, erotic, speed
	Green	Bio, nature, environment, envy
China	Black	Fun, party, power, thinking, money
	White	Purity, truth, death, grieving, erotic
	Yellow	Health, silence, power, money, respect, kingdom
	Red	Wedding, fertility, luck, success, happiness
	Green	Growth, life, defending the evil
India (Hindi)	Black	Sin, anger, money
	White	Truth, death, grieving, intelligence, peace
	Yellow	God, sickness, personal energy, defending the evil
	Red	Erotic, heat, energy, passion, wedding
	Green	Life, sympathy, the evil, attention, religion, love

1. The simplification itself is a source of misinterpretations.
2. The selection of objects serving as a base.

A particular source of irritation may be the selection of gestures as a basis for icons. And finally symbols like stars or crosses have a strictly cultural connotation.

For automotive HMIs gesture recognition plays a growing role. An upfront analysis is required which gesture work at all, and which gestures should be avoided were. Icons often reflect local habits, technologies, or interactions. This may lead to irritations when used globalized.

2.5 Legal, Religion, and Politics

Legal differences will lead to intercultural HMI differences [20]. Saudi Arabia for example requires automotive companies to provide Arabic navigation maps. Ignoring intercultural phenomenon may also have a political or religious component [12]. Frictions between countries may for example influence navigation map data. Vehicles shipping to China should represent Taiwan as a Chinese province, in Taiwan as an independent country. The same is valid for the West Bank and the Gaza strip in Arabic countries and Israel.

In Arabic countries green and violet are colors with a strong religious connotation. So, they should not be used for HMI designs. Due to the restrictions on alcoholic beverages, these should not be used on icons to indicate restaurants or bars. Similar restrictions occur when showing human limbs [12].

2.6 Interaction Devices

Over the years various interaction devices were used in the automotive industry. Traditional automotive user interfaces had been buttons, knobs, and sliders until BMW and Audi started using central controllers to manage the growing functionality and complexity on dashboards.

Other carmakers started using touchscreens earlier than others. Both kinds of interaction devices have pros and cons, which is not the focus of this paper. Japanese users preferred touchscreens earlier than US and European drivers [26]. Today the standard solution is a touchscreen. In almost every car, particularly in budget cars, touchscreen-based interactions are established. Luxury vehicles use multimodal devices, integrating controllers, voice, gesture, and touchscreens.

The ubiquity of smartphones and tablet led to a global acceptance of touchscreens in the past 10 years. The intercultural differences faded over the years; consumer devices set the pace for the automotive industry.

2.7 The Challenge of Complexity

For drivers until today driving is the primary task. This may change when higher levels of autonomous driving will be introduced, but for the foreseeable future the driver will remain in the control loop of driving [21]. This means that the interaction with in-vehicle devices is a secondary (horn, wind screen wipers, lights, etc.) or tertiary (media, audio, navigation, phone, etc.) task.

An optimal level of arousal needs to be realized in drivers, so the cognitive load of the over tasks, primary, secondary, and tertiary, needs to be on an appropriate level. The cognitive load from secondary and tertiary tasks shall be a low as possible to allow focusing on the primary task. The complexity of secondary tasks shall potentially be as low as possible.

Although there is a tendency in Asian, particularly Japanese, user groups to accept higher levels of complexity in secondary and tertiary tasks [1, 3], it shall in traditional driving without higher levels of automation, be reduced to a minimum to provide maximum safety. For higher levels of support of the driver by the vehicle this paradigm may change to reach the optimal level of cognitive load and with that arousal and attention.

2.8 Information Density

Asian users tend to accept higher levels of information density on screens. This often leads to screen designs European users rate as crowded, confusing, and overloaded [1, 3, 19]. Due to safety reasons, driving is and remains for the foreseeable time the primary task in a car, automotive HMIs shall be focused on the required information, present it in an easy to detect, read, and process manner.

3 Culture and the Automotive Industry

The very first vehicles were developed in the late 19th Century in Europe, followed by activities in North America. The status remained for over 100 years, only Japan became a vehicle developing and producing country. Asia outside Japan was a market, development and production of cars happened in Europa. After the year 2000 Korea appear on the global automotive landscape. China entered a few years later. Today we face a shift of automotive activities from Europa and the US to China and Korea.

Over many years the mindset of American and European brands was, that every European car will be sold easily in China [22]. Today most car maker have special car meeting the requirements of Chinese users. An obvious example is the offer of longer version of the vehicle to allow more space on the rear seats. This is an intercultural issue: in China, a larger number of people are driven by professional drivers. The owner of a cars sits on the rear seat.

For the future, an ongoing shift of the automotive industry towards China is expected [22]. In the end, this will lead to the mindset of Chinese car makers, that you can sell every car to the Europeans…

3.1 The Role of Consumer Devices and Applications

The ubiquity of smartphones and tablets with their touchscreen-driven interactions, has two major consequences for automotive HMIs:

- Users not only accept but expect touchscreens in vehicles [23]. The easy and direct interaction leads to fast and easy interactions. It is easy to transfer interaction patterns like swiping, zooming, scrolling, pressing from the consumer devices into the automobile.
- Smartphones offer high functionality, flexibility, and personalization. The use of apps and app stores allows the installation of features and functions according to personal preferences, use cases, use contexts.

The smartphone, consumer, and PC market is dominated by three operating systems [24, 25], representing larger software packages and HMI rules. This includes the icon-based interaction on smartphones and tablets and location and functionality of back-buttons and main screen buttons.

The HMIs are not localized by a larger extent. The producers shape the interaction between users and their technical devices globally. The prominent use of these HMIs leads to a unification of expectations, abilities, and preferences.

Other domains are hit by this as well, including the automotive HMIs. This trend leads to a domination of any HMI, including automotive applications, towards the smartphone paradigms. The sentence: “Make it like a smartphone” is often heard in HMI discussions.

3.2 The 3-Layer Approach

A row of definition exists for internationalization, globalization, and localization [18]. In this paper the definitions of Goldsmith [12] and Heimgärtner [20] are used, since it very well supports the development and decision processes. Other authors [eg. 27] promote similar approaches.

Localization of an HMI means that more than one variant of an HMI will be developed. These variants reflect local differences in culture. Internationalization means that one HMI fits many cultures. Globalization is the creation of an HMI that has market potentials worldwide. Internationalization is the basis of a pyramid, globalization the middle layer, and localization the capstone [18].

The decision which strategy shall be used for an intercultural HMI development should be made early in the development process. Later changes of strategy cause high costs and lead to suboptimal results. A real globalization, meaning one HMI for the whole globe works for quite simple devices only.

Literature [12] mentions five aspects relevant for the localization of HMIs and products:

1. Technical aspects
2. Linguistic aspects
3. Cultural aspects
4. Political aspects
5. Esthetic aspects

Goldschmidt [12] suggests considering these five main aspects with a 3-layer approach (Fig. 3).

In layer 1 the technical basis for localization is set. Appropriate tools and processes must be selected, and the right data formats defined. In layer 2 the "hard" localization is performed, like correct units, characters, and formats. Layer 3 represents the cultural and esthetic aspects like taste, technology orientation and habits of users.

The arrows on the left of Fig. 3 indicate the direction of activities for thinking and acting during the development process. Thinking shall start at layer 3, with the "softer" parameters of the process and then lead to decisions on the more technical levels. The action, the technical development process, itself shall start with the selection of the required tools and technical boundaries to allow the realization of the core cross-cultural usability and user experience.

3.3 Process of Internationalization and Localization

Basically, two mechanisms of handling intercultural issues are possible [20]. Internationalization describes a procedure where large parts of a technological artefact are unified. Local differentiations are put on top of them. Localization is the adaption of a product to local cultural, technical, and HMI standards.

General strategies of internationalization and localization need to be defined early in the process. In a project the author conducted, sales and head of engineering required a unified HMI for US and EU markets. Development time and cost should have been limited to a minimum. The task of the HMI development was to find a solution fitting both markets. After about 50% of the development time, it was clear, that is will never be possible to serve both markets, both having different features, HMIs, and marketing

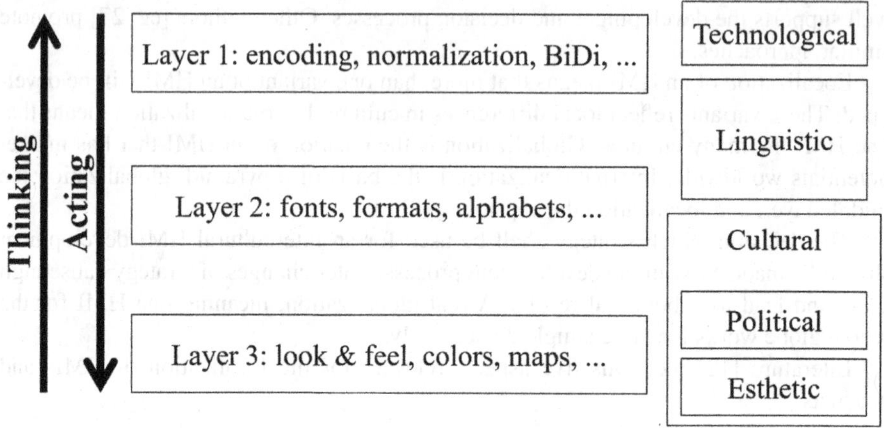

Fig. 3. The 3-Layer Approach for Intercultural HMI Developments

strategies with one solution. Enormous efforts were required to separate the development activities.

The strategy of internationalization and localization needs to be fixed before project start. Keeping an eye on cultural differences is required for any product sold ion international markets, particularly for car with their high functionality, high complexity, high levels of safety, and high prices.

4 Conclusion

Culture plays a significant role in the future of automotive HMI solutions. Intercultural differences shall be reflected in graphic design and interaction design. Based on the 3-layer model the following parameters need to be considered when designing for different markets:

- Parameters on layer 1: all technological and linguistic parameters. This is the basis for an implementation of HMI solutions. If they are not given, localization will not be possible.
- Parameters on layer 2: fonts, formats, units, alphabets. These are the basics of localization in HMI design.
- Parameters on layer 3: critical political and religious parameters like the use of colors and borders on maps shall be localized.

Other parameters on layer 3 need a closer analysis before localizing them. Depending in the expected user group interaction designs, colors, screen layouts and similar may be kept global, oriented towards the usual suspects in the smartphone and tablet area. Detailed research work is required to get an in-depth understanding on how far the global unification of HMI recognition, of user interactions is. It is not 100% but is larger than it was 20 years ago.

Localization of HMIs is required for vehicles, different markets need adaptions. The unification of user behavior based on the use of PCs, smartphones, and tablet computers plays a certain role, but does not eliminate the issues.

References

1. Rößger, P.: Fahrer-informations-systeme: situation und perspektive. In Karrer, G.S. (Hrsg.) Mensch-Maschine-Systemtechnik aus Forschung und Praxis. S149ff. Symposion (2005)
2. Young, K., Regan, M., Hammer, M.: Driver Distraction: a Review of the Literature. Monash University, Australia (2003)
3. Roessger, P., Rosendahl, I.: Intercultural differences in the interaction between drivers and driver-information-systems. SAE Technical Paper 2002–01–0087 (2002)
4. Heimgärtner, R.: Towards a Model of Culturally Influenced Human Machine Interaction. In: IWIPS 2010, Proceedings (2010)
5. Marcus, A, Baumgartner, V.J., Chen, E.: User-interface design vs. Culture. Evers et al. (Hrsg.): Designing for Global Markets 5. In: Proceedings of the 5th International Workshop on Internationalization of Products and Systems (2003)
6. Honold, P.: Culture and context: an empirical study for the development of a framework for the elicitation of cultural influence in product usage. Int. J. Hum. Comput. Inter. **12**(3), 327–345 (2000)
7. Scupin, R.: Cultural Anthropology: a Global Perspective. Prentice Hall (2011)
8. Hall, E.T.: Beyond Culture. Anchor Books (1976)
9. Hofstede, G., Hofstede, G.J.: Lokales Denken, globales Handeln. dtv Verlag (2009)
10. Gardt, A.: Beeinflusst die sprache unser denken? ein überblick über positionen der sprachtheorie. In: Andrea, L., et al. (Hg.) Sprache im Alltag. Beiträge zur neuen Perspektiven der Linguistik. Herbert Ernst Wiegand zum 65. Geburtstag gewidmet. Berlin/New York, De Gruyter (2001)
11. Karlgren, B.: Schrift und Sprache der Chinesen. Springer (2008)
12. Goldschmidt, D.: Workshop Arabic HMI. Internal Presentation (2009)
13. Wikipedia (as of 25 January 2021). Datumsformat. https://de.wikipedia.org/wiki/Datumsformat
14. Russo, P., Boor, S.: How fluent is your interface? designing for international users. In: Proceedings of the 1993 International Conference ACM CHI S 342ff (1993)
15. McCandless, D.: Das Bilderbuch des nützlichen und unnützen Wissens. Knaus (2010)
16. Broschinzky-Schwabe, E.: Interkulturelle Kommunikation: Missverständnisse – Verständigung. VS Verlag (2011)
17. Grosse, J., Reker, J., Bong-Kil, F.: Versteh mich nicht falsch! Gesten weltweit. Piper (2012)
18. Sikes, R.: Localization: the global pyramid capstone. MultiLingual, October/November 2012
19. Rößger, P.: Designing for world markets: HMI and cross cultural usability. Presentation at the conference Automotive Cockpit HMI. Bonn, Germany, 29 September 2012
20. Heimgärtner, R.: Intercultural user interface design. Springer (2019)
21. Kolrep, H., Röse, K., Gruhlke, F., Jürgenssohn, T.: Mobile anwendungen im kraftfahrzeug – mensch-maschine-interaktion und Akzeptanz. In: INFORMATIK 2003 - Innovative Informatikanwendungen, Band 2, Beiträge der 33. Jahrestagung der Gesellschaft für Informatik e.V. (GI), 29. September - 2. Oktober 2003 in Frankfurt am Main (2003)
22. Dudenhöffer, F.: China auf dem Weg zur Industrieführerschaft in der Autoindustrie (2018). https://docplayer.org/75781391-China-auf-dem-weg-zur-industriefuehrerschaft-in-der-autoindustrie.html. Accessed 25 Jan 2021

23. Franz, B.: Entwicklung und Evaluation eines Interaktionskonzepts zur manöverbasierten Führung von Fahrzeugen (Doctoral Dissertation, Technische Universität Darmstadt) (2014)
24. Statista (2020). Marktanteile der führenden Betriebssysteme weltweit von Januar 2009 bis September 2020. https://de.statista.com/statistik/daten/studie/157902/umfrage/marktanteil-der-genutzten-betriebssysteme-weltweit-seit-2009/. Accessed 25 Jan 2021
25. Statista. Marktanteile der führenden Betriebssysteme am Absatz von Smartphones weltweit vom 1. Quartal 2009 bis zum 2. Quartal (2020). https://de.statista.com/statistik/daten/studie/73662/umfrage/marktanteil-der-smartphone-betriebssysteme-nach-quartalen/. Accessed 25 Jan 2021
26. Rößger, P.: An international comparison of the usability of driver-information-systems. In: Proceedings of the Fifth International Workshop on Internationalization of Products and Systems. Berlin (2003)
27. Sturm, C.: TLCC – Towards a framework for systematic and successful product internationalization. Presentation at the IWIPS 2002, Austin TX (2002)

Hybrid Kansei Research of Product's Interactive Design Experience Based on "Sensing" Technology

Min Shi and Cheng-wei Fan(✉)

School of Art and Design, Fuzhou University of International Studies and Trade, Fuzhou, China
fanchengwei@fzfu.edu.cn

Abstract. Along with the rapid development of 5G technology, a new economic output model based on information technology services has emerged in microelectronics technology, optical technology, biotechnology, new materials, new energy technology, and artificial intelligence. The interactive economic model in interactive design mainly emphasizes the unique experience brought by commercial activities to consumers, with themed experience design as its core. Therefore, the study on Kansei of interactive design with "Sensing" technology is essential to improve user "Acceptance" of the product while developing, designing, and using the product. The Kansei cognition of interaction design can improve the quality of product design and help enhance designers' originality in the process of interactive design with "Sensing" technology, achieving economic benefits and experience value, which is also the indispensable pursuit of "Depth Experience" for people who live in modern times. This article mainly studies the product's interactive experience with "Sensing" technology based on hybrid Kansei engineering. The design and research of factors based on user Kansei image for future products' interactive design with "Sensing" technology is the analysis of Kansei factors used in the control and use of somatosensory perception, such as touch, pressure, temperature, pain, and proprioception (feelings related to muscles, joint position and movement, body posture and movement, and facial expressions), which focuses on physical product design and product service in the product's interactive design with "Sensing" technology, covering hardware, software, and service design. Whether users are satisfied or not depends on the study on the product's interactive experience with "Sensing" technology based on hybrid Kansei engineering, and the product design in terms of intelligent interaction, multimedia interaction, virtual interaction, and human-machine interaction depends on Kansei analysis of factors that conform to user experience. This article analyzes the existing products without clear description by using interactive design with "Sensing" technology based on "Depth Experience" and verify if the product satisfies users' demands according to their Kansei appeals after using the product. The study focuses on "Human-centered" human-machine interaction technology, which enables products to meet customer service demands by creating and improving user satisfaction, and users can expect the result of the operation and express this behavior through the senses. Therefore, interaction design is the most effective way to express this behavior in behavior planning and expression.

C. Stephanidis et al. (Eds.): HCII 2021, LNCS 13094, pp. 597–607, 2021.
https://doi.org/10.1007/978-3-030-90238-4_42

Keywords: "Sensing" technology design · "Depth Experience" · Interactive experience · Hybrid Kansei engineering

1 Introduction

1.1 Background and Purpose

In the context of the rapid development of 5G technology, the AI of products has completely changed our life. A new economic output model based on information technology services has emerged in microelectronics technology, optical technology, biotechnology, new materials and new energy technology [1]. Product designers are also in the midst of significant technological infrastructure and rapid digital transformation. Therefore, in the design innovation in the future, the area of digital products boasts a unique advantage. The interactive economy form of interactive design focuses on the unique experience brought by business activities to the consumers, which has enabled us to adopt new ways of thinking and working. These ways will be reflected in the trend of US design [2]. To enhance the user's "acceptance" of the products, the Kansei research of product "Sensing" technology interactive design and the Kansei cognition of user and product's interactive design are essential to the product technology development, design and use, which can enhance the quality and interest of product design and help enhance the designer's originality in the process of interactive design with "Sensing" technology, achieving economic benefits and experience value, which is also the indispensable pursuit of "Depth Experience" for people in the modern times.

1.2 Scope and Methodology of Research

This article mainly studies the Sensing technology of sematosensory perception of five senses and mechanisms such as physical, chemical and biological effects when the Sensing technology works and serves as a "window" to capture the information of the research object. It provides the product design system with the essential information for people's unconscious Kansei in the products, explores the user's Kansei appeal in the products and analyzes whether the products meet the user's demands.

2 Literature Review

Sensing system is one of the main areas of AI development, which has become popular in the world since the 1980s when many advanced industrial countries began to introduce the Sensing technology and study and apply such technology and produce intelligent products through further research and development [4]. At that time, the Sensing technology was a leading industrial technology in the world. The application and development of sensing technology in products became an important symbol of industrial development at that time.

2.1 The Concept of Sensing Technology

Conventional sensing technology can collect the needed measure information and convert it into input signal, equivalent to human body's sensory organs. With the development of science and technology, the function of modern sensing technology has become more powerful, which not only can collect characteristic information, but also can read information, and even conduct analysis, decision-making and control. In other words, modern sensing technology not only has the functional map of sensory organs such as eyes, ears and nose [1], but also is equipped with a "brain" that can make decisions [5] (Fig. 1).

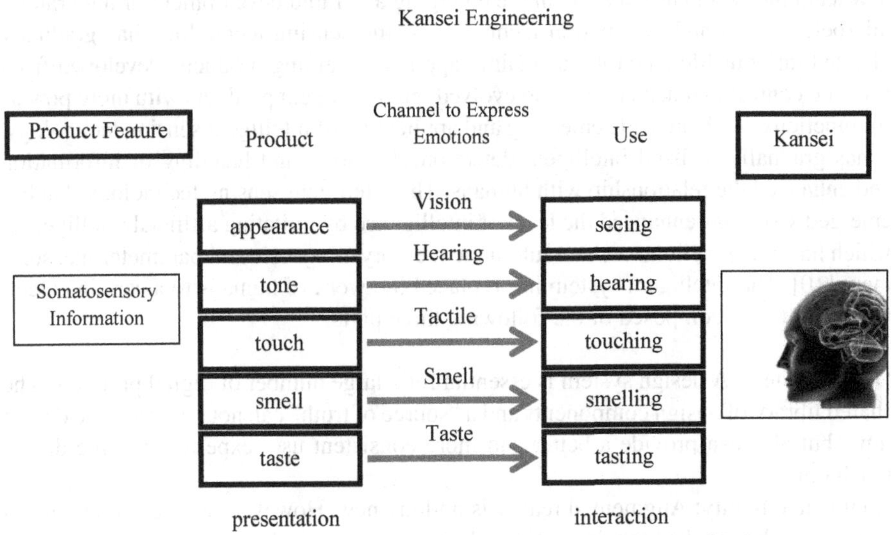

Fig. 1. Relationship between interactive design and Kansei engineering

The main role of sensing technology in the electromechanical automation system is to detect the working environment of the system, so as to provide information support for the operation of the system and equipment and improve the reliability of its operation. There are many types of sensing technology, which have different sensing functions. Therefore, in the electromechanical integrated system, a suitable type of sensing technology should be selected according to the detected object. Sensing technology can be divided into two categories by the nature of the measured object, namely internal information sensing technology and external information sensing technology. The former detects the internal information of the system such as position, pressure and temperature changes, while the latter mainly detects the external environment status such as temperature and humidity. The latter category can also be divided into more types, like tactile sensing technology, non-tactile sensing technology, laser rangefinder and ultrasonic rangefinder [6].

2.2 Kansei Development History of Sensing Technology

The core and foundation of sensing technology is computer, which uses network technology to collect, transmit and analyze information. It has been very common that sensing technology and devices are applied in the daily production and life of people [7, 8]. Figure 2 With the development of the times and the progress of science and technology, the functions of sensing technology keep innovating and improving and developing in the trend of integration, automation and digitization [9]. Viewed from its development history, the structure form of sensing technology has gradually developed from structural sensing technology such as the resistance strain type towards integrated and intelligent sensing technology, which has mainly benefited from the development of new technology. In particular, with the emerging and rapid development of integration, microelectronics and automation technologies, the sensing technology has gradually blended into our life when it was mainly applied in gaming products, developed from acoustic control to touch control and evolved into intelligent products with more powerful functions. With the wide emerging and application of intelligent sensing technology, it has gradually realized intelligent detection, diagnosis and handling of information and enhanced the relationship with humans. The intelligent sensing technology that has emerged so far has enhanced the level of intelligence by imitating artificial intelligence, which has new functions such as automatic memory, diagnosis and parameter measurement [10]. The intelligent system has replaced the work of humans to a certain extent, which is mainly composed of the following three parts.

Design system: A design system is essential for a large number of digital products. The shared library of design components and a "source of truth" can not only save the design time, but also can provide a better and more consistent user experience at the digital touch point.
Augmented reality: Augmented reality is nothing new. However, in recent years, it has been the widening the boundary of mobile user experience design.

3 Kansei Factor Analysis of product's Interactive Design Experience and Image Based on "Sensing" Technology

The future society will be a world full of sensing technology. Some argue that by dominating the sensing technology, one will grasp the new era. Therefore, the sensing technology is a commanding height scrambled for by people in the high-tech development in the 21st century. Starting from the 1980s, Japan has prioritized the development of sensing technology among high and new technologies. Western countries like America have also listed the technology among the key contents of national high-tech and national defence technology. The design and research of factors based on user Kansei image for future products' interactive design with "Sensing" technology is the analysis of Kansei factors used in the control and use of somatosensory perception, such as touch, pressure, temperature, pain, and proprioception (feelings related to muscles, joint position and movement, body posture and movement, and facial expressions).

Fig. 2. Application of interactive products

With the development of AI concept, wearable devices have become a hot field of intelligence after smart hand-held terminals. Applied in wearable devices, the inertia sensing technology can identify human's behaviors, which not only makes human-machine interaction faster and more convenient, but also can solve the problem of the weak interaction of smart wearable devices.

3.1 Generation of Kansei Factors in the Application of Sensing Technology

It focuses on physical product design and product service in the product's interactive design with "Sensing" technology, covering hardware, software, and service design. Whether users are satisfied or not depends on the study on the product's interactive experience with "Sensing" technology based on hybrid Kansei engineering, and the product design in terms of intelligent interaction, multimedia interaction, virtual interaction, and human-machine interaction depends on Kansei analysis of factors that conform to user experience (Fig. 3).

Humans are the subject of activities. Under this model, the solid square represents the users with the three abilities. The more it goes outward, the more serious the lack of such ability is. For example, different languages, cultures and gender identity, simply speaking, can be defined as: Design is made for different communities based on a conscious consideration of these communities. In the application of virtual technology, with the development of technologies such as artificial intelligence and multimedia technology, the sensing technology has realized an integrated application of Kansei, optimized the products and simplified the process of Kansei technology [12]. In actual production, combined with microelectronics technology and communication technology, Kansei has been enhanced effectively. The application of integrated Kansei technology has unified the quality of products and service.

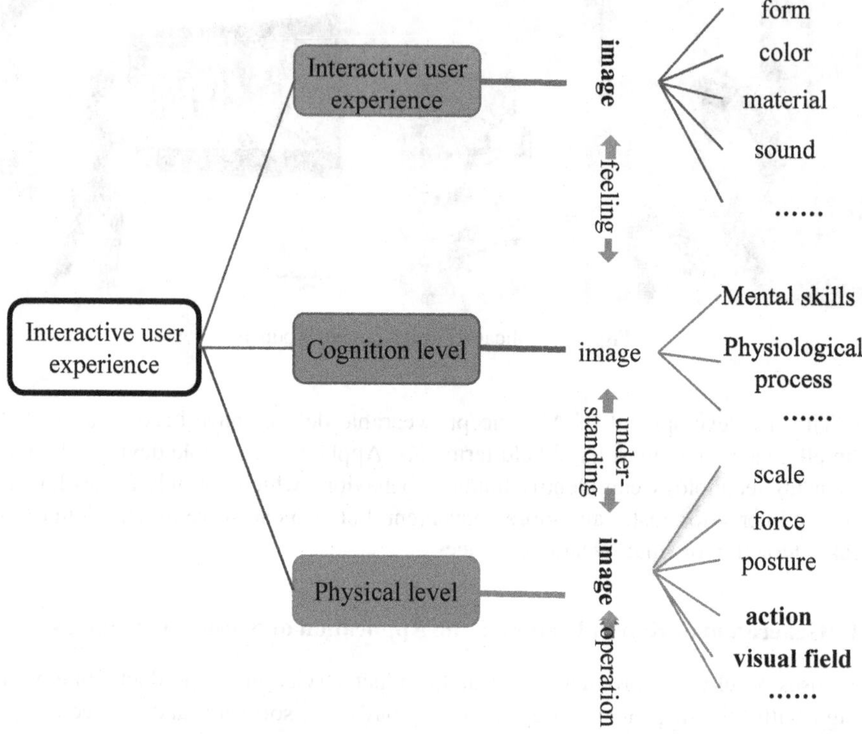

Fig. 3. Kansei factor analysis of interactive users

3.2 Research on Kansei Engineering for product's Interactive Design Based on "Sensing" Technology

The main application of sensing technology is bionic technology, whose mechanism is to collect and perceive external information with the internal construction and process and output the information to realize the purpose of control. This way of working, as one of the basic work of sensing technology, has played an increasingly important role in the field of sensing technology. Usually, the sensing technology in robot automatic control is divided into two types. One is internal sensing technology, the other is external sensing technology. Through the labor division and cooperation of the two types, it can achieve an effective operation. Through their application in the field of production, it has greatly enhanced the production efficiency. The internal sensing technology in robot is mainly applied in robot monitoring. The external sensing technology refers to external intelligent sensing technology and multifunctional sensing technology, which has a promising prospect of application in robots, such as vision, touch, force and proximity [13]. In particular, intelligent products need to identify, judge and decide based on the collected information. The intelligent sensing Technology, like the five sensory organs of humans, can make intelligent products have a perceptual functional map (Fig. 4).

Now some countries are studying and developing tactile sensing technology that can identify the shape of objects and olfactory sensing technology that can distinguish

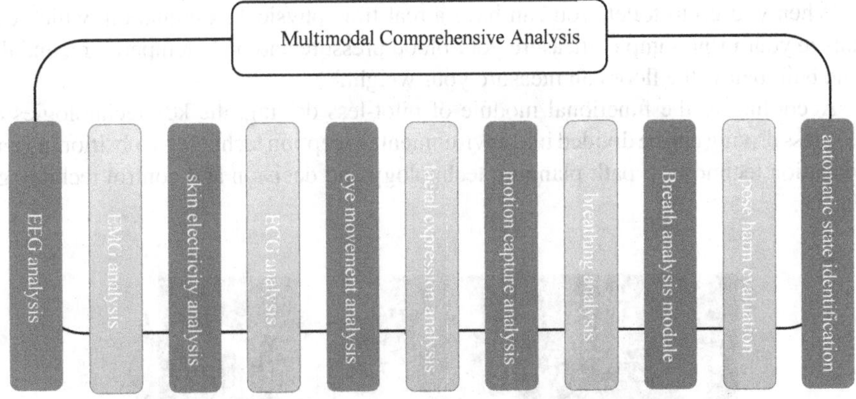

Fig. 4. Synchronized comprehensive analysis of multimodal data

different gases. With the development of intelligent sensing technology and multimedia sensing technology, they will be more widely used in various sectors such as industry, science and technology, and national defense.

4 Hybrid Sensing Research of the Interactive Experience of "Sensing" Technology in Product Design

In 21st century, we have fully ushered in the era of information. As one of the three pillars of modern information technology, sensing technology will surely have a long-term development. As the "electrical five senses" lack behind "the computer", we need to further develop and apply the new type of computers. The development of many new products with competitiveness and outstanding technical transformation are inseparable from sensing technology, like smart toilets (Fig. 5).

Fig. 5. Interior of smart toilets

When you go to toilet, you can have a real-time physical examination, which can analyze your urine sample, measure your blood pressure and body temperature, and the scale built below the floor can measure your weight.

According to the functional module of pilot-less driving, the key technologies of pilot-less driving can be divided into environment perception technology, positioning and navigation technology, path planning technology and decision and control technology (Fig. 6).

Fig. 6. Pilot-less autonomous vehicle

The comprehensive application of "The sixth sense" gesture identification technology, artificial intelligence, 3G and projection technology (Fig. 7).

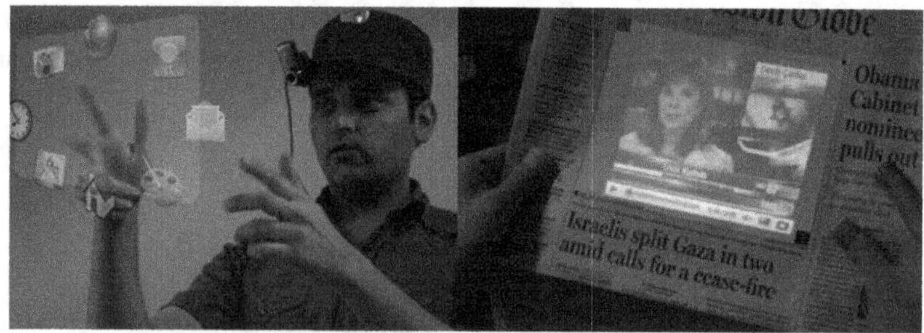

Fig. 7. The commanding system of multifunctional equipment

The development of spaceships and various probes, etc. is closely related with the sensing technology. The Kansei research has advanced at the same time. The application of sensing technology has improved the automation degree of product design, enhanced the fitness between products and people and yielded economic benefits [14].

The interactive sensing technology is widely applied in each area of the society, ranging from space to oceans, from various complex engineering systems to daily life necessities, which will create promising commercial prospects. All of them are a strong driving force for the development of interactive sensing technology. With the rapid development of modern science and technology, especially large-scale integrated electrical circuit technology and popularization of computers, the interactive sensing technology's status and role in the new technology revolution will become more prominent (Fig. 8).

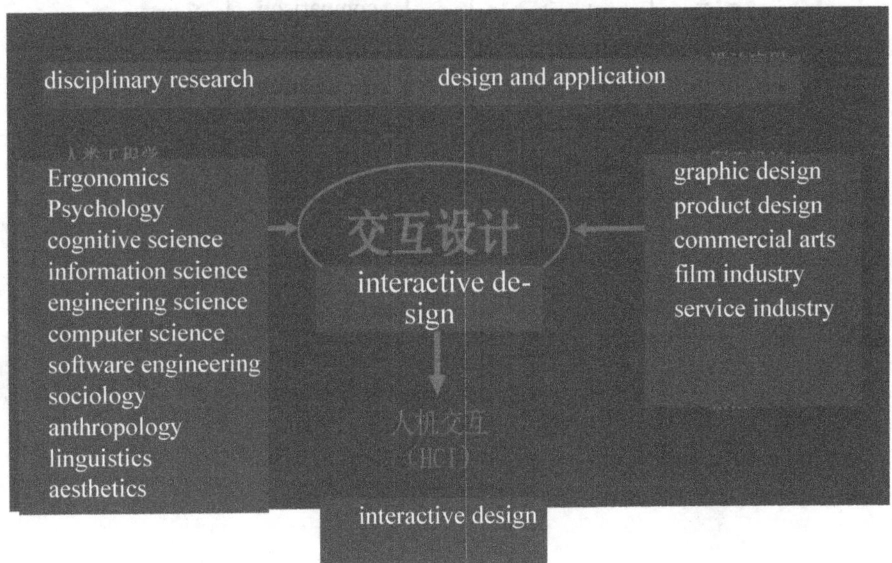

Fig. 8. The relationships between interaction design and other disciplines

A heat of development and application of interactive sensing technology has swept the world. At present, the sensing technology, whether in quantity, quality or function, is far from meeting the requirements for the development of society in various areas. Currently, while making full use of advanced electronic technology conditions, studying and adopting suitable external electrical circuits and maximizing the cost performance of existing sensing technology, people are seeking hybrid Kansei evaluation for the development of interactive sensing technology (Fig. 9). Especially, the development of technologies such as Electronic Design Automation (EDA), Computer Assisted Manufacture (CAM), Computer Assisted Test (CAT), Digital Signal Processing (DSP), Application-specific Integrated Circuit (ASIC) and Surface Mounting Test (SMT) has greatly accelerated the development and trend of sensing technology [15].

Product's interactive design based on "Sensing" technology and "in-depth experience" analyzes the existing Kansei appeal of users for products which has not been expressed clearly, thereby analyzing whether the products meet the user demands (**Fig.** 10).

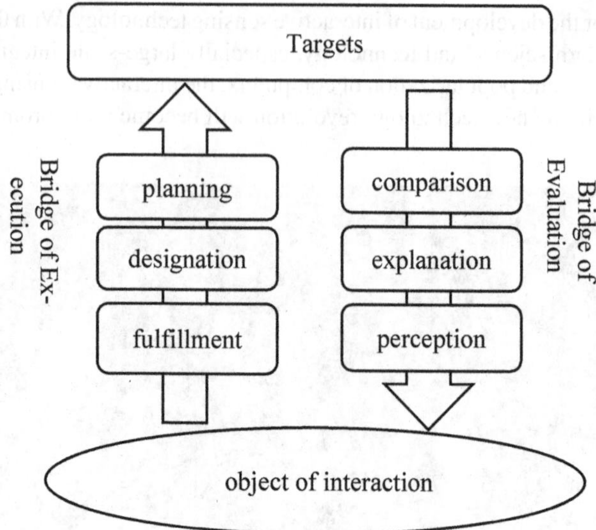

Fig. 9. Hybrid Kansei evaluation of interaction

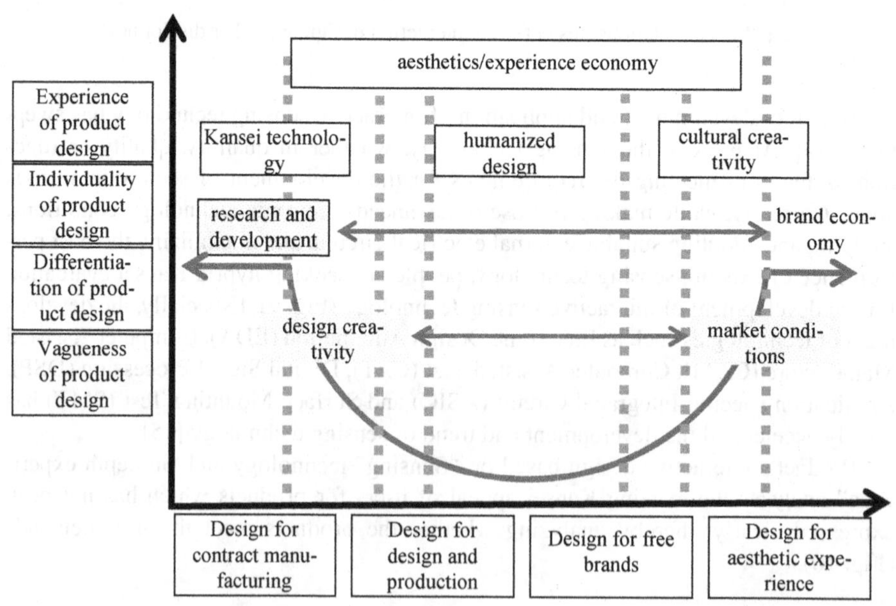

Fig. 10. Schematic diagram of Kansei demand in product's interactive design

5 Conclusion

In conclusion, this paper studies the hybrid Kansei technology in the product's interactive design. By creating and supporting the user experience, the product satisfies the users' appeal for service. Through their sensory organs, the users anticipate the results of operation. The planning and mode of interactive design is conveyed in human-machine interaction technology "centered around people". The sensing technology is widely applied in electromechanical automation control. With the continuous development and maturing of modern technology, the applied field of sensing technology has been further expanded, which has greatly enhanced the production and living efficiency and changed the traditional lifestyle, and boosted the progress of social productivity. At present, the sensing technology has been widely applied in the fields such as robot, machinery processing, monitoring, automotive production and digital medical care. Therefore, Kansei engineering will be applied in increasingly wide fields of sensing technology application and better promote the social development. The application of Kansei engineering in the sensing technology can effectively and continuously improve the intelligence of products.

References

1. Shan, C.: Theory of Sensing Technology and Design Foundation and Application, p. 45. National Defence Industry Press, Beijing (1999)
2. He, X.: Sensing Technology and Its Application Circuit, p. 12. Electronic Industry Press, Beijing (2001)
3. Huang, C.: Machinery Engineering Test Technology Foundation, p. 142. China Machine Press, Beijing (2001)
4. Yuanqing, W.: Mechanism and Application of New Type Sensing Technology. China Machine Press, Beijing (2002)
5. Zhanyou, S.: Mechanism and Application of Intelligent Integrated Temperature. China Machine Press, Beijing (2002)
6. Dix, et al.: Human-Machine Interaction. Electronic Industry Press (2006). Translated by Cai, L., et al. (2006)
7. Cooper, A.: About Face 3 Essence of Interactive Design. Electronic Industry Press, America (2008). Translated by Liu, S. (2008)
8. Jenifer, T.: Designing Interfaces. Electronic Industry Press, America (2008)
9. Luo, S., Zhu, S.: User Experience and Product Innovation Design. China Machine Press, Beijing, pp. 23 (2010)
10. Li, S.: Experience and Challenge--Product's Interactive Design. Jiangsu Fine Arts Publishing House (2008)
11. Jones, M.: Interactive Design of Mobile Devices. Electronic Industry Press, America (2008). Translated by Xi, D. (2008)
12. Coopeer, A., Reimann, R., Cronin, D.: Essence of Interactive Design. Electronic Industry Press, Beijing (2008). Translated by Liu, S., et al. (2008)
13. McRoberts: Arduino from Fundamental to Practice. Electronic Industry Press, Beijing (2013). Translated by Yang, Z. (2013)
14. Liu, W.: Approach Interactive Design. China Building Industry Press, Beijing (2013)
15. Mginno, J.P.: Arduino C Language Programming Practice. Posts and Telecom Press, Beijing (2013)

Author Index

Abraham, Muth Mary 481
Adhikary, Biswajit 16
Ahlers, Stefan 179
Almeida, João 326
Arconada-Alvarez, Santiago 63

Belavadi, Poornima 179
Beltrão, Gabriela 495
Bock, Sven 198
Braz Junior, Geraldo 326
Budree, Adheesh 310
Burbach, Laura 179

Calero Valdez, André 179
Castanho, Carla D. 367
Chakraborty, Joyram 554
Chen, Lingxi 509
Chen, Xuanyi 3
Choe, Pilsung 481
Choi, Woo Jin 217
Choi, Young Mi 63
Chowdhury, Ashraf Ferdouse 198
Coşkun, Aykut 564

de Bont, Cees 342
de Moraes, Iago L. R. 367
Delfino, Donatella 147, 387
Dishman, Samantha 229
dos Santos, Eduardo A. 367
Duffy, Vincent G. 229, 249, 451, 539
Dugar, Sumesh 16

e Silva, Tiago B. P. 367
Eduardo, José 326
Eloiriachi, Aya 564
Enebechi, Chidubem Nuela 249
Enebechi, Monica Okwuchkwu 249

Fan, Cheng-wei 597
Fernandes, Eduardo 326
Ferreira, Victor 326
Flores-Rivas, Víctor Ricardo 357
Fu, Bo 424

Galda, Rahul 24
Gamboa, Edwin 24
Gui, Yuanlong 165, 424

Hall, Margeret 280
Hamada, Yuri 270
Hatami, Zahra 280
Hawryluk, Bohdan 147
Hirth, Matthias 24
Huang, Kuo-Liang 522
Huang, Linda 509
Huang, Wentong 40

Ishii, Hirotake 54, 412
Ito, Kyoko 54

Jacobi, Ricardo P. 367
Jansen, Bernard J. 127
Jia, Chen 397
Jiang, Jinchen 522
Jung, Soon-gyo 127

Kim, Ted 63
Koratpallikar, Priyanka 539
Kundu, Anirudh 72
Kwon, Hyosun 342

Le, Chang 397
Liang, Xin 89
Liao, Naizheng 424
Lim, Chang Joo 217
Lin, Hsuan 522
Liu, Zhen 101
Lu, Hui-Ping 116

Manusuriya, Madhav 16
Mator, Janine D. 298
Mayas, Cindy 24
Merino Flores, Irene 357

Mitra, Abhishek 16
Miyazaki, Daisuke 54
Mochizuki, Rika 412
Momen, Nurul 198
Monteiro, Eliana 326

Nagata, Atsuya 270
Nandi, Shweta 16
Nogueira, Hugo 326
Ntumba, David 310

Ohkura, Michiko 72
Oladapo, Helina 554

Paiva, Anderson 326
Paiva, Anselmo 326
Paul, Sonit 16
Probst, Freya 342

Ramírez Galleguillos, María Laura 564
Reyna González, Julissa Elizabeth 357
Rivero, Luís 326
Rocha, Simara 326
Rosa, Marcos P. C. 367
Rössger, Peter 584

Sakamoto, Yoshiki 54
Salminen, Joni 127
Santos, Ítalo 326
Saravanos, Antonios 147, 387
Sarmet, Mauricio M. 367
Serdar, Büşra 564
Sheng-nan, Guo 397
Shi, Min 597
Shimoda, Hiroshi 54, 412
Shoji, Hiroko 270
Silva, Aristofenes 326
Soares, Marcelo M. 397, 465
Sousa, Sonia 495
Still, Jeremiah D. 298
Stott, Neil 147

Takahashi, Naoki 270
Takashima, Yuki 412
Tang, Jiaqi 465
Thorne, Neil 280
Tian, Renran 451

Ueda, Kimi 54, 412
Uotani, Takumi 412

Watanabe, Masahiro 412
Wu, Junchi 465

Xiao, Yuxuan 509
Xu, Na 522
Xu, Xiangrong 424
Xue, Chengqi 3

Yamamoto, Rieko 54
Yamawaki, Mizuki 54
Yang, Junyu 434
Yang, Zulan 101
Yu, Wei 89

Zeng, Jiayu 397, 465
Zervoudakis, Stavros 147, 387
Zhang, Ke 101
Zhang, Mu 434
Zhang, Xi 165
Zhang, Zhengming 451
Zhao, Tianjiao 434
Zhao, Xueqing 89
Zheng, Dongnanzi 147, 387
Zheng, Yawen 434
Zhou, Kun 165
Zhou, Yiqing 465
Zhu, Yanfei 3
Ziefle, Martina 179

Printed in the United States
by Baker & Taylor Publisher Services